THE ROUTLEDGE HANDBOOK OF LANDSCAPE ARCHITECTURE EDUCATION

In this handbook, 60 authors, senior and junior educators, and researchers from six continents provide an overview of 200 years of landscape architecture education. They tell the stories of schools and people, of visions, and of experiments that constitute landscape architecture education heritage.

Through taking an international perspective, the handbook centers on inclusivity with an appreciation for how education develops in different political and societal contexts. Part I introduces the field of education history research, including research approaches and international research exchange. Spanning more than 100 years, Parts II and III investigate and compare early and recent histories of landscape architecture education in different countries and schools. In Part IV, the book offers new perspectives for landscape architecture education. Education research presents a substantial opportunity for challenging studies to increase the pedagogic and didactic, the academic and historic, and the disciplinary knowledge basis.

Through a boundary-crossing approach, these studies about landscape architecture education provide a reference to teachers and students, policymakers, and administrators, who strive for innovative, holistic, and interdisciplinary practice.

Diedrich Bruns is Professor Emeritus at the University of Kassel, Germany. His research expertise is in planning history, landscape planning, and communication methods. His academic appointments include universities in Toronto, Canada; Stuttgart, Germany; and California and Minnesota, USA. Dr Bruns has published several peer-reviewed journal papers, book chapters, and books. He is the founder of the consulting firm Landscape Ecology & Planning, and a past president of ECLAS.

Stefanie Hennecke is Professor of Open Space Planning at the University of Kassel, Germany. She has a doctoral degree from the University of the Arts, Berlin. Until 2013 she was a junior professor of history and theory of landscape architecture at the Technical University of Munich. Her research topics are the history of urban green spaces in the 19th and 20th centuries and the development of adaptable urban green spaces for people and wildlife. In 2021, she edited a book on the impact of Covid-19 on public open spaces.

"This book provides welcome perspectives on the education of landscape professionals that generate both a starting point for a more profound understanding of the historical position of landscape architecture, as well as ideas of where the profession might take us in view of the contemporary environmental and social challenges facing humanity."

Jan Woudstra, Department of Landscape Architecture, University of Sheffield, UK

"This comprehensive handbook shares insights into landscape architecture education that are unprecedented in extent and depth. It describes and compares concepts and practices of education that are applied during the last two hundred years and addresses a wide range of perspectives by examining cases from around the world. The book lays the foundation for education studies into the character, commonalities and future of landscape architecture education."

Hiroyuki Shimizu, Emeritus Professor, Nagoya University, Japan

"This handbook makes an invaluable contribution to the present and future university education of landscape architects. It offers fascinating insights of both the discipline's historical development and of its current teaching and learning praxis in many countries around the world: a panorama of today's remarkable international diversity in landscape architecture education that unveils its implicit learning potential, not only for landscape architecture but also for other planning disciplines and environmental sciences."

Joachim Wolschke-Bulmahn, Professor Emeritus, Instutute for Landscape Architecture, Leibniz University Hannover, Germany

THE ROUTLEDGE HANDBOOK OF LANDSCAPE ARCHITECTURE EDUCATION

Edited by Diedrich Bruns and Stefanie Hennecke

NEW YORK AND LONDON

Cover image: Collaborative learning in the design studio at ETH Zurich.
Source: ETH (Swiss Federal Institute of Technology) Zurich,
Professorship of Günther Vogt, Thomas Kissling (2019).

First published 2023
by Routledge
605 Third Avenue, New York, NY 10158

and by Routledge
4 Park Square, Milton Park, Abingdon, Oxon, OX14 4RN

Routledge is an imprint of the Taylor & Francis Group, an informa business

© 2023 selection and editorial matter, Diedrich Bruns and Stefanie Hennecke; individual chapters, the contributors

The right of Diedrich Bruns and Stefanie Hennecke to be identified as the authors of the editorial material, and of the authors for their individual chapters, has been asserted in accordance with sections 77 and 78 of the Copyright, Designs and Patents Act 1988.

All rights reserved. No part of this book may be reprinted or reproduced or utilised in any form or by any electronic, mechanical, or other means, now known or hereafter invented, including photocopying and recording, or in any information storage or retrieval system, without permission in writing from the publishers.

Trademark notice: Product or corporate names may be trademarks or registered trademarks, and are used only for identification and explanation without intent to infringe.

Library of Congress Cataloging-in-Publication Data
Names: Bruns, Diedrich, editor. | Hennecke, Stefanie, editor.
Title: The Routledge handbook of landscape architecture education / edited by Diedrich Bruns and Stefanie Hennecke.
Description: New York, NY : Routledge, 2023. | Includes bibliographical references and index. | Identifiers: LCCN 2022023948 (print) | LCCN 2022023949 (ebook) | ISBN 9781032080413 (hardback) | ISBN 9781032080420 (paperback) | ISBN 9781003212645 (ebook)
Subjects: LCSH: Landscape architecture—Study and teaching.
Classification: LCC SB469.4 .R68 2023 (print) | LCC SB469.4 (ebook) | DDC 712.076—dc23/eng/20220614
LC record available at https://lccn.loc.gov/2022023948
LC ebook record available at https://lccn.loc.gov/2022023949

ISBN: 9781032080413 (hbk)
ISBN: 9781032080420 (pbk)
ISBN: 9781003212645 (ebk)

DOI: 10.4324/9781003212645

Typeset in Bembo
byApex CoVantage, LLC

Dedicated to Karsten Jørgensen (1953–2021), for his many contributions to the book, his enthusiastic support, and advice whenever we sought it.

CONTENTS

List of figures, tables, and boxes	xi
Biographies	xvii
Foreword	xxvii
Inspirations and acknowledgments	xxviii

1 Broadening the outlook, expanding horizons   1
   Diedrich Bruns and Stefanie Hennecke

PART I
Challenges and perspectives   11

2 Introducing the field of landscape architecture education research: challenges and perspectives   13
   Diedrich Bruns and Stefanie Hennecke

3 "A thing in movement": landscape history in professional curricula   16
   M. Elen Deming

4 Tracing discourses: learning from the past for future landscape architecture   31
   Mattias Qviström and Märit Jansson

5 European cooperation between educators and landscape architecture schools   40
   Richard Stiles

6 Building up historical continuity: landscape architecture archives in education   54
   Lilli Lička, Bernadette Blanchon, Luca Csepely-Knorr, Annegreth Dietze-Schirdewahn, Ulrike Krippner, Sophie von Schwerin, Katalin Takács, and Roland Tusch

7 Joining forces: landscape architecture and education for sustainable
   development   62
   Ellen Fetzer

8 Creating vital teaching communities through curriculum development   71
   Anne Katrine Geelmuyden

9 Conceptual thinking and relational models in landscape architecture pedagogy 81
   Juanjo Galan Vivas

10 Pedagogy for sustainability in landscape architectural education   91
   Dan Li

PART II
Agendas and standards   99

11 Chronicling education history: agendas and standards   101
   Diedrich Bruns and Stefanie Hennecke

12 Early landscape architectural education in Europe   109
   Barbara Birli, Diedrich Bruns, and Karsten Jørgensen

13 Landscape gardening, outdoor art, and landscape architecture: the
   beginning of landscape architecture education in the United States,
   1862–1920   121
   Sonja Dümpelmann

14 A hundred years of landscape architecture education in Ås, Norway:
   the pioneering work of Olav Leif Moen   135
   Karsten Jørgensen

15 German garden design education in the early 20th century   143
   Lars Hopstock

16 Landscape architecture university education under National Socialism
   in Germany   155
   Gert Gröning

17 'To broaden the outlook of training': the first landscape course in Manchester 164
   Luca Csepely-Knorr

18 Landscape architecture education history in Portugal: the pioneering
   roles of Francisco Caldeira Cabral and Francisco Simões Margiochi   176
   Ana Duarte Rodrigues

19 Dutch landscape architecture education in the first half of the twentieth century: the pioneering roles of Hartogh Heys Van Zouteveen and Bijhouwer   187
Patricia Debie

20 Early history of landscape architecture education initiatives in Romania: the pioneering work of Friedrich Rebhuhn   198
Alexandru Mexi

21 Landscape architectural education in Croatia   211
Petra Pereković and Monika Kamenečki

22 Landscape architectural education in Hungary: the pioneering work of Béla Rerrich, Imre Ormos and Mihály Mőcsényi   223
Albert Fekete

23 Landscape architecture education history in Slovakia and the Czech Republic   233
Attila Tóth, Ján Supuka, Katarína Kristiánová, Jan Vaněk, Alena Salašová, and Vladimír Sitta

24 A long, yet successful, journey: one hundred and fourteen years to implement a landscape architecture programme in Austria, 1877–1991   243
Ulrike Krippner and Lilli Lička

25 Landscape architecture education history in the German-speaking part of Switzerland   253
Sophie von Schwerin

26 The history of higher landscape architecture education at ETH Zurich, Switzerland   261
Dunja Richter

PART III
Broadening the common ground   273

27 Broadening the common ground: education for the design of human environments   275
Diedrich Bruns and Stefanie Hennecke

28 Landscape architecture education in Italy: fragmented patterns   278
Francesca Mazzino and Bianca Maria Rinaldi

29 The training of landscape architects in France: from the horticultural engineer to the landscape architect, 1876–2016   292
Bernadette Blanchon, Pierre Donadieu, and Chiara Santini, with Yves Petit-Berghem

30 Reflections on landscape and landscape architecture education in the
    Arab Middle East   303
    Jala Makhzoumi

31 Landscape architecture education in China: the pioneering work of
    Sun Xiaoxiang   315
    Lei Gao and Guangsi Lin

32 Landscape design education in Japan: the Meiji, Taisho, and
    Showa Periods   326
    Chika Takatori

PART IV
Aiming for justice, reconciliation, and decolonization   335

33 Innovating education policy: justice, reconciliation, and decolonization 337
    Diedrich Bruns and Stefanie Hennecke

34 Landscape architecture education in Albania after the fall of the Iron Curtain 340
    Zydi Teqja

35 Landscape architecture education in Poland   350
    Katarzyna Rędzińska and Agnieszka Cieśla

36 Landscape @Lincoln: place and context in the development of an
    antipodean landscape architecture programme   365
    Simon Swaffield, Jacky Bowring, and Gill Lawson

37 Learning to practice creatively: emergent techniques in the climate emergency 378
    *Alice Lewis, Sue Anne Ware, Martin Bryant, Jen Lynch, Penny Allan,
    and Katrina Simon*

38 Landscape architecture education in Africa   390
    Graham A. Young

39 Educational ecosystem on landscape in Latin America   409
    *Gloria Aponte-García, Cristina Felsenhardt, Lucas Períes, and Karla María
    Hinojosa De la Garza*

40 Conclusions and hopeful perspectives   421
    Diedrich Bruns and Stefanie Hennecke

Index   *429*

FIGURES, TABLES, AND BOXES

Figures

1.1	Landscape architecture education research is part of education studies and includes education history research.	5
3.1	Petworth House, England (1690; renovated Capability Brown ca. 1750).	22
3.2	"Servant's house T": dwelling for an enslaved family (1793–ca. 1830; reconstructed in 2014). Monticello, VA.	23
3.3	Thomas Jefferson estate at Monticello, VA.	24
4.1	Lorensborg, a neighbourhood developed in the 1950s in Malmö.	32
4.2	Field studies in Smedsby, Upplands Väsby, one of many modernist neighbourhoods where densification is underway.	34
4.3	Solskiftet, a neighbourhood in Landskrona, was developed in the 1970s. It was planned with open, green housing yards and also placed next to a park.	36
5.1	Group photograph of the participants at the 2019 ECLAS conference in Ås/Oslo at which 100 years of university education in landscape architecture in Europe was celebrated at the Norwegian University of Life Sciences, where the first programme began in 1919.	45
5.2	Members of the LE:NOTRE committee and international advisers representing some 14 countries – mainly from Europe but also North and South America – discussing informally in a Paris park during a break in the proceedings of the Versailles Spring Workshop in April 2009.	48
5.3	Excerpt from a folder made to present an early version of the LE:NOTRE web site to the broad project membership, illustrating the many interactive features it contained, including subject-related workspaces and user-editable databases.	50
6.1	Students gathering information for the seminar "Design in the Historical Context", Archive of Swiss Landscape Architecture (ASLA), Rapperswil, 2018.	56
6.2	Detail of a student poster on a historical site in Hungary.	56
6.3	Discussing historical drawings in ANLA in order for students to copy them during a course.	58

Figures, Tables, and Boxes

6.4	Isometric drawing of Dos Santos garden by Albert Esch (1930), transposed into a digital drawing.	59
7.1	The European integration pathway in landscape architecture education.	64
7.2	Key Competences for Sustainable Development according to Wiek et al. (2015).	65
9.1	Development and application of relational models for urban-nature concepts in a studio course on Green Area Planning.	83
9.2	Top: relational model for urban-nature concepts proposed by the teacher to activate conceptual discussions in the Green Area Planning course (adapted from Galan 2020). Bottom: relational model for urban-nature concepts proposed by the team of students working in the city of Oulu in 2019.	86
10.1	Pedagogy for teaching sustainability endorsement and effectiveness.	94
10.2	Teaching method for teaching sustainability endorsement and effectiveness.	94
10.3	Involvement of rating systems in teaching sustainability.	95
13.1	Two pages illustrating the horticultural and gardening training of Native American and African American students at Hampton Normal and Agricultural College, ca. 1909.	124
13.2	Two pages of a leaflet issued by Hampton Institute illustrating landscape planting and design guidelines, 1917.	125
13.3	Photograph by Frances Benjamin Johnston showing the Pennsylvania demonstration kitchen and flower garden at the Pennsylvania School of Horticulture for Women in Ambler, PA in 1919.	128
13.4	Surveying class at the Lowthorpe School in Groton, MA.	129
14.1	The original caption of this picture is: "Course for horticultural officials in garden art and garden architecture" at NLH in January 1931.	138
14.2	Students working in the NLH park during the 1940s.	139
14.3	Olav L. Moen: the 1924 park plan in perspective from ca. 1935.	140
15.1	Plan for the grounds of the *Höhere Gärtnerlehranstalt* in Dahlem, designed by Theodor Echtermeyer, director of the institute.	144
15.2	Drawing hall of the *Höhere Gärtnerlehranstalt* in Dahlem.	145
15.3	Jubilee Exhibition at Mannheim in 1907.	149
15.4	Page from Max Laeuger's *Kunsthandbücher* on which he explains that the laws of building are also valid for arranging trees.	151
16.1	Certificates of Honour for especially good achievements in the first Reich achievement competition of German universities and colleges (*Ehrenurkunden für besonders gute Leistungen im 1. Reichsleistungskampf der deutschen Hoch- und Fachschulen*) for Gert Kragh, Werner Lendholt, Gerhard Neef, Gerhard Prasser, Dietrich Roosinck, Ulrich Schmidt.	158
17.1	Course structure in the 1935–1936 School Catalogue.	170
17.2	Course structure in the 1937–1938 School Catalogue.	171
17.3	Diagram of the courses taught at the Manchester Municipal School of Art on the front page of the school's 1946–1947 Prospectus.	174
18.1	Children playing at the Estrela Garden with gardening utensils, 1927.	180
18.2	Engraving of the Froebel School at the Estrela Garden in 1883.	180
18.3	The building of the former Froebel School at the Estrela Garden in Lisbon, in 2015.	181
18.4	Plan of the monastic enclosure of the Monastery of Jerónimos, Lisbon, in which the premises for the School of Gardening was developed.	182

19.1	Advertisement at the plant nursery Moerheim Ruys at Dedemsvaart.	188
19.2	An example of a detailed garden design (1899) for the insane hospital Grave, containing a floor plan, a plan of construction and profiles.	190
19.3	Drawing instructions with good (*goed*) and wrong (*slecht*) examples of plant schemes to create natural plant communities.	191
19.4	A sample sheet with several examples of how to sketch the shape of a tree.	194
20.1	Romanian workers on the worksite of Bibescu Park in Craiova, ca. 1900.	200
20.2	Friedrich Rebhuhn's article in the Romanian Horticulture Magazine, 1913.	202
20.3	The picture Friedrich Rebhuhn shows in his manuscript "Theory, practice and aesthetics of green spaces in urban planning" (1942), portraying students learning outdoors.	206
20.4	One of the final drafts of the curriculum of 1998, translated in French.	208
21.1	Many pioneers of education in the field of garden architecture have been educated and/or have continued their educations abroad.	213
21.2	Participants and teachers of a gardening course held at the Botanical Garden in 1932, Zagreb.	214
22.1	Eyebird sketch of the Hungarian Royal School of Horticulture, designed by Béla Rerrich, around 1930.	225
22.2	Professor Ormos and his students during a seminar in the 1960s.	226
22.3	Landscape assessment seminar nowadays (2019).	228
23.1	Some characteristic hand drawings by Milan Kodoň, whose landscape drawings were instrumental in teaching.	238
23.2	Outdoor drawing classes belong to important learning practices, especially within the subjects of Drawing, Architectural Drawing, and Landscape Drawing at SUA Nitra.	240
24.1	According to Abel, gardens in the urban context should show a symmetrical and architectonic design (figure to the left), whereas the layout of suburban villa gardens combines regularity with nature-like forms (figure to the right).	247
24.2	In the 1950s, Friedrich Woess taught primarily garden design, following a simple and traditional style, which referred to the Wohngarten of the interwar period.	249
24.3	These two undated teaching charts by Friedrich Woess explain the design principle "contrast": left, contrasting lines; right, contrasting light.	249
25.1	Illustrations of garden designs which Albert Baumann used during lectures at the Oeschberg Horticultural School in the 1920s.	255
25.2	Admission requirements for the degree programme in the 1970s and early 80s.	256
25.3	Studying in Rapperswil in the early 1990s.	257
25.4	Professor Christian Stern during a presentation in the design studio.	258
26.1	Final presentation: students present their results of the design studio "Designing a Dynamic Alpine Landscape, Bondo/Switzerland" in collaboration with Gramazio Kohler Research.	265
26.2	The master's degree programme in landscape architecture at ETH comprises four semesters and a vocational internship of six months between the second and third semester.	267
26.3	Capture the acoustic dimension of the landscape: workshop "3D Landscape Mapping, Sounds and Point Clouds" in collaboration with Kyoto Institute of Technology, Shosei-en Garden in Kyoto/Japan.	268

26.4	Collaborative learning in the design studio: students are engaged in the analysis and study of different design variants in the project "Marseille – Maritime and Alpine Landscape".	269
28.1	Cover of the 1984 issue of the book series *Quaderni di Architettura*, edited by the School of Architecture of the University of Genoa and focusing on the two-year programme in Landscape Architecture for graduates, called *Scuola di Specializzazione in Architettura del Paesaggio*, launched in 1980.	285
28.2	Second-year studio final presentations for students of the Interuniversity Master's Programme in *Progettazione delle Aree Verdi e del Paesaggio* (Green Open Space Design and Landscape Architecture) at Politecnico di Torino, AY 2019–2020.	287
29.1	ENH in Versailles, early XXth century.	294
29.2	ENH in Versailles, around 1900. Students doing practical exercises in the *Potager du Roi*.	295
29.3	SPAJ. The design studio (*ateliers*) in 1965–1967.	297
29.4	ENSP, Versailles, 2019. Students doing ecological field exercises at the *Potager du Roi* (the King's Kitchen Garden).	298
30.1	Landscape architecture is conspicuously absent in countries of the Middle East and North Africa in a map of national associations of the International Federation of Landscape Architects.	307
30.2	Winning landscape design by first-year student Zeina Salam for the faculty garden.	310
32.1	Three design education systems developed side-by-side in Japan with links between them, the apprenticeship, the in-house, and the academic system.	328
34.1	The relationship between overall democracy index (a), Freedom house score (b) and number of landscape architecture programmes for 10 million people in former communist countries.	347
35.1	The frequency of the clusters of topics related to landscape architecture education, professional practice and landscape in the spatial planning system during the Landscape Forum, 1998 to 2019.	362
36.1	Design studios ca. 1970.	369
36.2	Landscape@Lincoln graduation numbers, 1971–2020.	373
37.1	Narratives of the future.	382
37.2	New cartographies to bring distance closer: the 2019 New Year's Eve bushfire (top), a drawing by Sam Clare; Deception Island in Antarctica (lower), a drawing by Faid Ahmad.	383
37.3	Explorations of the performance of sand, using dye and water to expose hidden flows and convergences, a drawing by Conrad J. Cooper MLA.	385
37.4	Rhizomatic mapping (top) and detail (bottom) exploring the interconnected communities of a practice of 'collecting', drawing by Louella Exton MLA.	387
38.1	Senior lecturer Graham A. Young illustrating urban design principles to third-year architecture and landscape architecture students at the University of Pretoria.	392
38.2	Supervising lecturers, from Ahmadu Bellow University, Dr Maimuna Saleh-Bala and Mr Bartho Ekweruo collaborating on addressing UNDP-Sustainable Development Goal 3 on health and well-being in Zaria City, Nigeria.	400

39.1	Landscape Architecture education in Chile.	413
39.2	Map of Landscape Architecture education in Argentina.	415
39.3	National coverage of Landscape Architecture education in México.	418

Tables

7.1	The key competences for SDG 11 (Sustainable Cities and Communities) as defined by the UN compared to the competences presented in three transnational frameworks for landscape architecture education	67
8.1	A sketch of curriculum development as an ongoing 'communal inquiry' amongst landscape architecture teachers with various disciplinary and practice backgrounds	78
9.1	External assessment by planners, academic experts and practitioners of the potential that a relational model and the subsequent academic proposals could have in sustainable urban planning (studio course on Green Area Planning, years 2016, 2017 and 2018)	87
9.2	Assessment of the evolution of conceptual and integrative thinking skills in the studio course Green Area Planning	88
11.1	The first landscape architecture programmes in North America and Europe	103
12.1	The first six landscape architecture programmes and educators at European universities	111
17.1	Student numbers registered to the architecture and landscape architecture courses in Manchester 1934–1952	168
21.1	The curriculum for the first Landscape Design study in Croatia – interfaculty postgraduate study held from 1968 to 1985 at University of Zagreb (Faculty of Agriculture – leading faculty, Faculty of Forestry and Faculty of Architecture)	218
21.2	The curriculum for the interfaculty graduate study 'landscape architecture' taught by University of Zagreb in 1996 (Faculty of Agriculture – leading faculty, Faculty of Forestry, Faculty of Architecture, Faculty of Geodesy, Faculty of Science, Faculty of Humanities and Social Sciences	219
24.1	History of landscape architecture education at BOKU and professionalisation in Austria	244
28.1	First graduate programmes in Landscape Architecture and undergraduate programmes with Landscape-Architecture related curricula established from the early 2000s in Italy following the three decrees issued by the Ministry of Education	286
31.1	The curriculum for the 1951–1953 landscape architecture programme, jointly established by Beijing Agriculture University and Tsinghua University	317
31.2	Table of contents of Sun Xiaoxiang's book *Garden Art and Landscape Design*	318
31.3	A comparison between the existing and the envisioned structures of landscape architecture-related institutions at Beijing Forestry University	322
34.1	SWOT analysis related to landscape architecture education in Albania and specifically at Agricultural University of Tirana	344
34.2	Democracy indexes and number of landscape architecture programmes for former communist countries	346

35.1 The stages of the development of landscape architecture education in Poland according to Wolski (2007), modified by authors based on Wolski (2015), Böhm and Sykta (2013), Niedźwiecka-Filipiak (2018), Sobota and Drabiński (2017) 351
35.2 Curricula and scientific disciplines mapping in 2020 355
35.3 Themes of the Landscape Architecture Forum, 1998 to 2019 359
36.1 Summary of key events and developments, 1960s to 2020 367
36.2 Evolving academic content of Landscape Architecture qualifications at Lincoln 371

Boxes

5.1 Objectives of European academic cooperation as defined in the ECLAS statutes 46

BIOGRAPHIES

Editors

Diedrich Bruns is Professor Emeritus at the University of Kassel, Germany. His research expertise is in planning history, landscape planning, and communication methods. His academic appointments include universities in Toronto, Canada; Stuttgart, Germany; California and Minnesota, USA. Dr Bruns has published several peer-reviewed journal papers, book chapters, and books. He is founder of the consulting firm Landscape Ecology & Planning, and a past president of ECLAS.

Stefanie Hennecke is Professor for Open Space Planning at the University of Kassel, Germany. She has a doctoral degree from the University of the Arts, Berlin. Until 2013, she was a junior professor of history and theory of landscape architecture at the Technical University of Munich. Her research topics are the history of urban green spaces in the 19th and 20th centuries and the development of adaptable urban green spaces for people and wildlife. In 2021, she edited a book on the impact of Covid-19 on public open spaces.

Authors

Penny Allan is Professor of Landscape Architecture at University of Technology Sydney, Australia. Her most recent design research projects, MOVED to Design, Earthquake Cities of the Pacific Rim, and Rae ki te Rae, deal with the relationship between environment, culture, resilience, and design.

Gloria Aponte-García is a Colombian landscape designer, lecturer, academic, and consultant researcher at Rastro Urbano, Universidad de Ibagué, Colombia. She holds a master's in landscape design, has run her firm Ecotono for 20 years, and initiated a master's in landscape design for U.P.B. Medellín, which she directed for eight years. Aponte-Garcia is the Colombian delegate to IFLA and a member of the Education and Academic Affairs Committee Americas Region.

Biographies

Barbara Birli holds a degree in landscape planning and management and a doctorate in regional planning. She currently works as environmental consultant at the Austrian Environment Agency. Dr Birli is involved in national and international projects concerned with land management, soil management, and education. Through her work, she provides a basis for decision making at local, regional, and international levels. She was responsible for the management of the LE:NOTRE network from 2004 to 2013.

Bernadette Blanchon is an architect, Associate Professor at *École Nationale Supérieure de Paysage* in Versailles, France, and research fellow at LAREP (*Laboratoire de Recherche en projet de Paysage*). Her teaching and research work focuses on landscapes in the urban environment of the post-war era. She has lectured at international conferences and various universities. A founding editor of the academic journal *JoLA, Journal of Landscape Architecture*, she headed the project review section *Under the Sky*.

Jacky Bowring has taught landscape architecture at Lincoln University, New Zealand, since 1997, and was Head of School in 2012–2014 and again in 2017. Her research interests are in memory aspects of landscape, design theory, and design critique, and she continues to practice as a registered landscape architect.

Martin Bryant is Professor of Landscape Architecture at University of Technology Sydney, Australia, and an architect, landscape architect, and urban designer with more than three decade's of experience in private practice. He has published, exhibited, and led design studios with a focus on resilience, indigenous knowledge in the Pacific, and form-making in landscape.

Agnieszka Cieśla is an architect and urban planner, a lecturer at the Warsaw University of Technology in Poland, and a specialist in issues related to demographic change and its spatial consequences. She holds a doctoral degree from the Bauhaus University Weimar and is a social activist, founder of the Despite the Age foundation (www.mimowieku.pl). In recent studies, Dr Cieśla has been concentrating on the living environment and the role of nature in maintaining its quality. She is a member of the Polish Association of Town Planners.

Luca Csepely-Knorr holds a doctoral degree from Corvinus University in Budapest, Hungary. She is a chartered landscape architect and art historian, working as Chair in Architecture at the Liverpool School of Architecture, UK. Her research centres on the histories of late 19th- and 20th-century landscape architecture. She is currently leading the Arts and Humanities Research Council (UK) funded research "Women of the Welfare Landscape" and is Co-Investigator of the project "Landscapes of Post-War Infrastructure: Culture, Amenity, Heritage and Industry".

Patricia Debie is a registered landscape architect and historian. Her certified firm Debie & Verkuijl examined and renovated over 550 Dutch heritage sites and won several awards: Oldenburger-Ebbers (2012), Garden of the Year (2013), ERM Historic Gardners (since 2019), and 100 years NVTL (2021). Her research projects are "Garden Art as Place Making with Dutch Avenue Systems" and "Influential Landscaping in Urban Planning (1900–1930)". She is an expert in heritage spatial quality committees and she guest lectures.

Biographies

M. Elen Deming is a professor of landscape architecture at North Carolina State University, USA, and Director of the Doctor of Design programme. She holds a doctoral degree from Harvard Graduate School of Design. She has taught design studios, design research, and topics in design history and theory for over 25 years. A former editor of *Landscape Journal* (2002–2009), Deming co-authored *Landscape Architecture Research* (with Simon Swaffield, 2011), and edited *Values in Landscape Architecture and Environmental Design* (2015) and *Landscape Observatory: The Work of Terence Harkness* (2017).

Annegreth Dietze-Schirdewahn is a professor in the School of Landscape Architecture at Norwegian University of Life Sciences (NMBU, Ås). She studied landscape and spatial planning at the University of Hannover, graduating in 1999, followed by history of garden art at the University of Bristol, UK, graduating in 2006, and she holds a doctoral degree from NMBU, also from 2006. She is a voting member of the International Scientific Committee on Cultural Landscapes (ICOMOS).

Pierre Donadieu is an agronomist, ecologist, and geographer. He holds a doctoral degree of Université Paris 7. He taught at the *École Nationale Supérieure de Paysage* in Versailles, France, from 1977 to 2017 where he founded the research laboratory and doctoral training programme. He is currently working on the school archives in order to reconstruct the history of the educational institutions (horticulture and landscape) based at the Potager du Roi in Versailles.

Sonja Dümpelmann is a professor at the University of Pennsylvania Weitzman School of Design, USA. She holds a doctoral degree from the University of the Arts, Berlin. Dümpelmann is the author and editor of several books, most recently *Seeing Trees: A History of Street Trees in New York City and Berlin* (YUP, 2019). She lectures internationally and has served as president of the Landscape History Chapter of the Society of Architectural Historians and as senior fellow in garden and landscape studies at the Dumbarton Oaks Research Library and Collection, Washington, DC.

Albert Fekete is a professor at the Institute of Landscape Architecture, Urban Planning and Garden Art Budapest (MATE), Hungary. He holds a doctoral degree and is licensed as a landscape architect and for heritage site restoration. Dr Fekete is contributor to over 60 landscape projects in Hungary, Romania, Germany, Netherlands, Spain, and China, and winner of the Landscape Architect of the Year award in Hungary in 2012 and in 2017, and the Europa Nostra Award in Research in 2014.

Cristina Felsenhardt holds a doctoral degree from *Universidad Politécnica de Cataluña*, Chile. She is Senior Lecturer at *Pontificia Universidad Católica de Chile*; architect at Royal Melbourne Institute of Technology, Australia; and the former head of the *Instituto de Estudios Urbanos y Territoriales at Pontificia Universidad Católica de Chile*. Dr Felsenhardt has worked on several research projects and is the author of books, chapters of books, papers, and other publications. She is member of ICHAP and IFLA.

Ellen Fetzer holds a doctoral degree from the University of Kassel, Germany. Since 2001 she has been working at the Faculty Environment, Design, Therapy at Nürtingen-Geislingen University (Stuttgart area). Dr Fetzer is primarily teaching master's students in landscape and urban planning. Her second focus is social innovation and transformative learning. She has been president of the European Council of Landscape Architecture Schools (ECLAS) during the publication period of this book.

Juanjo Galan Vivas is Associate Professor in the Department of Urbanism at the Polytechnic University of Valencia, Spain. Between 2015 and 2020 he held a similar position at the Department of Architecture of Aalto University, Finland. His research focuses on landscape planning, landscape design, sustainable development, regional and urban planning, and, on a more general level, on the study of the intersections between social and ecological systems as well as in the pedagogics of landscape architecture.

Lei Gao is an associate professor in the School of Landscape Architecture at Norwegian University of Life Sciences. She holds a doctoral degree from the University of Sheffield. Her main research interests include landscape architecture history, historical landscape management, values of private green space, rural landscape regeneration, and others.

Karla María Hinojosa De la Garza holds a master's degree in design, planning and conservation of landscapes and gardens from the *Universidad Autónoma Metropolitana*, Mexico. She is an architect at the *Universidad Autónoma de Tamaulipas*, and professor, researcher, and head of the Landscape Architecture Research Area and Laboratory at UAM-Azc., vocal of education in Landscape Architecture of the Society of Landscape Architects of México (SAPM), and member of the Standing Committee on Education and Academic Affairs of IFLA Americas.

Anne Katrine Geelmuyden is professor at the Norwegian University of Life Sciences (NMBU), where she currently heads the landscape architecture programme. She has also headed the department during two periods. Dr Geelmuyden holds a doctoral degree from NMBU, has taught a wide range of landscape architecture courses, and has tutored doctoral students for over 30 years. Her research regards the conceptualization of landscape, landscape aesthetics, landscape analysis, landscape criticism, and, more recently, the didactics of landscape architecture.

Gert Gröning is a retired professor of garden culture and open space development at the Universität der Künste in Berlin, Germany. Dr Gröning was a visiting professor at Shanghai Jiao Tong University, senior fellow in Garden and Landscape Studies at Dumbarton Oaks, USA. He is a long-standing member of ISHS and edited several *Acta Horticulturae* volumes. His recent book related to landscape architecture education in Germany is *From Dangast to Colorado Springs. Irma Franzen-Heinrichsdorff 1892–1983. Notes on the Life and Work of the First Woman Graduate in Landscape Architecture.*

Lars Hopstock is Junior Professor for Landscape Architecture in the architecture programme at the University of Kaiserslautern. He holds a doctoral degree from the University of Sheffield, UK. He worked for many years in both design practice and academia. His main research is about 20th-century landscape at the intersection of architectural, garden, and art history, most recently with a focus on the Bauhaus context. His interests also cover theory of designing, aesthetics of nature, and critique of urban open space.

Biographies

Märit Jansson is an associate professor of landscape planning in the Department of Landscape Architecture, Planning and Management at the Swedish University of Agricultural Sciences. Her research and teaching concerns mainly the management and planning of landscapes and urban open spaces for various user groups, with a particular focus on children. She takes an interest in the variations of landscape qualities, practices, and uses over time, lately through the lenses of the welfare landscape.

Karsten Jørgensen (1953–2021) was Professor of Landscape Architecture at the Norwegian University of Life Sciences. He earned his Dr. Scient. Degree in 1989, and was appointed a full professor in 1993. The Latvian University of Life Sciences awarded him an honorary doctorate in 2014. He actively contributed to the development of new landscape architecture programmes in Norway, Latvia, Estonia, and Palestine. He was a long-standing member of the executive committee of ECLAS, the European Council of Landscape Architecture Schools. He was founding editor of *JoLA, Journal of Landscape Architecture*, and one of the co-authors of the ECLAS Guidance on Landscape Architecture Education. He published regularly in national and international journals and books, and co-edited several books, including *Mainstreaming Landscape through the European Landscape Convention*, *Defining Landscape Democracy – A Path to Spatial Justice*, the *Routledge Handbook of Teaching Landscape*, and *Teaching Landscape: The Studio Experience*.

Monika Kamenečki is a landscape architect and assistant professor at the University of Zagreb, Faculty of Agriculture, Study of Landscape Architecture, Croatia. She had worked in private practice for more than a decade. Her work includes design, implementation, and construction supervision. She has a licence to work on cultural heritage projects. In teaching, her subjects are landscape construction and material science, sustainable technical planning, detailing, and plant use.

Ulrike Krippner is a senior researcher at the Institute of Landscape Architecture at BOKU in Vienna, Austria. She holds a doctoral degree in landscape architecture and teaches landscape history. Her research and writings are on the profession's history during the 20th century, with a special focus on women in landscape architecture and on post-World War II landscape architecture. She established the LArchiv, Archive of Austrian Landscape Architecture, and manages it together with Lilli Lička and Roland Tusch.

Katarína Kristiánová is an architect and urban designer with a focus on the management of urban green spaces. She holds a doctoral degree in landscape architecture and is an associate professor at the Institute of Urban Design and Planning in the Faculty of Architecture and Design of the Slovak University of Technology in Bratislava. She was the head of the former Institute of Garden and Landscape Architecture at the same institution and is ECLAS contact person at the Slovak University of Technology in Bratislava.

Gill Lawson is an associate professor and was head of the School of Landscape Architecture from 2018 to 2022 at Lincoln University in Christchurch, Aotearoa, New Zealand. She is originally from Australia and holds a doctoroal degree from Queensland University of Technology. Her research interests are in landscape pedagogy, landscape visualization, and landscape sociology in Australia, New Zealand, and other Asia-Pacific countries, and on water and plants as catalysts for improving the adaptation of our cities to climate change.

Alice Lewis is a lecturer in landscape architecture at Royal Melbourne Institute of Technology, Australia, where she also earned her doctoral degree. Her design research explores ways of engaging humans as caregivers for landscapes. Dr Lewis' approach to pedagogy builds directly on this research, which is often collaborative and interdisciplinary, and aims to cultivate practices of care between and for human and non-human environments.

Dan Li is a doctoral degree candidate in the Architecture and Design Research programme at Virginia Polytechnic Institute and State University where she has been a teaching and research assistant in the Department of Landscape Architecture. She previously was an adjunct lecturer of landscape architecture at Pennsylvania State University. Her work focuses on education for sustainability in landscape architecture, resiliency design, and research methodology.

Lilli Lička is a professor of landscape architecture at BOKU in Vienna, Austria. Her projects focus on public spaces, housing, heritage sites, urban parks, and green space justice. She co-curates *nextland*, an online collection on contemporary Austrian landscape architecture, and heads the LArchiv, Archive of Austrian Landscape Architecture. She is member of international design boards and academic commissions. She was principal of koselicka from 1991 to 2016 and opened LL-L Landscape Architecture in 2017.

Guangsi Lin is a professor and the head of the Department of Landscape Architecture at South China University of Technology (SCUT). He earned his degree in landscape architecture from Beijing Forestry University (BFU). He was a postdoctoral fellow in the Department of Landscape Architecture at Tsinghua University, and a visiting scholar in the Department of Landscape Architecture at the University of Pennsylvania.

Jen Lynch is a lecturer in the landscape architecture programme at Royal Melbourne Institute of Technology, Australia, and an AILA registered landscape architect. Her research and teaching explore the commons as an alternative reading of landscape and design practice as a form of commoning.

Jala Makhzoumi is an adjunct professor of landscape architecture at American University of Beirut, Lebanon, and president of the Lebanese Landscape Association. She studied architecture in Baghdad, Iraq; received her master's in environmental design at Yale University, USA; and earned her doctoral degree in landscape architecture at Sheffield University, UK. She is the recipient of the 2019 European Council of Landscape Architecture Schools (ECLAS) and the 2021 IFLA Sir Geoffrey Jellicoe Award for her outstanding contribution to education and practice.

Francesca Mazzino is Full Professor of Landscape Architecture at the University of Genoa and Director of the LA Master Programme. She holds a doctoral degree, is vice president of the Italian Scientific Society of Landscape Architecture, and a member of the scientific committee for the journals *Projects de paysage* and *RI-VISTA*. Dr Mazzino is Director for the series Landscape Studies and Researches at the University of Genoa.

She authored more than 100 publications on landscape planning and cultural landscape rehabilitation.

Alexandru Mexi is a landscape architect with a master's degree in cultural studies and a doctoral degree related to 19th-century municipal parks in Romania. He is a researcher at the National Institute of Heritage and he teaches cultural heritage and garden history at the University of Bucharest and at the University of Agronomic Sciences and Veterinary Medicine in Bucharest, Romania. He has been involved in cultural, editorial, and restoration projects dedicated to historic gardens and cultural landscapes.

Petra Pereković is a landscape architect and an assistant professor at the University of Zagreb, Faculty of Agriculture, in Croatia. Her research and professional interests are focused in the field of landscape and urban open space design, perception of landscape, and green infrastructure related with city development. At the time of writing the book, she was the head of the undergraduate study of landscape architecture in the Faculty of Agriculture at the University of Zagreb.

Lucas Períes holds a doctoral degree in architecture from the Universidad de Buenos Aires, Argentina, and a graduate degree in landscape architecture from Universidad Católica de Córdoba (UCC). He works as an architect at Universidad Nacional de Córdoba (UNC), and as professor and researcher at UNC and UCC. Dr Períes is executive co-director of LALI Latin American Landscape Initiative, director of Landscape Institute UCC, and a member of the Standing Committee on Education and Academic Affairs in IFLA Americas. He is the author and co-author of numerous books and papers in scientific journals.

Yves Petit-Berghem is a geographer and a professor at the *École Nationale Supérieure de Paysage* in Versailles, France. He manages the Theories and Approaches to the Landscape Project master's programme and uses a cross-disciplinary approach embracing ecology, geography, and landscape project design. His research focuses on the ecological and socio-spatial dynamics of landscapes as well as the processes which underpin them. His recent work centres on evolutions in the practices of landscape professionals.

Mattias Qviström is a professor in landscape architecture, especially spatial planning, in the Department of Urban and Rural Development at Swedish University of Agricultural Sciences. He holds a doctoral degree in landscape architecture and his research uses landscape theory, relational geography, and history to explore the interplay between landscape and planning. Thematically, his research concerns landscape and planning history, planning for urbanization and densification, urban nature, mobilities, and methods and theories for landscape planning.

Katarzyna Rędzińska holds a doctoral degree from Warsaw University of Life Sciences in Poland. She is a landscape architect and works as an assistant professor at Warsaw University of Technology, Faculty of Geodesy and Cartography, Department of Spatial Planning and Environmental Sciences. Her research focuses on integrated landscape planning. She is an author/co-author of 18 projects in landscape architecture, two of them awarded. She is a member of the Association of Architects of the Polish Republic.

Biographies

Dunja Richter is a landscape architect and a senior researcher at the Institute of Landscape and Urban Studies at ETH Zurich, Switzerland. From 2017–2021, she was significantly involved in the development of the new Master in Landscape Architecture, since 2019, she has been the programme manager. She holds a PhD concerning the plant trade of the Zurich Botanical Garden in 19th century, which was honored by ETH Zurich. Her research and teaching focus on garden history, especially on exchange processes, the meanings, and uses of plants.

Bianca Maria Rinaldi is Associate Professor of Landscape Architecture at *Politecnico di Torino*, Italy. She is the recipient of a fellowship in garden and landscape studies from Dumbarton Oaks Research Library and Collection, a fellowship for experienced researchers from the Alexander von Humboldt Foundation, and of a J.B. Jackson Prize from the Foundation for Landscape Studies. She served as an editor of *JoLA, Journal of Landscape Architecture* and currently is a member of its advisory board.

Ana Duarte Rodrigues is a professor in the Department of History and Philosophy in the Faculty of Sciences of the University of Lisbon, Portugal. She is the coordinator of the Interuniversity Centre for the History of Science and Technology (CIUHCT). She is also the coordinator of several research projects and has published intensively on gardens and landscape studies through the perspective of history of science and technology. She is the editor of *Gardens and Landscape Journal*, published by Sciendo.

Alena Salašová is an associate professor and dean of the Faculty of Horticulture and head of the Department of Landscape Planning at the Mendel University in Brno. Her main research focus is the identification and protection of historical cultural landscapes, landscape character assessment, adaptation to climate change, and the implementation of the European Landscape Convention in Czech Republic. She teaches landscape planning, landscape ecology, and master planning.

Chiara Santini is a full professor in history of gardens and designed landscapes at the *École nationale supérieure de paysage* in Versailles, France. Trained as an archivist-paleographer and historian, her research activity focuses on the design of gardens and public promenades in France between the 17th and 19th centuries. She is a member of the Administrative Board of Villa Adriana and Villa d'Este at Tivoli (UNESCO), and the master plan commissioner of the Potager du Roi in Versailles.

Katrina Simon is a designer and visual artist with a background in landscape architecture, architecture, and fine art in Australia. She gained her doctoral degree from the University of Sydney in Australia, and her research interests focus on the expression of memory and its loss in landscapes, cartography and landscape representation, design research methods, the history and design of cemeteries, and the impacts of earthquakes and disasters on the urban landscape. These topics are explored through a blend of traditional and non-traditional creative research methods.

Vladimír Sitta is Head of the Department of Landscape Architecture and Head of the Studio Sitta. He studied landscape architecture at MZLU in Brno, Czech Republic. He holds a master of science degree. He worked as a landscape architect in Czechoslovakia, Germany, and Australia. He founded Terragram in 1986, and was professor at the University of Western Australia.

He won the Peter Joseph Lenné Prize in 1981 and the President's Award of the Australian Institute of Landscape Architects in 2002.

Richard Stiles studied botany and landscape design at the universities of Oxford and Newcastle upon Tyne before working as a landscape architect in England and Germany. He joined Manchester University as Lecturer in Landscape Design, from where he was appointed Chair of Landscape Architecture at Vienna University of Technology, Austria, where he is now Professor Emeritus. He is a past president of ECLAS and led the LE:NOTRE European network project from 2002 to 2013.

Ján Supuka is Professor Emeritus of Landscape Architecture at the Slovak University of Agriculture in Nitra, where he was ECLAS contact person for many years. He worked as a researcher in dendrology, park and landscape design in Arboretum Mlyňany of the Slovak Academy of Sciences. His research focus is woody vegetation structures in urban and rural landscapes and the design of recreational spaces. He taught landscape planning and design, culture and perception of landscapes, and design of open spaces for recreation.

Simon Swaffield holds a doctoral degree from Lincoln University, New Zealand, where he taught landscape architecture from 1983 to 2018, served as Head of Section and Department, and was appointed New Zealand's first Professor of Landscape Architecture in 1998. He is now Professor Emeritus. His research and publications focus on rural landscape change, landscape assessment, landscape architecture theory, and research methodology.

Katalin Takács is a landscape architect in Hungary, with conservation expertise in the garden heritage field. She holds a post-graduate degree with work regarding the conservation and restoration of architectural heritage. Her research topics are the management of former manorial estates and the inventory of historic gardens in Hungary. Dr Takács is currently working as an associate professor at the Institute of Landscape Architecture, Urban Planning and Garden Art (MATE, Budapest), and as a freelance researcher and designer.

Chika Takatori is an associate professor at the Graduate School of Design, Kyushu University, Japan. She completed her doctoral studies at the University of Tokyo. Her professional background is urban planning and landscape ecology. Dr Takatori has conducted several empirical studies for sustainable landscape planning in collaboration with local stakeholders, and has been a committee member for more than 40 local governments. She has won several awards, including the encouragement award at the Japan City Planning Society.

Zydi Teqja is a professor at the Agricultural University of Tirana, Albania. He holds a doctoral degree and has many years of teaching experience in environmental sciences, green areas, and landscape architecture. Dr Teqja has broad knowledge of sustainability and green space planning, design, and management. His main research interests are the impact of green space on the health and well-being of a population and the impacts of climate change on the distribution of horticultural plants.

Attila Tóth holds a doctoral degree from the Slovak University of Agriculture in Nitra where he works as associate professor and head of the Institute of Landscape Architecture. His main research focus is green infrastructure in urban and rural landscapes from a planning and design perspective. He is chair of LE:NOTRE Institute, ECLAS contact person, and IFLA Europe delegate of Slovakia. He is a member of the Committee for Landscape Architects in the Slovak Chamber of Architects. He teaches master's and bachelor's design studios focusing on public open spaces and landscapes.

Roland Tusch is a senior scientist at the Institute of Landscape Architecture at BOKU, Vienna, Austria. He studied architecture and holds a doctoral degree from the Vienna University of Technology. He does research in the field of landscape and infrastructure. One focus is on the UNESCO World Heritage Site Semmering Railway, where he is a member of the ICOMOS monitoring team. He is also on the team of the LArchiv, Archive of Austrian Landscape Architecture.

Jan Vaněk is an associate professor and head of the Department of Garden and Landscape Architecture at the Faculty of Agrobiology, Food and Natural Resources at the Czech University of Life Sciences in Prague. He is an architect by training and owner of the W/4 Architects design studio. He teaches project design, urban green space design, landscape creation and introduction to garden and landscape architecture, as well as the history of architecture and art.

Sophie von Schwerin is a trained gardener and landscape architect who graduated from Technical University of Berlin. She worked for the historic gardens of Baden-Württemberg and holds a doctoral degree from Leibniz University in Hanover, Germany, on the botanical garden of the Herrenhäuser Gärten in 2011. She has been working at OST, Eastern Switzerland University of Applied Sciences Rapperswil, in the field of garden history since 2012 and became curator of the *Archiv für Schweizer Landschaftsarchitektur* in 2014.

Sue Anne Ware is a design research practitioner in Australia. Her projects are concerned with contesting the public realm, conflating intersectionality, provoking social engagement, and embracing radical hope. Her most recent work explores regimes of care and notions of feral; resulting in a wild, unrestrained, or uncultivated state as they pertain to contemporary and historic practices in ecology, landscape architecture, and gardening.

Graham A. Young is a professional landscape architect and academic who taught in the Department of Architecture, University of Pretoria, South Africa, for over 30 years. He has published broadly on landscape architectural issues, including cultural narratives in landscape design. He is a practising landscape architect whose projects have received local and international design awards. He is currently the president of the African Region of the International Federation of Landscape Architects.

FOREWORD

Today, I had a conference call with Landscape Architecture Institute staff members of the Ukrainian National Forest University in Lviv. Not an easy conversation; until we started to speak about ландшафт, landscape. Minds began detaching themselves from horrors of the present and shifted towards the future as soon as we moved our conversation to landscape, towards a future in a better landscape to which we all will contribute. People immediately started to make plans: Which knowledge and skills do we need? How might we build capacity? Which research is necessary to understand the many facets of the Ukrainian landscape? How might we prepare landscape architecture for the role it should play in the reconstruction of not only physical landscape, but also the mental one that gives a shared foundation to a disrupted society?

With a terrible war raging at the time of this writing, the content of this book appears in a new light to me. Its chapters present multiple stories from a discipline that has always believed in the idea of a better future, and that will continue to do so. While landscape design education unfolded, during the past 200 years, the conditions for achieving positive impact have grown and become more challenging. Our individual share of expanding bodies of knowledge is becoming proportionally smaller, day-by-day, as science in all fields is advancing rapidly. At the same time, the impact of our individual behaviour on this world becomes larger, day-by-day. What is it that a landscape architect really needs to know? We will continue to ask this question. However, one aspect seems to be certain: landscape architects have to be able to build trust.

My special thanks go to Stefanie Hennecke, Diedrich Bruns, and the editorial team at Kassel University for their fantastic work on this book. I also thank the landscape architecture academic community, both European and international, for sharing their valuable knowledge about landscape architecture education. This knowledge will become the basis for envisioning a better future also for our own educational practice. There is much to do. Let us do it!

Ellen Fetzer
ECLAS President since 2018
April 2022

INSPIRATIONS AND ACKNOWLEDGMENTS

It was in September 2019, during the conference *Lessons From the Past, Visions for the Future: Celebrating One Hundred Years of Landscape Architecture Education in Europe*, when the idea was born to prepare a book on education. The Norwegian University of Life Sciences hosted the annual conference of European Council of Landscape Architecture Schools (ECLAS) in Ås at the campus where Europe's first landscape architecture programme began enrolling students, in 1919. Aiming for an international view, the conference organizers invited scholars from several countries to report and discuss education research. Inspired by stimulating studies, by sharing new insights, and by comparing many stories, the editors of this book became fascinated with striking similarities and, at the same time, puzzled by a multitude of differences. We decided to weave together threads and stories of personal ambition and of social and political development and mechanisms under which landscape architecture education emerges and thrives.

The organizing committee of the 2019 ECLAS conference provided helpful comments and offered support in this project, for which we are grateful. We also would like to acknowledge the encouragement we received from participants of the special conference session *Bridging National and Disciplinary Boundaries: Concepts of Sustainability in Landscape and Urban Planning Education. Case Studies From Different European Countries*. To widen the view, the group that formed at Ås took a snowball approach to collecting reports from scholars who had not attended the conference. The group of authors continued to grow when four book proposal reviewers recommended widening the geographic scope and the range of cases beyond Europe and North America. We placed a call to invite contributions from Latin America, Africa, Australia, the Middle East, and Asia. The response was very encouraging, an expression, indeed, of the growing importance of the discipline on all continents. Together, 60 authors decided to collaborate in offering inspiring insights into and about education from around the globe.

Inspirations and acknowledgments

We would like to thank Mira Engler, Avigail Sachs, Terry Clements, and Adri van den Brink for their supportive reviews and suggestions. They all agreed this book would be useful to students and educators, to administrators involved in education policy and decision making. The sudden death of Adri van den Brink during the writing of this book is a great loss for us all.

At Taylor & Francis, we appreciate how the Routledge Planning, Landscape, and Urban Design publications committee supported the book project from the start. In particular, we are grateful to Kathryn Schell and Megha Patel for their continuous support. Reviewing the correspondence we had with the two of you over the course of well over one year, we see how you have responded to dozens of questions, always in a most friendly and supportive manner.

At Kassel, we thank Beatrice Pardon, the landscape architecture student who worked as editorial assistant, with great diligence and endless patience, to organize hundreds of documents into one coherent file, and who greatly supported the project. We also thank Chanda Hess, research assistant and great aide in reviewing manuscripts, and Margarete Arnold for generating graphic illustrations. Any remaining mistakes are the responsibility of the editors.

Finally, our thanks go to all authors who, each in their own way, committed themselves to developing the field of education studies. In the wonderful collaboration we are fortunate to have enjoyed since the fall of 2019, contour and structure of the field emerged that provide the context for studies of planning education in general and of landscape architecture education in particular. The field is on a good way of acquiring a high degree of legitimacy as an exciting area of research.

Diedrich Bruns and Stefanie Hennecke

1
BROADENING THE OUTLOOK, EXPANDING HORIZONS

Diedrich Bruns and Stefanie Hennecke

Landscape architecture education – a promising field of study

Today, in 2023, 200 years after the Prussian *Gärtnerlehranstalt* (Gardener Academy) started offering formal design education in 1823, professional landscape architecture programmes are available to students in all parts of the world. The time has come to talk of lessons learned and about shaping the future. Aiming to offer a diverse international voice, this handbook reports on education studies by way of weaving together societal and political context and the emergence of education where landscape architecture programmes exist, and where they are new or currently planned. As the discipline is expanding, there is much to add to the recent educational tiering that is trying to make sense of the older (European and North American) models. As academics in different areas of the world are developing education, explicit discussions of their stories would prove helpful in being more inclusive than we have been. As they are building and justifying new programmes with landscape emphasis, educators and academic administrators in Africa, Asia, Australia, the Middle East, and Latin America may find an interest in the reports compiled in this volume. In the same vein, education and education research should be moving along the global sustainability agenda, aiming for an inclusion of people and places that are most effected by landscape sustainability challenges.

During the 2019 Conference at Ås, Norway, ECLAS, the European Council of Landscape Architecture Schools, presented two books, *The Routledge Handbook of Teaching Landscape*, and *Teaching Landscape: The Studio Experience* (Jørgensen et al. 2019, 2020). These books provide a "wide-ranging overview of teaching landscape subjects (. . .) reflecting different perspectives and practices at university-level landscape curricula. Focusing on the didactics of landscape education", the books present and discuss "pedagogy, teaching traditions, experimental teaching methods and new teaching principles" (Jørgensen et al. 2019). While these books describe learning processes and provide examples that help understand how learning takes place, the *Routledge Handbook on Landscape Architecture Education* reports on research into higher education, including politics and policy, schools and people, ideas and visions, and experiments that constitute the past, present, and future of education. It covers four levels (or cycles) of learning; undergraduate (bachelor's) and graduate (master's) programmes, post-graduate or doctoral studies, and continuing education.

Fifty years ago, in 1973, Gary O. Robinnette had published *Landscape Architectural Education*, arguably the first scholarly treatment of education in the field.[1] In the introduction, he reports

DOI: 10.4324/9781003212645-1

on "glaring gaps" in information available about education and expresses hopes that "this compendium" will "spin-off or generate many other studies, dissertations and books". Education studies have been gaining prominence ever since. Practitioners rate education as one of the most "useful" research domains for the discipline (Meijering et al. 2015, p. 91). Scholars are interested in stories of individual schools that left educational legacies, for example in Britain (Lancaster 1986; Roe 2007; Woudstra 2010), in Germany (Nothhelfer 2008; Hennecke 2021), Norway and Sweden (Jørgensen and Torbjörn 1999), and the USA (Zube 1986). The field of education studies expanded into taking international views (for example Birli 2016; Fischer and Wolschke-Bulmahn 2016; Gao and Egoz 2019), discussing questions of specializations such as landscape planning (for example Steinitz 1986; Ogrin 1994), teacher-student interaction (for example Austerlitz et al. 2002; Smith and Boyer 2015), and the role of science in design education (for example Nassauer 1985; Corner 1997).

The *Routledge Handbook of Landscape Architecture Education* aims for landscape architecture to join the ranks of disciplines that systematically engage in education research, including education history research, defining education as a special subject of study distinct from but, from case to case, linked to research into professional practice. Primary purposes of the book are to document the state and to advance the study of landscape architecture education, and to articulate what constitutes high-quality research in this field. Taking a wide geographic scope, the reports collected in this volume bear witness to the events of education unfolding over a wide space and a long time. Readers may compare stories, reported from six continents that occurred over the course of two centuries. Examining landscape architecture education history, policy, programmes and practice in diverse learning environments, chapter authors explore the roles that educators, institutions, and methods play in shaping educational outcomes.

Tying many stories together, this book offers a multifaceted narrative. Reading its chapters together affords exploring parallels and differences among countries, schools, and people with regard, for example, to programme-building, pedagogic development, institutional framework, genealogy of ideas, and learning resources such as textbooks. Comparisons at the international level are possible, building on synchronic and diachronic juxtapositions of cases, for a number of purposes (Tilly 1984). First, everyone benefits from looking into different parts of the world and at happenings at different periods: By bringing out the singular features of each particular case makes understanding of one's own education practice easier. We can assess how much of it might be unique, or not, and benchmark our performance. The second purpose being to reveal divergences and variations within the greater realm of landscape architectural education. We can draw new insights, generalizations, demonstrating how some assumptions about landscape architecture education hold good from case to case and others do not. Third, only by including more than just a handful of cases, can we reliably use comparison also to develop educational typologies and models with considerable generality and wide range of applicability. Attention is called for to limitations associated with adopting comparison as a mode of analysis. Reliability of a comparison depends on the selection of the cases included (for example from the southern and northern parts of the world) and context considered (time, societal structure, size of country, and other). To reach acceptable levels of reliability, more research and cases are needed. The premise is that both education and education research should be evidence-based and theory-based.

Ambitions and opportunities

Landscape architecture education research is part of the branch of higher education research dealing with professional forms of learning (Jarvis 1983). Categories of inquiry are research *about* education, *for* education, and *through* teaching and learning practice. Contributions to this compendium mainly belong to the first category and partly to the second category. They are about the history of landscape architecture education, the institutes of higher learning and their staff who facilitate professional learning. They include, for example, studies into programmes and curricula, schools and resources, learning methods and formats, and so forth. Education history, a branch of historical research that studies disciplinary history, acquaints us with genealogies of individuals and ideas, visions, experiments, and lessons that constitute education in the field. Research for and in support of teaching and learning includes, for example, case studies where educators and education studies focus on studio, and on teacher-learner interaction.

Aiming to becoming part of the wider field of Higher Education Research, landscape architecture scholars are looking to reach new horizons, including opportunities to publish findings in more than one of the 80 academic journals of global distribution in the field (Tight 2018)[2]. A challenging vision, for, at least in the past, educators did "not write extensively" (Robinette 1973, p. vii). Searching content inventories and analysing thematic coverage of *Landscape Journal*, *Landscape Research*, and *Landscape and Urban Planning* reveals how, for a long time, education and pedagogy were among "the least conferred subjects", while, more recently, numbers of articles started to grow (Powers and Walker 2009, p. 104; Gobster et al. 2010; Gobster 2014; Vicenzotti et al. 2016). When landscape architecture scholars did venture into publishing findings from education studies in peer-reviewed journals, only a small number of papers included empirical research based on sound sampling of reliable evidence (Meijering et al. 2015). Doing research was mainly part of educational practice (Vroom 1994). Studies are of the research-through-education type; they are practitioner-led and localized in focus. Educator-researchers are looking into their own teaching and the learning experiences of their students. Occasionally, educators conduct evidence-based research, such as studies aimed to determine effective means of teaching, learning, and effective structural and instructional designs of university courses (for example: Brown et al. 1994; Stoltz and Brown 1994). The early education research can be categorized as "insider research" where it is easy to recruit (and obtain consent from) study participants, and the researcher has a deep understanding (opinion and bias) of the context and culture in which the study is conducted (Trowler 2014).

Since it is difficult to extract generally applicable evidence from a few singular cases that engage short-lived projects, the number and the time span of investigations must increase. Two approaches are to (1) engage in trials and longitudinal research to collect empirical evidence, and (2) generate an overview of and connect past and current research activities around the globe. Following up on Jørgensen et al. (2020) who made the start and collected substantial numbers of existing cases on studio teaching, the *Handbook of Landscape Architecture Education* goes a step further and links hitherto disparate educational studies from different countries and perspectives. To realize greater research and reporting ambitions, in the future researchers might pool resources needed for larger investigations and collaborate internationally and interdisciplinarily in pursuing long-term studies, such as repeated surveys and responding, for example, to challenges of developing sustainability education.

To comply with research ethics, the potential tensions between the educator's professional role (e.g. as a lecturer) and their role as "insider researcher" need addressing (Israel and Hay 2006). In the same vein, education research needs to consider and promote principles of inclusiveness as stipulated for all qualitative research (Czymoniewicz-Klippel 2010). Scholars will become aware of their ethical responsibilities and take inclusive research approaches serious. Taking inclusion a step further, they also must find ways to capture the diversity of landscape architecture education reflecting the cultural (and other) richness it exhibits in the many countries in which it practices. Together, the authors of this handbook take the initiative to compare, as many experiences and models as possible, making the case that taking international perspectives is indispensable not only from ethical points of view, but also for enabling comparative analysis that includes the breadth and complexity of the world's landscape architecture education. The ambition, then, is to prepare a foundation of the solidity required for theory building.

From professional introspection to interdisciplinary research

Working at a university requires of academic staff to engage in systematic research, a challenging task for an originally non-academic discipline. In addition, landscape architecture, in particular, is a field researched by people working within and by way of looking inward, examining one's own achievements, and one's own values, even opinions. The same holds for higher education where learning and teaching is the sector of activity that researchers study. In the future, the discipline will require significant advances in performing, reporting and discussing education research. This research will be basic and applied; it will combine disciplinary and interdisciplinary approaches.

Research expertise can, but need not reside in a single scholar. For conducting "interdisciplinary" studies, in the past, individual scholars acquired expertise in, for example, education research, historic research, and so forth. However, it might prove advantageous to distribute expertise across research teams in which researchers from different disciplines collaborate (Singer et al. 2012, p. 2). Academics who move between disciplines need a particular set of skills and knowledge if they should wish to undertake structured enquiry into learning and teaching practice (Cleaver et al. 2014). For many landscape architecture scholars, the focus and aims of higher education research will be different to their disciplinary and subject-based research. It may require developing new research approaches, and using new methodologies. There would be great interest and scope in learning about a variety of approaches to doing research. Interdisciplinary teams are able to raise questions or point at issues that are overlooked, or taken for granted, within one discipline alone. Both education and history research developed specializations, for example language education and history, business education and history, science education and history; now also landscape architecture education and history. Both education and history can be viewed as "exporter" disciplines with educators and historians contributing to other disciplinary areas.

Engaging in interdisciplinary research opens up opportunities for landscape architecture scholars to publish outside of discipline-based journals. Discipline-specific education research is of concern not only to the discipline itself but is of interest also to education in general, and to history, policy, and other areas. Good work will gain the attention for example of historians who are interested in histories of education as a way of discussing professional history. In return, education, pedagogy and history researchers might take the opportunity and publish findings in professional journals including landscape architecture. Filling the reporting gap between education- and profession-based research; landscape architecture might take architecture and planning as examples. Journals such as the *Journal of Architectural Education* (*JAE*) and the European Association for Architectural Education bring research in the field of professional education from different scholars and countries together. In planning, scholars publish reports on education in *Planning Perspectives* (Hennecke et al. 2018), the *Journal of Planning History* (Hise 2006), and in books on planning history (Hein 2018).

Spanning across disciplinary boundaries, applying mixed methods

Staying focussed on education is a major challenge to anyone disciplinarily rooted. The curious scholar is tempted to stray into the depths of design and landscape history each time we study education. We get all excited about the wealth of information gained by reading old books and journals, and through site visits, expert interviews, and from documents found in archives and museums. Repeatedly, while discovering the ages through words, images, artefacts and the built environment, we need to remind ourselves how our subject is education. We define the study of education as a sub-field of its own right within the disciplinary field of landscape architecture (see Figure 1.1). As landscape architecture history is set within the wider field of planning history, landscape architecture education history contributes to the history of planning education.

Education researchers tend to make use of generic research methods, such as literature and documentary analysis, surveys, interviews, and mixed methods of biographical studies. Landscape architecture education research also uses conceptual, observational and experimental methods that scholars have specifically developed in the context of teaching and learning practice. They are of special interest to the education field as few of the kind are reported in education research journals (Tight 2020). Overcoming discipline-specific boundaries, the scholars presenting research findings in this book apply a mixed methods approach. They took no small investment of energy to learn the ground rules of historic and education research; compiled and analyzed a great variety of sources; and put diverse content into regional, historic, and typological context. Methods that chapter authors primarily used include:

- Citation searching and searching relevant electronic databases and Internet sources for books, journal articles, surveys on study programmes, curricula, staff and student rosters, and so forth.
- Handsearching: Papers and reports which are not indexed in electronic databases and that appear in printed-only periodicals and conference proceedings of professional associations and publications of organizations, interviews printed in journals and university periodicals, documentations of research projects, documentation of special events such as anniversaries, inauguration and retirement, documentation of curriculum vitae, obituaries.

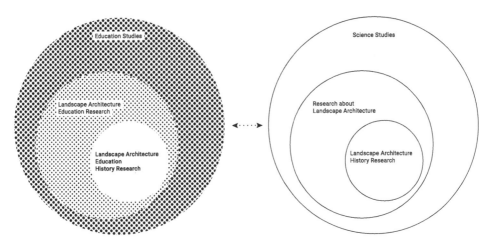

Figure 1.1 Landscape architecture education research is part of education studies and includes education history research. As Education Studies correspond to general Science Studies, research into professional education corresponds to research about Landscape Architecture as a professional field, including its history.

Source: Visualized by Diedrich Bruns with credit to Margarete Arnold.

- Archival searches: Authors delved into institutional archives, sampled and assessed what they found, such as student files and student's works, minutes of political and administrative meetings, study programme planning, correspondence, and much more.
- Conversations: Several researchers engaged in correspondence and conducted and recorded interviews with education administrators and practitioners, and analyzed the content, compared statements.
- Experience reports: Some authors used their own experience as administrators and managers, as teachers, as networkers, as people who are active in day-to-day teaching, in the conception and establishment of study programmes.
- Quantitative studies: Making use of social science methods, some researchers conducted empirical studies in class; invited educators and administrators to respond to surveys and questionnaires.

For landscape architecture education research to become truly interdisciplinary, however, the field would in the future need to include scholars from education, history, sociology, library science, and others. Research methods from a number of different fields, combined effectively with frameworks and techniques from planning, would provide a more robust understanding of education in the field of landscape architecture.

Structure and content of the book

The book has four parts framed by introduction and conclusions. Part I introduces the field of landscape architecture education research and links eight contributions that address three central challenges: (1) historical awareness and responsibility in the profession, (2) international networking, and (3) education for sustainable development. At (1), Deming analyzes observations from several years of teaching history. By collecting and comparing ideas and paradigms that appear in landscape architecture debates and practice, she traces how they evolve over time and discusses what they mean in the world today. In the same vein, Qviström and Jansson employ student projects, looking into densification as an example of ideas that designers pursue in a dogmatic manner. Combining landscape and planning history to examine contemporary debates, they argue for reflexive and critical examinations of discourses or taken-for-granted practices. At (2), Stiles discusses the role of international exchange in the development of landscape architecture education. Lička et al. as members of the International Network of European Landscape Architecture Archives (NELA) explain the role of archival material. At (3), seeking explanations for some of the complex und multifaceted challenges of teaching and learning, four scholars are presenting examples of learning for sustainability. Fetzer discusses the emerging role of landscape architecture in the light of a global sustainability agenda. Geelmuyden is investigating how programme managers and educators may find a balance between theory and practice. Galan makes the case for conceptual models and integrative thinking by analyzing a new educational format. Li discusses pedagogy and didactics in the context of Education for Sustainable Development.

Parts II, III, and IV report on past and current education development. Part II "Agendas and Standards" compiles findings on early study programmes. Fifteen chapters tell stories of how university education began in Europe and North America. Taking 19th-century vocational schools and colleges as their point of departure, Birli et al. present education pioneers who were involved in establishing the first schools and educational requirements in Europe. Dümpelmann discusses beginnings of education in the USA. Taking biographic approaches, Jørgensen, Hopstock, Csepely-Knorr, Duarte Rodrigues, and Debie present findings about instructors who were the first to offer landscape architecture programmes in Norway, Germany, the United Kingdom, Portugal, and the Netherlands. Employing methods of document and literature analysis, Gröning traces ideologies of National Socialism that affect landscape architecture education, in this case. Mexi chronicles the education history of Romania, a country located at the crossroads of Central, Eastern, and Southeastern Europe. Krippner and Lička present studies from Austria, and von Schwerin and Richter from Switzerland. Kamenecki and Pereković explore the education history of Croatia, Fekete that of Hungary. Tóth et al. compare educational developments in Slovakia and the Czech Republic as an example of countries that experienced waves of unity and separation.

Part III "Broadening the Common Ground", includes examples for regions where garden and landscape design cultures have developed for centuries, but the introduction of study programmes was conducted mainly or partly independently from antecendents of landscape architecture. In five chapters from China, Japan, the Middle East, Italy, and France, their authors invite us to follow the path of ideas, which, in the history of garden design, first emerge in Asian and Arabian cultures, then in Mediterranean and other European countries and beyond. Takatori, Lin, and Gao present studies on Japanese and Chinese education. With reference to schools of the Arabian world, Makhzoumi reports on landscape architecture education in the Lebanon. Representing the Mediterranean Region, Rinaldi and Mazzino highlight the Italy case. For examples from north of the Alps, Blanchon et al. report on cases from France.

In Part IV "Aiming for Justice, Reconciliation, and Decolonization", 17 authors discuss insights into the search for models in the introduction of study programmes in regions that are experiencing effects of social and political transition that landscape architecture education addresses. By implementing a practice of educational policies of justice, reconciliation and decolonization, we are advancing the discipline as a whole. Pertinent research includes, for example, comparing educational and design traditions and developments east and west of the (former) Iron Curtain, and south and north of the Equator, also comparing education in different societal and political contexts, different institutional settings, and different schools of thought. Regarding concepts of design and landscape, we learn about considering different design cultures, how different landscape words meaning different things, and how many peoples use different words altogether for expressing design and landscape ideas. European cases include examples from countries where universities developed landscape architecture education in the context of political changes for example after the fall of the Iron Curtain. In this group, Teqja investigates the case of Albania. Cieśla and Rędzińska are looking for a balance between scientific knowledge and political realities by examining educational programmes in Poland. In a second group, authors discuss examples from Africa, Latin America, and Oceania. They show how important it is to include identities and self-determination of ancestral cultures in the development of study programmes. Swaffield et al. explore how education developed in New Zealand, Aponte et al. present cases from Latin America, Young reports on Pretoria in South Africa, and Bryant et al. discuss education in Australia.

The conclusion discusses perspectives for future education and research. We review subjects and research findings compiled in the handbook, draw up a list of crosscutting themes, and discuss opportunities for comparative analysis. We are hoping for an interested and diverse readership, for everyone to be curious about the richness of stories, eager to compare different concepts and developments, initiatives and methods that this book holds in store. Most of all, we wish everyone much enjoyment in discovering, in the great matrix of findings compiled in this volume, parallels and cross-references, contradictions and inspirations to develop new ideas in order to further develop the field of landscape architecture education research.

Notes

1 Data collection began earlier. The Committee on Education of the (ASLA) started to collect information on landscape architecture higher education in 1911, and in 1968, the International Federation of Landscape Architect's (IFLA) Committee on Education published a first report on survey of educational institutions in the world. In 1972, Albert Fein presented "A Study of the Profession of Landscape Architecture, the Education Study" including a description of the curriculum of ASLA-accredited schools (published by the American Society of Landscape Architects, Virginia).
2 For example, the *Journal of the Learning Sciences, Educational Researcher, The Journal of Educational Research, The Journal of Research in Science Teaching*, and other journals regularly publish scholarly articles that are of interest to the education research community and that come from a wide range of areas of education research and related disciplines.

References

Austerlitz, N., Aravot, I. and Ben-Ze'ev, A. (2002) 'Emotional Phenomena and the Student-Instructor Relationships', *Landscape and Urban Planning*, 60(2), pp. 105–115.

Birli, B. (2016) *From Professional Training to Academic Discipline: The Role of International Cooperation in the Development of Landscape Architecture at Higher Education Institutions in Europe*. Doctoral thesis, Technical University of Vienna, Vienna.

Brown, R. D., Hallett, M. E. and Stoltz, R. R. (1994) 'Student Learning Styles in Landscape Architecture Education', *Landscape and Urban Planning*, 30(3), pp. 151–157.

Cleaver, E., Lintern, M. and McLinden, M. (2014) *Teaching and Learning in Higher Education: Disciplinary Approaches to Educational Enquiry*. London: SAGE.

Corner, J. (1997) 'Ecology and Landscape as Agents of Creativity', in Thompson, G. F. and Steiner F. R. (eds.) *Ecological Design and Planning*. New York: John Wiley & Sons, pp. 81–108.

Czymoniewicz-Klippel, M., Brijnath, B. and Crockett, B. (2010) 'Ethics and the Promotion of Inclusiveness within Qualitative Research: Case Examples from Asia and the Pacific', *Qualitative Inquiry*, 16(5), pp. 332–341.

Fischer, H. and Wolschke-Bulmahn, J. (2016) *Travels and Gardens*, symposium report. Hannover: Druckerei Hartmann.

Gao, L. and Egoz, S. (eds.) (2019) *Lessons from the Past, Visions for the Future: Celebrating One Hundred Years of Landscape Architecture Education in Europe*. Ås: School of Landscape Architecture, Norwegian University of Life Sciences.

Gobster, P. H. (2014) 'Mining the LANDscape: Themes and Trends Over 40 Years of Landscape and Urban Planning', *Landscape and Urban Planning*, 126, pp. 21–30.

Gobster, P. H., Nassauer, J. I. and Nadenicek, D. J. (2010) 'Landscape Journal and Scholarship in Landscape Architecture: The Next 25 Years', *Landscape Journal*, 29(1), pp. 52–70.

Hein, C. (ed.) (2018) *The Routledge Handbook of Planning History*. New York and London: Routledge.

Hennecke, S. (2021) 'Reflections on Landscape Architecture Education in Germany in 2019, with Reference to the Bologna Process', in Wolschke-Bulmahn, J. and Clark, R. (eds.) *From Garden Art to Landscape Architecture. Traditions, Re-Evaluations, and Future Perspectives* (CGL-Studies 28). München: Akademische Verlagsgemeinschaft (AVM), pp. 141–152.

Hennecke, S., Kegler, H., Bruns, D. and Reinert, W. (2018) 'Centre for Urban & Landscape Planning History (CUL), established at Kassel University, Germany', *Planning Perspectives*, 33(3), pp. 449–453.

Hise, G. (2006) 'Teaching Planners History', *Journal of Planning History*, 5(4), pp. 271–279.

Israel, M. and Hay, L. (2006) *Research Ethics for Social Scientists. Between Ethical Conduct and Regulatory Compliance*. London: SAGE.

Jarvis, P. (1983) *Professional Education*. London: Routledge.

Jørgensen, K., Karadeniz, N., Mertens, E. and Stiles, R. (eds.) (2019) *The Routledge Handbook of Teaching Landscape*. London: Routledge.

Jørgensen, K., Karadeniz, N., Mertens, E. and Stiles R. (eds.) (2020) *Teaching Landscape. The Studio Experience*. London: Routledge.

Jørgensen, K. and Torbjörn, S. (1999) 'Om etableringen av landskapsarkitektutdanningen i Norge og Sverige', in Eggen, M., Geelmuyden, A.-K. and Jørgensen, K. (eds.) *Landskapet vi lever i*. Oslo: Norsk Arkitekturforlag, pp. 251–267.

Lancaster, M. L. (1986) 'Education for Landscape Architecture in Britain', in Jellicoe, G. Jellicoe, S., Goode, P. and Lancaster, M. (eds.) *The Oxford Companion to Gardens (Oxford Companions)*. Oxford: Oxford University Press, p. 324.

Meijering, J. V., Tobi, H., van den Brink, A., Morris, F. and Bruns, D. (2015) 'Exploring Research Priorities in Landscape Architecture: An International Delphi Study', *Landscape and Urban Planning*, 137, pp. 85–94.

Nassauer, J. I. (1985) 'Bringing Science to Landscape Architecture', in Stoltz, R. R. (ed.) *CELA Forum 1985, Issues of Teaching and Instructional Development in Professional Education*. Council of Educators in Landscape Architecture, pp. 41–44. Available at: https://catalog.libraries.psu.edu/catalog/2050052

Nothhelfer, U. G. (2008) *Landschaftsarchitekturausbildung – zwischen Topos und topologischem Denken*. Dissertation, Universität Kaiserslautern, Tönning.

Ogrin, D. (1994) 'Landscape Architecture and Its Articulation into Landscape Planning and Landscape Design', *Landscape and Urban Planning*, 30(3), pp. 131–137.

Powers, M. N. and Walker, J. B. (2009) 'Twenty-Five Years of Landscape Journal: An Analysis of Authorship and Article Content', *Landscape Journal*, 28(1), pp. 96–110.

Robinette, G. (1973) *Landscape Architectural Education*, 2 vols. Dubuque and Iowa: Kendall/Hunt Publishing Company.

Roe, M. (2007) 'British Landscape Architecture: History and Education', *Urban Space Design*, 20(5), pp. 109–117.

Singer, S. R., Nielsen, N. R. and Schweingruber, H. A. (eds.) (2012) *Discipline-Based Education Research. Understanding and Improving Learning in Undergraduate Science and Engineering*. Washington, DC: The National Academies Press.

Smith, C. A. and Boyer, M. E. (2015) 'Adapted Verbal Feedback, Instructor Interaction and Student Emotions in the Landscape Architecture Studio', *International Journal of Art and Design Education*, 34, pp. 260–278.

Steinitz, C. (1986) 'World Conference on Education for Landscape Planning', *Landscape and Urban Planning*, 13(5–6), pp. 329–332.

Stoltz, R. R. and Brown, R. D. (1994) 'The Application of a Pedagogical Framework to the Design of University Courses', *Landscape and Urban Planning*, 30(3), pp. 159–168.

Tight, M. (2018) 'Higher Education Journals: Their Characteristics and Contribution', *Higher Education Research and Development*, 37(3), pp. 607–619.

Tight, M. (2020) 'Higher Education: Discipline or Field of Study?', *Tertiary Education and Management*, 26, pp. 415–428.

Tilly, C. (1984) *Big Structures, Large Processes, Huge Comparisons*. New York: Russell Sage Foundation.

Trowler, P. (2014) *Doing Insider Research in Universities: Volume 1 (Doctoral Research into Higher Education)*. Scotts Valley, CA: CreateSpace Independent Publishing.

Vicenzotti, V., Jørgensen, A., Qviström, M. and Swaffield, S. (2016) 'Forty Years of Landscape Research', *Landscape Research*, 41(4), pp. 388–407. https://doi.org/10.1080/01426397.2016.1156070

Vroom, M. J. (1994) 'Landscape Architecture and Landscape Planning in Europe: Developments in Education and the Need for a Theoretical Basis', *Landscape and Urban Planning*, 30(3), pp. 113–120.

Woudstra, J. (2010) 'The "Sheffield Method" and the First Department of Landscape Architecture in Great Britain', *Garden History*, 38(2), pp. 242–266.

Zube, E. H. (1986) 'Landscape Planning Education in America: Retrospect and Prospect', *Landscape and Urban Planning*, 13(5–7), pp. 367–378.

PART I

Challenges and perspectives

2
INTRODUCING THE FIELD OF LANDSCAPE ARCHITECTURE EDUCATION RESEARCH

Challenges and perspectives

Diedrich Bruns and Stefanie Hennecke

Acting in the awareness of history, international networking and alignment with sustainable development goals (SDGs) offer three promising perspectives in future landscape architecture education. Part I of this volume discusses how the three are interwoven.

History is the first. Why is it relevant, even interesting, to scholars to study the history of landscape architecture education? Responding, the first three book contributions consider the "double role" of history, that is, history as (a) a subject that students take, an important part of any curriculum, and (b) a field of knowledge, a subfield of educational research. Three questions arise repeatedly throughout the book: How do we learn (about) history? Who learns which part of the story? What do we learn from historical research? Every time we arrive at some answers to these questions the conclusion is, we must learn to act in the awareness of and take responsibility for our own history. This is true, for example, when we consider the evolution of educational agendas, the construction and presentation of the professional and educational history, such as select numbers of ideas threading along canonical collections of books,[1] and design examples considered relevant from a Western point of view (Deming, this volume). This is true also realizing what we learn through and in the context of education (and its *schools*) about the role of different ideas and ideologies (Qviström and Jansson, Gröning, this volume), about *elitist-* and *Eurocentric*-dominated teaching (Deming, Dümpelmann, this volume), about long-hidden ideas and groups (indigenous, migrant, low income, etc.) (Swaffield et al., Young, Aponte et al., this volume), about developments (outside the mainstream landscape architecture) in the relationship between professional history and educational history.

Elen Deming uses the example of teaching history to illustrate how changeable the historical point of view has and can be, and how politically and culturally problematic any canonization of knowledge is. Deming develops her analysis based on many years of experience teaching professional and landscape design history, and her findings apply to educational history as well: Who are the celebrated role models? Why are curricula designed as they are and implemented as they are? Which social contexts have determined the formation of any of the institutions of higher education, the curricula, the appointment of teachers, and the setting of mission statements? What can we learn from trying to answer these questions for the further development of education in our disciplinary field? Which elements, features, and traditions can and must we question critically? For example, Mattias Qviström and Märit Jansson illustrate how to "crystallize" any history of planning by identifying evidence thereof in the built environment, to make history readable and debatable, and how this critical and sympathetic reading of history can become part of any timely education.

Moving on to the second major educational challenge, we discuss international networking. Several chapter authors demonstrate how educational institutions for landscape architecture have been established worldwide at different times and in different contexts. However, they also make it clear that decisive progress in the design of study programmes was achieved by educators conveying ideas and methods via exchanges through international networks. The establishment of the International Federation of Landscape Architects (IFLA), the European Federation for Landscape Architecture (EFLA), the European Council of Landscape Architecture Schools (ECLAS), and the Council of Educators in Landscape Architecture (CELA) are important milestones. Richard Stiles describes beginnings of international networking and also emphasizes the dependence of networking on infrastructure, such as travel opportunities, on communication possibilities across time zones, on correspondence, and finally on the opportunities that opened up with email, the internet, and video conferencing. The contribution of Lilli Lička and seven co-authors, all members of NELA (the international Network of European Landscape Architecture Archives that was founded at the ECLAS conference in 2019), shows how fruitful the international exchange can be, about the professional history, about traditions and how to assess and evaluate them. The NELA chapter, written by eight authors from different countries (who contribute also individual educational studies to this volume), exemplifies the practice of international exchange on ideas and teaching methods using the example of the integration of archives into landscape architecture education.

A central challenge of higher education is its alignment with visions and principles of sustainable development. It is important to consider Sustainable Development Goals (SDG) and the basic features of the Education for Sustainable Development (ESD) concept. ESD is a learning process that is based on the ideals and principles that underlie sustainability. Individual studies conducted so far offer insights into ESD in design education. Systematic research is needed beyond single cases to provide the basis for landscape architecture educators to align on pedagogic approaches and methods. As curricula include subjects pertaining to ecology, economy, and the socio-cultural-political domain (including governance), landscape architecture education connects with several SDGs and integrates several elements of ESD. Its mission is for students to comprehend how planning determines the future course of action regarding sustainable landscapes, and how all landscape design must meet, as spelled out in the Brundtland Report of 1987, the needs of the present without compromising the ability of future generations to meet their own needs.

Applying a set of ESD specific skills and competencies, Ellen Fetzer assesses the degrees to which landscape architecture study programmes are able to align with ESD learning objectives. Based on her findings, the author emphasizes how the discipline is strong in systems thinking, while detecting a certain vagueness regarding the analysis and assessment of human needs. Perspectives for programme development include strengthening social dimensions of landscape design and exposing students to co-designing experiences, such as community action, strategy building, and collaborative transformation. Schools might also work together with professional organizations to make ESD terminology explicit in educational discourses and practice, and to agree on domain-specific sets of sustainability principles and standards. According to Fetzer, it is important for the discipline to present a joint internationally developed and respected framework for landscape architecture education and to align it with sustainability education, including SDG. In her contribution, Ann-Katrin Geelmyden formulates how teachers would have to work together and jointly develop a curriculum for sustainability education. Considering the academization of education and the roles that practitioners play in education, she reflects on the development of education practice towards a system of scholarly competition. Based on educational theory, she sketches a vision of an educational community working in the spirit of eudaemonia, which is oriented towards knowledge exchange for knowledge generation. Juanjo Galan presents an example of how to implement Fetzer's vision of integrating principles of sustainable development into landscape architecture teaching in a transdisciplinary approach. Galan addresses the question of how education might address the complex requirements of sustainable development goals by fostering conceptual and integrative thinking, working, and teaching in transdisciplinary and interdisciplinary ways. By enhancing and refining planning models, in exchange with practitioners, the students themselves contribute developing contents of study programmes in an interdisciplinary and transdisciplinary manner. Aiming to help educators to gain insights into sustainable teaching, Dan Li reports findings from literature reviews and a survey conducted among 209 educators in the USA. Accordingly, landscape architecture educators appear to be strongest and most effective in pedagogic approaches they use most frequently. Included are methods that have shown to be successfully associated with sustainability teaching such as, among others, problem-based and place-based (or community-based) learning, scenario thinking and networked learning, together with multidisciplinary collaboration, community outreach, and using sustainability rating systems. Opportunities exist for linking landscape architectural fieldwork and community engagement with sustainability teaching and sustainability rating systems.

Note

1 Probably the first textbooks written for design teaching are Gustav Meyer's *Lehrbuch der schönen Gartenkunst* (*Teaching Book of Fine Garden Art*) of 1859–60, and Jean Darcel's *Etude sur l'architecture des jardins* (*Study in Garden Architecture*) of 1875. Of the early scholarly history books, Marie L. Gothein's *Geschichte der Gartenkunst* (*A History of Garden Art*) of 1914, published in English in 1929, is still in print today. Later titles covering much of the same canon include, among others, *An Introduction to the Study of Landscape Design* (Hubbard and Hubbard 1917), *Design on the Land: The Development of Landscape Architecture* (Newton 1971), *A History of Landscape Architecture. The Relationship of People to Environment* (Tobey 1973), *The Landscape of Man. Shaping the Environment from Prehistory to the Present* (Jellicoe and Jellicoe 1975), *The History of Gardens* (Thacker 1979), *The World Heritage of Gardens* (Ogrin 1993), *Landscapes in History* (Pregill and Volkman 1999), *Landscape Design: A Cultural and Architectural History* (Rogers 2001), and *Garden History: Philosophy and Design* (Turner 2005).

3
"A THING IN MOVEMENT"
Landscape history in professional curricula

M. Elen Deming[1]

Introduction

This chapter explores the place of landscape design history in the education of landscape architects. Accreditation standards in the United States require all professional programmes to offer at least one introductory history survey; for many landscape architects, this one class may comprise the whole of their exposure to design history. The challenge is to select and curate case studies from the cornucopia of historical scholarship in such a way that opens beginning students to important landscape values without disorienting and overwhelming them.

Over centuries, scholarly research in design history has mirrored broader social and environmental positions. Given the evolution of professional practice today, inclusive cultural landscape histories are arising to meet the moment, effectively helping designers develop empathy. In attending to themes of race and injustice, landscape scholars acknowledge that oppression of Black and Indigenous people, women, and workers is marbled through the history of landscape architecture. How then shall design history courses adapt to acknowledge and address this sorry reality?

Written from the perspective of a North American academician, this chapter comprises four linked topics: contemporary issues in teaching history, institutions and the knowledge base, principles of restorative history, and formats for future history. Such issues are reciprocal: as social and professional values shape the historical narratives we teach, historical narratives also shape the future direction, values, and thus relevance of the profession. Historians from many disciplines, including landscape architecture, are marshaling exciting new teaching perspectives, tools, and stories. It is now up to educators to integrate these perspectives and reframe the history(s) we tell.

Landscape history now

For two decades I've taught a comprehensive landscape history survey course at three institutions, for both undergraduate- and graduate-level students, in a combination of face-to-face, hybrid, and online formats. Each year, it is far and away my most difficult teaching assignment. I profess that learning the history of the designed landscape is important, as Peirce Lewis writes, because it is society's "unwitting autobiography" (Lewis 1979, p. 12). Despite that pearl of wisdom, many design students do not regard history as intrinsically valuable – at least, not until they begin to understand how landscape shapes their personal autobiography as well (Deming 2019).

History is a social construction and thus a living discipline. Before it is anything else, history is a story, an interpretive narrative about the past. As a noun, the term *history* is understood as something we make – a cultural artefact. Historical narratives can be selective, even designed, to be monumental or stabilizing. This, of course, explains why history is never synonymous with "the past."[2] Landscape history is also analogous to science, referring simultaneously to *a body of knowledge* as well as the *methods* engaged to build and temper that knowledge. But as a method for inquiry, history is also an action – an investigative and constructive operation in constant motion – and thus powerful, positional, and potentially destabilizing.

Feminist cultural historian Aurora Levins Morales puts it this way: "History is the story we tell ourselves about how the past explains our present, and the ways we tell it are shaped by contemporary needs" (2019, p. 71). Compelling historical narratives can shape professional, institutional, social, and political identities as well. It is no wonder that historical constructs are so often deployed to justify professional claims to specialized mission, expertise, and leadership. But no history – indeed, no knowledge at all – is fixed, hermetic, immune to currents of contemporary insight. At best, every history is simply an incomplete project. Or, as Marc Bloch (1886–1944) explains:

> [Research in] history is neither watchmaking nor cabinet construction. It is an endeavor toward better understanding and, consequently, a thing in movement. To limit oneself to describing a science just as it is will always be to betray it a little. It is still more important to tell how it expects to improve itself in the course of time.
>
> (Bloch 1953, p. 12)

For the sake of every landscape architect in training who, in effect, is preparing to construct the historical data and narratives of their own future, we take to heart Bloch's challenge to improve the histories we write. The bigger question is "How?" How nimble and responsive are our history courses to professional and societal changes? How do the institutions that support landscape architecture programmes affect the way histories of the profession are taught? Or, as landscape historian Thaïsa Way writes: "It is clear that history is increasingly at the center of public discourse in powerful ways. So where are landscape architects in this discussion?" (Way 2020a, n.p.).

Teaching history in the culture wars

To tease out the links between historical patterns and practices and their consequences in contemporary environments is obviously important work. For beginning designers, a working knowledge of professional history is useful to help grasp the range and scale, diversity, and magnificence of landscape design achievement. But the kind of environmental humanism that most benefits and transforms design students and helps them understand the deep sources of their own tastes, experience, privilege, and identities has always been difficult to teach. Part of the value of teaching and learning history is to have difficult conversations and to co-construct new historical knowledge with others. Critiquing and naming practices, seeing other cultures, developing empathy: all of it sheds light on one's own condition (Deming 2019).

To question the darker histories of professional institutions, aesthetics, and ideologies, however, has lately become contentious. Following the social paroxysms of the 1960s, college classrooms enjoyed a relatively privileged autonomy, where scholars largely were insulated from external agendas by their expertise – as well as the "academic freedom" bestowed by tenured appointments. We note this privilege is historically contextual; for many reasons it has not been enjoyed equally in all regions. In general, however, popular perception of a left-leaning professoriate, comfortably ensconced in publicly funded institutions, has long enraged conservatives and other groups on the right. Since the 1980s, along with the systematic dismantling of state funding for many public services, the professoriate too has been gradually underfunded, with many tenured lines replaced by contract teaching positions. To make ends meet, increasingly corporate or entrepreneurial models of higher education have been inculcated to reward grantsmanship and commercialization of intellectual property.

Of late, state interference – recently driven by private manipulations of popular antagonism towards expertise (with its whiff of elitism) – has become noticeably strident. Near-daily news reports on the politicization of teaching racial history, for instance, sketch the contours of how the "culture wars" have invaded classrooms (Redman 2021; Romero et al. 2022). Across the United States (as well as in Europe and elsewhere), a sweeping tide of conservatism impugns educators for teaching "threatening" and "uncomfortable" histories of racial and social justice. Books are banned, teachers are hounded, school board meetings descend into uncivil chaos, and, in several statehouses, new legislation aims to prohibit anti-racist teaching in secondary schools and publicly funded universities (Robertson 2021; Young 2022).

Distinguished university academicians have been condemned in the press and social media for teaching certain subjects and forms of history. The saga of Nikole Hannah-Jones, Pulitzer Prize–winning journalist and founder of *The 1619 Project*, famously denied tenure at UNC Chapel Hill, is only the most visible example (Hannah-Jones et al. 2021; Hannah-Jones 2021). But then this is part of a political playbook, is it not?

> One of the first things a colonizing power, a new ruling class, or a repressive regime does is attack the sense of history of those they wish to dominate by attempting to take over and control their relationships to their own past. . . [and] by interfering with the education of the young.
>
> (Levins Morales 2019, pp. 69–70)

As with all forms of knowledge, historical narrative in landscape architecture responds to the intellectual schema and values of its composers. Historical narrative is a mirror we hold up to ourselves: it shows us who we think we are, how far we judge we have come, and where we hope we may be going. Critical historical perspectives are cast as "revisionist" only when they shine unflattering light on a cherished set of myths or undermine justification for the claims of the powerful.

Two recent spasms in public opinion resulting from critical historical scholarship make this plain: widespread and accelerating removal of Confederate monuments from public spaces, and anti-intellectual attacks on critical race theory (CRT).[3] Both issues have parallels and implications for landscape architecture. Many iconic designed landscapes – involving owners and labourers, wealth production and extraction, aesthetic and political expressions of power, mass oppression and cultural erasure – have deep histories implicated in race and class oppression. So let us go ahead and say the quiet part out loud: for over a century, the modern professional teaching of landscape architecture's history has been, both generally *and* specifically, a reinforcement of White privilege.

There is no better evidence for this than the traditional Eurocentric historical canon of case studies selected for professional students – a canon in which my own teaching has been grounded for two decades. Ibram X Kendi asks us to acknowledge that the Western canon of Enlightenment Europe is, to a significant degree, a racist construct (2016, pp. 80–81). American designers have worshipped at the altar of Eurocentric high culture for over three centuries; it suggests, by extension, that the design history of our field may be rooted in racist world views. This is not to say that designers are racist; it *is* to suggest that we really need to examine our assumptions.

Such a recognition compels scholars to "improve" – as in, acknowledge, examine, critique, and expand – the histories we construct and teach. To rebalance our teaching is not necessarily to cancel the teaching of Villa d'Este, Stowe, Central Park: it is only to stop teaching them naïvely, narrowly, and uncritically. The recent robustness of historical scholarship in landscape architecture is already lighting our way. So while Whiteness and Western-ness have all but defined the canon of landscape architectural history as we once taught it, they cannot and will not continue to do so.

Institutional settings and the knowledge base

Professionalization of a field requires, among other things, the development of a specialized knowledge base (theory, case studies, best practices, etc.) as well as a formal curriculum (technical, processual/applied, and theoretical coursework). Professional curricula should be living configurations of knowledge, evolving in response to changing needs and knowledge of the society they serve and the institutions where they are based. The study of curricula thus offers a distinctive double portrait – juxtaposing the aspirations of the profession with the achievements of an institution.

One can readily surmise how the complex priorities and values of a profession may be embedded in the evolution of its history courses. So it is in landscape architecture, where the role of history has shifted from a typological catalogue of formal precedent and *parti* towards analysis of humanistic, ethical, and environmental dimensions of design (Ward Thompson and Aspinall 1996; Harris 1997). By extension, the changing content of history courses can signify two things: first, how the profession encompasses *designed landscapes*, as well as changes in the range and types of designed landscapes that it values.

What do we mean by designed landscapes? Landscape architects have followed historians and cultural landscape geographers in delimiting the designed landscape with great elasticity. J.B. Jackson famously takes issue with "The Word Itself," complaining that the English language privileges aesthetic appreciation of landscape over its other more expressive and performative dimensions. Landscape, he argues, is actually a system, a collective achievement: "a composition of man-made spaces on the land... not a natural feature of the environment but a *synthetic* space... functioning and evolving not according to natural laws but to serve a community" (1984, pp. 7–8).

Jackson, along with his well-published disciples (e.g. John Stilgoe at Harvard, Robert Riley at Illinois, and Paul Groth at Berkeley), fundamentally changed institutional perspectives as well as methods for the study of cultural landscapes. Vernacular and amateur landscapes are now valued as part of the same historical record as elite design with equal (or greater) interest for the profession (Deming 2015). Such broadly interdisciplinary definitions leave history courses with highly porous boundaries and myriad ways to "cut the cake" – to identify relevant thematic transects. In 1920 the history survey may have been a candy box of iconic design projects; in 2020 landscape history comprises nearly every aspect of the built environment – from Neolithic ritual centers to regional infrastructure, and from palaces to national parks.

A distinctive institutional pedagogy, instantiated by equally distinctive historic practices and projects, may generate a history class as unique as the scholar who curates it. Maintaining that range and diversity actually seems terribly important for the social, intellectual, and professional health of the discipline. For if, indeed, the profession's own practices encompass public monuments to community gardens to carceral landscapes to coastal zone disaster resilience, the kind of history needed to provide relevant context must, arguably, also exhibit equivalent breadth.

Schools of thought

Naturally, the efficacy of any course in professional history relies on institutional factors, including curricular objectives. Student respect for historic precedents, practices, and principles usually takes root only when ideas are recognized and reinforced in other classes, especially design studios. Other factors may pertain to a prevailing school of thought, or pedagogical tradition (e.g. emphasis on vernacular, design-build, sustainability, systems-thinking, or innovation), as well as the type of university within which landscape architecture programmes are found.

American programmes traditionally formed in two types of institutions. In (typically rural) public land-grant universities, landscape architecture was associated with schools of horticulture, agriculture, forestry, or natural resources; landscape architects were prepared for land development, recreation, and resource harvest/management practices. Land-grant institutions were intended to build educational programmes for technical, agricultural, and/or military training, and thus stimulate national economic development.[4] In (typically urban/suburban) private universities, landscape architecture was most often associated with schools of architecture or fine arts; landscape architects were taught to provide highly skilled design services for elite private or progressive public clients. Curricular history thus parallels and navigates the inherent dualism in landscape architecture's professional identity.

Virtually concurrent with the formation of the American Society of Landscape Architects, the first comprehensive programme of professional study in the United States was founded at Harvard University. Traditionally, a shared understanding of the scope of the field falls within the magisterial purview of textbook writers. But in 1900, no comprehensive textbooks as yet existed for the emergent field.[5] The earliest and, for six decades, *only* scholarly textbook integrating history with design principles appeared in 1917 as *An Introduction to the Study of Landscape Design*, by Theodora Kimball Hubbard and Henry Vincent Hubbard, both staff at Harvard (1917).

During her tenure as librarian of the Harvard School of Landscape Architecture (1913–1924), Kimball Hubbard was responsible for organizing collections of visual and textual materials. Despite Harvard's failure to admit women to its professional programme until 1942, Kimball Hubbard gave Harvard an astounding gift – the first genuine structure of knowledge for landscape architecture *and* its then-subfield of city planning. She recognized that "landscape architecture and city planning were not just design arts. . . but were also design sciences, defined by research" (Hohmann 2005, p. 174). In preparing an "encompassing theory of landscape design based in aesthetics and history," Kimball Hubbard's textbook "codified a 'Harvard' approach to landscape design" (Hohmann 2005, p. 176). Given the stature and influence of the "Harvard approach," Kimball Hubbard's structure of knowledge would thereafter be reproduced in the curricula of many other design programmes.

Structures of knowledge

Rapid social changes following World War I, together with changing relationships among the design disciplines, meant that history books, along with the design curriculum, needed to be overhauled. Harvard institutional legend has it that when Walter Gropius arrived from the Bauhaus in the mid-1930s, he ordered all history books to be tossed out. In actuality, "the removal of history courses from graduate design education [at Harvard] began during the 1920s. . . to make room for improved design courses" (Alofsin 2002, p. 13). It would not be the last time history was compelled to shift around other curricular exigencies.

The enterprise of writing a new English-language landscape history book would not be taken up again seriously at Harvard until Norman Newton's *Design on the Land: The Development of Landscape Architecture* (1971), quickly followed by Geoffrey and Susan Jellicoe's *The Landscape of Man: Shaping the Environment from Prehistory to the Present Day* (1975) and Christopher Thacker's

The History of Gardens (1979).⁶ As production of new historical case studies accelerated in the 1990s, so too did historical surveys. Philip Pregill and Nancy Volkman's *Landscapes in History* (1999) was quickly supplanted by Elizabeth Barlow Rogers's *Landscape Design: A Cultural and Architectural History* (2002). European authors (some having studied in the United States) chose to write landscape histories in their own languages: for example Dušan Ogrin's *World Heritage of Gardens* (1993) is translated from Slovenian. The most recent entry is Christophe Girot's *The Course of Landscape Architecture* (2016). The field is fortunate to be so enriched.

Despite the unique curatorial distinctions of all these works, however, their coverage circles around a similar canon of cultures and precedents. Each addresses a slightly different audience and scholarly standard, yet all represent a distinctly Western perspective, often even pertaining to non-Western landscapes. The good news is that, since the 1990s, hundreds of historical dissertations and monographs on landscape history have been produced by gifted scholar-specialists from a variety of environmental, humanistic, and design disciplines, addressing a wide diversity of other cultures, periods, designers, and projects. This is not to mention the thousands of conference and symposia proceedings, transcripts of seminars and webinars, cultural landscape inventories, oral histories, technical reports, and semi-scholarly formats in constant production.

Given the plethora of new publications across several related disciplines, there appears to be almost unlimited potential for the growth of subaltern histories. After a generation of work, there is a robust and growing literature on the contributions of women to landscape architecture, as historical actors, practitioners, researchers, and critics. Today our resources for teaching the history of women in the profession are deep.⁷ Around 2000, scholarly examinations on the intersections of race, diversity, cultural landscape, and design justice began to flourish.⁸ Over the past two decades, Diane Jones Allen, Craig Barton, Kofi Boone, Clifton Ellis, Rebecca Ginsburg, Dianne Harris, Walter Hood, and many others have produced valuable case studies for a new canon – vernacular practices and cultural landscapes designed by African Americans.

And this is just the beginning. With increasing numbers of graduates from doctoral programs specializing in landscape history, it is increasingly common to see trained landscape historians now teaching surveys and advanced history-theory seminars in top-ranked programmes. Of course, navigating the vastness of such quickly growing literature can easily overwhelm students as well as instructors. The surge of new historical scholarship, while glorious and welcome, challenges the very notion of the history survey – and compounds the challenge of choosing what, and whose, history to teach. The development of an explicit agenda for landscape design history remains an open question.

Restorative teaching

In the midst of his editorial tenure at *Landscape Journal*, Robert Riley issued a series of provocations to readers, notably "What History Should We Teach and Why?" (1994). Catherine Ward Thompson and Peter Aspinall rose to the bait with "Making the Past Present in the Future: The Design Process as Applied History" (1996). Dianne Harris countered with "What History Should We Teach and Why? An Historian's Response" (1997). The series of three essays remains an important resource for opening thoughtful discussion (Deming 2023).

A generation later, Riley's question has never been more relevant. Scholars in proximate fields may pursue compelling histories of landscapes and their impact on society and the environment. Yet many studies grounded in exquisite historical research will, for complex reasons, never make it to the survey courses that anchor professional curricula. The range and scale of landscapes the profession values forms a sort of ontological boundary for our curricula. What comprises a relevant history for landscape architecture depends almost entirely on the profession's (or the professor's) definitions of merit. But then who does that serve?

Historical empathy and curiosity are culturally interdependent. Teaching new histories first requires, and then supports, the formation of new values. It demands, and also generates, new imaginative and critical capacities. Above all it elicits and also innovates new research questions and investigative methods. Thus our understanding of professional history and, by extension, the field itself is bounded by the kinds of historical questions we are capable of asking. . . or even imagining.

Consider one illustration. The emergence of Naturalism may be interpreted as an aesthetic foil to the infinite axes of the *ancien regime* and a formal metaphor for networked liberal systems of shared power (Figure 3.1). However, as W.J.T. Mitchell explains, the imperialist canon obscures its own origins (1994). One need only consider the pathway of capital supporting the newly liberalized British ruling oligarchy in the late 17th and early 18th centuries to realize how much wealth was extracted (directly or indirectly) from industrial-scale agro-colonial enterprises on Caribbean and North American soils.

In the ubiquitous history of English landscape gardens, discussion about the wealth generated by a vast population of abducted and enslaved Africans all but disappears – along with African lives, languages, knowledge, and landscape practices. Also unacknowledged is the disappearance of English landscapes, local knowledge, and folkways as well, as that small island was deforested to build the fleets defending wealth-generating trade as well as tenements that re-homed evicted cottagers. After cleared landscapes were depopulated during the Enclosures Acts, they were reconstituted through new pastoral industries and mythically reinvented as the English Landscape Garden. To answer Riley's question then, the holistic interdependence and agency of *all* parties involved in the 18th-century transformation of lives and landscapes on three continents is a history we should try to teach (Figures 3.1–3.3).

Figure 3.1 Petworth House, England (1690; renovated Capability Brown ca. 1750).
Source: Photo by M. Elen Deming, September 1990.

"A thing in movement"

Figure 3.2 "Servant's house T": dwelling for an enslaved family (1793–ca. 1830; reconstructed in 2014). Monticello, VA.

Source: Photo by M. Elen Deming, May 2019.

Figure 3.3 Thomas Jefferson estate at Monticello, VA.
Source: Photo by M. Elen Deming, May 2019.

Medicinal history

To improve the "science" of landscape history, scholars recognize that changing social values create new mirrors – opening space for a different kind of historical framework. Work is well underway: adding new regional emphases, alternate voices, and nuanced stories. In *Medicine Stories* (2019) feminist historian Aurora Levins Morales speaks of "medicinal history" as a strategy for cultural and ecological restoration, and presents the metaphor of the historian as *curandera* – as healer. Drawn from her experiences as a Jewish Puerto Rican woman, "A Curandera's Handbook of Historical Practice" contains guidance for an alternative telling of history grounded in the lived experience of those who are not, or have not yet been, represented. What follows is my tactical digest of her principles:

1. *Tell untold or undertold histories.* By seeking out the stories of members of an underclass or outgroup, the poor, the colonized, the workers, we uncover better historical questions.
2. *Centering women changes the landscape.* There is a far greater chance that overlooked subjects, for example women, can provide valuable new historical perspectives, scales, and positions.
3. *Identify strategic pieces of misinformation and contradict them.* The historian's first task is authentic diagnosis, to see through myths and occlusions and prevent further harm.
4. *Make absences visible.* A lack of historical evidence may be as significant as its presence; a hole in the archival record may hold the shape of a decision, a practice, a value, or an erasure.
5. *Asking questions can be as good as answering them.* Speculative (abductive) "what if?" questions may reframe historical potentialities, leading to new evidence and relationships.
6. *What constitutes [historical] evidence?* Even archives may be biased in privileging textual documents; other types of non-literate material evidence may be equally "vocal."
7. *Show agency.* False narratives about the apparent passivity of oppressed people is consolidated by repetition; by explaining their agency and resistance, "medicinal history" can flip the script.
8. *Show complexity and embrace ambiguity and contradiction.* Rather than caricaturing historical actors, reveal their shadings; this gives greater dimension to the human heart and its affects.
9. *Reveal hidden power relationships.* Expose and denaturalize asymmetrical power relations by tracing financial, geographic, and environmental impacts (especially in landscape history).
10. *Personalize.* To particularize, rather than generalize, augments the development of historical imagination and empathy – a particularly useful skill for "medicinal landscape historians."
11. *Show connection and context.* Historical actors (artists/designers) are not anomalous (unicorns); always acknowledge partners/mentors in the society, community, or studio supporting them.
12. *Restore global meaning.* While working on alternative histories, place the focus on linkages across time, space, materials, and media, and show the interdependence of parts of the globe.
13. *Provide access and digestibility.* *Curandera* historians make stories accessible. "[H]istory is a form of healing. . . it has to work like a network of community clinics, not a luxury resort spa" (Levins Morales 2019, p. 86).
14. *Show yourself in your work.* Objectivity is a residual myth of positivism; in humanist scholarship owning a position and being transparent meets scholarly obligations to research integrity.
15. *Cross borders.* Be broadly curious and transgress boundaries of "discipline, geography, and historical period. . . [that] don't necessarily serve the projects of medicinal history" (Levins Morales 2019, p. 88).

Landscape architecture needs the kind of historical landscape scholarship that, as Levins Morales puts it, "shifts the landscape in which the questions are asked and makes a different kind of sense out of existing information" (2019, p. 74). In particular, medicinal landscape history may prove valuable for "reestablishing a sense of the connectedness of world events to one another" (2019, p. 86). Landscape historians would do well to take these tactics to heart.

Future landscape histories

Currents now in play across society demand alternative and oppositional histories with currency for professional education. For example, even as the Morrill Act is celebrated by American educators in landscape architecture, it forces reckoning with its imperialist and racist roots. In slow-dawning recognition of their historical debt to America's old continental aggression, land-grant universities today offer ancestral land acknowledgements (similar to other postcolonial nations). And while it is not commonplace to find history courses in landscape architecture that explicitly address decolonization or reparations, several notable exemplars are reframing "anti-racist" landscape histories (Harris 2013; Allewaert et al. 2019; Dalla Costa 2020; Boone 2021; Bierbrauer 2020; Hirsch 2021). More efforts are, undoubtedly, under vigorous development.[9]

As Way writes: "History is the foundation for arguments about the future" (Way 2020b, n.p.). New historical perspectives(s) are needed to relate a trove of valuable new scholarly material to landscape architectural history. What kind of scholarship might make it possible to teach a just or equitable landscape history course that represents the invisible designers of built landscape and the constituent communities served by the profession? What kind of landscape histories might make it possible for educators to teach honestly about the painful ways in which landscape architecture has been complicit in social and institutional crimes against these very same contributors and communities? And what kind of learning outcomes emerge from a just historical scholarship?

Twelve points for a just historical scholarship

- *documents vernacular and ordinary landscapes*
- *focuses on under-studied cultures, regions, and actors*
- *promotes landscape literacy for understanding deep site history*
- *analyzes economic trade-offs in urban surplus and rural residual*
- *is anti-racist and anti-imperial*
- *shows how wealth is created, extracted, and stolen*
- *explains how wealth is symbolized and concentrated through design*
- *maps separation from and proximity to valued landscapes*
- *studies environmental beauty, health, depletion, and degradation*
- *gathers stories of resilience in public health and climate adaptation*
- *moves beyond binaries of culture and nature*
- *is volunteered, drawn from citizen-science, crowdsourced, multivalent*

Then to make a just historical scholarship more accessible, what kind of teaching and learning formats might help disseminate new knowledge in landscape architectural history?

The integrated curriculum

For history to actually matter to students (typically eager to gain skills and find a place in practice), history must be thoroughly integrated as part of the main curricular sequence. This includes the design studio, of course; but history may also enter into technical courses on representation, or fabrication, or maintenance. Each of these topics has its own social, technical, and professional history (for examples, see Qviström/Jansson in this volume).

Alternatives to the textbook

Alternative formats for teaching landscape architecture may dispense with magisterial textbooks in favour of other strategies. Immersive alternatives for print media abound: blogs, YouTube videos, documentaries. Academic organizations are busy producing curated content online (e.g. The Cultural Landscape Foundation, www.tclf.org/, and the Center for Cultural Landscapes, www.arch.virginia.edu/ccl).

The curated critical seminar

To accelerate alternate historical conversations, thematic seminars (preferably interdisciplinary) could focus on history vis-à-vis selected contemporary issues such as climate change, domestic and vernacular landscapes, sustainability, design justice and equity, construction practices, commemoration, and so on. Visiting designers, lecturers, and studio critics may introduce community-based content that extends critical themes for a semester or longer, perhaps outside the regular academic calendar.

Student historians

Students are key to assembling new narratives and new sources. Success is reported with student-run and -edited wikis, as well as interactive student-led online forums and reading circles. Repositories of case studies and stories may be collected and curated by individual students and scholars corresponding to curricular themes.

Landscape history in the news

Special events, thought patterns, and movements are regularly featured in mainstream and alternative press, blogs, social media, and so on. Attending to non-academic sources can excite students about the politics, economics, and social impacts of landscape histories.

"A thing in movement"

Historical research is always emerging – a "thing in movement" – inviting new stories to be told, new voices to be heard, new layers to be investigated. As we expand the breadth of our landscape histories, surely the horizon of our ignorance grows as well. The old histories of concentrated wealth and power once provided landscape architecture with its inheritance, with its surplus of meaning. Yet it also maintains a surplus of Whiteness – so systemic and so *invisible* that it strikes at the moral and intellectual foundations of the field. Lest we "betray it," as Bloch warns, we must work harder to improve the science of landscape history, by telling its stories fully, accurately, critically, and compassionately.

Landscape architects have so much to learn. There is so much at stake in today's political climate for historians and history teachers at all levels. Yet we find reason enough for restorative teaching in the professional aspirations of landscape architecture itself. Innocuous as it may seem to the general populace, the history of designed landscape is core to all human history; it indexes the evolution of human values. Landscape is *the* living matrix that connects past and present, here and there, "us" and "them," I and thou. Naturally, the history of landscape architecture, like all history, contains the same spectrum of sin and grace as humanity itself.

The imperative is to continue connecting and learning new aspects of history, as steadily as we can. For just as historical education tends to mirror professional values, historical questioning guides future insights and decisions. And whether or not our institutions support this work, our students and the future of landscape architecture certainly demand it.

Notes

1 Author's Note. Extending my deepest gratitude to the editors for their patient expertise, I dedicate this chapter to our gentle friend, Karsten Jørgensen (1953–2021).
2 Mississippi writer William Faulkner (1897–1962) famously puts it: "The past is never dead. It's not even past" (1951 n.p.).
3 CRT is an academic framework developed by legal scholars Derrick Bell (1930–2011), Kimberlé Crenshaw, and others, to examine how racial bias is "baked into" American legal codes, financial institutions, and education (Delgado and Stefancic 2017).
4 During the American Civil War, the federal Morrill Act (1862) provided each state with extensive sites and/or capital from seized, stolen, or federalized lands for public colleges – thus the *land grant* (see Nash 2019).
5 In 1900 a library of landscape design history would have included architectural mainstays – from Vitruvius to Alberti to Pugin to Ruskin – as well as treatises by European landscape practitioners Humphry Repton, Joseph Paxton, Jean-Marie Morel, and J.C.A. Alphand, among others. American landscape practitioner Frederick Law Olmsted (1822–1903) was a prolific and widely published essayist. The works of landscape gardener Andrew Jackson Downing (1815–1852) and progressive landscape architect Charles Eliot, Jr. (1859–1897) published in *Forest and Garden* magazine, would also have been studied. H.W.S. Cleveland's (1814–1900) sweeping *Landscape Architecture as Applied to the Wants of the West* (1873) foreshadowed the mature profession. Urban historians such as Camillo Sitte (1843–1903), and contemporaries documenting canonical European gardens, such as Charles Platt (1861–1933) and Edith Wharton (1862–1937), would also be read.
6 This discounts Christopher Tunnard's idiosyncratic chapter on history in *Gardens in the Modern Landscape* (1938).
7 In 1994, *Landscape Journal* published a special issue titled Women | Land | Design (vol 13:2 (Fall), with papers drawn from the eponymous Radcliffe Seminars Symposium (April 1993). A second wave of interest resulted in *Women in Landscape Architecture: Essays on History and Practice* (2011) edited by Louise Mozingo and Linda Jewell, as well as Thaïsa Way's *Unbounded Practice: Women and Landscape Architecture in the Early Twentieth Century* (2013). Taking a global perspective, Sonja Dümpelmann and John Beardsley edited *Women, Modernity and Landscape Architecture* (2015).
8 Craig Barton's *Sites of Memory: Perspectives on Architecture and Race* (2001) opened the door to a wider set of landscape designer-authors. Cultural landscape geographers such as Richard Schein contribute foundational work in *Landscape and Race in the United States* (2006). Architectural historians Dell Upton and Dianne Harris featured in *Race and Space*, a special issue of *Landscape Journal* edited by Harris (2007). Most recent is *Black Landscapes Matter*, edited by Walter Hood and Grace Tada (2020).
9 Professional advocacy organizations support anti-racist teaching with reading lists, websites, videos, and model syllabi. Landscape-specific sites are maintained by the Society of Architectural Historians (2020), the Landscape Architecture Foundation (2022), the University of Virginia's Center for Cultural Landscape Studies and University of Virginia School of Architecture (n.d.), among others.

References

Allewaert, M., Gomez, P. and Mitman, G. (2019) 'Interrogating the Plantationocene', unpublished course syllabus for English 817/History of Science 921 (Spring). Madison: University of Wisconsin.

Alofsin, A. (2002) *The Struggle for Modernism: Architecture, Landscape Architecture and City Planning at Harvard*. New York: W.W. Norton.

Barton, C. (2001) *Sites of Memory: Perspectives on Architecture and Race*. New York, NY: Princeton Architectural Press.

Bierbrauer, A. (2020) 'History of Landscape Architecture', unpublished course syllabus for LDAR 5521 (Fall). Denver: University of Colorado.

Bloch, M. (1953) *The Historians Craft: Reflections on the Nature and Uses of History*. Intro. J. Strayer, trans. P. Putnam. New York: Vintage Books.

Boone, K. (2021) 'Environmental and Social Equity in Design', unpublished course syllabus for LAR 535–001 (Fall). Raleigh: North Carolina State University.

Dalla Costa, W. (2020) 'Indigenous Architecture, Planning and Construction', unpublished course syllabus for DSC/CON/AIS-598/494. Tempe: Arizona State University.

Delgado, R. and Stefancic, J. (2017) *Critical Race Theory: An Introduction*. 3rd ed. New York: New York University Press.

Deming, M. E. (2015) 'Value-Added: An Introduction', in Deming, M. E. (ed.) *Values in Landscape Architecture and Environmental Design: Finding Center in Theory and Practice*. Baton Rouge: LSU Press, pp. 1–29.

Deming, M. E. (2019) 'Values and Transformative Learning: On Teaching Landscape History in a Community of Inquiry', in Jorgenson, K. et al. (eds.) *The Routledge Handbook of Teaching Landscape*. New York: Routledge, pp. 177–190.

Deming, M. E. (2023) 'More Important Questions', in Brown, B. (ed.) *Landscape Fascinations and Provocations: Robert B. Riley*. Baton Rouge: LSU Press.

Dümpelmann, S. and Beardsley, J. (eds.) (2015) *Women, Modernity and Landscape Architecture*. New York: Routledge.

Faulkner, W. (1951) *Requiem for a Nun*. New York: Random House.

Hannah-Jones, N. (2021) *Nikole Hannah-Jones Issues Statement . . .* [Online]. NAACP Legal Defense and Educational Fund, Inc. Available at: www.naacpldf.org/press-release/nikole-hannah-jones-issues-statement-on-decision-to-decline-tenure-offer-at-university-of-north-carolina-chapel-hill-and-to-accept-knight-chair-appointment-at-howard-university/ (Accessed 29 January 2022).

Hannah-Jones, N., Roper, C., Silverman, I. and Silverstein, J. (eds.) (2021) *The 1619 Project: A New Origin Story*. New York: New York Times.

Harris, D. (1997) 'What History Should We Teach and Why? An Historian's Response', *Landscape Journal*, 16(2), pp. 191–196.

Harris, D. (2007) 'Race, Space, and the Destabilization of Practice', *Landscape Journal*, 26(1), pp. 1–9.

Harris, D. (2013) 'Race and Space', unpublished course syllabus for Landscape Architecture 587. Urbana and Champaign: University of Illinois.

Hirsch, A. (2021) 'Global Histories of Designed and Cultural Landscapes', unpublished course syllabus for ARCH 565 (Fall). Los Angeles: University of Southern California.

Hohmann, H. (2005) 'Theodora Kimball Hubbard and the "Intellectualization" of Landscape Architecture', *Landscape Journal*, 25(2), pp. 169–186.

Hood, W. and Tada, G. (eds.) (2020) *Black Landscapes Matter*. Charlottesville, VA: University of Virginia Press.

Hubbard, H. V. and Hubbard, T. K. (1917) *An Introduction to the Study of Landscape Design*. New York: The Macmillan Company.

Jackson, J. B. (1984) 'The Word Itself', in Jackson, J. (ed.) *Discovering the Vernacular Landscape*. New Haven: Yale University Press, pp. 1–8.

Jellicoe, G. and Jellicoe, S. (1975) *The Landscape of Man: Shaping the Environment from Prehistory to the Present Day*. New York: Van Nostrand Reinhold Co.

Kendi, I. X. (2016) *Stamped from the Beginning: The Definitive History of Racist Ideas in America*. New York: Bold Type Books.

Landscape Architecture Foundation. (2022) *Read/Watch List to Learn to be Anti-Racist* [Online]. Available at: www.lafoundation.org/resources/2020/06/anti-racist-read-watch-list (Accessed 20 January 2022).

Levins Morales, A. (2019) 'The Historian as Curandera', in *Medicine Stories: Essays for Radicals*. Durham: Duke University Press, pp. 69–88.

Lewis, P. (1979) 'Axioms for Reading the Landscape', in Meinig, D. W. (ed.) *The Interpretation of Ordinary Landscapes: Geographical Essays*. New York: Oxford University Press, pp. 11–32.

Mitchell, W. J. T. (1994) 'Imperial Landscape', in Mitchell, W. J. T. (ed.) *Landscape and Power*. London: University of Chicago Press.

Mozingo, L. and Jewell, L. (eds.) (2011) *Women in Landscape Architecture: Essays on History and Practice*. Jefferson, NC: McFarland & Co.

Nash, M. A. (2019) 'Entangled Pasts: Land-Grant Colleges and American Indian Dispossession', *History of Education Quarterly*, 59(4), pp. 437–467.

Newton, N. (1971) *Design on the Land: The Development of Landscape Architecture*. Cambridge: Belknap Press of Harvard University Press.

Redman, H. (2021) 'Wisconsin Senate Republicans Move School Culture War Bill Forward', *Wisconsin Examiner* [Online], 30 November. Available at: https://wisconsinexaminer.com/2021/11/30/wisconsin-senate-republicans-move-school-culture-war-bill-forward/ (Accessed 30 November 2021).

Riley, R. (1994) 'What History Should We Teach and Why?' *Landscape Journal*, 14(2), pp. 220–225.

Robertson, K. (2021) 'Nikole Hannah-Jones Denied Tenure at University of North Carolina', *New York Times* [Online], 19 May. Available at: www.nytimes.com/2021/05/19/business/media/nikole-hannah-jones-unc.html (Accessed 19 May 2021).

Rogers, E. B. (2002) *Landscape Design: A Cultural and Architectural History*. New York: Abrams.

Romero, D., Caputo, M. and Finn, T. (2022) 'Florida School District Cancels Professor's Civil Rights Lecture Over Critical Race Theory', *NBC News* [Online], 24 January. Available at: https://nbcnews.my.id/2022/01/24/florida-school-district-cancels-professors-civil-rights-lecture-over-critical-race-theory-concerns/ (Accessed 24 January 2022).

Schein, R. (ed.) (2006). *Landscape and Race in the United States*. New York, NY: Routledge.

Society of Architectural Historians. (2020) *Resources for Learning and Teaching about Race and the Built Environment* [Online]. Available at: www.sah.org/about-sah/news/sah-news/news-detail/2020/06/10/resources-for-learning-and-teaching-about-race-and-architecture (Accessed 20 January 2022).

Tunnard, C. (1938) *Gardens in the Modern Landscape*. London: The Architectural Press.

University of Virginia School of Architecture. (n.d.) *Educate Yourself: Inclusion & Equity Resources* [Online]. Available at: www.arch.virginia.edu/about/jedi/educate-yourself (Accessed 20 January 2022).

Upton, D. (2007) 'Sound as Landscape', *Landscape Journal*, 26(1), pp. 24–35.

Ward Thompson, C. and Aspinall, P. (1996) 'Making the Past Present in the Future: The Design Process as Applied History', *Landscape Journal*, 15(1), pp. 36–47.

Way, T. (2013) *Unbounded Practice: Women and Landscape Architecture in the Early Twentieth Century*. Charlottesville, VA: UVA Press.

Way, T. (2020a) 'Why History for Designers (Part 1)', *Platform* [Online], 2 March. Available at: www.platformspace.net/home/why-history-for-designers-part-1 (Accessed 27 January 2021).

Way, T. (2020b) 'Why History for Designers (Part 2)', *Platform* [Online], 9 March. Available at: www.platformspace.net/home/why-history-for-designers-part-1 (Accessed 27 January 2021).

Young, C. (2022) 'We Need an Honest Conversation about Teaching Social Justice in Public Schools', *Daily Beast* [Online], 24 January. Available at: www.thedailybeast.com/we-need-an-honest-conversation-about-teaching-social-justice-in-public-schools (Accessed 24 January 2022).

4
TRACING DISCOURSES
Learning from the past for future landscape architecture

Mattias Qviström and Märit Jansson

Introduction

The urban landscape is a contested terrain in contemporary planning. With sustainability and improved "urban qualities" used as main arguments for densification, green space that exists as the result of former planning eras is being transformed into mainly new residential units. This is the case in several parts of the world, and clearly shown in built environments in Sweden (see Figure 4.1). Studies on Stockholm show how both planners and parents, the latter experiencing the negative effects of densification for their children, see these ideals of densification as unavoidable (Cele 2015). Even landscape architecture students might take the densification approach to planning for granted, as it employs a mainstream urban sustainability rhetoric, and then use it as a base for practice. Engagement with environmental concerns is an important foundation for landscape architecture and also runs deep amongst students today, and to find ways to contribute to sustainability is therefore of the essence. In this situation, however, educational institutions have a responsibility to place current discourses in a wider context and, in this case, to problematise and broaden ideas of sustainability within landscape architecture practice.

This chapter argues for the use of landscape and planning history in landscape architecture education to unpack the taken-for-granted ideals and ideologies prevalent in contemporary practices, and support students to find a way to be open for potentially "other" futures for the urban landscape as well as for landscape architecture. While the use of history as a source of inspiration has a long tradition within the field of landscape architecture, its use to scrutinise planning ideologies is less pronounced today. History, we argue, provides a means to examine decisions people have taken in the past and analyse subsequent rationalisations, which make such studies serve well as a learning context for landscape architecture students in general. We are also in particular lifting the possibilities for using history to compare and review current discourses, revealing alternative ideas and ideals, without launching yet another normative agenda on to the students. However, there are many more reasons for, and ways to use, landscape and planning history in education, as we will come back to.

Discussing landscape and planning history studies as a "tool", and using the case of modernist landscape architecture as an example, we illustrate education practice by presenting how graduate students have been able to scrutinise contemporary landscape architecture through historical studies of the post-war planning era in their recent student projects.

Figure 4.1 Empty space or an invaluable resource for recreation and play? The spacious design of modernist neighbourhoods are under question in current planning: historical studies offer a way beyond defending or overlooking this legacy. Lorensborg, a neighbourhood developed in the 1950s in Malmö.

Source: Photo by Alexander Stigertsson.

Landscape and planning history

The history of our discipline is out there, in full view. Take a walk in any city and you will encounter, as Schein (1997) illustrates using a case of a North American neighbourhood, a meshwork of "discourses materialized". Some discourses are specifically about landscape architecture, while others have made significant imprints in how, where and to what extent landscape architecture has affected the urban environment. History, in this sense, is not the domain of the past but something which affects what we do, our everyday actions and viewpoints, and thereby opens up or closes down our horizons in terms of thinking and acting.

Also, not all plans and policies are tucked away in dusty archives. Some of the "historic" plans, policies and investigations, outdated as they may seem at first glance, are still at work and are used as a base for contemporary ones. With a cut and paste tradition observable in planning, contemporary plans become palimpsests, reiterating decades of planning (Qviström 2010, 2013). Abbot and Adler (1989, p. 470) bring up one particular aspect of how old plans and planning linger when they notice that

> people remember wins, defeats, promises, and broken promises, and they develop a trust or distrust of institutions based on these experiences. They harbour grudges (or, more rarely, warm feelings) long after the offending individuals have moved to other jobs. Thus, a planner who walks into a situation without knowing the expectations that other participants have built up or how they are invested in an issue can easily be blind-sided.
>
> (see Bickerstaff 2012, for a pertinent example)

History, then, is perceptible almost anywhere; in historical documents, in the places created. There are also traces of history in contemporary discourses – and in planning education. On the one hand we are "stuck" with history; on the other hand we can find new futures through history (Qviström 2013). However, *learning* from history is not the same as simply copying it, or as using a design or style as a norm or for aesthetical inspiration. There is no reason to be naïve: history has always been used to mobilise the landscape for a cause, from national politics down to individual design proposals. Landscape theory also offers a useful base for further critical examinations of this use of history (see e.g. Schein 1997; Crang 2000; Wylie 2007). Yet, with this in mind, we argue for the need to try to walk in the shoes of the architects of the past, to better understand their ideas and the landscapes they shaped – and as a way to gain awareness of taken-for-granted norms in contemporary landscape architecture practice.

The history of landscapes and planning can be studied with, broadly speaking, a discursive understanding of the practice of landscape architects and planners, informed by a social constructivist approach which acknowledges how the way of talking, thinking, representing and acting in planning expresses different ideas which are translocal in character (see e.g. Schein 1997; Pries and Qviström 2021), and which have at least partly materialised in the landscape. Yet, concepts of *discourse* and *social* constructivism do not entirely capture the landscape and planning history approach we have in mind. There is more.

While landscape representations and planning visions can be deleted "with a press on a button" as Daniels and Cosgrove (1988) famously suggested, the customary use of a place, and its material features, are obdurate and messy to govern. Furthermore, landscapes are not merely governed by the actions of human beings but also by other events, such as growth, erosion, invasive species and so on. Therefore, as landscape and environmental historians have shown, in order to capture the combined landscape and planning history one needs to go beyond the conventional historical focus on actions and intents to also include events, some of which are driven by non-humans (Sörlin 2020). With this in mind, we need to go beyond a conventional discourse analysis (which primarily focuses on texts) to include a "material semiotics" to capture how we relate to the world (Law 2009). This could, however, be to ask too much of the students if a historical and empirical (rather than theoretical) exploration is supposed to be centred upon.

Fortunately, landscape architecture students are skilled at thinking through space and place. They usually take on a new site by sketching, measuring, capturing its scale and spatial qualities, scrutinising maps and renderings and so on, but also by trying to capture the character of the place, through field observations, interviews and so on. They literally bring in the spatial and material aspects of their cases, supported not the least by the heterogeneous toolbox of landscape analysis (see e.g. Stahlschmidt et al. 2017). It is in relation to such examinations of space and place that we argue the usefulness of studies of materialised discourses. A discursive approach provides a useful complement to the spatial and place-based examinations the students are often familiar with.

Taking on the modernist landscape

The so-called record years of especially the 1960s and 1970s set a clear mark on the urban and suburban landscape. Highways, high-rise suburban development and traffic separated centres are well-known examples of this modernist planning, but there are also the leisure facilities, such as public baths and beaches, ice-hockey arenas, indoor tennis courts, outdoor recreation centres and so on, which taken together materialised certain ideas of modernity, progress and welfare (Pries and Qviström 2021). Equally important, however, is the role the planning during the record years plays in the current planning debate, as the "base line" for a critique, or as a rhetorical figure, which motivates current development. As the neighbourhoods from that era serve as a main target for critique and as suggestions for "improvements", for example via intense densification, such areas and their history can be fruitful cases for students to examine.

Modernist planning has been severely critiqued for decades (see e.g. Natrasony and Alexander 2005), including its reliance on functional divides (e.g. separated land-use, as a way to remediate conflicts), its belief in standardisation and thus its inability (or even disinterest) to listen to place and the local community, and the naïve faith in technology and progress. Relph refers to the period from the 1950s to the early 1970s as the "high age of placelessness" (2016).

There are several reasons why we encourage students to go beyond this taken-for-granted narrative concerning the legacy and critique of the modernist era. First, the critique, as devastating as it may be, echoes the language of the modernists themselves in commenting on planning conducted during the decades before their own practice, and which they wanted to go beyond in the name of progress. The similarities in this approach are more than just superficial, as it usually expresses the same progressive ideal and attitude. Ironically, the critique could thus show how planners and architects are still part of the modernist (or at least modern) project (c.f. Qviström and Bengtsson 2015).

Second, modernist planning is what has shaped much of today's cities. Its materialised discourse is our legacy and the base for coming planning decisions. Landscapes planned or designed during the record years have endured (and matured) for half a century or more. They have been the objects of emotional and physical investments, of development and of everyday maintenance. There are no frozen layers in the landscape: these areas have evolved, sometimes into lush neighbourhoods with their own identities, which also need to be assessed based on their current qualities (see Figure 4.2).

Third, the rhetoric of modernist planning should not be confused with the actual practice of the time. For instance, the amenities of the landscape, and local practices, have in some projects had a much more pronounced role than the modernist programme suggested; it does not make sense to take for granted that site-specific assets or local practices did not play a role in the planning.

Figure 4.2 Field studies in Smedsby, Upplands Väsby, one of many modernist neighbourhoods where densification is underway. The modernist neighbourhoods of today are often lush and green, a quality that is not always acknowledged in the public debate or in planning.

Source: Photo by Mattias Qviström.

Fourth, in countries such as Sweden, high modernism is entangled with the development of a social welfare society. Even if under pressure today, the general idea of a welfare society is still widely acknowledged: if we aim to preserve the idea and maintain its practice despite limited resources, it might be worth paying heed to the remains and ruins of the kind of planning conducted during the time when the welfare society flourished.

To be clear, we raise the preceding points not in defence of modernist planning. Our main argument is that we need to *learn* from the modernist era – and by doing so gain a critical perspective on the current debate, and perhaps also constructive ideas on how to move forward. By gaining a deeper understanding of the past, we usually get surprised, disappointed, fascinated and inspired at the same time: in their educational progress landscape architecture students benefit from engaging in that interpretive phase.

Students learning from history

Starting in the early years in the 1960s, the landscape architecture programme at the Swedish university of agricultural sciences has included aspects of garden history. While garden and landscape history studies expanded for a long period (e.g. with the new complement of cultural heritage), they have, in recent years, been given slightly less attention and also been integrated in other courses, partly to avoid history being a "compartmentalized knowledge" (Abbott and Adler 1989) in the education. While garden history and cultural heritage are acknowledged as important subjects in landscape architecture education, urban environmental and planning history receives less attention and rarely goes beyond classes on design styles and morphologies of different epochs of city building. The master thesis, however, opens up for spending time on such historical themes, and for methodological experiments. This chapter is based on general experiences of using history as a tool in our educational approach, focusing specifically on two recent student projects.

One of the studies was a joint study by Stigertsson and Palmqvist (2020), who developed their master's thesis around shifting planning ideals for child-friendly environments in Sweden, aiming to understand the focus on children in planning and its consequences in the built environments. They placed specific focus on the possibilities for children's outdoor recreation, and how well providing these have been prioritised within landscape planning. In their study, they are looking both at discourses and at their material consequences for children. They made the choice of analysing the modernist planning of the mid 1900s as a comparison to what they called the neoliberal planning of the 2000s, in studies of three residential areas, each in two cities in southern Sweden (see Figures 4.1 and 4.3). The other study was done by Stryjan (2020), who focused on shifting ideas within landscape planning for physical activity, which she examined with the case of the neighbourhood of Hallonbergen in Sundbyberg municipality, focusing primarily on the initial planning of the area in the late 1960s and the recent development and ongoing plans for densification. Both studies were based on the students using case studies of selected built environments, with a combination of qualitative interviews with planners, field studies in built environments (walking around and taking photos, sometimes with protocols for assessment of various aspects) and document studies (primarily historical and contemporary planning documents, comprehensive and detailed development plans), in combination with literature reviews. Thus, the students did not engage in archival studies, nor in documenting the oral histories of the sites, but settled with an analysis of the main planning documents. This light touch on history would nevertheless consume a substantial part of their work, as taking on such a historical approach was novel to them as was the study of modernist planning. Given the time limit of about 20 weeks of full-time studies, there is a delicate balance between a rich history and a detailed analysis of the contemporary landscape.

Figure 4.3 Solskiftet, a neighbourhood in Landskrona, was developed in the 1970s. It was planned with open, green housing yards and also placed next to a park.

Source: Photo by Alexander Stigertsson.

While focusing equally on history and today as an outset, the critique of current planning ideals and the built environments developed into a main concern in both studies. By comparing the two ideals and eras and looking specifically at environments planned and built in the modernist era on the one hand, and those after 2000 on the other, the students were able to unveil shortcomings of one-sided approaches aimed at densification. Experiences of the historical study clearly showed silences in contemporary planning, and made the specific ideology of the time stand out. Stryjan (2020) would bring to the forth the notion of paradigm shifts to characterize the difference between the ideals which had formed the original design of Hallonbergen compared with the compact city ideology of the contemporary plans. Using a child-friendly perspective (based on theory by e.g. Kyttä 2004), Stigertsson and Palmqvist (2020) found that in the neighbourhoods from the early 2000s, planning has affected children in a particularly negative way, hindering their independent mobility and providing few affordances for play and activity, while the modernist areas support these child-friendly qualities much more, providing both spaces for play and structures for mobility. Such critique is however only part of the learning outcome, as a general knowledge of past planning will be of use next time the students engage in a neighbourhood from the modernist era – or in another study of landscape and planning history.

Findings from the student studies show that landscape and planning history is a Pandora's box. It takes time to gain a rich understanding of even the local history of planning, especially if the period studied is a new experience. Furthermore, even if the study is clearly defined from start, a historical study nevertheless opens up new paths and nurtures new questions, which on the one hand might be worth pursuing, and on the other hand takes time. The challenge is to engage with history and yet not get stuck in it, as a comparison with contemporary planning or lessons for future landscape architecture are set in focus. An external examiner might not accept a project which "only" presents a landscape/planning history analysis as a way to scrutinise contemporary planning, but require that the student apply it in a proposal.

And yet our teaching at the Swedish university of agricultural sciences also includes experiences of shorter student essays which draw on landscape and planning history, without getting caught into the rich stories of the past. For instance, examples can be found from the master's course "Landscape Planning in Theory and Practice", in which the students are asked to spend four weeks writing a shorter essay in which landscape theory is used as a lens to critique contemporary planning. In this course, the theoretical lens helps the students to compare one aspect from the modernist epoch with the current plans for densification. While such an analysis only touches on the complexity of past planning, it illustrates how historical material can be used to bring in alternative ways of understanding the contemporary planning discourses – and perhaps also function as a gateway to more elaborative historical examinations. It also captures how an understanding of history and theory can evolve hand in hand.

History in future landscape architecture education

Saab (2007) notices the widespread application of nostalgia in urban design. Nostalgia, she argues, is history without guilt. The use of modernism as a scapegoat is hardly nostalgic in the way 19th-century towns or medieval villages are portrayed, and it is all about guilt. Yet, in the way it is used, a demarcation line is drawn between "them" (the modernists) and "us" which frees us of guilt. This is an odd way of treating history, given that we are ensnared by it. While a feeling of guilt is most likely not what students need, they do need the insight that the history of our discipline has *consequences* (as they materialise specific ideas of society, nature, and justice), which are out there in the landscape and "in" here in our taken-for-granted practices, techniques and ways of thinking. This comes close to Thomas' (2006) call for history as a means to make students grasp the social justice implications of their coming practice. The reason why Thomas argues for history is, however, primarily due to its narrative qualities:

> Carefully crafted and compelling historical accounts. . . have the power to allow readers (and listeners or viewers of alternative media) to access and interpret fairly complex situations at a very basic, intuitive level of comprehension.
>
> (Thomas 2006, p. 317)

Thus, history can inspire and explain without compromising on its situated qualities. It does so by bringing new questions as well as answers. It brings to the forth uncertainty rather than dogmas. This is a reason as such for letting students dwell on historical studies, whether they manage to apply the insights in a planning or design proposal or not (c.f. Harris 1997). However, to dwell on history takes time and can be overwhelming, especially if done without a pronounced theoretical lens. If theory is used in a reflexive manner, that is, not as a dogma but as a preliminary understanding that is tested throughout the work with the historical material, then theoretical and historical engagement can be reached simultaneously.

There are certainly other arguments for introducing history to landscape architect students. For instance, as Harris (1997) argues, it is crucial to engage with history as such, without focusing on its applications. Riley (2015) offers a discussion on different approaches to history, and argues the need for an idea of why history is taught, a question which needs to be answered in each course rather than for an entire programme, as history can (and should) serve different purposes in different courses.

Our experiences from teaching suggest that landscape and planning history studies can provide a fruitful tool allowing new perspectives on current discourses. The master's thesis in particular provides opportunities for critical reflections as well as attention to the values of past discourses. This approach to history is as demanding as it is fruitful as a tool to gain specific knowledge concerning the case and more general knowledge concerning the time period studied: such an analysis needs to be acknowledged as a study as such, not just as a background for a design or plan proposal. Specific courses in landscape/planning history could be one solution.

So what about the use of landscape/planning history as a tool? More than 30 years have passed since Abbott and Adler called for "making historically based analysis a regular part of planning practice" (Abbott and Adler 1989, p. 467). Perhaps the challenge lies precisely in the usefulness and importance of history for developing such a tool. As history is everywhere, and as there are several reasons to study history, there are also different ways of doing it. This goes for the use of landscape and planning history as a tool too. What is needed, then, is to differentiate ways to combine landscape history and planning history as an integrated tool, such as the rapid time-layer analysis that primarily focuses on land-use change, theoretically inspired examinations with a narrow searchlight into history, and studies with a stronger focus on providing a rich historical account of a case. These different approaches are united in the use of history as a tool, that is: the *main* reason for the historical endeavour is to understand the present day landscape and to find ways to move forward. Yet, their different potentials and constraints need to be discussed with the students. Future landscape architecture education requires then, not only an improved interplay between history and theory, but also more explicit discussions on how history is used in different ways in different courses, and in the development of several landscape/planning history tools.

References

Abbott, C. and Adler, S. (1989) 'Historical Analysis as a Planning Tool', *Journal of the American Planning Association*, 55, pp. 467–473.

Bickerstaff, K. (2012) ' "Because We've Got History Here": Nuclear Waste, Cooperative Siting, and the Relational Geography of a Complex Issue', *Environment and Planning A*, 44, pp. 2611–2628.

Cele, S. (2015) 'Childhood in a Neoliberal Utopia. Planning Rhetoric and Parental Conceptions in Contemporary Stockholm', *Geografiska Annaler Series B*, 97, pp. 233–247.

Crang, M. (2000) 'Between Academy and Popular Geographies: Cartographic Imaginations and the Cultural Landscape of Sweden', in Cook, I., Crouch, D., Naylor, S. and Ryan, J. (eds.) *Cultural Turns/ Geographical Turns*. Hoboken: Prentice Hall, pp. 88–108.

Daniels, S. and Cosgrove, D. (1988) 'Introduction: Iconography and Landscape', in Cosgrove, D. and Daniels, S. (eds.) *The Iconography of Landscape: Essays on the Symbolic Representation, Design and Use of Past Environments*. Cambridge: Cambridge University Press, pp. 1–10.

Harris, D. (1997) 'What History Should We Teach and Why? A Historian's Response', *Landscape Journal*, 16, pp. 191–196.

Kyttä, M. (2004) 'The Extent of Children's Independent Mobility and the Number of Actualized Affordances as Criteria for Child-Friendly Environments', *Journal of Environmental Psychology*, 24(2), pp. 179–198.

Law, J. (2009) 'Actor Network Theory and Material Semiotics', in Turner, B. S. (ed.), *The New Blackwell Companion to Social Theory*. Chichester: Blackwell Publishing, pp. 141–158.

Natrasony, S. M. and Alexander, D. (2005) 'The Rise of Modernism and the Decline of Place: The Case of Surrey City Centre, Canada', *Planning Perspectives*, 20, pp. 413–433.

Pries, J. and Qviström, M. (2021) 'The Patchwork Planning of a Welfare Landscape: Reappraising the Role of Leisure Planning in the Swedish Welfare State', *Planning Perspectives*, 36, pp. 923–948.

Qviström, M. (2010) 'Shadows of Planning: On Landscape/Planning History and Inherited Landscape Ambiguities at the Urban Fringe', *Geografiska Annaler Series B, Human Geography*, 92, pp. 219–235.

Qviström, M. (2013) 'Searching for an Open Future: Planning History as a Means of Peri-Urban Landscape Analysis', *Journal of Environmental Planning and Management*, 56, pp. 1549–1569.

Qviström, M. and Bengtsson, J. (2015) 'What Kind of Transit-Oriented Development? Using Planning History to Differentiate a Model for Sustainable Development', *European Planning Studies*, 23, pp. 2516–2534.

Relph, E. (2016) 'The Paradox of Place and the Evolution of Placelessness', in Freestone, R. and Liu, E. (eds.) *Place and Placelessness Revisited*. New York, NY: Routledge, pp. 20–34.

Riley, R. B. (2015) *The Camaro in the Pasture: Speculations on the Cultural Landscape of America*. Charlottesville: University of Virginia Press.

Saab, J. (2007) 'Historical Amnesia: New Urbanism and the City of Tomorrow', *Journal of Planning History*, 6, pp. 191–213.

Schein, R. (1997) 'The Place of Landscape: A Conceptual Framework for Interpreting an American Scene', *Annals of the Association of American Geographers*, 87, pp. 660–680.

Sörlin, S. (2020) 'Hägerstrand as Historian: Innovation, Diffusion and the Processual Landscape', *Landscape Research*, 45, pp. 712–723.

Stahlschmidt, P., Swaffield, S., Primdahl, J. and Nellmann, V. (2017) *Landscape Analysis: Investigating the Potentials of Space and Place*. London: Routledge.

Stigertsson, A. and Palmqvist, R. (2020) *Samhällsplanering – ett barn av sin tid? Modernismens och nyliberalismens påverkan på barns utemiljöer i staden*. Alnarp: Swedish University of Agricultural Sciences.

Stryjan, N. (2020) *Planering för fysisk aktivitet: en analys av Hallonbergen, 1967–2020*. Uppsala: Swedish University of Agricultural Sciences.

Thomas, J. M. (2006) 'Teaching Planning History as a Path to Social Justice', *Journal of Planning History*, 5, pp. 314–322.

Wylie, J. (2007) *Landscape*. Oxon: Routledge.

5
EUROPEAN COOPERATION BETWEEN EDUCATORS AND LANDSCAPE ARCHITECTURE SCHOOLS

Richard Stiles

Introduction

This chapter investigates the phenomenon of international cooperation in landscape architecture education from a European perspective. This is in part a reflection of the author's close involvement with many of the developments over recent decades, but also due to the fact that accidents of history and geography have meant that the potential, and indeed the necessity, for cooperation between Europe's schools is arguably greater than is to be found elsewhere.

Another important factor is that, within Europe, recent political and policy developments have provided a unique impetus for strengthening cooperation in many fields, including landscape architecture education and scholarship, above all the establishment of the European Single Market. This has had a significant impact on international cooperation in the field, while the European Landscape Convention calls explicitly for both relevant education and international cooperation on the part of the signatory states, which include 41 of the 46 Council of Europe members (Sarlöv Herlin and Stiles 2016).

Today, sitting in internet meetings with colleagues to discuss involvement in Erasmus or Horizon programme projects, or to discuss a book project, it would seem that at the beginning of the third decade of the 21st century, international cooperation between European landscape architecture schools and their academics has become commonplace, yet it was not always so.

The evolution of landscape architecture schools and their missions

For the first eight centuries in the history of European universities, starting with the foundation of the University of Bologna in the late 11th century, international exchange became an accepted part of study and scholarship. The early European universities were religiously inspired communities of scholars and students. Their subject matter was initially largely ecclesiastical, while including some civil law as well as medicine and philosophy.

It was not until the early 19th century that the concepts of the unity of teaching and research, and of academic freedom were established, and subjects of study expanded to include the natural sciences. This involved the pursuit of knowledge for its own sake to educate thoughtful and responsible citizens with sound judgement and an ability for critical reflection. As the 19th century progressed, there began to be added a second purpose: that of offering specialised *professional* education and training for an increasingly broad public.

Initiatives for the establishment of the first university landscape architecture programmes aspired to profit from both traditions. These efforts came not from the universities themselves but from professional organisations. Their immediate goal was that of ensuring the necessary flow of trained young people to enter the profession. There was, no doubt, also a secondary motivation: namely a desire to acquire some of the kudos of an elite university liberal education for the profession of landscape architecture. This could confer on it a status similar to that of the other university educated professions, which by then included architecture and engineering.

Combining the profession's goal of training technically competent graduates to meet the needs of public and private landscape architecture offices, with the liberal educational ethos of the European university resulted in an inherent tension which persists to this day. The consequence was that both schools and educators became proverbial servants of two masters. Their roles combine providing clearly circumscribed professional training with academic freedom and commitment to both teaching and research, as well as to the open-ended intellectual enquiry embodied by a liberal education. And this can in a modest way be seen as having had an indirect influence on the development of international cooperation between landscape architecture schools. These dual missions of today's landscape architecture schools in Europe are broadly reflected in the functions of teaching and research. In defining the potential fields for cooperation, to these must be added a third area of what might be termed societal engagement or public service. If the potential fields for cooperation were clear, what about the motives and opportunities for cooperating, and what possible forms might it take?

What is cooperation and why it does and doesn't happen?

Cooperation is a rather diffuse concept: at its most basic, it merely suggests some form of voluntary working together. In practical terms, it can encompass a wide spectrum of activities and organisational forms. In the context of teaching, it can range from more or less spontaneous informal cooperation between individual teachers, who may decide to run a joint studio project, to formal institutional arrangements at university level involving the award of joint degrees. Forms of research cooperation can be equally varied, from joint publications to large-scale funded research projects.

Whatever form it may take, cooperation is beneficial to all involved. Given its obvious advantages, such as broadening horizons and acquiring insights into other's viewpoints, one could be excused for thinking that international cooperation was just a specialised instance of a much wider phenomenon. Counterintuitively, however, cooperation at the national level is not common, despite the fact that factors such as shared ways of thinking, geographic proximity and the absence of communication problems would encourage it. In practice, however, schools within the same country often tend, rightly or wrongly, to view each other as competitors rather than cooperation partners. In addition to the general benefits of cooperation, international cooperation, carries with it echoes of repute and recognition, and, is therefore the norm rather than the exception.

There are, however, considerable practical barriers, and these help explain why for most of the 20th century international cooperation between higher education institutions was not the default condition. Reasons include the inverse of factors which would theoretically seem to favour national cooperation, such as the lack of geographic proximity and the prevalence of communication problems. While passive barriers such as these have been gradually broken down over the years, as the historical evolution of cooperation in Europe reveals, it was the advent of active support for European integration that has had the greatest impact on promoting academic cooperation between landscape architecture schools. Schools were then able to make use of this support to develop their own institutions to strengthen international cooperation still further.

The limited nature of early cooperation

The role that professionals played in instigating international academic cooperation has been a central one. This can be traced back to the first international meetings involving both professionals and academics held during the 1930s. Initiatives to form international organisations came to fruition after World War II with the establishment of IFLA, the International Federation of Landscape Architects, in 1948 (Imbert 2007). The founding of IFLA can also be said to mark the end of the first phase of the international cooperation story, which had begun even before the establishment of the first degree programmes. The increasingly tense geopolitical situation of pre–World War II Europe was, however, not the most fertile ground for international cooperation, with the exception of some limited exchange between the Nordic countries (Jørgensen et al. 2020b), although by 1939 there were still only four full programmes in existence across the continent as potential cooperation partners (Birli 2016, p. 147).

In 1948, there were seven professional degree programmes in Europe. From 1949, the number of programmes in Europe almost doubled every ten years, such that by the end of the 1970s, some 51 landscape architecture programmes at higher education institutions are recorded (Birli 2016, p. 147) and a critical mass of landscape architecture schools existed in Europe as a basis for cooperation. In addition to its biennial and later annual meetings, at which educators had the opportunity to meet one another, in 1959 the IFLA Grand Council established its Committee on Education. This committee was charged to undertake a worldwide survey of the state of education (Birli et al. in this volume). During the post-war period the professional focus was on responding to the needs of post-war reconstruction and so international cooperation was perhaps not a primary concern. Later the rise in environmental awareness, in the context of the general expansion of higher education helped drive the increase in the number of programmes and indeed the number of countries in which they were located. But, despite the fact that environmental issues do not respect national borders, as far as we can be aware, cooperation remained at best sporadic during this whole period, and where it took place was largely the result of individual personal initiatives.

Two possible explanations can be put forward for this. From today's perspective, it is easy to forget that during the 1970s international travel was not straightforward or as affordable as it is today; furthermore, travel funding tended to be scarce. Also, Europe was still divided by the 'Iron Curtain'. International communication mainly involved writing letters and towards the end of the decade, sending faxes, while widespread access to e-mail and the World Wide Web was still a further decade away. With hindsight, it can be said that both individually and together, shortcoming in communications infrastructure must be seen as impediments to international cooperation not outright barriers. From today's perspective we may view the conditions surrounding international travel and communication as inconvenient, but then they were the norm, and had there been an imperative to cooperate more closely at an international level, it would certainly have happened, albeit at a less intense level than we now take for granted.

This leads us to the second possible explanation, namely that there was little in the way of supporting structures for cooperation and no strong incentive for it to happen. Other than the 'normal' frictions affecting international travel and communication, there were certainly no hindrances to cooperation between educators or landscape architecture schools, but because there are few records of cooperation projects during the 1960s and 1970s it can be assumed that such examples as there were resulted from personal initiatives and had limited lasting impact.

However, as soon as a supporting framework began to be established in the early 1980s, corresponding initiatives began to appear.

European integration as a key driving force for academic cooperation

When international cooperation did begin to take place in a more structured form during the 1980s, the main driving force took the form of initiatives developed by the European Community (from 1993 the European Union). The passing of the Single European Act in 1986, in particular, paved the way for international cooperation between European landscape architecture schools to step up a level based on a stable organisational structure. The planned establishment of the single European market indirectly supported activities leading to closer international cooperation in higher education and professional organisations.

The European Community first discussed educational initiatives in the 1970s,[1] but it was not until the early 1980s that pilot student exchanges in the form of a precursor to the Erasmus Programme began. In 1986, the first involvement of landscape architecture schools started with the establishment of ELEE – European Landscape Education Exchanges[2] – initiated by Roger Seijo of Thames Polytechnic (now Greenwich University, London, UK). This was a group of 12 landscape architecture schools from 11 of the then 12 member states of the European Community (Luxembourg had and still has no landscape architecture school). The main activities receiving funding support centred on what became known as 'Intensive Programmes' (IP). These involved small groups of teachers and students from several schools working jointly on design projects usually on-site. While the focus was on joint student design projects, and although only small groups of students were able to participate, the initiative was significant because of its multilateral nature, bringing together schools from several countries and allowing teachers to develop shared pedagogic approaches and get to know each other's teaching programmes.

As part of the preparations for the introduction of the European Single Market in 1993, bodies representing the landscape architecture profession in the then 12 member states of the European Community began talking to each other (independently of IFLA) about how to respond to the coming freedom of movement and opportunities to offer professional services throughout the European Community. The result was the establishment of two bodies, EFLA – the European Foundation for Landscape Architecture, and HKL – the *Hochschulkonferenz Landschaft*. EFLA's task was that of representing the interests of the landscape architecture profession 'in Brussels'. The structure of EFLA largely mirrored that of the UK's Landscape Institute (LI) whose role included defining the educational syllabus. As a result, EFLA – like the LI – had two committees: professional practice and education. With the help of funding from the European Commission's fledgling ERASMUS Bureau, EFLA undertook a survey of *Teaching Landscape Architecture in Europe: A Comparative Study of Course Contents, Durations and Emphases with Comments and Statements of Aims and Objectives and Definitions*, which was published in September 1992, in time for the inauguration of the European Single Market.[3] Apart from providing a first, if incomplete, overview of comparative data, the survey concluded that it would take a lot more work to achieve "harmonious integration of the education systems".[4]

Also in the spirit of the forthcoming Single European Market, and on the 60th anniversary of the first landscape architecture programme in Germany, a meeting of European schools was hosted by Berlin Technical University in September 1989, with the title 'HKL Europe', and a second HKL Europe meeting in Vienna in 1990. HKL is short for *Hochschul-Konferenz Landschaft*, a joint standing conference jointly established in 1979 by professional organisations and the higher education institutions in German-speaking countries. These developments did not occur independently, the HKL Europe meetings taking place in parallel with EFLA's survey of European landscape schools, while the majority of the EFLA sub-committee responsible were landscape architecture academics, including Helmut Weckwerth of Berlin Technical University, which was also a member of ELEE. Michael Downing, the chair of the EFLA Education Committee played a key role on the survey. The third meeting of European landscape architecture schools took place at the invitation of another member of the EFLA survey committee, Meto Vroom of Wageningen University in the Netherlands. At this meeting in September 1991, it was unanimously decided to establish a new organisation to take over the responsibility of organising future meetings. This was given the name European Conference of Landscape Architecture Schools (ECLAS).[5]

ECLAS — a new vehicle for European cooperation

In recognition of his work with the EFLA education committee and on the education survey, Michael Downing was elected as chair of ECLAS, however not before a discussion had taken place about whether ECLAS should in fact be identical with the EFLA Education Committee. The decision to retain a clear separation between the two organisations, with their bases in the profession and the academic world respectively, was driven by considerations of the dual role of the universities with their academic goals as well as their professional training role. But it was also inspired by the wish not to limit membership of ECLAS to the narrow definition of Europe of the 12 European Community members comprising EFLA.

Participation in the inaugural 1991 ECLAS meeting at Wageningen underlined the validity of this broader geographical remit, as colleagues from non-EC countries including Norway, Israel and the newly independent Slovenia did not just attend the meeting but also became part of the new ECLAS committee. Indeed, it was agreed that the second ECLAS conference the following year should be hosted by the University of Ljubljana.

While this new phase of European cooperation was clearly stimulated by the moves towards European integration, it was still largely reactive in character with self-selected groups of schools taking advantage of the opportunities offered by the European Community – rather than being proactive in terms of creating their own cooperative structures. Now, with the foundation of ECLAS as a permanent multi-lateral and inclusive institution, it can be argued that a new cooperation phase had begun.

The ECLAS Conference, which now became an annual event, provided the opportunity for landscape architecture academics to meet once a year and discuss aspects of education and report on research (see Figure 5.1). However, cooperation was still limited to these once-yearly exchanges, and even the schools taking part in the annual meetings usually sent at most a couple of representatives. Thus, there remained a considerable scope to enhance cooperation. Between the ECLAS conferences, international cooperation still had largely to rely on the ERASMUS Programme. A new network run from Manchester University was established in 1992 with student exchange as the focus. Although these arrangements only allowed for a small number of students to study at another network member university, a vital benefit was provided by regular funded opportunities for representatives from member schools to meet and learn about each other's programmes. Given the limitations of the ECLAS conferences, it was thanks to the ERASMUS Programme's financial support for ELEE and the Manchester network, that the process of European cooperation could be intensified, although still overall only some two dozen schools were involved.

As to what proportion of the schools in Europe this represented, no one really knew. At this stage, ECLAS was still an *ad hoc* collection of schools defined largely by those which took part in the conference. There was no formal membership, and its size and composition thus varied

Figure 5.1 Group photograph of the participants at the 2019 ECLAS conference in Ås/Oslo at which 100 years of university education in landscape architecture in Europe was celebrated at the Norwegian University of Life Sciences, where the first programme began in 1919. The event also marked 40 years of European cooperation in landscape architecture education since a first informal meeting of schools in Berlin.

Source: Photo by Ellen Fetzer.

from year to year. This situation continued through the 1990s until changes in the ERASMUS Programme expanded the opportunities for student exchange, but reduced the role of network coordination to an administrative task undertaken at university level. This instantly removed the previous opportunities to strengthen academic cooperation in the form of meetings between ERASMUS contact persons. At the 1998 conference in Vienna, Michael Downing stepped down from his role as first ECLAS president, and the necessity for a new ECLAS team provided an unexpected opportunity to take a fresh look at the state of international cooperation within the European context. This can be said to have marked the end of the first, pioneering, phase of ECLAS and the start of what might be described as 'ECLAS 2.0'.

Conference – Council – Community: taking international cooperation to the next level

The new ECLAS team felt that despite the success of the initial conferences, the potential for international cooperation reached well beyond a once yearly meeting, and that the organisation ought to have some kind of existence between the conferences too. To symbolise this ambition, at the conference in 2000 it was agreed that ECLAS's name should be changed from the European *Conference* . . . to the European *Council* of Landscape Architecture Schools. The challenge was now to fill out this symbolism with substance.

It has already been remarked that limitations on communication acted as a brake on the development of international cooperation in the 1970s. By the late 1990s, much had changed in this respect, but it was still not the world which we take for granted today. This is illustrated by a coincidence: simultaneously with ECLAS coming under 'new management', in September 1998, a new internet company by the name of 'Google' was incorporated in California. Despite the growth of the World Wide Web during the 1990s, ECLAS did not yet have a presence. A web site would, if nothing else, make potential members aware of its existence. Creating a web site called for two things: a minimal amount of funding and some content, to begin to substantiate the new aspiration to take cooperation beyond the annual conference. ECLAS had no financial resources, so the obvious response was to begin to charge a modest membership fee, but ECLAS also had no formal membership. As a precondition to rectifying this situation, it was necessary to define a set of aims and objectives for the achievement of which the fees were to be levied, and to encode these in statues to govern the operation of the organisation. As well as defining the general goal of furthering the landscape architecture academic community they attempted to set out in some detail what was entailed in international cooperation and this became the explicit *raison d'être* of ECLAS (see Box 5.1).

Box 5.1   Objectives of European academic cooperation as defined in the ECLAS statutes

Article 3

The Association's objectives are to foster and develop scholarship in landscape architecture throughout Europe by strengthening contacts and enriching the dialogue between members of Europe's landscape academic community, by representing the interests of this community within the wider European social and institutional context and by making the collective expertise of ECLAS available, where appropriate, in furthering the discussion of landscape architectural issues at the European level.

Article 4

In pursuit of this goal the European Council of Landscape Architecture Schools seeks to build upon the rich European landscape heritage and intellectual traditions to:

a. Further and facilitate the exchange of information, experience and ideas within the discipline of landscape architecture at the European level, stimulating discussion and encouraging cooperation between Europe's landscape architecture schools through, amongst other means, the promotion of regular international meetings, in particular an annual conference, publishing an academic journal and maintaining a website;
b. Foster and develop the highest standards of landscape architecture education in Europe by, amongst other things, providing advice and acting as a forum for sharing experience on course and curriculum development, and supporting collaborative developments in teaching and learning;
c. Promote interaction between academics and researchers within the discipline of landscape architecture, thereby furthering the development of a Europe-wide landscape academic community, through, amongst other things, the development of common research agendas and the establishment of collaborative research projects;
d. Represent the interests of scholarship in landscape architecture within Europe's higher education system, encourage interdisciplinary awareness and enhance the overall standing and the public understanding of the discipline;
e. Stimulate dialogue with European bodies, institutions and organisations with interests in landscape architecture and with other international organisations furthering landscape scholarship.
f. Undertaking any other lawful activity as may be conducive to achieving the objectives of the Association as set out above, either singly or in conjunction with other organizations and/or individuals.

Source: Statutes and Standing Orders, ECLAS: www.eclas.org/statutes-and-standing-orders/ (accessed 30.11.2021).

After the creation of a first simple static web site, it became clear that to be useful it would need to contain information about each member school and its staff. An academic community cannot be created from the top down but is dependent on the interaction of individual staff members, yet it was soon clear that to provide this information in sufficient detail and keep it up to date was a task beyond the resources of the non-existent ECLAS staff. The obvious solution to this was to build in the possibility for members to add and update their own information. In this way, the web site became an early example of a Web 2.0 site, characterised by user-generated content, even before the term had been coined. Although an academic community could not be created, with virtual infrastructure such as this it could be nurtured.

This level of ambition was, however, far beyond the scope of the resources made available by the new membership fee. The solution to this challenge was to come from a new part of the European Union's ERASMUS Programme: 'Thematic Networks'. These discipline based projects had objectives broadly similar to ECLAS's own: furthering European cooperation in higher education, and they offered an annual budget several orders of magnitude higher than ECLAS's annual fee income. In line with ECLAS's goal of furthering the landscape architecture community as a whole, a project involving all of Europe's known 73 landscape architecture schools, as well as 10 further partner universities from 'non-eligible' countries together with 10 non-university partner organisations, was conceived and submitted to Brussels.

LE:NOTRE – a European experiment in international cooperation

The LE:NOTRE Project (Landscape Education: New Opportunities for Teaching and Research in Europe) started in October 2002 with a budget of €125.000 for its first year, at least 25 times the then annual ECLAS budget, and the annual LE:NOTRE budget increased as the project progressed. It offered an unrivalled potential for supporting international collaboration. LE:NOTRE was a large and complex project[6] which eventually ran for some 11 years, had an overall budget of some €5 million, including contributions of time from project members, and at its peak involved at least 125[7] universities worldwide (see Figure 5.2).

One important way to look at the project is in terms of its outputs.[8] Perhaps the most high-profile of these was the establishment in 2006 of the peer-reviewed *Journal of Landscape Architecture (JoLA)*, which has since become an important communication and cooperation platform in the field. Other outputs included ECLAS-endorsed book publications. The first was an international collection of essays by members of related disciplines *Exploring the Boundaries of Landscape Architecture* (Bell et al. 2011), investigating the relationship between the discipline and its academic neighbours and was intended to lay the groundwork for cooperation across disciplinary frontiers. Several more ECLAS endorsed books have appeared, including those on research and education (van den Brink et al. 2017) (Jørgensen et al. 2016). The formal incorporation in 2006 under Netherlands law of ECLAS itself was a further important result of the project, drawing a line under the informal existence of the organisation since its first official meeting in 1991.

On the education side, the LE:NOTRE network became one of the participants in the EU funded 'Tuning Project' – an initiative to harmonise higher education programmes in many disciplines as a precursor to establishing the 'European Higher Education Area' as a means of operationalising the so-called Bologna Process. This engaged the broad membership of the sub-disciplinary 'working groups' set up within the context of the project in order to define what graduates of landscape architecture programmes should be capable of, in terms of both subject-specific and generic competences. The output of the working groups resulted in the publication in 2010 of the *ECLAS Guidance on Landscape Architecture Education* (ECLAS 2010).

Figure 5.2 Members of the LE:NOTRE committee and international advisers, representing some 14 countries – mainly from Europe but also North and South America – discussing informally in a Paris park during a break in the proceedings of the Versailles Spring Workshop in April 2009.

Source: Photo by Diedrich Bruns.

All EU projects are expected to make a web site to present information about the project and disseminate the results. In the case of LE:NOTRE the web site began to build on the beginnings made with the ECLAS site, and from the start was understood as far more than just an electronic noticeboard. The World Wide Web was still at a relatively early stage in its development as the project began: Wikipedia had only just started (2001) and Facebook was still just a twinkle in Mark Zuckerberg's eye (since 2004), and yet, with hindsight it can be said that LE:NOTRE.org independently pioneered features now found on many of these. From this point of view, it could be said that the LE:NOTRE web site represented the most ambitious attempt to enhance international cooperation by both broadening and deepening it. It became, not a way of presenting the results of the project, but a central part of those results.

The development of ECLAS and of LE:NOTRE were highly encouraging and entirely in line with the ECLAS's stated goals of 'strengthening contacts' and 'enriching dialogue' between members of the academic community. In practice, however, the group of people participating actively in this dialogue was mainly limited to those few, often senior, academics from each school who were able to regularly attend the annual conferences and LE:NOTRE workshops. An interactive web site, which was continuously accessible and open to all members of every school, was felt to offer a way to democratise cooperation and build the academic community from the bottom up.

The rise and fall of LE:NOTRE.org

In seeking to establish an academic community in virtual space, it was first necessary to enable academics with similar areas of interest to identify each other. Two features of the site aimed to facilitate this: first, individuals had to be registered to gain access, their names being listed on screen whenever they logged in alongside their national flag. This sought to establish the feeling of a community, linking people using the site at any one time. Secondly, 12 working groups or channels were created reflecting the main sub-disciplines of landscape architecture. Members were encouraged to 'join' at least one so that anyone could easily identify colleagues with similar areas of interest.

Working group members were encouraged to share information of common interest using a series of web-databases, in order to build up a common cumulative resource, such as a shared literature database, which could also help to increase awareness of literature that had, perhaps, previously only been locally familiar. Another joint resource was a database of design projects. Case studies of designed open spaces are a staple of all teaching and research, and with its international membership, the network was ideally placed to collect and share first-hand information on and images of local projects which could be used by everyone.

Projects, images and references could be tagged according to the relevant sub-disciplines (working groups), and thereby shared on the respective working group pages. Tagging entries according to location, made it possible to generate pages for specific cities, containing relevant information from all databases, which also included, the design offices responsible for individual projects. Such pages, based on information entered by local academics, could provide an ideal basis for planning field visits. Other databases covered landscape plans, plant material, research projects and web links. A sophisticated multilingual thesaurus of technical terms was also developed to ensure clarity and accuracy of international communication of specialist concepts within the discipline.

This brief outline only captures a fraction of the features and facilities which the web platform came to offer. It was conceived as an interactive and open-ended resource to be used and added to by all project members (see Figure 5.3). Yet, anyone searching for this information today or wishing to contribute to it will be disappointed. Although it was maintained and further developed for some time after the project finished, eventually it had to be taken off-line due to the lack of resources available to maintain it.

During the 11 years of developing the web platform, the main focus was on adding functionality rather than maximising usability. This perhaps resulted in users who had become familiar with newer sites such as Facebook and Wikipedia being less willing to invest the effort necessary to tag and geo-reference each entry. Hindsight suggests this was perhaps a significant hurdle to project members adding information, as was the increasing pressure on all academics living in the 'publish or perish' environment of modern universities, which did not give credit for co-creating community resources such as the LE:NOTRE web platform. As a result, the many thousands of entries made during the course of the project, together with the information on 'who's who' within the landscape architecture academic community and the range of 'in-house' tools available for all to collaborate have been lost.

International cooperation in Europe since LE:NOTRE

As an EU co-funded project, LE:NOTRE ended in 2013. Its 11 years of funding far exceeded the regular three-year lifespan of a Thematic Network Project. At the centre of the project was not just the aim of furthering international cooperation within the discipline but the desire to

Richard Stiles

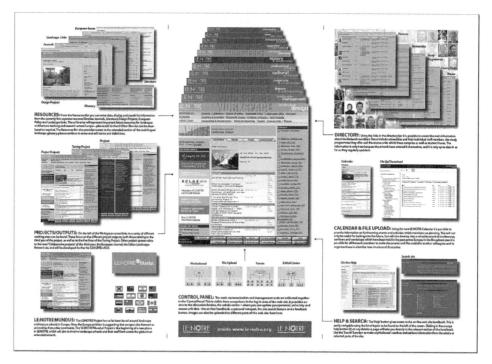

Figure 5.3 Excerpt from a folder made to present an early version of the LE:NOTRE web site to the broad project membership, illustrating the many interactive features it contained, including subject-related workspaces and user-editable databases. The site continued to develop during the course of the project, gaining yet further functionality as well as a more friendly user interface.

Source: Richard Stiles, LE:NOTRE Project.

create a sustainable basis for its further development in the future. It demonstrated, with the help of co-funding from the European Union, that international cooperation could indeed be advanced. The question remains, how sustainably?

The foundation of the LE:NOTRE Institute[9] was the tangible form that the intention took, aiming to put the project outcome on a sustainable footing. The Institute continues to thrive in an adapted form. EU funding programmes continue to provide possibilities for cooperation projects, and many of Europe's landscape architecture schools have taken advantage of these. This confident culture of European cooperation between landscape architecture schools is something which can indeed be seen as a positive outcome of the LE:NOTRE Project.

Furthermore, the LE:NOTRE Institute has found a role both as a consortium member and in helping to manage some of these projects. By participating as a consortium member, the Institute is able to indirectly involve and represent the interests of the broader academic community. In doing so at least partly to compensate for the absence of a large scale network in a small-scale consortium of five or six partners and at the same time can make links between projects in which it is involved, thereby also performing a valuable task of helping to network individual projects together.

While some of these projects have been focussed on specific landscape issues, such as Landscape Education for Democracy – LED, or COLAND – Inclusive Coastal Landscapes, or been run by and for a small group of schools, such as for developing a 'European Master in Landscape

Architecture', or promoting continuing professional development in the Baltic Region, several others have had a specific European or collaborative theme. These include EU-Teach (looking at European policy issues in landscape architecture teaching), EULand21 (Trans-European Education for Landscape Architects), or INNOLand – preparing a European Common Training Framework for landscape architects. Involvement in the management of some of these projects has also provided a degree of financial stability for the Institute.

Conclusions

It may appear a paradox that, despite the prevailing competitive orthodoxies as embodied in university league tables, departmental rankings and assessment exercises, the level of cooperation between Europe's landscape architecture schools has grown successively. Indeed, the greatest growth has occurred in parallel with the rise of the neoliberal ideology of the competitive market-place as embodied by the European Single Market, while the very policies intended to minimise barriers to European competition have been the main driving force behind the warming climate of academic cooperation between landscape architecture schools.

There are, however, signs that Neoliberalism has run its course, and with this historic reassessment lies hope for further strengthening European cooperation in landscape architecture education and scholarship. The cooperative ethos has been further reinforced by the recognition of the 'free market's' inability to address the Corona pandemic and the climate crisis. Even economists are now acknowledging that it was not through simplistic Darwinian competition that *Homo sapiens* rose to global prominence, but rather due to the human ability to cooperate (Collier and Kay 2020).

We should also not forget that while academic teaching and scholarship have historically always been collaborative rather than competitive, the *laissez faire* approach to international cooperation – leaving it to take care of itself – resulted in only slow progress for most of the 20th century. Even with the support of European Union programmes, academic cooperation remained largely opportunistic. Only with the establishment of ECLAS did the discipline began to organise itself at the European level and to achieve significant progress. It was this combination of the political push for greater European integration and the financial support this offered, together with the creation of a supportive institutional structure by the discipline itself, that resulted in the current level of cooperation between Europe's landscape architecture schools.

Cooperation can ultimately only be the outcome of the activities and commitment of individual academics. The question is whether, despite this fact, the wider academic community can continue to promote further international cooperation? As we have seen, what can be done is to create an environment which is as supportive of and conducive to cooperation as possible. Yet more effective future cooperation will require more opportunities to meet and exchange ideas, locations where this can take place, and knowledge of potential cooperation partners. Such expectations may seem ambitious but, following the coronavirus pandemic, the outlines of possible solutions have become apparent.

The elimination of travel time through increased virtual working has provided the first piece of the jigsaw, a potential surplus of time. Virtual work spaces and meeting rooms bringing people together across international borders have now become commonplace, only the challenge of finding potential research partners has not been made easier by the pandemic. This was a task addressed by the online directory of 'who is who' within the landscape architecture academic community, created as part of the LE:NOTRE Project. Although the LE:NOTRE web platform is no longer online, the data which it contained still exist on various storage media, making the possibility of restoring and updating them something which might be seriously considered.

Whatever the merits of this specific initiative, the potential of online communication platforms can be the basis for the next stage of European or indeed international cooperation in landscape architecture teaching and research. This might involve international team teaching, which would not just be more stimulating and dynamic for all concerned but also release more time resources for international research collaboration (Jørgensen et al. 2019, 2020a). A cooperative future beckons!

Notes

1 Origins of the Erasmus programme – interview with Hywel Ceri Jones | Erasmus+ (Origins of the Erasmus programme – interview with Hywel Ceri Jones | Erasmus+ (erasmusplus.org.uk)) accessed 18.03.2021
2 http://elasa.org/archive/archive1/YB94/YB94-24.html accessed 18.03.2021
3 https://issuu.com/ifla_publications/docs/efla_blue_book_1992
4 This cannot strictly be classed as an instance of educational cooperation, as it took the form of an external overview carried out on behalf of the profession, rather than a cooperative activity undertaken by the universities themselves.
5 Birli (2016, p. 52) reports that the name ECLAS was suggested by Professor Dušan Ogrin as a way of differentiating the organisation from the American CELA (Council of Educators in Landscape Architecture) to emphasise the role of schools rather than individual teachers.
6 In fact, it comprised eight separate projects, each requiring its own funding application, separate outputs and reporting requirements.
7 Many non-European schools also joined as associate members of the network as the project was encouraged to extend its reach worldwide as part of a separate 'LE:NOTRE Mundus' project. At its peak some 190 universities were registered in the LE:NOTRE web site.
8 The EU expects to assess the success of an ERASMUS project in terms of the concrete outputs promised in the proposal. In the worst case, this can risk the definition of outputs which are safe and predictable, risking a project degenerating into a simple box-ticking exercise. International cooperation, by contrast, is an abstract 'outcome' and is therefore much harder to pin down. This is not to suggest that the fruits of cooperation will not take the form of concrete outputs as in the case of LE:NOTRE, but that cooperative outcomes – if they are to be truly valuable – are the result of open-ended creative processes and therefore cannot necessarily be predicted in advance.
9 The decision not to simply reabsorb LE:NOTRE back into ECLAS, which had been the original intention, can be said to be the result of the success of the project. Although it started as a project of European landscape architecture academia, it gradually expanded to include landscape architecture schools from outside Europe to engage with related disciplines and to involve students, none of which had a natural home within ECLAS. The LE:NOTRE Institute is therefore seen as a kind of external affairs arm of ECLAS.

References

Bell, S., Sarlöv Herlin, I. and Stiles, R. (eds.) (2011) *Exploring the Boundaries of Landscape Architecture*. Abingdon: Routledge.
Birli, B. (2016) *From Professional Training to Academic Discipline: The Role of International Cooperation in the Development of Landscape Architecture at Higher Education Institutions in Europe*. Unpublished doctoral dissertation, Technische Universität Wien, Vienna.
Collier, P. and Kay, J. (2020) *Greed Is Dead: Politics after Individualism*. London: Allen Lane.
European Council of Landscape Architecture Schools (ECLAS). (2010) *ECLAS Guidance on Landscape Architecture Education*. Available at: www.eclas.org/eclas-education-guide (Accessed 16 March 2022).
Imbert, D. (2007) 'Landscape Architects of the World Unite: Professional Organisations, Practice and Politics 1935–1948', *Journal of Landscape Architecture*, 2007, pp. 6–19.
Jørgensen, K., Clementsen, M., Halvorsen Thorén, K. and Richardson, T. (eds.) (2016) *Mainstreaming Landscape – Through the European Landscape Convention*. Abingdon: Routledge.
Jørgensen, K., Karadeniz, N., Mertens, E. and Stiles, R. (eds.) (2019) *Teaching Landscape: The Studio Experience*. Abingdon: Routledge.

Jørgensen, K., Karadeniz, N., Mertens, E. and Stiles, R. (eds.) (2020a) *The Routledge Handbook of Teaching Landscape*. Abingdon: Routledge.

Jørgensen, K., Stiles, R., Mertens, E. and Karadeniz, N. (2020b) 'Teaching Landscape Architecture: A Discipline Comes of Age', *Landscape Research*, pp. 167–178. https://doi.org/10.1080/01426397.2020.1849588.

Sarlöv Herlin, I. and Stiles, R. (2016) 'The European Landscape Convention in Landscape Architecture Education', in Jørgensen et al. (eds.) *Mainstreaming Landscape – Through the European Landscape Convention*. Abingdon: Routledge, pp. 175–186.

van den Brink, A., Bruns, D., Tobi, H. and Bell, S. (eds.) (2017) *Research in Landscape Architecture: Methods and Methodology*. Abingdon: Routledge.

6
BUILDING UP HISTORICAL CONTINUITY
Landscape architecture archives in education

Lilli Lička, Bernadette Blanchon, Luca Csepely-Knorr, Annegreth Dietze-Schirdewahn, Ulrike Krippner, Sophie von Schwerin, Katalin Takács, and Roland Tusch

Introduction

In 2019, European Archives of Landscape Architecture started to collaborate on joint research projects to help integrate archives into landscape architectural education and achieve high standards for archives in technical and scholarly terms. The Network of European Landscape Architecture Archives (NELA) was founded at the 2019 conference of the European Council of Landscape Architecture Schools (ECLAS) in Ås, Norway. International collaboration in the field of landscape architecture has always been important (see Stiles in this volume). Malene Hauxner, writing about icons on the move, has demonstrated how styles, trends, and theoretical ideas in Europe have spread (Hauxner 2006, p. 45ff.). The recognition of cross-border stylistic influences on styles and ideas, the regional understanding of landscape architecture, and, in particular, the changing political landscape and borders in twentieth-century Europe have highlighted the need for international research and collaboration.

In this chapter, we present examples of the many ways that educators employ archives in landscape architecture education. Drawing on sources and experience from NELA, we provide thoughts and scholarly reflections on how archives serve to foster an awareness of landscape architecture's continuous development in higher education over the space of decades and centuries. These reflections are based on the teaching experiences of NELA members (IFLA 2021) in various academic surroundings and beyond academia.

Past–present–future

It seems that landscape architecture has hitherto lacked a sense of historical continuity directed toward the future. On the one hand, the profession's historical narrative is particularly sketchy when it comes to more recent designs, works, and practitioners. John Dixon Hunt stressed that both landscape architects and historians of landscape need to understand "the conceptual principles of [the] profession" (Hunt 2000, p. 207). However, tracing links between garden heritage and contemporary design is, so far, not very robust (see Qviström and Jansson in this volume). A reason might be that the narrative of garden heritage is strongly rooted in art history; many scholars and writers are historians or art historians. Looking forward into the future is not usually part of their occupation (Harris 1999). On the other hand, designers and scholars of contemporary landscape architecture often have little to do with history of art and gardens unless they are specialized in heritage sites or heritage conservation. This discrepancy may explain how a certain gap remains between the past and the future – garden heritage and contemporary landscape architecture – which calls for efforts to increase awareness of continuous timelines between the two with all their disruptive breaks and subtle frictions.

Archives play a key role in developing this understanding of continuity, through the processes of collecting, preserving, curating, and teaching, and through the preservation and restoration of landscape architectural heritage. The question to be investigated is how, in education, the ties linking to both directions on the timeline can be enhanced. Most universities have stipulated a form of research-guided education, which includes learning by research as well as learning research and learning about research outcomes. This has resulted in the formulation of research-guided teaching and research-guided learning, which can be connected through specific contexts (Hughes 2005, p. 15).

Archives are a means of closely relating research to teaching, because archival material is subject to analysis, reflection, and contextualization. Both designed landscapes and designers are the foci of research based on archival material. Together with the designers' approaches and techniques, they form a "complex epistemic structure", as Mareis (Mareis et al. 2010, pp. 11–12) puts it. To make use of archival materials, therefore, means furthering an approach with respect to history as well as to the future, relating design tasks and processes "to wider social, political and cultural force" (Dümpelmann 2011, p. 627). These aims can be achieved through teaching that is focused on archives, archival work, or archival material as well as history classes and classes that go beyond the traditional scope of didactic methods. Archival material can serve well to exemplify other topics, where contemporary problems are complemented by historical approaches and examples. This cross-referencing in time can help to make students confident about working in a continuous timeline, integrating past experiences in their future projects.

Enhancing the use of archives

Work in and with archives of landscape architecture is usually connected to the history of the profession, to art history, to heritage sites, to design theory and to biographical studies. It has been used in educational settings to impart knowledge about the historical development of landscape architecture and to learn and illustrate different stages, projects, and personalities (see Figure 6.1). In any traditional curriculum, the history of garden art is a subject in which the teaching is based on texts, images, and field trips. Archival material is mostly integrated through those sources.

Some engaged teachers take students to analogue archives to introduce them to archival work and make them familiar with original material. For example, at the Hungarian University of Agriculture and Life Sciences, Budapest (Institute of LA and Urban Planning and Garden Art), students learn how to fill out a detailed garden inventory sheet while researching archive data. In Budapest – in lectures, exercises, and complex design work on an existing historical site – a wide range of archival sources are employed, including written sources, drawings, images, and cartographic materials (see Figure 6.2).

Universities that have a landscape architecture archive can make use of their own sources and easily provide access for students. These sources may be extremely varied in their origins. At the École Nationale Supérieure de Paysage (ENSP) in Versailles, as in other schools, the archival material has derived from the archives either of designers or of commissioning clients, owners, and people in charge of the maintenance of the sites. They provide a wide range of information.

Lilli Lička et al.

Figure 6.1 Students gathering information for the seminar "Design in the Historical Context", Archive of Swiss Landscape Architecture (ASLA), Rapperswil, 2018.

Source: Photo by Simon Orga.

Figure 6.2 Detail of a student poster on a historical site in Hungary.

Source: Alexandra Bátki (student), Katalin Takács and Imola G. Tar (teachers) – Szent István University, Faculty of Landscape Architecture and Urbanism, 2019. Map sources: https://maps.arcanum.com/hu/map/cadastral/, Cadastral map of Oroszlány (1885) and Google Maps (2019).

They also widen the scope of study cases to a wider history of landscape architecture. When dealing with history and heritage, the use of archives is obvious. However, archives have not often been used as a basis for investigating the contemporary aspects of the field that can come up in academic education.

Research in education: learning from the past for the future

The relationship between research and teaching can take various forms. In his article "The Link between Research and Teaching in Architecture", Andrew Roberts (2007) defines four different possible relationships. Research-tutored pedagogy encourages students to "apply or interpret research content and [the] ideas of tutors through their project work". In the course of research-based education, "students use the process of designing as a means to advance and develop knowledge". Research-oriented teaching encourages students "to develop research skills and design-enquiry and related information-gathering skills through focussed teaching", while during units delivered using a research-led methodology, students learn about the findings of other researchers through lectures and seminars (Roberts 2007, p. 16). The integration of archives into research-led education is yet to be expanded beyond the core history courses. There is an ongoing discussion about the "research-by-design" approach and the extent to which it creates new global insights – as well as what the nature of these insights might be, as elaborated thoroughly by Martin Prominski in his essay "Design Knowledge" (2008) and by other authors (Nijhuis and de Vries 2019; Lenzholzer et al. 2013). Clemens Steenbergen seeks to incorporate plan and topological analysis into design research based on a study of the drawings of a large number of Italian gardens. He thereby establishes the relation between context and object as a given (Steenbergen 2008, p. 20). In archives, we find designs which may be placed in a given context, when the author, the time, and the socio-political circumstances are known. In many cases, however, this context is still to be elaborated. In an empirical manner, when using material from the archive as a starting point, establishing context is at the centre of investigations, making it possible to refer to contemporary questions within the profession.

At the Manchester School of Architecture, the postgraduate elective unit Arch.Land.Infra Research Methods aims to use a mix of methodologies to introduce students to the recent histories of British architecture and landscape architecture. The overarching aim of this unit is to introduce a range of approaches and methods for understanding, interrogating, and researching the built environment. Within this framework, Arch.Land.Infra relies on archives and archival research as a crucial part of the teaching process. The project is delivered within an ongoing live research project that brings together historic research and contemporary policymaking in a multidisciplinary network. Through research-led and research-tutored elements, students benefit from current research results. The collaboration with archives contributes to the research-oriented elements of the unit, where students are asked to activate the methodologies learned in the context of archival research. Students are asked to gather historical data from a variety of archives related to case studies. This data is examined using a variety of design analysis and design research techniques to create new interpretations of the primary sources in a visual format. This analysis and visualization of archival data and the contextualisation of it within a socio-political framework is not just a useful method of analysing the spatial implications of design decisions but also helps to depict the complexities of the design and implementation process, a crucial area of knowledge for contemporary professionals. Physical models enable students to reconstruct now altered landscapes, while virtual models allow them to reconstruct change over time. Interactive models based on the deconstruction of the design process as an outcome of the thorough analysis of archival documents recreate the debates and choices made by designers.

A close relation between research and teaching is established in the students' final bachelor's and master's dissertations. Following a research-based strategy, LArchiv at the University of Natural Resources and Life Sciences (BOKU), Vienna, operates as a pool to generate research questions for theses on several levels. At a post-graduate level, students contribute significantly to research when they uncover material that is so far unedited. This is of mutual benefit to both the archive and the student: the student adds a thoroughly inventoried and analyzed set of archival material to the archive and contributes new findings to landscape history, while also gaining experience with archival work, which has increasing relevance in landscape practice and research (Powers and Walker 2009, p. 105). Three learning outcomes of this research-based process are achieved almost as a side effect: thresholds for consulting an archive are lowered considerably and can more easily be overcome; an awareness of historical continuity is generated by the student and enhanced by publications; and the student becomes a skilled archival worker.

Learning by drawing

The starting point for generating new insights is the analysis of the design itself with the drawing as its representation. Catherine Dee (2001) uses drawings of contemporary projects to learn and specify the nature of the design and its formal aspects. It is well-known that physical activities such as drawing by hand support a deeper understanding than mere thinking (Hasenhütl 2009; Hufendiek 2012; Pallasmaa 2009). The interplay of thinking and doing is essential here, because both hemispheres of our brains are mobilized. The first step is to copy a design drawing from an archival source, such as a plan. Steenbergen (2008, p. 24) calls copying "the simplest manner of research by drawing". Repeating lines that a predecessor had skilfully put on paper involves tracing their original production (see Figure 6.3).

Figure 6.3 Discussing historical drawings in ANLA in order for students to copy them during a course.

At BOKU, both hand drawing and digital drawing are used for copying in order to acquaint students with old material, earlier design principles, and different modes of representation, while allowing them to exercise their technical skills. Archival material is firstly decontextualized and used as a practical example. By reading the design in combination with a physical drawing, new insights are generated into skills and techniques as well as content. This also allows students to reflect on their own intentions and techniques. At the Institute of LA, Urban Planning and Garden Art (Budapest), for instance, small-scale design or analysis tasks are integrated into history lectures. The productive outcome of the first master semester is thus a garden-art-style sketchbook for each student. At the Historical Archive of Norwegian Landscape Architecture (ANLA) history is deliberately included in a number of courses, using archival material as a starting point for discussion rather than overloading students with a long introduction to the history of landscape architecture beforehand. Historical plans from student projects drawn in the 1920s, 30s, 50s, and 60s are selected as a way to practise freehand drawing. Students are given several tasks while studying these plans – they are asked, for example, to look at composition principles and the use of symbols in the plans. They are also given the task of interpreting a 2D drawing to create a 3D perspective, which helps them develop spatial awareness. Historical drawings and plans are used at BOKU as a basis for teaching hand drawing as well as digital drawing. In digital drawing classes, students have to vectorize and digitize historical hand-drawn plans and overlay them with site plans using current open-source data maps (see Figure 6.4). The decoding of graphic style and the meaning of the symbols requires precise analysis as well as interpretation of the archival documents. This thorough examination leads to an expanded understanding of the project's historical context.

The experience with the group of undergraduate students at ANLA has shown how archival material can be an innovative and effective learning tool. By decoding historical hand-drawn plans, students acquire a good knowledge of the underlying design principles. Comparing and analyzing plans from different periods leads to an expanded understanding of the projects' contexts. The process of reinterpreting and redesigning enables students to get a better sense of the designers' intentions than oral explanations. Students understand various creative ways of presenting a project in a plan or perspective. They see in detail how forms and shapes are created in historical drawings. By going through materials from different periods, they can also trace the different trends and fashions at work in visualizing project ideas. Such experiences have brought students a different understanding of space and form and encouraged them to reflect on the role of the designer or landscape architect in the project.

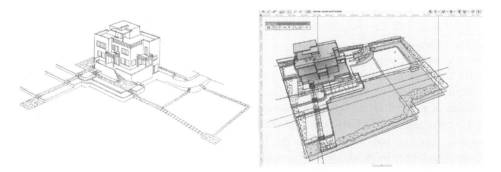

Figure 6.4 Isometric drawing of Dos Santos garden by Albert Esch (1930), transposed into a digital drawing.

Source: Albert Esch, LArchiv BOKU Vienna, student Vigil Peer, teacher Roland Wück, 2017, BOKU Vienna.

Exploring sites, learning ideas

As a case study, students taking part in design classes at the Eastern Switzerland University of Applied Sciences in Rapperswil, Switzerland, are encouraged to create new ideas for different types of existing sites with heritage value but a lack of heritage protection. For this purpose, they use resources available in the Swiss Landscape Architecture Archives (ASLA). The process includes a deep analysis of the history, the area, and the place itself. The original garden style and the use of materials and plants are determined as traces of the past and provide inspiration for the new designs. Photos, plans, drawings, and written documents from the archive are the basis for understanding the original ideas of the earlier designs. The study remains close to professional practice, as it constitutes one of the main tasks in landscape design studios and gives more depth to the designs. One goal is for students to learn and understand the importance of the material kept in the archives and encourage them to come back to check on the resources when they are working as professionals. Within this class, connections between past, present, and future reveal themselves. The need for, and the benefits of using, archival material become obvious.

Deciphering historical and contemporary design achievements is a means of understanding the translation of ideas into concrete projects in different periods of time. In the history course at the ENSP in Versailles, case studies are presented, including the use of archives, to illustrate the evolution of a project and place from conception to its current state. In the tutorial, students work their way backwards. They start from the current situation (which they have observed on-site) and decipher the different stages hidden beneath it, until they arrive at the original project and archive, which includes intermediate stages and photos. The work is based on representations in plans and sections complemented by views and texts. Approaching history through case studies allows the student to apply to a specific site a number of concepts presented during the lectures, which focus on iconic cases. The scope it opens up ranges from historical examples to common cases such as social housing districts.

Conclusions

The preceding examples illustrate parts of the wide range of educational benefits that archives offer. They reveal a number of creative ways of introducing both archival and historical content and methods into landscape architecture courses and curricula at all levels, from undergraduate to graduate and post-graduate studies. Some tutors seek to underpin their courses with historical contexts, while others dive straight into project histories. In order to trigger the use of archival material for contemporary sites and design questions, it is helpful to show students around an analogue archive, where they can find and explore real plans and drawings. Drawing, or copying, represents a "light" way of approaching history and serves well as a complement to the landscape history courses, which students sometimes perceive as "heavy". It tends to be difficult for students to obtain useful insights or in-depth knowledge from history courses alone because, as a discipline and mode of analytical thinking, history differs from the constructive and creative approach of landscape architecture. In addition to 'learning-by-drawing', archive material offers easy ways into understanding history as it contains local and regional projects that students have knowledge of, are familiar with, or feel close to. Therefore, learning about such projects not only improves students' skills but also increases their understanding of the history and heritage values of a place and may even build up a sense of how the projects emerged. It also supports students in developing skills in critical reading. Furthermore, getting to know the way in which archives can be used and how the history of sites and designs can be experienced helps lower the threshold to implementing archival work in future design projects. If students learn to use archives at early stages of their educations and start to implement this skill as a working tool, it is likely that they will make greater use of archives in their future careers. In the European Network of Landscape Architecture Archives, NELA, the use and application of archival material in teaching is one of the foci of the exchange activities. Building up a sound basis of archives is a prerequisite to make the material accessible. Both enhancing archives and furthering the use of materials in teaching is an ongoing process.

References

Dee, C. (2001) *Form and Fabric in Landscape Architecture: A Visual Introduction*. London and New York, NY: Spon Press.
Dümpelmann, S. (2011) 'Taking Turns: Landscape and Environmental History at the Crossroads', *Landscape Research*, 36(6), pp. 625–640. https://doi.org/10.1080/01426397.2011.619649.
Harris, D. (1999). 'The Postmodernization of Landscape: A Critical Historiography', *Journal of the Society of Architectural Historians*, 58(3), pp. 434–443. https://doi.org/10.2307/991537.
Hasenhütl, G. (2009). 'Zeichnerisches Wissen', in Gethmann, D. and Hauser, S. (eds.) *Kulturtechnik Entwerfen*. Bielefeld: transcript, pp. 341–358.
Hauxner, M. (2006) 'Either/Or, Less and More – An Exchange of Forms and Ideas', in LAE (ed.) *Fieldwork, Landschaftsarchitektur Europa*. Basel, Boston, Berlin: Birkhäuser, pp. 4–53.
Hufendiek, R. (2012) 'Draw a Distinction, Die vielfältigen Funktionen des Zeichnens als Formen des Extended Mind', in Feist, U. and Rath, M. (eds.) *Et in imagine ego: Facetten von Bildakt und Verkörperung*. Berlin: Akademie Verlag, pp. 441–446.
Hughes, M. (2005) 'The Mythology of Research and Teaching Relationships in Universities', in Barnett, R. (ed.) *Reshaping the University: New Relationships between Research, Scholarship and Teaching*. Maidenhead: Society for Research into Higher Education and Open University Press, pp. 14–26.
Hunt, J.D. (2000) *Greater Perfections: The Practice of Garden Theory*. London: Thames and Hudson.
IFLA Europe. (2021) NELA – European Network of Landscape Architecture Archives, https://iflaeurope.eu/index.php/site/general/nela-european-network-of-landscape-architecture-archives (Accessed 29 October 2021).
Lenzholzer, S., Duchhart, I. and Koh, J. (2013) 'Research Through Designing in Landscape Architecture', *Landscape and Urban Planning*, 113, pp. 120–127.
Mareis, C., Joost, G. and Kimpel, K. (eds.) (2010) *Entwerfen – Wissen – Produzieren: Designforschung im Anwendungskontext*. Bielefeld: transcript, pp. 9–32.
Nijhuis, S. and de Vries, J. (2019) 'Design as Research in Landscape Architecture', *Landscape Journal*, 38(1–2), pp. 87–103.
Pallasmaa, J. (2009) The Thinking Hand: Existential and Embodied Wisdom in Architecture. Hoboken: Wiley.
Powers, N. and Walker, J. (2009) 'Twenty-Five Years of *Landscape Journal*: An Analysis of Authorship and Article Content', *Landscape Journal*, 28(1), pp. 96–110. https://doi.org/10.3368/lj.28.1.96.
Prominski, M. (2008) 'Design Knowledge', in Seggern, H., Werner, J. and Grosse-Bächle, L. (eds.) *Creating Knowledge, Innovation Strategies for Designin Urban Landscapes*. Berlin: Jovis, pp. 276–289.
Roberts, A. (2007) 'The Link between Research and Teaching in Architecture', *Journal for Education in the Built Environment*, 2(2), pp. 3–20. https://doi.org/10.11120/jebe.2007.02020003.
Steenbergen, C. (2008) *Composing Landscapes, Analysis, Typology and Experiments for Design*. Birkhäuser: Basel.

7
JOINING FORCES
Landscape architecture and education for sustainable development

Ellen Fetzer

Introduction

This chapter reflects the mainstreaming of landscape architectural education in Europe against the process of Education for Sustainable Development (ESD) that is currently advancing globally with accelerating speed. This reflection aims to assess how aligned landscape architecture education is to ESD goals and which efforts would make the discipline more effective in this regard.

Landscape architectural education has developed in response to diverse socio-economic, cultural and political conditions in which universities have established study programmes since about 1900. In Europe, a continent that was and to a certain degree still is fragmented politically and socially, processes towards integration were set in train that will probably continue for a while. The Council of Europe, with its explicit Human Rights agenda, has been a promoter of integration from the start. Economy, however, became the strongest integrative motor, and the idea of driving transnational integration using market forces worked well in overcoming borders and boundaries leading to what we know as the European Union. In the following, I will focus on (a) when and where landscape architecture education appeared in the European integration story and (b) on the role the profession can play in the future, (c) particularly with regard to the global Education for Sustainable Development agenda.

A brief history of the European integration of landscape architecture education

Setting the stage for discussing European integration of landscape architectural education, two political developments are probably important that appeared at the same time (see Stiles in this volume). First, the Single European Act of 1987 provided, in title VII on the environment, the first legal basis for a common environmental policy (EU 1987). Landscape is included and started to become a European policy issue. Second, also in 1987, the EU introduced the ERASMUS programme to support both the cooperation of European institutes of higher education and student mobility (EU 1987). European landscape architecture educators and practitioners began to collaborate more closely. They established ECLAS, the European Council of Landscape Architecture Schools, and EFLA, the European Foundation for Landscape Architecture (now

IFLA-Europe). ECLAS and IFLA-Europe cooperate on finding answers to the question: what do we expect landscape architects to be able to do after graduating from a study programme?

Discussing this question is somehow a mix of bottom-up and top-down processes that, similar to various other development lines in European policy, advance one step at a time. The ERASMUS programme, encouraging exchanges of staff and students across EU member states, created the basis for building an academic community. However, it also gives rise to new challenges such as degree recognition, comparability of curriculum contents and performance assessment. In addition, professional organizations are concerned about cross-border mobility of European graduates. How would offices know which graduates meet professional requirements? Could assessment and degree requirements somehow be standardized? In response to these questions, EFLA set up an Education Committee and charged it with formulating common standards and with the registration, inspection and recognition of schools, their curricula and their diplomas (Vroom 1994, p. 113). In 1992, EFLA published its first Education Policy Document (revised in 1998), the so-called Blue Book (EFLA 1992). The process of school recognition started. Programme reviewers use a standard template that each school fills out, supported by ECLAS and national professional associations. This convention-based recognition model worked well until recently, in particular so as a European accreditation body for landscape architecture does not exist.

The process of educational integration evolved dynamically in the early 2000s (see Figure 7.1). Again, political developments play a role. The European Landscape Convention (ELC) was adopted (Council of Europe 2000) and, all of a sudden, the Council of Europe appeared on stage as a landscape actor, involving EU and many non-EU states, and calling, amongst others, for the enhancement of landscape-related professions and their education. The ELC is the most holistically framed landscape policy statement that exists. Paradoxically (or shall I say typically European?), the ELC does not have any synchronicity with EU policy. The 2004 enlargement of the European Union and the participation of accession states, including Turkey, in the ERASMUS programme doubled or even tripled the scope of European academic cooperation within a few years. The Bologna Process started in 1999 (European Ministers of Education 1999), which aimed at building the European Higher Education Area, today involving 48 states. For the EU, the Professional Qualifications Directive (EU 2005) and the Services Directive (EU 2006) substantially improved labour mobility across the Union. In addition, IFLA, the global representative of landscape architecture, also became increasingly concerned about the issue of professional qualification and published its own guidance document (IFLA 2008), revised it in 2012, including a UNESCO label (IFLA 2012).

Landscape architecture educators in the EU were lucky, in a way, to be able to use the ERASMUS programme for a broad and inclusive discussion on the knowledge, skills and competences that students may expect to acquire when they enrol in any European study programme. In the light of the ongoing reforms the Bologna process initiated 1999, the greatest impact it had was the introduction of three distinct study cycles (undergraduate, graduate, and post-graduate) leading to bachelor's, master's and doctorate degrees. What seems normal today was a challenging and laborious transformation during the early years after the turn of the twenty-first century. The method the ERASMUS-funded LE:NOTRE thematic network projects used derived from the so-called Tuning Project (Tuning Educational Structures in Europe n.d.). Tuning is a Europe-wide approach towards aligning the diversity of study programmes and to meeting shared Bologna standards. Using the Tuning approach resulted in the adoption and publication of the first ECLAS Education Guidance in 2010 (ECLAS 2010). This document still is the main ECLAS reference in the field. It presents a well-selected and carefully formulated set of subject-specific and generic competences. It is respectful in the way it considers the diversity of landscape architecture study programmes. The reason for this diversity is primarily that European study programmes exist in different disciplinary environments, ranging from affiliations with areas ranging from forestry to fine arts. Following EU Council recommendations, the European Qualifications Framework (EQF) appeared in 2012 (EU 2008, 2017), only two years after the ECLAS Education Guidance. (This is why the guidance does not yet fully comply with the EQF standardization format.)

More recent initiatives are two: the ERASMUS projects EU-Land21 and INNO-Land. Both aim at developing curriculum frameworks for undergraduate and graduate programmes in compliance with the EQF standardization. The EU-Land21 model curriculum for undergraduate programmes serves as complement to the ECLAS Education Guidance. INNO-Land is on the way to developing the same for graduate programmes. In addition, the project aims at developing a Common Training Framework for Landscape Architecture in Europe (CTF) in response to the amendment of the EU Professional Qualifications Directive (EU 2013). The amendment allows EU Member States to decide on a common set of minimum knowledge, skills and competences required to pursue a given profession through a Common Training Framework (CTF). This process is currently considered a good chance to revisit the competence profile of landscape architecture. A very important question to be addressed is whether the profile needs amending or modifying in order to respond to emerging requirements, especially in the light of the many grand societal challenges and of sustainable development goals. I will elaborate on both in the following.

Education for Sustainable Development and the Global Sustainability Goals

Landscape architecture education integrates many relevant elements of Education for Sustainable Development (ESD), and it seems relevant that the community of educators and practitioners work towards furthering this integration. The United Nations (UN) are the global catalyst in this field. The starting point was the Brundtland Report from 1987 that explained sustainability as a "development that meets the needs of the present without compromising the ability of future generations to meet their own needs" (United Nations General Assembly 1987). Today, this definition serves worldwide as a conceptual basis. It integrates the idea of the future as a reference for our actions today. In addition, the concept builds on three pillars: the natural resources and ecology, the economy and the socio-cultural-political domain. All areas must be understood, assessed

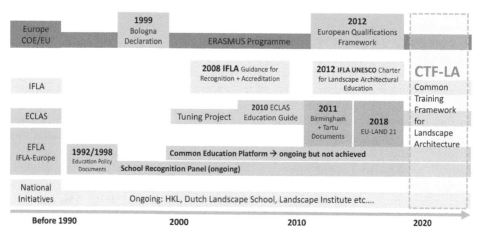

Figure 7.1 The European integration pathway in landscape architecture education.

Source: Graphic by Ellen Fetzer.

and designed in an integrated manner. A recent milestone was in 2015, when all United Nations Member States adopted 17 Global Goals, officially known as the Sustainable Development Goals, SDGs, and when they set out a 15-year plan to achieve the Goals and their related targets (UN 2015). Education is the key transversal force for achieving the goals, which is why ESD gets particular attention worldwide, endorsed by UNESCO, the UN's Educational, Scientific and Cultural Organisation. In addition, Quality Education (SDG 4) represents a goal in its own right.

Education for Sustainable Development is a lifelong process aimed at ensuring that people of all ages acquire knowledge and skills in order to be able to live and act in the interests of sustainability. This includes understanding the consequences of one's own actions and for everyone making responsible decisions. It is also about actively shaping the future. ESD promote the ability for dialogue and orientation, creative and critical thinking as well as holistic learning that considers religious and cultural values. It aims at the willingness to deal with uncertainties and contradictions, to solve problems and to participate in the creation of a democratic and culturally diverse society.

UNESCO has been promoting ESD since 1992. The first step was the UN Decade of Education for Sustainable Development (2005–2014), followed by the Global Action Programme (GAP) on ESD (2015–2019). The GAP is currently in process of implementation at national level. In Germany, for example, GAP translates into the *National Action Plan Education for Sustainable Development*. Thirty goals and 349 specific recommendations for action in all of the individual areas of education are intended to ensure that ESD is structurally anchored in the German educational landscape. Each UN member state is following up similar parallel processes. This covers all educational domains, starting from kindergarten all the way through to higher education, including also professional and adult education, as well as informal learning.

In recent years, various educational scientists (de Haan 2002; Wiek et al. 2011, 2016) have dealt with a definition of skills and competencies that could serve as a target framework for training in this context. Since the models do not substantially differ, which a very recent Delphi Study has reconfirmed (Brundiers et al. 2021), I will focus in the following on the key competences for sustainability as presented by Wiek et al. (2016) (see Figure 7.2).

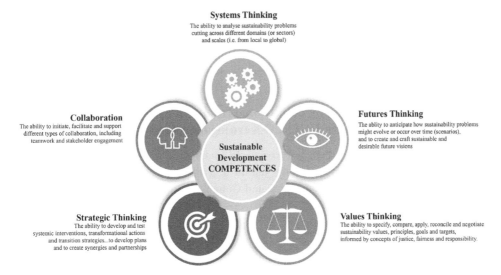

Figure 7.2 Key Competences for Sustainable Development according to Wiek et al. (2015).
Source: Visualized by Ellen Fetzer.

Even if some of these key competence definitions seem rather generic, they are quite compatible with the identity of landscape architecture as a profession focussing on changing any given state of landscape towards a better, more sustainable future, taking normative dimensions into account. In the following, I will draw the observational grid a bit tighter for us to see if this general observation still holds true if we look at specific learning goals pertaining to sustainable planning and design in a community context.

Learning objectives for SDG 11, Sustainable Cities and Communities

In a report from 2017, UNESCO started specifying learning objectives in relation to the 17 SDGs. As a holistic discipline, landscape architecture is involved in many of the SDGs. A most relevant one is certainly SDG 11, Sustainable Cities and Communities. One of the ten targets specifying this goal refers in particular to landscape architecture key activities. Target 11.7 reads: "By 2030, provide universal access to safe, inclusive and accessible, green and public spaces, in particular for women and children, older persons and persons with disabilities."[1]

The UNESCO report "Education for Sustainable Development – Learning Objectives" (UNESCO 2017, p. 32) suggests the following learning goals for SDG 11 and presents them in three categories, briefly summarized here:

(A) **Cognitive learning objectives:** This section refers to the understanding of basic physical, social and psychological human needs, the evaluation of the sustainability of a settlement system and knowledge of the basic principles of sustainable planning and building.

(B) **Socio-emotional learning objectives:** This part underlines the ability to connect to local communities and show advocacy, to reflect on regional and local identities and to reflect roles and responsibilities.

(C) **Behavioural learning objectives:** This category speaks about the ability to plan, implement and evaluate community-based sustainability project with a specific emphasis on advocacy and co-creation.

Taking into account how leading educational scholars have written this report in the light of a global sustainability agenda, I suggest setting the learning goals specified for SDG 11 as sustainability benchmarks and applying them to existing landscape architecture education frameworks.

Comparing existing educational frameworks for landscape architecture with UNESCO learning objectives for SDG 11

In the following, I consider three internationally leading landscape architecture education frameworks and compare them against the full list of learning objectives presented by UNESCO (2017) for SDG 11, Sustainable Cities and Communities. The following documents are included in this comparison:

1. The ECLAS Education Guidance (ECLAS 2010), a document resulting from a pan-European tuning process that presents core competences of landscape architecture.
2. The IFLA-UNESCO Charter for Landscape Architecture Education (2012).
3. Output 2 of the EU-Land21 ERASMUS Project (EU-Land21 2019) Trans-European Education for Landscape Architects: peer learning methods on the development of a curriculum.

Table 7.1 The key competences for SDG 11 (Sustainable Cities and Communities) as defined by the UN compared to the competences presented in three transnational frameworks for landscape architecture education.

Learning objectives for SDG l 11 Sustainable cities + communities	ECLAS Guidance Report 2010	IFLA + UNESCO 2012	EU Land 21 2019
as defined by the United Nations (2017)	EU tuning process	Charter for LA Education	Output 2/ Block of Competences
Cognitive learning objectives			
The learner understands basic physical, social and psychological **human needs** and is able to identify how these needs are currently addressed in their own physical urban, peri-urban and rural settlements.	+	+	+
The learner is able to **evaluate** and compare the sustainability of their and other settlements' systems in meeting their needs particularly in the areas of food, energy, transport, water, safety, waste treatment, **inclusion and accessibility**, education, **integration of green spaces** and disaster risk reduction.	++	++	++
The learner understands the **historical reasons for settlement patterns** and, while respecting cultural heritage, understands the need to find compromises to **develop improved sustainable systems**.	++	++	++
The learner **knows** the **basic principles of sustainable planning and building**, and can identify opportunities for making their own area more sustainable and inclusive.	+	+	+
The learner understands the role of **local decision makers** and **participatory governance** and the importance of representing a sustainable voice in planning and policy for their area.	+	+	+
Socio-emotional learning objectives			
The learner is able to **use their voice** to identify and **use entry points** for the public in the local planning systems, to call for the investment in sustainable infrastructure, buildings and parks in their area and to debate the merits of long-term planning.	+	+	+
The learner is able to **connect** with and **help** community groups locally and online in developing a sustainable future vision of their community.	+	+	+
The learner is able to **reflect** on their region in the development of their own identity, understanding the roles that the natural, social and technical environments have had in **building their identity and culture**.	++	++	++
The learner is able to **contextualize their needs** within the needs of the greater surrounding ecosystems, both locally and globally, for more sustainable human settlements.	++	++	++

(Continued)

Table 7.1 (Continued)

Learning objectives for SDG l 11 Sustainable cities + communities	ECLAS Guidance Report 2010	IFLA + UNESCO 2012	EU Land 21 2019
as defined by the United Nations (2017)	EU tuning process	Charter for LA Education	Output 2/ Block of Competences
The learner is able to feel **responsible** for the **environmental and social impacts** of their own individual lifestyle.	+	+	+
Behavioural learning objectives			
The learner is able to **plan, implement and evaluate** community-based sustainability projects.	+	+	+
The learner is able to participate in and **influence decision processes** about their community.	–	–	–
The learner is able to **speak** against/for and **to organize their voice** against/for decisions made for their community.	–	–	–
The learner is able to **co-create** an inclusive, safe, resilient and sustainable community.	–	–	–
The learner is able to **promote low carbon approaches** at the local level.	+	+	+

Source: Ellen Fetzer.

The comparison was performed as a content analysis. Starting from the learning objectives defined in the UNESCO report, all documents were reviewed in order to identify the learning objectives that they include. The findings were synthesized in the form of short descriptions and organized on a comparative matrix; both are presented here in an abbreviated version. The short descriptions are clustered according to three categories:

- ++ = The learning objective is well covered and specified.
- + = The learning objective is recognizable, but lacks specification.
- – = The learning objective is not clearly identifiable.

Discussion

The comparative analysis reveals both strengths and weaknesses of landscape architecture. The field is, obviously, strong in the field of systems thinking, mainly since landscape is per definition a metasystem, a combination of natural, social, physical, cultural and economic systems, each with their respective subsystems. Thinking about and understanding systems of high complexity includes understanding the history and evolution of landscapes as a basis for envisioning their future. However, the preceding educational frameworks remain somewhat vague regarding the analysis and assessment of human needs. This finding reveals to some extent the limits of one single discipline dealing with landscape. Deeper insights into the systems require cooperation with neighbouring disciplines, or other systems, including the non-academic world, as suggested and required by transformative science. Many landscape architecture faculties include teachers and researchers from disciplines such as ecology, architecture, fine arts and sociology. This multidisciplinary input certainly deepens the systems thinking competence as one key component of ESD. Nevertheless, our study programmes do not necessarily offer interdisciplinary educational experience when it comes to co-designing, strategy building and collaborative transformation, which are the other key components outlined in the UNESCO report. A clearly strong point is the persistent context-orientation of most landscape architecture education programmes, a quality that allows learners to explore important sustainability dimensions such as identity, culture, behaviour, and the wider regional and global linkage of an area or settlement.

What seems to be missing in general is a domain-specific definition of sustainability principles in the context of landscape architecture practice. A reason for this shortcoming might be that professionals have developed their own "implicit" understanding of sustainability, a belief that has prevailed as an element of professional identity ever since the field established more than 100 years ago, even though different words were in use back then. The global, cross-sectoral discussion on sustainable development started at the end of the 1980s when the profession had been operational and effective for many decades. Distilling the contributions of landscape architecture for sustainable development, ideally in line with the SDGs, should not be too difficult, and it would make the profession more visible in the global sustainability discourse.

However, there still seems to be room for improvement. A truly ESD-based approach to landscape architecture education requires cooperation with local communities and exposure to controversial stakeholders with converging landscape interests. This includes the collaborative mapping of these stakeholders' assets, problems, dreams and power structures; and the involvement of the community into the analysis, assessment, design and evaluation of whatever alternative future might be desirable. This is exactly where the preceding comparison reveals gaps. Collaborative design or co-creation does not appear in any European landscape architecture educational framework: neither does power mapping, collective visioning, activism, advocacy or citizenship.

The analysis also shows how ESD competence terminology has, so far, not been mainstreamed in the existing transnational frameworks for landscape architecture education. One explanation might be that the educators have yet to include ESD into their discourse and practice, a process that appears to be lagging behind of their political development and advancement. The current revisiting of educational documents as part of the development of the Common Training Framework is a chance for aligning landscape architectural learning goals and competence definitions more closely with ESD terminology. This aligning could become a constructive process contributing to highlighting the transformative power of landscape architecture and the disciplinary capacity for sustainable development.

Note

1 www.un.org/sustainabledevelopment/cities/

References

Brundiers, K., Barth, M., Cebrián, G., Cohen, M., Diaz, L., Doucette-Remington, S., Dripps, W., Habron, G., Harré, N., Jarchow, M., Losch, K., Michel, J., Mochizuki, Y., Rieckmann, M., Parnell, R., Walker, P. and Zint, M. (2021) 'Key Competencies in Sustainability in Higher Education – Toward an Agreed-Upon Reference Framework', *Sustainability Science*, 16, pp. 13–29.
Council of Europe (2000) *European Landscape Convention*. Details of Treaty Nr 176 [Online]. Available at: www.coe.int/en/web/conventions/full-list/-/conventions/treaty/176 (Accessed 16 June 2021).
de Haan, G. (2002) 'Die Kernthemen der Bildung für eine nachhaltige Entwicklung', *Zeitschrift für internationale Bildungsforschung und Entwicklungspädagogik*, 25(1), pp. 13–20.
EFLA. (1992) *Teaching Landscape Architecture in Europe* [Online]. Available at: www.iflaeurope.eu/assets/docs/EFLA_Blue_Book_1992.pdf (Accessed 16 May 2021).

EU-Land21 ERASMUS Project. (2019) *Trans-European Education for Landscape Architects, Output O2: Peer Learning Methods on the Development of a Curriculum* [Online]. Available at: https://lnicollab.landscape-portal.org/goto.php?target=cat_1305&client_id=main (Accessed 16 May 2021).

European Council of Landscape Architecture Schools (ECLAS). (2010) *ECLAS Guidance on Landscape Architecture Education*. Available at: www.eclas.org/eclas-education-guide (Accessed 16 March 2022).

European Ministers of Education: Bologna Declaration (1999) [Online]. Available at: www.magna-charta.org/resources/files/BOLOGNA_DECLARATION.pdf (Accessed 16 May 2021).

European Union (1987a) *Council Decision of 15 June 1987 Adopting the European Community Action Scheme for the Mobility of University Students (Erasmus)* [Online]. Available at: https://eur-lex.europa.eu/legal-content/EN/TXT/HTML/?uri=CELEX:31987D0327 (Accessed 16 May 2021).

European Union (1987b) *Single European Act* [Online]. Available at: https://eur-lex.europa.eu/resource.html?uri=cellar:a519205f-924a-4978-96a2-b9af8a598b85.0004.02/DOC_1&format=PDF (Accessed 16 May 2021).

European Union (2005) *Directive 2005/36/EC of the European Parliament and of the Council of 7 September 2005 on the Recognition of Professional Qualifications* [Online]. Available at: https://eur-lex.europa.eu/eli/dir/2005/36/oj (Accessed 16 May 2021).

European Union (2006) *Directive 2006/123/EC of the European Parliament and of the Council of 12 December 2006 on Services in the Internal Market* [Online]. Available at: https://eur-lex.europa.eu/eli/dir/2006/123/oj (Accessed 16 May 2021).

European Union (2008) *Recommendation of the European Parliament and of the Council of 23 April 2008 on the Establishment of the European Qualifications Framework for Lifelong Learning (Text with EEA Relevance)* [Online]. Available at: https://eur-lex.europa.eu/legal-content/EN/ALL/?uri=CELEX:32008H0506(01) (Accessed 16 May 2021).

European Union (2013) *Directive 2013/55/EU of the European Parliament and of the Council of 20 November 2013 Amending Directive 2005/36/EC on the Recognition of Professional Qualifications and Regulation (EU) No 1024/2012 on Administrative Cooperation Through the Internal Market Information System ('the IMI Regulation')* [Online]. Available at: https://eur-lex.europa.eu/eli/dir/2013/55/oj (Accessed 16 May 2021).

European Union (2017) *Council Recommendation of 22 May 2017 on the European Qualifications Framework for Lifelong Learning and Repealing the Recommendation of the European Parliament and of the Council of 23 April 2008 on the Establishment of the European Qualifications Framework for Lifelong Learning* [Online]. Available at: https://eur-lex.europa.eu/legal-content/EN/TXT/?uri=uriserv:OJ.C_.2017.189.01.0015.01.ENG (Accessed 16 May 2021).

International Federation of Landscape Architects (IFLA) (2008) *Guidance Document for Recognition or Accreditation* [Online]. Available at: www.eclas.org/wp-content/uploads/2020/09/E2_IFLA-Guidance-Document-for-Recognition-or-Accreditation_2008_with_IFLA-Europe-addenda_2017.pdf (Accessed 16 May 2021).

International Federation of Landscape Architects (IFLA) (2012) *IFLA/UNESCO Charter for Landscape Architectural Education* [Online]. Available at: www.eclas.org/wp-content/uploads/2020/09/E1_IFLA-Charter-for-Landscape-Architectural-Education_2017_with_IFLA-Europe-addenda_2017.pdf (Accessed 16 May 2021).

Tuning Educational Structures in Europe [Online] (n.d.) Available at: www.unideusto.org/tuningeu (Accessed 16 May 2021).

UNESCO (2017) *Education for Sustainable Development – Learning Objectives* [Online]. Available at: www.unesco.de/sites/default/files/2018-08/unesco_education_for_sustainable_development_goals.pdf (Accessed 16 May 2021).

United Nations General Assembly (1987) *Report of the World Commission on Environment and Development "Our Common Future"* [Online]. Available at: https://sustainabledevelopment.un.org/content/documents/5987our-common-future.pdf (Accessed 11 July 2022).

United Nations General Assembly (2015) *Resolution Adopted by the General Assembly on 25 September 2015: 70/1 – Transforming Our World: The 2030 Agenda for Sustainable Development* [Online]. Available at: https://daccess-ods.un.org/access.nsf/Get?Open&DS=a/res/70/1&Lang=E (Accessed 16 May 2021).

Vroom, M. J. (1994) 'Landscape Architecture and Landscape Planning in Europe: Developments in Education and the Need for a Theoretical Basis', *Landscape and Urban Planning*, 30(3), pp. 113–120.

Wiek, A., Bernstein, M. J., Foley, R. W., Cohen, M., Forrest, N., Kuzdas, C., Kay, B. and Withycombe Keeler, L. (2016) 'Operationalising Competencies in Higher Education for Sustainable Development', in Barth, M., Michelsen, G., Rieckmann, M. and Thomas, I. (eds.) *Routledge Handbook of Higher Education for Sustainable Development*. New York, NY: Routledge, pp. 241–260.

Wiek, A., Withycombe, L. and Redman, C. L. (2011) 'Key Competencies in Sustainability: A Reference Framework for Academic Program Development', *Sustainability Science*, 6, pp. 203–218.

8
CREATING VITAL TEACHING COMMUNITIES THROUGH CURRICULUM DEVELOPMENT

Anne Katrine Geelmuyden

Introduction

One question has persistently preoccupied my mind during my time as head of department/school (2006–2011 and 2014–2016) and programme director (2003–2006 and 2018–now) at Norwegian University of Life Sciences, NMBU: How does one, in a professional university programme, create a balanced and cooperative environment between members of the teaching and research staff, when some have various kinds of academic backgrounds and others mainly experience as practitioners. At first glance, the challenge is one of coming to terms with conflicting knowledge cultures and interests. However, regardless of backgrounds, all staff members of a landscape architecture department share the common practice of landscape architecture education. Our *purposes* converge in our combined contributions to the whole of one educational curriculum. I shall approach the question of creating a rewarding working environment for different teachers from the perspective of a programme director and curriculum developer.

The 'academisation' of practical professions in the last three decades

Taking the case of the landscape architecture programmes at NMBU, their example serves to illustrate the challenges of 'academisation' in professional education (for the history of the programmes, see Jørgensen in this volume). The academisation process can be traced in employment lists provided by the current faculty administration, as well as in the department's yearbooks in the Norwegian Archive of Garden History at NMBU (Norsk hagearkitektlag, Årsskrifter 1952–1956, Institutt for hagekunst (landskapsarkitektur), Årsmeldinger 1970–1998). Based on the department's employment history and records of the various major curriculum evaluations and revisions since the programme started in 1919, the following can be stated: But for one exception in the year 1974–1975, there were no landscape architecture instructors with a doctoral degree until 1984. Until the mid-1980s, professors were men[1] who had practical training in (mostly) gardening, sometimes architecture and art, as well as years of professional practice. Formal academic training was on the whole uncommon at NMBU (formerly NLH) until the 1970s[2] (Olsen 2000). Until 1971, students were accepted to the garden architecture programme on the basis of a minimum of two years of practice in horticulture, in addition to one year's education in the same subject and a baccalaureate or equivalent (Norges landbrukshøgskole

1859–1959, p. 200). Until 1971, completing the programme took three years; then the practice requirement was dropped and the programme was extended to five years, of which a first preparatory year was taught at an agricultural or horticultural school.

Today, the programme at NMBU still retains a practice-based, practice-oriented and natural science-biased character.[3] However, since the 1980s, landscape architecture has, not unlike other professional educations, experienced an 'academic shift', in applied disciplines also known as 'academisation' (e.g. Laiho 2010; McEwen and Trede 2014; Ek et al. 2013). Doctorate studies and research projects in landscape architecture began to emerge during the 1970s and their number increased, first slowly, then more rapidly after the 1990s (Institutt for hagekunst (landskapsarkitektur), Årsmeldinger 1970–1998). For the first time in 2001, the five years' programme was taught in its entirety at the university. A third cycle programme leading to doctoral degrees in landscape architecture was in place in Norway by the end of the decade. One doctoral degree was awarded in 1968, seven in the 25 years between 1970 and 1995, and the number increased to 21 during the following 25 years (1995–2020) (Institutt for hagekunst, Årsmeldinger 1970–1998, plus author's own records).

When a Norwegian university now announces a new teaching position, the norm is to require a doctoral degree. In practice-oriented fields, however, we often find ourselves arguing against such rigid requirements. First, the number of applicants who have the sought-after teaching qualification *and* a doctoral degree is (still) very small. Secondly, and more importantly, the requirement is irrelevant in an unconditional form. One argument is that the very *combination* of disciplinary perspectives and practical as well as theoretical approaches to what and how to teach is a precondition for educating students to become *good* landscape architects. While on the one hand both self-evident and true, this argument, however, on the other hand implies an instrumentality in our conception of teaching that has some problematic aspects. In the following, I argue that too much of a utilitarian attitude to knowledge and teaching obstructs our view towards a different and better way of conceiving of the purpose and character of teaching landscape architecture.

The origin of the mentioned instrumentality lies in the historical development that leads to the modern understanding of how we acquire knowledge primarily through research, notably in and by science. Both Hannah Arendt (1958) and François Lyotard (1984) have shown the development of a utilitarian attitude to knowledge in the West and its consequences as motivated by the need for tools that make production more efficient. One of these consequences is that the standing of practically gained insight is diminished compared with knowledge from 'basic research'. The price for this is paid by the practitioners that are employed and enter a workplace whose renumeration system grants advancement primarily on the grounds of publication numbers and successful research applications. However hard working and dedicated, they tend to be regarded as second-rate citizens in the academic community. Overloaded with teaching obligations as they usually become, there are few development opportunities if they stay on. Their other option is to take on part-time teaching jobs beside their work in landscape architecture offices or public agencies, with the consequence that they never really become part of their department colleagues' intellectual and social exchange. We lose sight of the fact that teaching is the one practice which we share and through which we may develop and thrive, as a team of colleagues at a landscape architecture department.

Our department at NMBU is now embarking on yet another revision of its five years' integrated master's programme of landscape architecture. The reasons for this have to do with the university structure and other external factors having changed. I shall address the task of developing the curriculum of a landscape architecture programme as a way to build a meaningful working environment despite teachers' and researchers' sometimes very different core interests and teaching backgrounds. I shall do this leaning on the virtue ethics of education offered by Chris Higgins in his book *The Good Life of Teaching. An Ethics of Professional Practice* (2011). Higgins' overall project concerns teaching, but what he offers is a more general critique of the *ethical* core of practices. Combining the philosophies and concepts of his fellow neo-Aristoteleans, Alasdair MacIntyre, Hannah Arendt, John Dewey and Hans-Georg Gadamer, Higgins (re-) posits ethical reflection (theorising) as intrinsic to any serious practical occupation or profession.

A key concept in Higgins' professional ethics is the Aristotelean term 'Eudaimonia', which he translates as "the good life", a meaningful life, a life where one can flourish through the use of one's given capacities. Eudaimonia is what any human being should ultimately strive for and what the reflection on human virtues revolves around. In Higgins' own words, his

> project [. . .] concerns the interplay of altruism and self-interest in the practice of teaching; it is a philosophical exploration of teacher motivation, identity, and development. To see that such questions are not the exclusive province of psychologists or sociologists, we need simply rephrase the familiar question 'Why teach?' as 'Why is the practice of teaching worth putting at the centre of my life?' It then turns out that we are dealing with one of the central questions of professional ethics.
> (Higgins 2011, p. 9)

How to make landscape architects – What is the recipe?

One of the external factors to take into account in our curriculum revision is that members of the European Council of Landscape Architecture Schools ECLAS, following European educational policies to harmonise the EU-member countries' various educational systems[4] (e.g. Winterton et al. 2006), have worked together to implement the European Qualification Framework (EQF 2008). ECLAS's formidable work has resulted in guideline documents such as the ECLAS Guidance on Landscape Architecture Education (ECLAS 2010) and EU-LAND 21 (2017, 2019). The guidelines coincide in that they adopt EQF's term 'Learning Outcomes' expressed as 'Competences' ('generic' and 'subject specific'), 'Knowledge' and 'Skills' for curriculum description and assessment. The given lists as well as the explanation of their content and methods are recognisable to anyone who has struggled with composing a landscape architecture curriculum that is suited for a period of up to ten semesters and for foreseeably available resources.

As I read these guidelines, they emphasise the curriculum's role in assuring students' acquisition of competences, knowledge and skills as a set of more or less assessable *prerequisites* for being authorised to enter and engage in professional practice. I find myself wondering whether the emphasis on measurable education items might be an unintended 'import' of the aforementioned utilitarian attitude to knowledge. Is there not an instrumentality concealed in the output-terminology that EQF and education guidance have adopted?[5] First, the prevailing terms reflect governmental needs for standardised policies and control mechanisms. Second, the policy documents and their terms place students in the role of customers. For good as well as for bad, students are provided with tools with which to complain about their education in terms of delivered products. Neither of these connotations reflect goals that coincide with the way the professions in question, teaching and landscape architecture, would themselves assess their own quality. In Higgins' (2011) words (leaning on MacIntyre), they exemplify an institution (in this case European educational authorities' university regulations) housing a practice (in this case the practice of educating landscape architects) threatening the integrity of that practice by focusing on a good (production and efficiency control) in a way that is *external* to it.

Part of our mandate as programme developers is to explicate what the purpose of landscape architecture is. What is the 'Why' which explains the 'How' of teaching landscape architecture? EU-LAND 21 (2017, p. 6) recommends that "when designing a new programme and also when evaluating an existing one, it is important to consider the main concept lying behind the programme structure. This may be related to a specific or pedagogical theory". Ellen Fetzer follows this up in EU-LAND (2019, pp. 11–16) by presenting a "constructive theory of learning", where the main tenet is that in the course of education, teachers and learners of landscape architecture co-construct a vision of the world (e.g. in the form of landscapes and landscape problems). They construct what Higgins (2011, with Dewey 1997) terms a "vocational environment",[6] which is a perceptive restructuring of students' world through the continuous purposive practices which a profession implies. A vocational environment constitutes the universe of one's possibilities. In Higgins' words (ibid., p. 123):

> When I ask myself, 'Is this occupation worth putting at the centre of my life?', I am considering what sort of environment I want to inhabit. I am determining, to some extent, the very aspects of the world I will notice and with which I will vary. Thus, to choose a vocation is to make a fundamental existential choice: it is to choose the very world I want to inhabit.

Like Fetzer, Higgins advocates a *transformative* model of teaching, an understanding of teaching where a student transforms as a person through learning. To my mind, however, Higgins delivers a more accurate exploration *of the grounds* on which we can assess whether this co-construction is good. The reason for this lies in his point of departure: practices are essentially evaluative, not morally, but in terms of the *internal values or good* of a profession. As stated by the ECLAS guidelines (2010, p. 12), the goods pursued in landscape architecture are articulated in policies aimed at "high quality of life and environmental quality" as substantiated through international conventions and agreements and referred to with a (long) list beginning with the European Landscape Convention (Council of Europe 2000). However, is this all there is to say about the good we are pursuing as *teachers and programme developers* in the profession?

It is within practices that ethical questions are truly posed, the quintessential one being "what should I do next?" (Higgins 2011, p. 9). Making good judgements on what to do next, we pursue the internal good of the profession we are practicing. Good judgement is the *virtue* by which we pursue this good. Professions are intrinsically value-laden, existentially, as argued earlier and aesthetically (in Dewey's terms), because of the *experiences* they afford through our practicing. 'Experience', here, has the connotation of the German *Erfahrung* and ties up with Gadamer's circle of understanding (Gadamer 1975 [1960]): In order to know how to increase sustainability through my work, for instance, I need a general idea of what sustainability means. Although this can be explained to me in a lecture or on the internet, it is only when I am confronted with a concrete planning or design task in a particular landscape situation that I will experience (sensually, emotionally and cognitively) what sustainability can be. Understanding requires ethical reflection, or judgement. It is the ability to connect concepts and generalities with concrete cases and to bring past experiences to bear on new situations.

"Practices are in fact our ethical sources: They are the sites where aspects of the good are disclosed to us as well as the primary scenes of our ethical education" (Higgins 2011, p. 10). Accordingly, ECLAS (2010, p. 14) lists "Generic competences" as "taught and acquired in a contextual way, above all in the context of studio and project work" (ibid., p. 16). Practice or case-based studio courses are typically the environment that is most conducive to students' ethical education, because we never educate directly, but indirectly, by means of the (vocational) environment, that is, what people in the profession have said and made before us *and*, most importantly, the thought-processes that lead them to do so. Labelling landscape architecture as an applied discipline, thus, "puts the ethical cart before the practical horse" – if I may paraphrase Higgins (ibid., p. 127). His conclusion (with Dewey) is that

> Education is not preparation for vocations; vocations themselves are (more or less) educative, preparing us for more complex vocations, wider experience, and a richer life. This is the insight that gets lost in the progressive slogan 'learning by doing'.
>
> (ibid.)

Having established what it means to teach landscape architecture, I shall now return to the question of creating beneficial co-working environments for landscape architecture teachers. The dominant vocation in any human life is self-formation or growth in experience. We achieve this by exercising the central virtue of good judgement and practical wisdom in our occupations: when a general good is reflected on and encountered in a particular practical situation.

What is (the) good (of) teaching in landscape architecture?

A practice's *internal* good is something that we simultaneously value *for its own sake* while also judging that it "contributes to or is partially constitutive of [our overall] well-being" (Higgins 2011, p. 49, quoting MacIntyre 1999, p. 64). For a further exploration of this, Higgins turns to Arendt's three-partite conceptual model of human practical activities or 'vita activa' as either 'labour', 'work' or 'action' (Arendt 1958). Any occupation may be pursued in more than one of these modes. The purpose of 'labour' is life itself, our day-to-day activities which are less than conscious, and so it is the latter two modes that are most relevant in Higgins' investigation of a virtue ethics for teaching. Whereas 'work' is performed *in order to* achieve a predefined or expected goal (for instance more efficient storm water management or less time-consuming evaluation of students' projects), 'action' is performed for *the sake of* something (for instance a harmonious coexistence between humans and animals or a student's development into a sound professional). The mode of 'work' is easily recognisable as a characteristic of occupations such as landscape architecture. We make things and are evaluated, including by ourselves, by these things' technical, biological, cultural or other worth. As teachers, we produce course outlines and PowerPoints.

However, it is Arendt's 'action' (in his transcription: 'deeds') that Higgins distinguishes as the mode that may help us live a good life as teachers. We perform deeds when we express values in a way that is unique to each one of us and in a way that affects others' lives. The 'deed' fuses the existential with the ethical/political: to become authors of deeds, Arendt (1998) holds, we need a public space, a 'polis'. We need to appear in front of others and disclose our personhood. Interestingly, Arendt reserves her concept of 'action' (Higgins' 'deeds') to situations when we speak with each other, because this is when we are most likely to come up with the unexpected. I read this as a confirmation of the importance of conversations between teachers as well as teachers and students on specific landscape situations, tasks and assignments if we are to achieve any real transformative learning.

Still following Arendt (ibid.), two conditions must be met for this public space of appearance (polis) to be ethically productive. First, our confrontation with a plurality of others needs to be centred on a *common object of concern*. Our common object of concern as teachers, and the sake for which we have decided on our profession, is to support the landscape architect student's development into becoming a landscape architect. As university teachers, researchers and curriculum developers, we are engaged in a 'para-practice' bordering on landscape architecture, whose role it is "to advance the purposes of its host-practice" (Higgins 2011, p. 72). We are needed because

> In trying to track down practices, we need descriptions from the doers and descriptions from those who, while keeping one foot in the role of observer, learn how to participate in the experience itself.

The reason for this, and as many of us in academia may have witnessed,

> we may not be able to rely solely on first-person reports about practice, [. . .] for we may expect that [. . .] the best practitioners are those whose knowledge will be shown more than told, [. . .] exemplified as lived experience in the specific medium of their practice.
>
> (ibid.)

However, the observation from outside, I would like to add, must go both ways: Researchers, theoreticians and practitioners of landscape architecture must mutually look to each other in the cards! That would be academisation in a fruitful, internally driven sense.

Building a community of inquiry through the 'polis' of curriculum development

The second condition that must be met for Hannah Arendt's 'polis' of plural voices to prevent calcification and endless repetition of one's habits is the virtue of being willing to *unsettle* one's self-formation. In discursive interplay with individuals who have a different practice background and perspective (for example that of a natural scientist or that of a municipal park manager), we can potentially renew ourselves, thereby becoming more experienced human beings. In Higgins' (2011, p. 132) referral to Gadamer

> to initiate a dialogue with the other (a text, a novel situation, a distant epoch, another person) means risking our prejudices by putting them in play, while trying to remain open to those moments in which we find ourselves noticing just a little more than we thought we knew how to notice.

With Gadamer, Higgins posits *questioning* as the means to widen one's experience. Not any question or inquiry will do. For an individual's perceptive world to be expanded, he or she must be open to experience in the normative sense that it transforms them. Significant experience occurs when our expectations or preconceptions are thwarted. A degree of alienation is necessary. Reconstructing Gadamer's investigation of the various modes of questioning, Higgins (2011, p. 137) concludes that

> The true open question does not sit there inscrutably before all answers, but exists because substantive, conflicting answers have already been developed. [. . .] In this way, Gadamer helps us move away from a notion of openness as freedom from assumptions to ask what assumptions can be productive.

In the context of landscape architecture education, concretisations of what this may mean are presented in Wall (2019), in Abbott and Bowring (2019) as well as by Deming under the label of a *community of inquiry* (2019). Just as our students do, our colleagues with varying disciplinary and occupational perspectives come from outside, either of the profession of landscape architecture or of academia. As teachers, and researchers, our 'vocational environment' and our potential polis for the meeting of these different horizons of understanding is the continuous re-evaluation and co-construction of the curriculum. Shared and mutual questioning may prove conducive to self-development and professional growth for teachers and, thus, indirectly for students, too. Table 8.1 is a sketch of how curriculum development can serve as an ongoing process of 'communal inquiry' amongst landscape architecture teachers with various disciplinary and practice backgrounds.

Conclusion

I have offered a reconceptualisation of what it means to teach a profession such as landscape architecture in a way that lays the ground for prioritising 'learning outcomes', not in the sense of predefined knowledge items and skills that are useful for future practice but rather as more processual (and probably less assessable) development of the practical *virtue of good judgement*, exercised in the co-constructions of vocational environments as well as landscapes between teachers and students. With a perspective on teaching as *the* shared *practical purpose* in university departments that house professional studies, the task of curriculum and course development becomes the arena where 1) the staff member's varying knowledge, experience and skills can be exposed, exchanged and appreciated and 2) conditions for experience-rewarding teaching environments can be born and develop. Curriculum development should be treated as the glue that can keeps departments of professional education together as vital communities that afford professional growth. For this to take place, a common understanding must be established that everyone's own professional well-being, as teachers, is conditional on one's involvement in teaching teams with a communal dedication to initiating students to the various aspects of the profession.

Notes

1 A few women held positions as teaching assistants. From when on, still needs to be researched.
2 The first doctoral degree at NLH/NMBU was awarded in 1927 and by 1968, 38 doctoral degrees had been awarded at the university as a whole to teachers from disciplines such as forestry, animal breeding, agricultural sciences and soil sciences, of which only a very few, if any, taught landscape architecture students (Norges landbrukshøgskole 1859–1959).
3 Until a decade ago, NMBU's landscape architecture programme was the only one in the country. Also, an important aspect is a close cooperation with the professional organisation of Norwegian landscape architects, NLA. After earning their degree, students automatically become members.
4 Even though Norway is not an EU member state, its educational policies follow those of the EU through governmental agreements.
5 In ECLAS (2010), the overarching "core competences" and "generic competences" offer intimations of the reasoning behind the curriculum development descriptors. A footnote (p. 13) discloses the rationale behind the core competences ("Knowledge, skills and understanding of planning, design and management" and "holistic knowledge and understanding of the nature of landscape"):

> The concept of 'core competences' comes from the world of business management. Here, it is used to describe the set of unique capabilities which a particular company is able to develop or acquire in order to give it competitive advantage in the marketplace. In the case of an academic discipline, the term 'core competences' can be used to refer to those distinctive capabilities which give it its specific characteristics and thereby distinguish it from other disciplines.

6 "Vocation" in Dewey is used interchangeably with "occupation" and "is simply 'a concrete term for continuity' or in another formulation, 'a continuous activity having a purpose'" (Dewey 1997, p. 307, cited in Higgins 2011, p. 112).

Table 8.1 A sketch of curriculum development as an ongoing 'communal inquiry' amongst landscape architecture teachers with various disciplinary and practice backgrounds.

Questions on curriculum/programme level	WHY? = For the sake of what value – or with what theoretical perspective?	WHAT/WHO? = Our common object of concern: The *Eudaimonic* student and practitioner of the profession.	HOW? = Through what pedagogical work can the profession's central virtue of practical wisdom be taught?

	For the sake of what internal values/goods do we *teach*? For the sake of what internal values/goods do we teach *landscape architecture*? Why do we compose our curriculum as we do now? What is its history? What tradition does it convey? Do we need to change it? Why? For example:
	• Is the curriculum suited to raise students' awareness of landscape architecture as **cultural tradition**? Do we as teachers agree on what it is? Is it e.g. "Related to stewardship – the protection, and enhancement of the conceptual, material and phenomenal relationships between humans and non-human nature"? (Deming and Swaffield 2011, p. 18)
	• Does the curriculum contribute to the growth of students into experienced self-developers? Do we encourage and value their exercise in good judgement as their main way to learn?
		What role in society do we educate our students for? The landscape architect as designer of outdoor environments at any scale. The landscape architect as expert of green cultural heritage. Park and garden history. The landscape architect as urban green space manager. Ecological steward. The landscape architect as strategic landscape planner.	Through an environment of lectures, exercises, demonstrations, exemplifications in case studies, ...? In what way do I, in my capacity as teacher with *my particular background* pursue the landscape architecture's internal good (or aspects of it) and provide a role model? What is necessary and how much is sufficient, e.g. of ecological concepts; the history of landscape architecture (material and conceptual); landscape drivers (cultural, natural, political, economic, Technological, ...); etc.? For example:
			• How can we concretise the profession's internal values? Through
			  o Exemplary planning or design tasks, historical and contemporary? Which ones?
			  o Exemplary works? historical and contemporary. Which ones?
			• What processes will raise students' competence in good judgements, in design, planning and management? Understood as awareness of aesthetic experience(s); as analytical acuteness (context awareness and responsive acquisition of knowledge); as conscious value orientation (relevant framing of site, and problem); as imaginative and forward-looking problem solving?

	• Does the curriculum encourage theorising during practical work? • Does the curriculum incite students to reflect on the **ELC's goals**? • Does the curriculum incite students to think about how landscape architecture may increase **biological diversity, social equity and justice** as well as **aesthetic experience** and **beauty** in the physical environment? • Does the curriculum and course work adequately contextualise the concrete case-studies and problems?	The landscape architect as project manager, social entrepreneur, influencer, etc.	• How and to what extent should we provide training in skills, such as: Techniques for creative project and design development; visualising skills such as hand drawing, sketching, 2D and 3D digital visualisation, etc.; cooperation skills; scholarly inquiry and argumentation; written, spoken and visual communication; etc. • How should we provide declarative knowledge? Lectures, integrated individual or group-wise investigation exercises, student-led colloquia and reading sessions?
Questions on course – or module-level	For example: Why/for the sake of what do we teach a special course in planting design? • Are plants our most important materials? • Is it through knowledge of/on plants that landscape architects can enhance some of the profession's internal values? • Do we teach planting design because it is a way of concretising landscape architecture's contribution to sustainable development?		Same questions as preceding, but adapted to the specific subject matter.

Source: Anne Katrine Geelmuyden.

References

Abbott, M. and Bowring, J. (2019) 'The DesignLab Approach to Teaching Landscape', in Jørgensen, K., Karadeniz, N., Mertens, E. and Stiles, R. (eds.) *The Routledge Handbook of Teaching Landscape*. New York: Routledge.

Arendt, H. (1998) [1958] *The Human Condition*. 2nd ed. Chicago: University of Chicago Press.

Council of Europe (2000) *Council of Europe Landscape Convention*. ETS NO 176. [Online]. Available at: www.coe.int/en/web/conventions/full-list?module=treaty-detail&treatynum=176

Deming, M. E. (2019) 'Values and Transformative Learning. On Teaching Landscape History in a Community of Inquiry', in Jørgensen, K., Karadeniz, N., Mertens, E. and Stiles, R. (eds.) *The Routledge Handbook of Teaching Landscape*. New York: Routledge.

Deming, M. E. and Swaffield, S. (2011) *Landscape Architecture Research. Inquiry, Strategy, Design*. Hoboken: John Wiley & Sons Inc.

Dewey, J. (1997) [1938] *Experience and Education*. New York: Simon and Schuster.

Ek, A.-C., Ideland, M., Jönsson, S. and Malmberg, C. (2013) 'The Tension between Marketisation and Academisation in Higher Education', *Studies in Higher Education*, 38(9), pp. 1305–1318.

EU-LAND 21. (2017) *Trans-European Education for Landscape Architects, Output 01 Guidelines on Revision and Developing Study Programmes in Landscape Architecture* [Online]. Available at: www.iflaeurope.eu/assets/docs/EU-Land21_Output_01_27-11-18.pdf (Accessed 7 February 2020).

EU-LAND 21. (2019) *Trans-European Education for Landscape Architects, Output 02 Peer Learning Methods on the Development of a Curriculum* [Online]. Available at: www.iflaeurope.eu/assets/docs/EU-Land21-O2-2018reportaapp-learning-lines.pdf (Accessed 7 February 2020).

European Council of Landscape Architecture Schools (ECLAS). (2010) *ECLAS Guidance on Landscape Architecture Education*. Available at: www.eclas.org/eclas-education-guide (Accessed 16 March 2022).

European Qualification Framework (EQF). (2008) *Revision 2017* [Online]. Available at: https://ec.europa.eu/social/BlobServlet?docId=15686&langId=en (Accessed 1 November 2020).

Gadamer, H.-G. (1975) [1960] *Wahrheit und Methode*. Tübingen: J. C. B. Mohr (Paul Siebeck).

Higgins, C. (2011) *The Good Life of Teaching. An Ethics of Professional Practice*. Chichester: Wiley-Blackwell.

Institutt for hagekunst (landskapsarkitektur). (n.d.). *Årsmeldinger 1970–1998*. Ås: Historical Archive of Norwegian Landscape Architecture at the Norwegian University of Life Sciences, NMBU.

Laiho, A. (2010) 'Academisation of Nursing Education in the Nordic Countries', *Higher Education*, 60, pp. 641–656.

Lyotard, J. F. (1984) *The Postmodern Condition: A Report on Knowledge*. Manchester: Manchester University Press.

MacIntyre, A. (1999) *Dependant Rational Animals. Why Human Beings Need the Virtues*. Chicago: Open Court.

McEwen, C. and Trede, F. (2014) 'The Academisation of Emerging Professions: Implications for Universities, Academics and Students', *Power and Education*, 6(2), pp. 145–154.

Norges landbrukshøgskole (1859–1959). (1959) *Festskrift til 100-årsjubileet*. Oslo: Grøndahl & Søns Boktrykkeri.

Norsk hagearkitektlag, Årsskrifter (1952–1956). Ås: Historical Archive of Norwegian Landscape Architecture at the Norwegian University of Life Sciences, NMBU.

Olsen, T. B. (2000) 'Norske doktorgrader ved årtusenskiftet. En oversikt over utviklingen i det 19. og 20. Århundre, med hovedvekt på 1990-årene', NIFU – Norsk institutt for studier av forskning og utdanning Rapport 14/2000 [Online]. Available at: https://nifu.brage.unit.no/nifu-xmlui/handle/11250/273796 (Accessed 9 March 2022).

Winterton, J., Delamare-Deist, F. and Stringfellow, E. (2006) 'Typology of Knowledge, Skills and Competences: Clarification of the Concepts and Prototype', Cedefop – European Centre for the Development of Vocational Training series 64. Official Publications of the European Communities [Online]. Available at: www.cedefop.europa.eu/files/3048_en.pdf (Accessed 10 September 2020).

9
CONCEPTUAL THINKING AND RELATIONAL MODELS IN LANDSCAPE ARCHITECTURE PEDAGOGY

Juanjo Galan Vivas

Introduction

This chapter explores how landscape planning studio courses can support conceptual and integrative thinking. It presents a study that builds on the development and use of relational models linking different urban-nature concepts and connecting them with sustainability and resilience goals. The study was conducted between 2016 and 2019 at the Aalto University master's degree programme in landscape architecture. Students were invited to develop strategic proposals for Green-Blue infrastructures in different Finnish and Baltic cities. A particular focus was on promoting and assessing the development of conceptual and integrative thinking capacities. The assessment of those two capacities used two tools. One was a self-assessment of levels of conceptuality and integration that students progressively achieved during the course. The other was an assessment, done by external experts, of the relational models students employed and of the potential impact that proposals based on those models could have in integrative and sustainable urban planning.

Background and rationale

Pedagogic approaches and methods designed to foster conceptual and integrative thinking in landscape architecture education are important for a number of reasons. Firstly, the role of conceptual thinking in learning for deep comprehension, meaningful understanding and higher-order thinking is widely recognized (Etkind 2011; Trygestad 1997). Secondly, planning and design are informed by different types of knowledge and by new and increasingly disconnected concepts emerging from several disciplines. This situation is particularly notable in landscape architecture, partly due to the nodal and highly connected position of the discipline in the web of knowledge, and partly due to the role that landscape architects can play in integrating hitherto disaggregated concepts and methods pertaining to the multiple dimensions of landscape (Galan 2020).

Conceptual thinking is the ability to understand a situation or problem by identifying patterns or connections between abstract and disparate ideas and by integrating issues and factors into a conceptual framework (Etkind and Shafrir 2013). Accordingly, conceptual thinking pedagogy focuses learners' attention on the identification of patterns, similarities and differences, and on underlying meanings. Conceptual thinking implies a certain level of abstraction and the creative and non-self-evident formulation of relations between concepts. These relations can be formalized through new concepts or 'code words' and procedures for their use and further development (Etkind et al. 2010). From a constructivist perspective, conceptual thinking implies the use of concepts as "fundamental building blocks of cognition" that "organize ideas into patterns for deeper cognitive understanding and faster cognitive processing" (Trygestad 1997, pp. 3–4). Consequently, knowledge would be arranged into patterns of clustered concepts that may be represented, for example, through concept mapping (Trygestad 1997). Since the definition of concepts and concept-clusters is based on the identification of patterns and relationships, conceptual thinking can be associated with the definition of *relational models* exploring hypothetical or factual connections between the considered concepts (Van Kamp et al. 2003).

The pedagogic study of conceptual thinking and relational models may be developed in several of the many complex and multifaceted challenges that landscape architects address in practice. For the purpose of this chapter, and following the spirit of the European Landscape Convention (ELC 2000), *landscape* is understood as the physical and cultural materialization of the interaction between people and the environment in evolving socio-ecological systems, including cities. *Sustainability and resilience* are conceptualized as system properties that can help manage the continuity and adaptation of those systems. *Sustainable and integrative urban planning* are both understood as processes and procedures that help facilitate a sustainable evolution of cities through synergic interactions between the many sectors involved.

In addressing urban growth, new visions and concepts of resilient and sustainable development increasingly serve as frameworks for analyzing and managing the evolution of socio-ecological systems (Fiksel 2006; Berkes and Folke 1998; Forman 1995). Helping to construct and implement these visions, different disciplines have defined concepts such as Green Infrastructure (GI), Urban Green Infrastructure (UGI), Ecosystem Services (ESS), Nature Based Solutions (NBS) or Natural Capital (NC), that for the purpose of this research on urban landscapes, are grouped under the term *Urban-Nature concepts*. These concepts have become part of the basic vocabulary framing current research, education and professional practice in landscape planning and design. To date, attempts to use and combine these concepts are few (Pauleit et al. 2017; Nesshöver et al. 2017; Lafortezza et al. 2013; Hansen and Pauleit 2014; Tzoulas et al. 2007). Relational models have served to link different concepts, specialized research with integrative practice (Hansen and Pauleit 2014), and conceptual thinking with applied planning (Kambites and Owen 2007; Mell 2010). However, any consistent attempt to define a relational model initially requires a critical reflection and definition of the concepts it is supposed to connect.

Addressing the issues mentioned earlier provides great pedagogical opportunities in landscape architecture education. Not only is conceptual thinking becoming important across highly dynamic and complex environments and in advanced education (Berman and Smyth 2015), but by encouraging students to reflect on human-nature relationships we can also create a fertile ground for learners to develop conceptual, integrative and transdisciplinary thinking skills. Moreover, due to their complexity, human-nature relationships in urban settings can be particularly challenging (White et al. 2016) and, consequently, pedagogically productive in landscape architecture.

Research question, study design and methods

To investigate how conceptual and integrative thinking might be promoted in landscape planning education using relational models, it was decided to design a study that introduces and assesses the acquisition of both of these capacities by learners. The study incorporated specific learning outcomes and assessment methods in a studio course on Green Area Planning (see Figure 9.1). In landscape architecture higher education, studio courses provide learning environments that are particularly useful to overcome or mirror the possibilities and constraints of professional practice (Armstrong 1999). Studio courses based on case studies can also effectively link research and practice by applying theory and conceptual reflection to real-world challenges (Baum 1997; Yin 1993). Additionally, studio courses can facilitate the joint engagement of students and educators through active learning, in which both process and outcome are equally important, and the multidisciplinary development of proposals can go beyond standard solutions (Higgins and Simpson 1997; Shepherd and Cosgriff 1998; Kumar and Kogut 2006; Higgins et al. 2009; Galan 2018; Galan and Kotze 2022). Therefore, studio courses offer adequate platforms to explore and test the practical potential of conceptual and relational models in planning. The use of conceptual models in the context of planning has previously been studied (Margoluis et al. 2009), and different authors have proposed a combination of various methods to assess their quality and utility. Oliver-Hoyo and Allen (2006) recommend their validation through a triangulation process combining interviews, surveys and journal entries. Alternatively, Mehmood and Cherfi (2009) use surveys to identify key factors that, according to users, would be critical to define and assess the quality of conceptual models.

In 2016, 2017, 2018 and 2019, the pedagogic potential of employing relational models to link different urban-nature concepts was tested in the Green Area Planning course of the Aalto University Landscape Architecture graduate programme. With a workload of seven credits, this is one of the programme's two compulsory courses. Organized in a studio format, the course

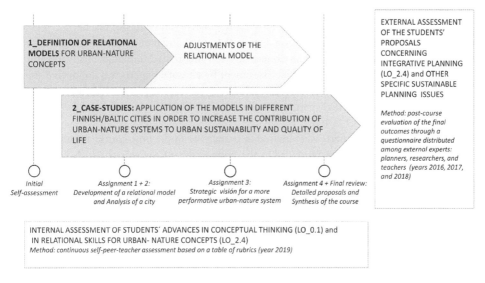

Figure 9.1 Development and application of relational models for urban-nature concepts in a studio course on Green Area Planning. This diagram emphasizes the specific learning outcomes and assessment methods that were used to evaluate the effect of the course on students' conceptual and integrative thinking skills.

Source: Juanjo Galan Vivas.

promotes active learning, connects theory and innovative practice, and engages students with real-world cases. Seventy-five per cent of the enrolled students were landscape architects and 25% of students came from other degrees such as architecture, urban studies and planning, creative sustainability and water engineering. Students worked in multidisciplinary teams of three to four members each. The size of the whole class fluctuated between 15 students in 2018–19 and 33 students in 2019–20.

The expected learning outcomes of the Green Area Planning course were defined in accordance with the specific goals of the course and the general goals of the programme. During 2019–20, the list of learning outcomes was expanded to incorporate soft skills such as conceptual thinking, collaboration skills and self-management or communication skills, among others. Soft skills are considered essential to increase employability, professional success and self-learning capacities of graduates (Heckman and Kautz 2012; Wats and Wats 2009). The learning outcome 'LO_0.1' concerned conceptual and critical thinking and the learning outcome 'LO_2.4' addressed the capacity to integrate urban-nature concepts in relational models supporting sustainable urban planning. By incorporating these two learning outcomes in a continuous process of self and peer assessment, students were invited to explicitly address them in all assignments.

The first part of the course (see arrow 1 in Figure 9.1) included the comparative study of different urban-nature concepts and the definition of a relational model that could link them all and support strategic urban planning. This exercise was catalyzed through the discussion of a model proposed by the teacher. Simultaneously, each team started analyzing one Finnish or Baltic city of their choice. Their analysis included general aspects such as the morphological and functional evolution of the cities. During the second part of the course (see arrow 2 in Figure 9.1), students explored how the studied urban-nature concepts and the developed relational models could help them to envision new urban-nature systems or Green-Blue infrastructures promoting urban sustainability and resilience. During this process, many teams iteratively adjusted their relational models to make them more operational for the ideas and challenges that they were progressively facing. Each year, the final proposals were presented in a set of posters and in a public presentation attended by local representatives and external teachers.

As displayed in the dotted boxes on Figure 9.1, two different assessment methods were devised to evaluate advances made in conceptual and integrative thinking. Firstly, during the years 2016, 2017 and 2018 a post-course assessment by external experts was conducted using a survey. Secondly, in the academic year 2019–20, an internal process was designed to help students self and peer evaluate their progress in all of the learning outcomes defined for the course, including LO_0.1 and LO_2.4.

In 2018, 75 external experts were invited to participate in the post-course assessment. They all received samples of the work produced by the students during the years 2016, 2017 and 2018, together with a survey containing Likert questions about the potential of the relational models and students' proposals to support sustainable urban planning. The experts came from different sectors, academic backgrounds and professional fields. Twenty-four answers were received and processed, 14 from planners and decision makers from the studied cities, five from educators, and five from practitioners from different Finnish landscape architecture offices.

During the year 2019, advances in conceptual thinking and relational skills were internally evaluated through the use of an assessment form that included all the learning outcomes of the course and a table of rubrics specifying the meaning of each grade for each learning outcome (ranging from one as a minimum to five as a maximum). Both the assessment form and the table of rubrics were used to conduct an individual self-assessment at the beginning of the course (diagnostic assessment) and a self, peer and teacher assessment after the review of each assignment (formative and summative assessments). The grades and comments from peers were anonymously shared after each review in order to provide the students with a continuously updated and multifaceted picture of their performance for each learning outcome during the whole course. The course consisted of 33 students organized into nine teams. The comparison between the scores obtained by each student and team on the learning outcomes LO_0.1 and LO_2.4 at the beginning and end of the course provided an indicator of the students' evolution in conceptual thinking and relational skills. In addition, the comparison between the grades obtained by each team in the learning outcomes LO_0.1 and LO_2.4, and their overall final grades, allowed analyzing potential correlations.

Findings

Starting with the relational model proposed by the teacher (see Figure 9.2, top), students were able to successfully develop their own models and use them in studying and developing proposals in the cities they had selected. In the years 2016, 2017 and 2018, several teams adopted the teacher's model, but from 2019 onwards, they were strongly encouraged to prepare their own models in order to promote deeper conceptual discussions. This last decision proved to be extremely useful to achieve a higher level of appropriation of the models by the students and a stronger connection between the models and the subsequent proposals. As displayed in one of the students' models (Figure 9.2, bottom), urban sustainability, resilience and quality of life often defined an overarching framework supporting a socio-ecological approach to urban areas. Within this framework, the interaction between people and urban nature was conceptualized through a layered model in which the Urban Green Infrastructure concept (UGI) incorporated the spatial dimension of the urban nature system (European Commission 2016) whereas the Ecosystem Services concept (ESS) included the functions and benefits (Millennium Ecosystems Assessment 2005) provided to humans by the same system. In addition, urban-nature was perceived as a key component of the urban metabolism by impacting positively in water, matter, air, waste and food flows (Baccini and Brunner 2012; Ferrão and Fernandez 2013). As displayed in the inner part of Figure 9.2 (bottom), students were also able to link the development of the urban-nature system with contextual challenges (both global and local) affecting their cities, with new citizens' needs, and with the understanding of the city as dynamic landscape collage. From an operational point of view, and as displayed in the lower part of Figure 9.2 (bottom), the models also helped students to organize their work in the studio and to move from the initial and proactive analysis to the definition of coherent strategies and specific actions. Overall, the elaboration of the relational model became the central element of the course and helped students to connect theory and practice.

Table 9.1 presents the results of the assessment by external experts of the works produced by the students during the years 2016, 2017 and 2018. External evaluations of conceptual models produced by students have been successfully implemented in different disciplines (Mehmood and Cherfi 2009; Maes and Poels 2007). Using Likert scales, 24 respondents scored how the relational model and the subsequent proposals performed in terms of sustainable landscape and urban planning. Experts considered the relational model that students used an adequate tool to support sustainable city/urban planning. All participants shared this opinion, although landscape architecture practitioners, architects, social scientists, engineers, city planners, landscape designers, and other public servants gave particularly high scores. Similarly, the relational model was considered adequate to integrate the studied concepts, although in this case, the scores were slightly lower. Finally, the assessment of the students' proposals by the respondents coincided almost completely with the assessment of the integrative potential of the relational model itself. Overall, the external assessment suggests that the academic use of relational models can promote

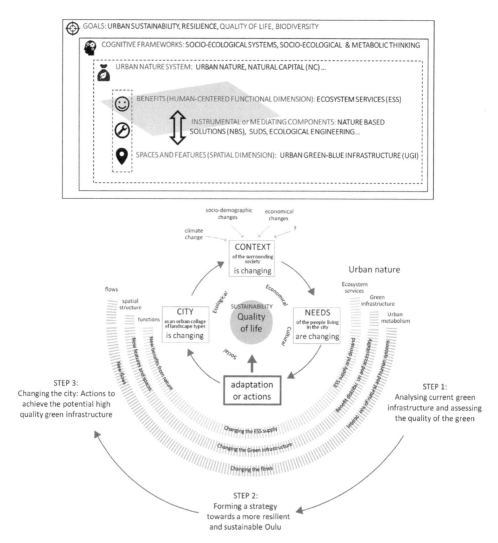

Figure 9.2 Top: relational model for urban-nature concepts proposed by the teacher to activate conceptual discussions in the Green Area Planning course (adapted from Galan 2020). Bottom: relational model for urban-nature concepts proposed by the team of students working in the city of Oulu in 2019.

Source: Students: H.Y. Lai, Y. Liang, S. Kangas, K. Rahkola, C. Yao; teacher: J. Galan, 2019.

the coherent integration of different concepts and their combined use in planning education. However, due to the small size of the sample and of some of the groups consulted, it would be recommendable to develop a more extensive study.

Following the changes made in the Green Area Planning course in 2019, Table 9.2 displays the results of the internal assessment of conceptual thinking (LO_0.1) and integrative thinking skills (LO_2.4). Regarding self-assessed progress in conceptual thinking skills, 61% of the students expressed important or very important advancements, while 39% expressed no change. The comparison of the peer-evaluation and teacher-evaluation of the first and last assignments

Table 9.1 External assessment by planners, academic experts and practitioners of the potential that a relational model and the subsequent academic proposals could have in sustainable urban planning (studio course on Green Area Planning, years 2016, 2017 and 2018). Mean Likert scale values are presented.

	AVERAGE TOTAL (24)	Sector (24)			Academic background (24)							Professional field (24)					
		CITY EMPLOYEES (15)	TEACHERS & RESEARCHERS (5)	PRACTITIONERS PRIVATE SECTOR (4)	LANDSCAPE ARCHITECTS (10)	ARCHITECTS (4)	ENGINEERS (2)	ENVIRONMENTAL SCIENTISTS (4)	GEOGRAPHERS (2)	SOCIAL SCIENTISTS (1)	PUBLIC ADMINISTRATION (1)	GREEN AREA PLANNING + LANDS PLANNING (5)	ENVIRONMENTAL PLANNING (6)	CITY & URBAN PLANNING (6)	LANDSCAPE DESIGN (5)	CIVIL ENGINEERING (1)	PUBLIC ADMINISTRATION (1)
Capacity of the model TO SUPPORT SUSTAINABLE CITY/URBAN PLANNING	3,7	3,6	3,6	4,3	3,5	4,3	4,0	3,1	3,8	5,0	4,0	3,2	3,5	4,0	4,0	4,0	4,0
Capacity of the model TO INTEGRATE THE STUDIED CONCEPTS	3,5	3,2	3,8	4,0	3,4	3,5	3,0	3,3	3,5	5,0	4,0	3,4	3,5	3,2	3,8	3,0	4,0
Capacity of the proposals TO SUPPORT SUSTAINABLE CITY/URBAN PLANNING	3,4	3,5	3,2	3,3	3,3	4,0	3,5	3,0	3,0	4,0	4,0	3,6	3,2	3,8	3,0	3,0	4,0

Source: Adapted from Galan (2020).

reveals how 89% (22 + 45 + 22%) of the teams improved their level of conceptual thinking during the course, whereas only 11% remained at the same level.

Concerning self-assessed progress made in integrative thinking skills through the use of relational models for urban-nature concepts, 85% (36 + 49%) of the students experienced important or very important individual advancements, and only 15% gave themselves the same score at both the beginning and end of the course. According to the assessment made by the teacher and peers of the first and final assignments, all teams experienced an increase in integrative thinking skills. This increase was moderate for one team (11% of the class) and moderately high or high for the other eight teams (78% and 11% respectively).

As displayed in the last column of Table 9.2, despite the low weight that the learning outcomes LO_0.1 (conceptual thinking) and LO_2.4 (integrative thinking through relational

Table 9.2 Assessment of the evolution of conceptual and integrative thinking skills in the studio course Green Area Planning.

	Individual evolution (based on initial and final self-assessment) ★ 33 students	Team evolution (based on peers' and teacher's assessment of first and final team work) Nine teams	Correlation with final grade (based on the comparison between the final score for a particular Learning Outcome [LO_XX] and the final grade) Nine teams
Conceptual thinking (learning outcome LO_0.1)	39% no change 37% increase (1–1,5 points) 24% increase (2–3 points)	11% no change (0 points) 22% increase (0–0,5 points) 45% increase (0,5–1 points) 22% increase (>1 point)	79% correlation between grade in 'conceptual thinking' and 'final grade' in the course
Integrative thinking: Integration of urban-nature concepts through relational models serving sustainable urban transitions (learning outcome LO_2.4)	15% no change 36% increase (1–1,5 points) 49% increase (2–4 points)	11% increase (< 0,5 points) 78% increase (0,5–1 points) 11% increase (>1 point)	75% correlation between grade in 'capacity to define relational models integrating different urban nature concepts' and 'final grade' in the course

Source: Juanjo Galan Vivas.

models) were given in the final grade, they both showed a strong correlation with the final grade (79% and 75% respectively). However, due to the limitations of the study, it is not possible to ascertain if a statistically significant causal relationship exists between the level of conceptual or integrative thinking and the final grades obtained by the students.

Conclusions

The content and character of the pedagogic approach tested in this study provides new support to the conception of landscape architecture as a discipline that brings together concepts and knowledge from different fields. Research findings suggest that the proposed pedagogy and methods help students increase their conceptual and integrative thinking skills. According to the external experts consulted, the relational models employed in this study were useful in creating coherent and fruitful connections between different urban-nature concepts and in supporting their combined use in planning. As tested in 2019–2020, the generation of conceptual models is a key exercise *per se* that can facilitate a deeper understanding of the studied concepts and their active use in practice. Research findings also reveal how the adequate definition and alignment of learning outcomes, teaching methods and assessment methods can specifically address and support the development of cognitive skills such as conceptual and integrative thinking.

In addition, the study unveils the pedagogical potential that the development and use of relational models might have in (1) linking research, education and practice, especially in complex systems (e.g., landscapes or cities) informed by multiple types of knowledge and conceptual frameworks; (2) generating a common ground for people with different academic or professional backgrounds; and (3) facilitating the coordinated and synergic use of complementary or overlapping concepts within integrated planning.

From a landscape architecture education perspective, the study can be an initial step to advance new pedagogies supporting innovative landscape planning and design through abstract, conceptual and relational thinking. These skills have been widely recognized as essential to operate in the complex, uncertain and porous conditions that characterize contemporary societies, and could give landscape architects a key role in the definition and coordination of the multifaceted projects that occur in the intersection between social and ecological systems. The findings of this study suggest how these cognitive skills can considerably enrich the students' palette for planning and design processes and, in addition, easily connect theory with practice. The most rewarding moments during the Green Area Planning course were those in which students enjoyed the feeling of being intellectually challenged and discovered how the construction of their own conceptual frameworks could open new, unexpected and solid possibilities for them.

References

Armstrong, H. (1999) 'Design Studios Research: An Emerging Paradigm for Landscape Architecture', *Landscape Review*, 5, pp. 5–25.

Baccini, P. and Brunner, P. H. (2012) *Metabolism of the Anthroposphere*. 2nd ed. Cambridge: MIT Press.

Baum, H. (1997) 'Teaching Practice', *Journal of Planning Education and Research*, 17, pp. 21–29.

Berkes, F. and Folke, C. (1998) *Linking Social and Ecological Systems for Resilience and Sustainability*. Cambridge: Cambridge University Press, pp. 1–24.

Berman, J. and Smyth, R. (2015) 'Conceptual Frameworks in the Doctoral Research Process: A Pedagogical Model', *Innovations in Education and Teaching International*, 52(2), pp. 125–136.

Etkind, M. (2011) 'Pedagogy for Conceptual Thinking: Certificate Program for Instructors in Innovative Teaching', in *EDULEARN11 Proceedings*. Barcelona: IATED, pp. 2729–2737.

Etkind, M., Kenett, R. S. and Shafrir, U. (2010) 'The Evidence-Based Management of Learning: Diagnosis and Development of Conceptual Thinking with Meaning Equivalence Reusable Learning Objects (MERLO)', in *Proceedings from International Conference on the Teaching of Statistics ICOTS-8 Data and Context in Statistics Education: Towards an Evidence-Based Society*. Ljublijana, Slovenia: IATED, pp. 1–18.

Etkind, M. and Shafrir, U. (2013) 'Teaching and Learning in the Digital Age with Pedagogy for Conceptual Thinking and Peer Cooperation', in *Proceedings of INTED2013 Conference*. Valencia: IATED, pp. 5342–5352.

European Commission. (2016) *The Forms and Functions of Green Infrastructure*. Brussels: European Commission.

European Landscape Convention (ELC). (2000) Strasbourg: Council of Europe.

Ferrão, P. and Fernandez, J. E. (2013) *Sustainable Urban Metabolism*. Cambridge: MIT Press.

Fiksel, J. (2006) 'Sustainability and Resilience: Toward a Systems Approach', *Sustainability: Science, Practice and Policy*, 2(2), pp. 14–21.

Forman, R. T. T. (1995) *Land Mosaics: The Ecology of Landscapes and Regions*. Cambridge: Cambridge University Press, pp. 480–505.

Galan, J. (2018) 'Pedagogical and Academic Reflections from the iWater Summer Schools: Storm Water Management in Urban and Landscape Planning', in Delarue, S. and Dufour, R. (eds.) *Landscapes of Conflict, Proceedings of the ECLAS-2018 Conference*. Ghent: University of Ghent, pp. 560–568.

Galan, J. (2020) 'Towards a Relational Model for Emerging Urban Nature Concepts: A Practical Application and an External Assessment in Landscape Planning Education', *Sustainability*, 12, p. 2465.

Galan, J. and Kotze, D. J. (2022) 'Pedagogy of Planning Studios for Multidisciplinary, Research-Oriented, Personalized, and Intensive Learning', *Journal of Planning Education and Research*, 1(13), pp. 1–13.

Hansen, R. and Pauleit, S. (2014) 'From Multifunctionality to Multiple Ecosystem Services? A Conceptual Framework for Multifunctionality in Green Infrastructure Planning for Urban Areas', *AMBIO*, 43, pp. 516–529.

Heckman, J. and Kautz, T. (2012) 'Hard Evidence on Soft Skills', *Labour Economics*, 19(4), pp. 451–464.

Higgins, M., Aitken-Rose, E. and Dixon, J. (2009) 'The Pedagogy of the Planning Studio: A View from Down Under', *Journal of Education in the Built Environment*, 4, pp. 8–30.

Higgins, M. and Simpson, F. (1997) *Work-Based Learning within Planning Education: A Good Practice Guide*. London: University of Westminster Press for the Discipline Network in Town Planning.

Kambites, C. and Owen, S. (2007) 'Renewed Prospects for Green Infrastructure Planning in the UK', *Planning Practice & Research*, 21, pp. 483–496.

Kumar, M. and Kogut, G. (2006) 'Students' Perceptions of Problem Based Learning', *Teacher Development*, 10, pp. 105–116.

Lafortezza, R., Davies, C., Sanesi, G. and Konijnendijk, C. C. (2013) 'Green Infrastructure as a Tool to Support Spatial Planning in European Urban Regions', *iForest*, 6, pp. 102–108.

Maes, A. and Poels, G. (2007) 'Evaluating Quality of Conceptual Modeling Scripts based on User Perceptions', *Data & Knowledge Engineering*, 63, pp. 701–724.

Margoluis, R., Stem, C., Salafsky, N. and Brown, M. (2009) 'Using Conceptual Models as a Planning and Evaluation Tool in Conservation', *Evaluation & Program Planning*, 32, pp. 138–147.

Mehmood, K. and Cherfi, S. S. S. (2009) 'Evaluating the Functionality of Conceptual Models', in Heuser, C. A. and Pernul, G. (eds.) *Advances in Conceptual Modeling – Challenging Perspectives. ER 2009. Lecture Notes in Computer Science volume 5833*. Berlin and Heidelberg: Springer.

Mell, I. C. (2010) *Green Infrastructure: Concepts, Perceptions and its Use in Spatial Planning*. Doctoral thesis, Newcastle University, Newcastle.

Millennium Ecosystems Assessment. (2005) *Ecosystems and Human Well-Being: Ecosystems and Their Services*. Washington: Millennium Ecosystems Assessment.

Nesshöver, C., Assmuth, T., Irvine, K., Rusch, G., Waylen, K., Delbaere, B., Haase, D., Jones-Walters, L., Hansm, K. and Esztern, K. (2017) 'The Science, Policy and Practice of Nature-Based Solutions: An Interdisciplinary Perspective', *Science of the Total Environment*, 79, pp. 1215–1227.

Oliver-Hoyo, M. and Allen, D. D. (2006) 'The Use of Triangulation Methods in Qualitative Educational Research', *Journal of College Science Teaching*, 35, pp. 42–47.

Pauleit, S., Zölch, T., Hansen, R., Randrup, T. B. and Konijnendijk van den Bosch, C. (2017) 'Nature-based Solutions and Climate Change – Four Shades of Green', in Kabisch, N., Korn, H., Stadler, J. and Bonn, A. (eds.) *Nature-Based Solutions to Climate Change Adaptation in Urban Areas. Theory and Practice of Urban Sustainability Transitions*. Berlin and Heidelberg: Springer, pp. 29–49.

Shepherd, A. and Cosgriff, B. (1998) 'Problem based Learning: A Bridge between Planning Education and Planning Practice', *Journal of Planning Education and Research*, 17, pp. 348–357.

Trygestad, J. (1997) 'Students' Conceptual Thinking in Geography', in *Proceedings from Annual Meeting of the American Educational Research Association*. Chicago: American Educational Research Association, pp. 1–33.

Tzoulas, K., Korpela, K., Venn, S., Yli-Pelkonen, V., Kazmierczak, A., Niemela, J. and James, P. (2007) 'Promoting Ecosystem and Human Health in Urban Areas Using Green Infrastructure: A Literature Review', *Landscape & Urban& Urban Plannning*, 81, pp. 167–178.

Van Kamp, I., Leidelmeijera, K., Marsmana, G. and Hollander, A. (2003) 'Urban Environmental Quality and Human Well-Being towards a Conceptual Framework and Demarcation of Concepts; A Literature Study', *Landscape and Urban Planning*, 65, pp. 5–18.

Wats, M. and Wats, R. K. (2009) 'Developing Soft Skills in Students', *International Journal of Learning*, 15(12), pp. 1–10.

White, D. F., Rudy, A. P. and Gareau, J. (2016) *Environments, Natures and Social Theory*. London: Palgrave Macmillan.

Yin, R. K. (1993) *Applications of Case Study Research*. London: SAGE.

10

PEDAGOGY FOR SUSTAINABILITY IN LANDSCAPE ARCHITECTURAL EDUCATION

Dan Li

Introduction to research into Education for Sustainability

Sustainability means "the capacity to be kept in existence or maintained indefinitely, in particular, the capacity to maintain the ability of social systems, economic systems, and environmental systems to support human life and well-being" (Portney 2015, p. 9). Sustainability in this chapter is related to global challenges such as climate change, food security, and water resources. Landscape architecture has the potential to help achieve sustainability.

In 1987, the United Nations proposed principles of sustainable development to address global environmental challenges, emphasizing how education would be important in achieving sustainability goals (Brundtland 1987). Commonly used concepts that help implement these goals are Education for Sustainability, EfS, and Education for Sustainable Development, ESD (UNESCO 2022). Both concepts promote holistic and interdisciplinary approaches to learning and multimethod research approaches (Singer et al. 2015; Van Lopik 2012; UNESCO 2005). Accordingly, research into EfS is (1) multidisciplinary, (2) participatory, and (3) constructivist in nature.

First, the multidisciplinary character of EfS is emphasized by a lot of scholars. Johnston and Johnston (2013) pointed out in "What's Required to Take EfS to the Next Level", that EfS requires "multidisciplinary collaboration" (p. 1). The broad definition of sustainability is the main reason that makes multidisciplinary fundamentally necessary in EfS (Christie et al. 2013; Wals and Blewitt 2010). In addition, EfS must be interdisciplinary because all concepts of sustainability bring together ecologic, economic, and sociocultural factors (Christie et al. 2013; Littledyke and Manolas 2010).

Second, the participatory and locally relevant character of EfS stems from the definition of sustainability relying on social and cultural norms as they vary from one region to another (Little et al. 2016; Pfister et al. 2016). The success of educational approaches to sustainability will depend on the degree and quality of considering social and political contexts (Corcoran and Wals 2004). In educational practice, educators and students will involve local communities in real-world planning assignments (Faurest and Ruggeri 2016).

Third, EfS is associated with post-positivist approaches to education. It should be subjective, value-laden, tentative knowledge recognized as subject to change due to context, challenge and social acceptability. For ideological influences, EfS should be using reconstructive theory emphasizing social change (Christie et al. 2013; Littledyke and Manolas 2010).

Studies of Education for Sustainability in design education

Research into EfS in design education, including landscape architecture, started after 2010. Most studies are in the category of applied research and pertain to the development of individual courses or projects (Li et al. 2018). They focus on examining different pedagogy and teaching methods as scenario-based planning, reality gaming, integration of theory courses and design studio, using integrated student teams, visualizing collaborative experiences, place-based approaches and so on (Albert et al. 2015; Ayer et al. 2016; Bozkurt 2016; Brncich et al. 2011; McMahon and Bhamra 2016; Nikezić and Marković 2015). Few cases of fundamental research exist. One is where Hakky (2016) used a mixed-methods approach to examine, from the educator's perspective, the status and future of design for sustainable behaviour in interior design education.

From individual studies and the few basic research studies existing to date, it is difficult to understand the current situation and potential of EfS in design education (Li et al. 2018). More research needs to be carried out to support both pedagogy and curricular development and gain insight regarding policy. In doing so, one step is to connect education to design practice where developing landscape sustainably is a major aim. Another step is to explore the views of educators of the pedagogy and methods used for EfS in landscape architectural courses.

Connecting education to design practice

The purpose of professional education is to help graduates to emerge as professionals, and programmes involving EfS engage students with professional responses to global, regional and local sustainability challenges. For students to learn from professional practice, they may refer to studies into the practice of sustainable landscape design focusing on design theories and strategies (Calkins 2011; Dinep and Schwab 2010; John 1992; Schwarz and Krabbendam 2013; Thompson and Sorvig 2007). Most studies take a case study approach. However, it is difficult for educators and students to draw general conclusions from individual examples. Single projects are also only partly useful as teaching material for sustainability.

To make a difference when schools are designing landscape architecture programmes that aim at preparing students for sustainable development, new research is needed, combining practice with education. One way of making this connection is to include practical evaluation tools for sustainability – rating systems such as LEED (Leadership in Energy and Environmental Design) and SITES (The Sustainable SITES Initiative). LEED supports practitioners and researchers in designing and examining the sustainability of buildings and neighbourhoods (LEED 2022). SITES is a comprehensive rating system designed to develop and measure sustainable landscapes (SITES 2022). Rating systems have served in education as part and content of newly developed courses (Ahn et al. 2009). Including sustainability rating systems in education increases the possibilities of systematically integrating sustainability parameters into landscape architectural curricula or even creating new sustainability-oriented programmes.

Research questions, approach and findings

The literature review just presented suggests that current studies on EfS in landscape architectural education in the USA are fragmentary. To fill the knowledge gap about the type and effectiveness of pedagogic methods used for sustainability teaching in professional programmes, the following research questions are raised: 1) what kinds of pedagogic methods are used for

EfS, 2) how effective are they, and 3) whether educators use sustainability rating as part of their pedagogic methods, and how they are using them.

A quantitative research approach is used by conducting a survey with educators in the landscape architecture programmes in the USA. The study aims at generating data that allows for a fundamental examination of EfS pedagogy and teaching methods employed in landscape architectural education. Based on a review of pertinent literature, the survey includes three parts. First, as a whole, the contemporary pedagogies are examined. The emphasis is on learner-centred pedagogy, networked learning, critical pedagogy, problem-based learning, multidisciplinary, collaboration, and outreach. These pedagogies have been shown, in several studies, to be associated with teaching sustainability or come from the features of EfS (Ayer et al. 2016; Bosselmann 2001; Breiting and Mogensen 1999; Christie et al. 2013; Cotton et al. 2007; Dawe et al. 2005; Thomas 2009; Tilbury 2007; Tilbury and Cooke 2005). Second, teaching methods are examined. Methods suggested for teaching sustainability include critical thinking, role plays and simulations, group discussions, debates, problem-based learning, case studies, reflective accounts, critical reading and writing, fieldwork, and the teacher and university modelling good practice (Brncich et al. 2011; Christie et al. 2013; McMahon and Bhamra 2016; Nikezić and Marković 2015). Third, the survey examines how sustainability rating systems are used as part of pedagogic methods and how educators use rating systems in their teaching practice. Finally, demographic data is collected from participating educators for comparative and descriptive information. This included questions about sex, race, and status of educators, such as title and rank.

The survey was administered with Qualtrics and sent to 951 educators. The educators were from all 69 professional programmes accredited by the Landscape Architecture Accreditation Board (LAAB) by October 2018. The survey was open for participation from 12.11.2018 to 12.12.2018. A total of 209 educators completed and returned the survey. The response rate was 22.0%. The raw data were exported from Qualtrics into a Microsoft Excel file (.xlsx) for analysis using Excel and SPSS 26 (IBM 2019). Data cleaning was conducted using SPSS 26 to evaluate data integrity, test assumptions, and visualize the data. Descriptive statistics included frequencies, parametric statistics, univariate statistics, and bivariate statistics. Inferential statistics were conducted and included χ^2 cross-tabulation tests, Pearson correlations, independent samples t-tests, analysis of variance (ANOVA), and simple linear regression. Data visualization was conducted using both Excel and SPSS 26.

Males accounted for 56.3% of the respondents, 37.1% were female, and the rest other (or provided no information); 81.5% of the respondents are white, 3.3% Asian, 2.0% Hispanic or Latino, 1.3% black or African American, 1.3% multiracial, 0.7% Native Hawaiian or Pacific Islander, and the rest other (or provide no information). Regarding status, 76.1% of the responders were full-time, and 23.9% were part-time educators; 45.8% of the responders were tenured professors, 13.4% tenure track without being tenured, and 26.8% not on the tenure track. The percentage of the demographic data excluded the missing answers of the demographic information.

Endorsement refers to the agreement or support of a concept or action. In the charts in Figure 10.1 and Figure 10.2, endorsement is the prevalence of survey respondents who engage in each pedagogy or teaching method.

First, for the frequency of pedagogy used for teaching sustainability in landscape architecture and their effectiveness, problem-based learning is the most commonly used pedagogy for teaching sustainability according to educators' self-report, with a percentage of 93. Place-based or community-based learning, multidisciplinary learning, and learner-centred pedagogy are also commonly used with percentages of 86, 82, and 79. According to study findings,

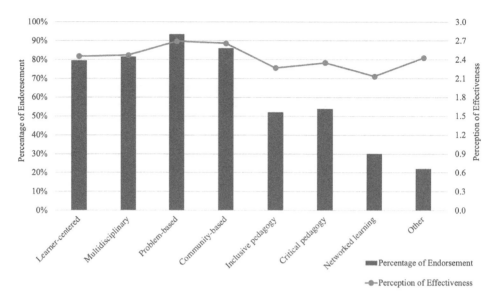

Figure 10.1 Pedagogy for teaching sustainability endorsement and effectiveness.
Source: Dan Li.

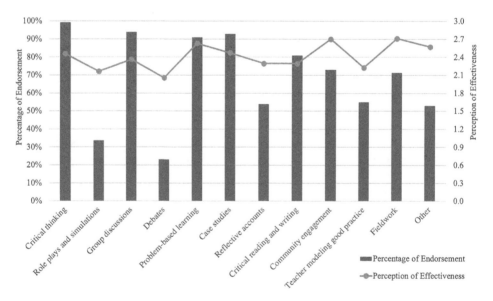

Figure 10.2 Teaching method for teaching sustainability endorsement and effectiveness.
Source: Dan Li.

networked learning is the least commonly used pedagogic method, while critical pedagogy and inclusive pedagogy are in the middle. When asked how effective the pedagogic methods are for sustainability teaching, the survey participants indicated that problem-based learning and community-based learning are the most effective ones, followed by multidisciplinary learning

Pedagogy for sustainability

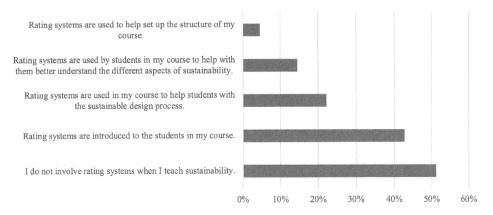

Figure 10.3 Involvement of rating systems in teaching sustainability.
Source: Dan Li.

and learner-centred methods. Other methods respondents used are experiential learning and culture awareness.

Second, for specific teaching methods, survey participants reported that critical thinking, group discussions, case studies, and problem-based learning were the most commonly used in teaching sustainability, with percentages of 99, 94, 93, and 91, while critical reading and writing, community engagement, and fieldwork were used less. Debates, role-play and simulations, reflective accounts, and practice modelling were the least used methods. When asked about the effectiveness of the methods they used, educators responded that fieldwork, community engagement, and problem-based learning were the most effective methods for teaching sustainability, while debates and role-plays and simulations were the least effective ones. Some other teaching methods indicated by the responders are systems design, team research, guest speakers, and so on (Figure 10.3).

Third, when asked about their familiarity with sustainability rating systems, educators reported that they are more familiar with LEED and SITES, and less familiar with Living Building Challenge and Envision.

Fourth, for the involvement of sustainability rating systems in teaching, 51.1% of the respondents reported that they did not involve rating systems while teaching sustainability; 48.9% of the educators involved rating systems in single or multiple ways; 42.7% of educators reported that they introduced rating systems to students; 23.7% of the educators indicated that rating systems were used to help students to better understand different aspects of sustainability; and 22.1% said the systems were employed to help students in sustainable design processes. Only 4.6% of the responders said that they used rating systems to help set up the structure of the course (Figure 10.3).

Conclusions

Educational programmes need to prepare graduates for the challenges of achieving sustainability through landscape architecture, as theories and strategies for sustainable landscape practice are important in landscape architecture (Calkins 2011; Dinep and Schwab 2010; John 1992; Schwarz and Krabbendam 2013; Thompson and Sorvig 2007). In preparing graduates to take leadership in achieving sustainability goals, EfS plays an increasingly important role. However, knowledge about the extent of educators applying EfS in landscape architectural education is sparse. Findings from literature reviews reveal that studies about EfS used in landscape architecture programmes are limited to examining single cases, leaving a fragmentary picture from which scholars cannot draw general conclusions. Furthermore, single cases provide little information for educators in structuring and preparing courses that focus on or relate to sustainability.

With the aim to provide more substantial support for methodological and curricular development and gain insight regarding policy implications, findings from a study are presented to explore pedagogic approaches and methods employed for sustainability teaching in landscape architectural education. Findings are limited to the USA. However, they provide the general substance needed to improve educational practice beyond pedagogy for EfS in landscape architecture. Findings suggest that the pedagogy and teaching methods that put students in the scenario engaging with the problem, the place, and the stakeholders are the most effective way to help students learn about sustainability. For effectiveness ranking, the different pedagogic approaches studied appear to match the frequency they are used in teaching sustainability. Specifically, problem-based learning, place-based or community-based learning, multidisciplinary learning, and learner-centred pedagogy are the top four most effective, and the top four commonly used approaches. On the other hand, some of the most effective teaching methods, fieldwork and community engagement, are not the most commonly used in sustainability teaching. This finding suggests opportunities for improvement. While the study shows how educators are familiar with some sustainability rating systems, these systems are largely missing as course content or teaching method. Where sustainability rating is involved in teaching, the utilization is limited to introducing the sustainability rating systems to students. Using sustainability rating systems to set up the course structure or integrating them into design processes might become another way of improving sustainability teaching practice. Including sustainability rating systems in education increases the possibilities of systematically integrating sustainability rating into landscape architectural curricula or even creating new sustainability-oriented programmes.

Considering the multidisciplinary and collaborative nature of sustainability concepts, educators might consider working together with colleagues from different departments inside their universities, with members of the design industry, and getting involved in community design projects to improve sustainability teaching in landscape architecture. First, educators might consider inviting students from different disciplinary fields to attend the same classroom (McMahon and Bhamra 2016). In turn, the learning experience and outcomes from this kind of boundary-crossing course would provide substance for new research, for example, in the form of experimental research, systematic pedagogic trials, and in the context of longitudinal studies. Second, connecting sustainable landscape education with current design practice will be important. Outreaching to the design industry or community design projects will open a new door for sustainable landscape education research. Using real design projects as the content of the design studio course, inviting designers into the design team with the students, and evaluating the learning outcomes will be a new kind of experimental research.

Learning about and utilizing effective teaching methods are not limited to the field of landscape architecture. All results presented in this chapter could be important to and inspire all educators interested in advancing EfS.

References

Ahn, Y. H., Kwon, H., Pearce, A. R. and Wells, J. G. (2009) 'The Systematic Course Development Process: Building a Course in Sustainable Construction for Students in the USA', *Journal of Green Building*, 4(1), pp. 169–182.

Albert, C., von Haaren, C., Vargas-Moreno, J. C. and Steinitz, C. (2015) 'Teaching Scenario-based Planning for Sustainable Landscape Development: An Evaluation of Learning Effects in the Cagliari Studio Workshop', *Sustainability*, 7(6), pp. 6872–6892.

Ayer, S. K., Messner, J. I. and Anumba, C. J. (2016) 'Augmented Reality Gaming in Sustainable Design Education', *Journal of Architectural Engineering*, 22(1), p. 04015012.
Bosselmann, K. (2001) 'University and Sustainability: Compatible Agendas?', *Educational Philosophy and Theory*, 33(2), pp. 167–186.
Bozkurt, E. (2016) 'Integration of Theory Courses and Design Studio in Architectural Education Using Sustainable Development', *SHS Web of Conferences* 26. Les Ulis, France: EDP Sciences, pp. 01102.
Breiting, S. and Mogensen, F. (1999) 'Action Competence and Environmental Education', *Cambridge Journal of Education*, 29(3), pp. 349–353.
Brncich, A., Shane, J. S., Strong, K. C. and Passe, U. (2011) 'Using Integrated Student Teams to Advance Education in Sustainable Design and Construction', *International Journal of Construction Education and Research*, 7(1), pp. 22–40.
Brundtland, G. H. (1987) *Report of the World Commission on Environment and Development: "Our Common Future"*. United Nations General Assembly Document A/42/427.
Calkins, M. (2011) *The Sustainable Sites Handbook: A Complete Guide to the Principles, Strategies, and Best Practices for Sustainable Landscapes*. Hoboken, NJ: John Wiley & Sons.
Christie, B. A., Miller, K. K., Cooke, R. and White, J. G. (2013) 'Environmental Sustainability in Higher Education: How Do Academics Teach?', *Environmental Education Research*, 19(3), pp. 385–414.
Corcoran, P. B. and Wals, A. E. (2004) *Higher Education and the Challenge of Sustainability: Problematics, Promise, and Practice*. New York, NY; Boston, MA; Dordrecht; London; Moscow: Kluwer Academic Publishers.
Cotton, D. R. E., Warren, M. F., Maiboroda, O. and Bailey, I. (2007) 'Sustainable Development, Higher Education and Pedagogy: A Study of Lecturers' Beliefs and Attitudes', *Environmental Education Research*, 13(5), pp. 579–597.
Dawe, G., Jucker, R. and Martin, S. (2005) *Sustainable Development in Higher Education: Current Practice and Future Developments*. Heslington: Higher Education Academy.
Dinep, C. and Schwab, K. (2010) *Sustainable Site Design: Criteria, Process, and Case Studies for Integrating Site and Region in Landscape Design*. Hoboken, NJ: John Wiley & Sons.
Faurest, K. and Ruggeri, D. (2016) 'Landscape Education for Democracy: A Proposal for Building Inclusive Processes into Spatial Planning Education', *Proceedings of the Fábos Conference on Landscape and Greenway Planning*, 5(1), Article 51. Available at: https://scholarworks.umass.edu/fabos/vol5/iss1/51 (Accessed 30 December 2019).
Hakky, D. (2016) *Examining the Status and Future of Design for Sustainable Behavior in Interior Design Education*. Doctoral dissertation, Virginia Tech, Blacksburg, VA.
IBM. (2019) *SPSS* (Version 26) [Computer software]. Available at: www.ibm.com/products/spss-statistics (Accessed 30 June 2019).
John, A. S. (1992) *The Sourcebook for Sustainable Design: A Guide to Environmentally Responsible Building Materials and Processes*. Boston, MA: Architects for Social Responsibility.
Johnston, D. D. and Johnston, L. F. (2013) *Introduction: What's Required to Take EfS to the Next Level. Higher Education for Sustainability: Cases, Challenges, and Opportunities from Across the Curriculum*. Abingdon, Oxon; New York, NY: Routledge, pp. 1–9.
LEED. (2022) 'How LEED Works', *LEED* [Online]. Available at: https://www.usgbc.org/leed (Accessed 10 July 2022).
Li, D., Kim, M. and Bohannon, C. L. (2018) 'Methodological Review of Sustainable Landscape Education Research', *Landscape Research Record*, 7, pp. 266–277.
Little, J. C., Hester, E. T. and Carey, C. C. (2016) 'Assessing and Enhancing Environmental Sustainability: A Conceptual Review', *Environmental Science & Technology*, 50(13), pp. 6830–6845.
Littledyke, M. and Manolas, E. (2010) 'Ideology, Epistemology and Pedagogy: Barriers and Drivers to Education for Sustainability in Science Education', *Journal of Baltic Science Education*, 9(4), pp. 285–301.
McMahon, M. and Bhamra, T. (2016) 'Mapping the Journey: Visualising Collaborative Experiences for Sustainable Design Education', *International Journal of Technology and Design Education*, 27(4), pp. 595–609.
Nikezić, A. and Marković, D. (2015) 'Place-based Education in the Architectural Design Studio: Agrarian Landscape as a Resource for Sustainable Urban Lifestyle', *Sustainability*, 7(7), pp. 9711–9733.
Pfister, T., Schweighofer, M. and Reichel, A. (2016) *Sustainability*. Abingdon, Oxon: Routledge.
Portney, K. E. (2015) *Sustainability*. Cambridge, MA: MIT Press.
Schwarz, M. and Krabbendam, D. (2013) *Sustainist Design Guide: How Sharing, Localism, Connectedness and Proportionality Are Creating a New Agenda for Social Design*. Amsterdam: BIS Publishers.

Singer, J., Gannon, T., Noguchi, F. and Mochizuki, Y. (eds.) (2015) *Educating for Sustainability in Japan: Fostering Resilient Communities After the Triple Disaster.* Abingdon, Oxon; New York, NY: Routledge.

SITES. (2022) *SITES Is the Most Comprehensive Rating System for Creating Sustainable and Resilient Land Development Projects* [Online]. Available at: www.sustainablesites.org (Accessed 10 July 2022).

Thomas, I. (2009) 'Critical Thinking, Transformative Learning, Sustainable Education, and Problem-based Learning in Universities', *Journal of Transformative Education*, 7(3), pp. 245–264.

Thompson, J. W. and Sorvig, K. (2007) *Sustainable Landscape Construction: A Guide to Green Building Outdoors.* Washington, DC: Island Press.

Tilbury, D. (2007) 'Monitoring and Evaluation During the UN Decade of Education for Sustainable Development', *Journal of Education for Sustainable Development*, 1(2), pp. 239–254.

Tilbury, D. and Cooke, K. (2005) *A National Review of Environmental Education and Its Contribution to Sustainability in Australia: Frameworks for Sustainability.* Canberra: Department for the Environment and Heritage and Australian Research Institute in Education for Sustainability.

UNESCO. (2005) *United Nations Decade of Education for Sustainable Development (2005–2014): International Implementation Scheme.* United Nations Educational, Scientific and Cultural Organization Document ED/DESD/2005/PI/01.

UNESCO. (2022) *The Concept of Sustainability and Its Contribution towards Quality Transformative Education: Thematic Paper.* United Nations Educational, Scientific and Cultural Organization Document ED/PSD/GCP/2022/02.

Van Lopik, W. (2012) *Keys to Breaking Disciplinary Barriers That Limit Sustainable Development Courses. Higher Education for Sustainability: Cases, Challenges, and Opportunities from across the Curriculum.* Abingdon, Oxon; New York, NY: Routledge, pp. 79–92.

Wals, A. E. and Blewitt, J. (2010) *Third Wave Sustainability in Higher Education: Some (Inter) National Trends and Developments. Sustainability Education: Perspectives and Practice Across Higher Education.* Abingdon, Oxon; New York, NY: Earthscan, pp. 55–74.

PART II

Agendas and standards

11
CHRONICLING EDUCATION HISTORY
Agendas and standards

Diedrich Bruns and Stefanie Hennecke

Part II offers overviews of and presents examples from the early history of education. Within a period of about 150 years, innovative people developed new educational agendas, and universities, professors, and professional associations negotiated terms of qualification requirements. Landscape architecture is part of the formative phase of higher professional education and its educational history, therefore, is not an isolated case, but rather representative of many professions that built university study programmes on foundations of vocational and trade education. At the turn of the 20th century, founding professional programmes proved advantageous for all parties involved. Young universities, recently established in a competitive field of higher education, strove to expand their palette of academic fields. In newly formed disciplines to be recognized as a university field of study was regarded as the ultimate conquest of the highest bastions of social appreciation.

Barbara Birli, Diedrich Bruns and Karsten Jørgensen seek to draw connections between the backgrounds and personal experiences of early landscape design teachers and their educational ideas and standards. The analysis includes the professors who led the first study programmes that European universities established, and members of the IFLA Education Committee who developed a process of school and programme recognition. Sonja Dümpelmann presents a survey of the beginnings of education in the United States of America. Considering the stories of the first universities that established professional programmes, she discusses findings about the need for and the success of schools of landscape architecture for women, and schools where African American students found educational opportunities. Her themes are the courses that agricultural colleges and architecture schools developed, the preference that educators and practitioners had for landscape architecture as the profession's name, and questions of equality and justice.

Karsten Jørgensen, Lars Hopstock, Gert Gröning, Ana Duarte Rodrigues, Luca Csepely-Knorr, and Patricia Debie discuss findings from studies about formative years of higher education. They tell the stories of the *Norges Landbrukshøgskole* in Ås, Norway; the *Landwirtschaftliche Hochschule* in Berlin, Germany; the *Universidade Técnica de Lisboa*, Portugal; Reading University and the Manchester School of Architecture, United Kingdom; and *Wageningen Universiteit*, the Netherlands. They explore the (different) reasons why governments and universities decided to develop study programmes, and the roles, ideas, and identities of professionals and professors who made these programmes come to life.

DOI: 10.4324/9781003212645-13

The seven chapters that follow present European examples of programme development in the 20th century. Their common background is the impact that societal transformation, and the initiatives and influence of individual educators and professional organizations had on changing affiliations and denominations of faculties and study programmes, oscillating between art and design, horticulture and agriculture, architecture and planning. Sophie von Schwerin chronicles the course of landscape architectural education in the German-speaking part of Switzerland and Dunja Richter highlights how the Swiss Federal Institute of Technology Zurich came to establish a graduate programme in the field. For Austria, Ulrike Krippner and Lilli Lička explain how research helps understand long traditions of landscape design education as a continuous process with periods of innovation, success, and setbacks. Providing an overview of education history in Croatia, Hungary, Romania, Slovakia, and the Czech Republic, Petra Pereković and Monika Kamenečki, Albert Fekete, Alexandru Mexi, and Attila Tóth, Ján Supuka, Katarína Kristiánová, Jan Vaněk, Alena Salašová, and Vladimír Sitta illustrate how prominent educators led landscape design education through times of change from monarchy through communism to democratic republic. (Educational developments in Albania, France, Italy, and Poland are included in parts three and four of this compendium.)

The authors of the 15 chapters contributing to the standards-and-agendas section demonstrate how exciting it is to explore and read the contouring of education in the discipline, as it is full of tension and surprises. Parallels become visible across national developments. Five different approaches to comparative analysis are:[1]

1 Classical chronology,
2 Structural models and equality issues,
3 Affiliation of departments and degree programmes (based on determining how far our discipline leans towards the arts or sciences, whether graduates see themselves more as generalists or as specialists),
4 Disciplinary priorities in study programmes and curricula,
5 Qualification of teachers.

Classical chronology

Recognizing four different periods of about 40 to 60 years each, they are the period of higher education institutions since the middle of the nineteenth century, the establishment of the first courses of study at universities in the first half of the 20th century, the phase of the strong growth of such courses in the second half of the 20th century, and the current period of diversification with adaption to regional specificity, further outlined as follows:

1 Beginning in the 1840s, gardener institutes, vocational schools, and agricultural colleges are offering landscape design education.
2 Since 1900, universities are offering degree courses in garden and landscape architecture.
3 Number and size of landscape architecture programmes are growing during the second half the 20th century; specializations develop.
4 Universities are offering education and conducting landscape architecture research around the world.

From the 1840s, prospective garden and landscape designers acquire their professional qualification in a dual system; they take up technical knowledge in practice and theoretical-conceptual

Table 11.1 The first landscape architecture programmes in North America and Europe.

University, city, founding year	Division/programme, starting year	Graduation degree	First director
Harvard, Cambridge, 1636	Architecture/Landscape Architecture, 1900	BS Landscape Architecture	F.L. Olmsted
Mass. Agricultural College, Amherst, 1863	Horticulture/Landscape Gardening, 1903	BS Landscape Gardening	F.A. Waugh
Cornell University, Ithaca, 1865	Agriculture/Landscape Design, 1904	BLA	B. Fleming
University of Illinois, Champaign, 1867	Horticulture/Landscape Gardening, 1907	BS Landscape Gardening	J.C. Blair
Lowthorpe School of Landscape Architecture, Groton, 1901	Landscape Architecture, 1901 (elementary professional course)	Landscape Architecture certificate	E.G. Low
Landwirtschaftliche Hochschule, Berlin, 1881 (1934 included into Berlin University, 1949 renamed Humboldt University)	Gartenkunst/ Gartengestaltung, 1929	*Diplom Gärtner*	E. Barth
Norges Landbrukshøgskole, Ås, 1897	Hagearkitektur/Linje for hagekunst, 1919	Hagearkitekt	O.L. Moen
Wageningen Universiteit, 1918	Tuin- en Landschapsarchitectuur, 1948	Agricultural Engineer	J. Bijhouwer
University of Reading, 1926	Landscape Architecture, 1930	Diploma in Landscape Architecture	A.C. Cobb
Instituto Superior de Agronomia, Lisbon, 1930 (formal approval in 1942)	Curso Livre de Arquitectura Paisagista, 1941	*Diploma de engenheiro agrónomo e arquiteto paisagista*	C. Cabral

Source: Diedrich Bruns.

knowledge at a secondary school, such as vocational schools, trade schools, and other career colleges that award degrees. The number of study opportunities increases in the course of the 19th century, laying the foundations for study programmes which universities established in the second period from 1900 onwards. Universities offer professional programmes initially for garden architecture and increasingly for landscape architecture (see Table 11.1). In the first and second periods, formal curricula and qualification requirements develop and trade, occupational, and vocational training becomes a formal system of higher disciplinary-specific education. Universities and professional organizations are driving this process forward. Universities are expanding their range of educational palette. Professional organizations are in need of young talent with both general and special qualification. Universities and organizations together are strengthening the discipline by building up a body of knowledge and by developing a disciplinary field beyond professional practice. In the third period, the number of educational institutions and students increases with a rapidly growing demand for design professionals. Supported by international communication and cooperation, the structure, content, and minimum requirements of landscape architecture courses of study are further standardized. During the current period, universities are offering education and conducting landscape architecture research in all parts of the world, each with adaptations of general standards to meet regional and cultural specificity, and policies of inclusion and justice, reconciliation and decolonization (see Parts III and IV).

Structural models and equality

Until the middle of the 19th century, aspiring garden artists acquired their professional qualifications with established practitioners. This apprenticeship model remained in place for a long time, even after the establishment of higher education institutions. In some countries, including Japan, the apprenticeship system remains in place until the present (Takatori in this volume). Formal courses were, in many countries, only available to a few applicants who could afford the costs and meet the required qualifications, such as school degree and several years of practice. Moving into the 20th century, in order to generate suitable employees for the increasingly wide range of professional tasks, garden designers such as Froebel and Mertens in Switzerland and urban planners such as Mawson in England, educate the next generation of designers in their own companies (Schwerin, Csepely-Knorr in this volume). The British ILA Education Committee lists this form of continuation of the long-established apprentice-journeyman system in its "Statement of Education" in 1945 and names it as one of three possible educational paths. The other two paths, the committee lists study programmes linked to an "allied profession" (Architecture, Town Planning, Horticulture), either undergraduate or postgraduate, and, their preference, an independent (five-year) landscape architecture study programme.

Eventually, in North American and European education policy, the founding of independent landscape architecture studies is gaining momentum. Three structural models develop. One, the consecutive model with academic degrees and professional recognition separate, as described by the ILA Education Committee for the UK. Two, the dual model with interlinking of separate education and training components, made possible by contracts between educational institutions and practitioners. Three, the dual model with integration of education and training components under the sole responsibility of the university.

In the consecutive model, students must, after achieving academic qualification, provide further proof of practical knowledge before receiving professional recognition (including admission to professional organizations). In the case of dual models, proof of academic and practice achievements provided together are prerequisite for academic and professional qualifications. Examples of the combination of dual and consecutive models exist at higher education institutions from the second half of the 19th century onwards. For instance, after completing his studies at the *Gärtnerlehranstalt* (Gardener Institute) in Potsdam, Erwin Barth underwent further training with garden architects and in municipal administrations before being appointed university professor in Berlin. Education at most universities in Northern, Western, and Central Europe from the turn of the century onwards is based on the dual-integrative model. Academic degree and professional qualification are a prerequisite for taking up professional activity, whereby in some countries any independent professional requires obtaining a state license, which is carried out, for example, in Germany by the professional chamber accredited for this purpose.

In many cases, professional practice is a prerequisite for admission to college and university. Practical components take up large portions in the curriculum and include, for example, construction and maintenance work on campus grounds. In addition, after successfully passing exams, some models require additional periods of practical training before graduates may enter into professional life. Those who do not meet these three conditions could not be admitted or graduated. Women are disadvantaged. In the first decades of education history, they could not even take up studies because they were denied the opportunity to gain practical experience in garden and construction companies. Women who nevertheless made their way into a successful professional life are mostly self-taught. Born into wealthy families, they receive their first commissions through societal connections. Some find access to the profession via detours, such as in Portugal as educators in orphanages or the Fröbel kindergarten, which also integrates horticultural elements (Rodrigues in this volume).

In order to open up better opportunities for women, special universities were created, before and around the turn of the century, first in Germany, but soon also in the USA. In Hexable, England, Swanley Horticultural College was continued in 1903 as a school for women, and, in 1913, the Higher Horticultural Institute for Women was founded in Austria. In Eisgrub (today Lednice, Czech Republic), where the Higher Horticultural Institute offered education for prospective garden architects of the Austro-Hungarian Empire, women were accepted to study from 1911. From 1934 also in Manchester, in 1942 at Harvard.

In the USA, African Americans and Native Americans experienced additional (mainly racist) difficulties to entering professional programmes. The Agricultural College Act of 1890 required states to establish a separate land-grant college for African Americans if they were being excluded from the existing ones. In some cases, for example the Hampton Institute, schools extended their education to Native Americans in the late 19th century, seeking to educate Native Americans so that they would "assimilate". However, establishing special schools for women, African Americans and Native people does not necessarily lead to higher equality, but rather manifests barriers. Equality requires equal opportunities for all to enter schools and the profession.

Until the turn of the 21st century, further educational policy decisions had led to the alignment of course structures and university degrees in Europe, within the framework of the European Qualification Framework, which regulates structures and degrees for universities that are comparable to those of other parts of the world, such as bachelor's, master's and doctorate degrees in particular. Degree programmes are (or are supposed to be) open to all applicants regardless of gender, race, and societal background.

Affiliation of institutes, titles of study programmes, specializations

To governments and educational administrators who thought of the "new" profession as one growing out of gardening, so to speak, it may have appeared a matter of logic when they identified horticultural departments as best suited to accommodate garden art and garden architecture. Thus, most of the early courses had their beginnings in the academic environment of natural-science-based agricultural schools. Not all parties, professional organizations in particular, believed that these were good affiliations. For example, the president of Harvard University, Charles W. Eliot, insisted that landscape architecture was a fine art and belonged to a school of design.

At the turn of the century, a broad spectrum of different views appear in discourses on the question of disciplinary and institutional affiliation. At one end of the spectrum is the view that architecture, itself considered one of the arts, has been responsible for the design of built human environments as a whole since ancient times, so exemplified in Italy where landscape architecture is still seen today as an appendage of architecture. The first university-recognized course of study for landscape architects in England was also set up at the School of Architecture of the Manchester College of Art, scientifically supported by the Botany Department of the University of Manchester. This example of cooperation between artistic-architectural and scientific institutions lies at the intersection of lines of discussion around the question of how far our discipline leans towards the arts or sciences and whether graduates should see themselves more as generalists or as specialists. In Germany, Norway, Poland, Portugal, the Netherlands, and the

USA, governments, ministries, and universities initially assign the new subject to agricultural institutions, with individual protagonists such as Barth in Berlin and Bijhouwer in Wageningen maintaining the connection to architecture and urban planning on their own. It was not until the second half of the 20th century that it was possible to re-establish close institutional links between art, architecture, and landscape architecture in Europe, partly supported by educational reforms, individual universities, for example in art academies in Copenhagen, Denmark, and Kassel, Germany.

When universities established new study programmes it took several decades before all of them decided to adopt "landscape architecture", emphasizing how students learn and graduates understand landscape as their medium of design. Some schools actively pursued establishing the new name from the start, while others adhered to traditional lines of garden art and garden architecture. For example, in Warsaw, Reading, and Lisbon, universities founded programmes that awarded degrees in *Architektury Krajobrazu i Parkoznawstwa* (landscape architecture and park science, est. 1930), landscape architecture (est. 1932), and *Arquitectura Paisagista* (est. 1941), respectively. Mirroring the time of transition, as it were, Berlin and Wageningen universities integrated garden and landscape. Their reasons, however, differed from each other. Berlin renamed its department from *Gartengestaltung* to *Landschafts- und Gartengestaltung* (landscape and garden design) in 1939 because *Landschaft* had become an ideologically charged subject of national importance (Gröning). Wageningen University united garden and landscape in one title, in 1946, as Bijhouwer, the first Dutch Professor of *Tuin en Landschapsarchitectuur* insisted students acquire an understanding of landscape as the decisive factor of urban and infrastructure development. Students initially earned a *Landbouw-Ingenieur* (agricultural engineering) degree, the diploma certificate specifying *Richting Tuinbouw, gespecialiseerd in Tuinarchitectuur* (gardening and garden architecture). Some decades later, the degree title became 'landscape architect'. In the course of time, several schools decided to adopt the new title. For example, Massachusetts Agricultural College introduced 'landscape architecture' in 1930, the University of Michigan in 1939; three decades after Harvard had launched the first programme of that name.

Different structural models and pedagogic approaches to general and specialist education include integrating design and planning in one curriculum, and offering separate educational stands. In landscape architecture, specializations include open-space and landscape planning. F.L. Olmsted and T. Mawson are examples of prominent historic figures who mastered design and planning competence, and for whom it was possible for one person to absorb most relevant sources of information. The growth in specializations and information has been enormous ever since, particularly for open-space planning in relation to town and urban planning, and for landscape planning in relation to countryside, regional, and infrastructure planning.

Curricula and subjects

During the first period at institutes of higher design education, study programmes initially took three and, towards the turn of the century, four years to complete. Curricula focussed on design training, garden design history, construction, and maintenance and included, among other subjects, draughtsmanship and painting, botany, horticulture and plant material, surveying, and ground levelling and grading.[2] As teaching staff was small, each instructor took on several different subjects and classes, and they themselves decided on content. With increasing structural alignment of study programmes, there was a canonization of content, but how far this actually went in view of the extensive autonomy the university teacher then had is difficult to determine. Course titles listed in curricula and syllabus do not easily reveal the true nature of what teachers believe students should learn, and what students really did learn.

From the studies reported on in this volume, all landscape architecture curricula had a design core and a number of design supporting subjects. Students spend a third or more of their time learning different forms of basic and advanced "landscape design", including studying design history and theory. Other important subjects are landscape engineering, also called landscape construction, and graphic and verbal communication. For engineering/construction, study programmes include surveying and topography. Graphic communication includes, among others, perspective and freehand drawing. Subjects that students might spend less time on, unless their university environment mandates otherwise, are general history, natural sciences, plant science and agriculture (and sometimes forestry), and social sciences. For natural sciences that agricultural universities emphasize, curricula include botany, vegetation, plants, and horticulture, chemistry, physics, geology, soil science, meteorology, and hydrology. Town and regional planning gained importance after the turn of the century, with open-space planning often remaining inside of the landscape architecture degree programme, and landscape and countryside planning developing as specializations for which several universities developed graduate studies.

Teachers and teaching qualification

With the establishment of new degree programmes, institutions of higher education face challenges of finding candidates to fill teaching, programme manager, and other academic leadership positions. In the first period, practitioners are responding to growing demands for trained professionals, create courses and take over teaching. Examples are Lenné in Prussia; Fuchs in Belgium; Margiochi in Portugal; Bergstrøm in Norway; and Culley, Hill, and Fleming in the USA. Even in the transition to the second period, experts became teachers who self-trained in leadership and pedagogy, and mostly came from neighbouring disciplines. Examples are the architects Rerrich and Abel in Hungary and Austria, the agronomist Olmsted in the USA, and the engineer Darcel in France (where the combination of architecture and engineering has a long tradition in garden design). Educational pioneers of the second period build on the pedagogic foundations of the first. Some of them, such as Misvær, Barth, and van Zouteveen, had received their educations in the first period and built teaching methods on what they had learned themselves. Teaching proficiency improved from one generation to the next where university study programmes developed from professional and vocational schools of the first period, including Illinois, Ås, Berlin, and Wageningen.

For university instructors to gain teaching competences, initially in-service training was the order of the day and it was largely autodidactic. In order to raise the pedagogical qualification and to align education to the new professional and academic profile, universities began to specifically promote teacher training and award scholarships for it. For example, the Norwegian garden designer Moen received a scholarship to qualify in Potsdam, and then take on a professorship in Ås, Norway. In Portugal, the agronomist Cabral was sent to Berlin on a scholarship to study landscape architecture in order to take up a leading position in the Lisbon City Garden Office. In Wageningen, Bijhouwer, who trained in garden architecture, became the first professor of landscape architecture in the Netherlands.

It was, and in some instances still is, common for members of the teaching staff to spend up to 50 percent of the official working hours outside of the university, particularly where instructors predominantly are practitioners. It was also common for students to work on design tasks that professors brought to school from their private firms, including design competitions. Vice versa, students would work in the ateliers of their professors and partly also on construction sites and in the maintenance of completed projects. In addition, the *training campus* becomes the object of practical activity for teachers and students at many institutions, for example in Norway,

England, and in the USA. In all of these ways, professional requirements arrive in everyday learning without any time delay and are immediately defined as relevant for education, training, and qualification. This is crucial, because the disciplinary range of tasks is constantly expanding. In addition to gardens and parks of different sizes, at the end of the 19th century, projects also included public open space of various uses such as urban squares, cemeteries, zoos; the redesign of canals, streets, and urban fortifications became part of the professional palette, as well as contributing to urban expansion projects.

The examples of Cabral and Bijhouwer mark the moment when, after around 100 years of educational history in Europe, university professors are appointed for the first time to teach landscape architecture with a university degree in their own subject. They are also the ones who begin to attract young scientists in their own field as part of their own research activities. Cabral and Bijhouwer (also the first university professor of the discipline to receive a doctorate in science) set up research institutes at their universities in order to generate disciplinary knowledge independent of practical activities. The further history of education shows how the number of persons and groups who are involved in the development and expansion of discipline-specific knowledge within landscape architecture is increasing, generating knowledge both from practical activity and as scholars through systematic research.

Notes

1 In the introduction to *Landscape Architectural Education* (1973), Gary O. Robinette lists nine approaches: History and chronology, the organizations (including their education committees), the cooperative programmes (such as student exchange, extension and community involvement), the personalities (educators, researchers, administrators), the cycles (growth and development of schools, programmes), the trends (changing opportunities and challenges), the schools (universities, departments, faculties), the periods (pioneering periods, maturing periods), the existing documents, and available literature (records, archives, etc.).

2 For example, the Prussian *Gärtnerlehranstalt* (Gardener Academy) of higher learning in Wildpark-Potsdam offered a four- to five-term (three-year) study programme. The Institute admitted applicants on the condition of them having gained two years of practical experience. Students worked from 6 am (summer term) or 8 am (winter term) to 6 pm, six days per week. First-year subjects included natural science (botany, dendrology, soil sciences, physics, and chemistry), technical subjects (surveying, grading, and mathematics), draughtsmanship and painting, agriculture and horticulture, and business administration. Second-year subjects included "Garden Art" including art history and theory, and design training (three to four different design assignments each term). Year 3 was for students to pursue special scholarly and pedagogic interest, aiming for a teaching career, and for additional practical training. (Land, D. and Wenzel, J., (ed.) (2005) Heimat, Natur und Weltstadt. Leben und Werk des Gartenarchitekten Erwin Barth. Leipzig: Koehler & Amelang: 40, 41).

12
EARLY LANDSCAPE ARCHITECTURAL EDUCATION IN EUROPE

Barbara Birli, Diedrich Bruns, and Karsten Jørgensen (†)

Introduction

Landscape architects today owe no small debt to their education pioneers. It was a progressive group of people who, through the programmes and courses they created, shared a responsibility that was for the very definition and maintenance of a form as close as possible to their vision of a professional realm ranging "from the domestic garden designer. . . to the planner of a whole countryside" (ILA 1973 [1945]). When they negotiated what eventually became the landscape architecture education agenda, agricultural, horticultural and gardening traditions met art, architectural, engineering and planning ambitions. Recognizing two periods, this chapter begins with a chronicle of the first and then focuses on examples from the second period. The first period starts in the nineteenth century when ambitious visionaries began offering education in a dual system combining specialist practice and schooling. The second period begins at the turn of the twentieth century when professional portfolios expanded to include town, recreational, industrial, infrastructural and regional planning. The stories of seven educators serve to illustrate how, building on and adding to first period foundations, a number of innovative individuals are the first in Europe to develop university programmes and examination requirements.

The review draws on chapters compiled in this volume; on biographies on Barth, Jellicoe and Colvin; on documents held at the archives of the International Federation of Landscape Architects (IFLA) and the Landscape Institute (LI); and on studies and the database prepared (from 2006 to 2009) as part of the EU-funded project 'Landscape Education: New Opportunities for Teaching and Research in Europe', LE:NOTRE. Included is information from expert interviews that generated insights on how landscape architecture programmes were established and how education practice initially developed (for details see Birli and Vugule 2007; Birli and Fetzer 2019). When analysing content of curricula and courses offered a hundred years ago and earlier, it is important to consider how titles and terms used then and today differ. For example, some courses called "rural planning" would later be included in the subject of *regional planning*, while "rural engineering", "engineering biology" and "grading and levelling" might today be part of general engineering or landscape architecture technology (Birli 2016, p. 158).

Acknowledgements

We are grateful to Olav Skage, Dušan Ogrin (1929–2019), Meto Vroom (1929–2019), Michael F. Downing (1934–2015), Ralf Gältzer (1931–2007) and Manuel Ribas Piera (1925–2013) who shared thoughts about their own educations and for their encouragement to engage in research investigating the development of landscape architecture education. Conversations with them and with the LE:NOTRE project leader Richard Stiles shed light on how international cooperation between landscape architecture schools evolved (for details see Birli 2016; Stiles in this volume). Studies that led to this chapter reveal how much there still is to learn about educational pioneers. For pointing out valuable sources and for sharing insights about Madeline Agar and Brenda Colvin, we thank Dr Luca Csepely-Knorr, Liverpool School of Architecture; Annabel Downs, chair of the Friends of the Landscape Archive at Reading; and Karen Fitzsimon (personal communication 2021), School of Architecture and Cities, University of Westminster, London. For sharing insights and information on René Latinne, we thank Ursula Wieser Benedetti, Belgian Landscape Architecture Archives, and Jef De Gryse (personal communication 2021), landscape architect, both Brussels.

Examples of early professional education

For professional training in the nineteenth century, practitioners took on articled pupils and apprentices who, during or after several years of practice, enrolled in professional schools to attend design and science courses. A dual education system developed where aspiring designers learned their trade on-the-job and at institutes of higher education (Jørgensen 2005; Holden 2015). Perhaps the earliest higher education in the field that would become landscape architecture was offered at the Prussian *Königliche Gärtnerlehranstalt* (Royal Gardener Academy) in Wildpark-Potsdam near Berlin (Brüsch 2010). This school of higher education was established 1823 under the leadership of Peter Joseph Lenné (1789–1866), a learned gardener who studied architecture in Paris (Hinz 1980, p. 57) and became a pioneer in urban design (Girot 2016, p. 238). Two of Europe's first landscape architecture professors earned their degrees studying 'Garden Art' at this school: Olav Moen who would lead the professional programme in Ås, Norway, and Erwin Barth who became director of the Institute for Garden Art at Berlin University.

In 1849, the *Belgian Tuinbouwschool* (Gardening School) was founded in Vilvoorde near Brussels where, from 1860 onwards, Louis Fuchs (1818–1904) taught architecture and drawing courses. As in Berlin, Vilvoorde students will have learned how their professional realm expanded beyond designing gardens, for, during the 1850s, Fuchs was working on public commissions such as cemeteries, a semi-public zoo, an urban expansion and remodelling former fortifications and historic gardens to be used as city parks (Duquenne 2005). Hendrik François Hartogh Heys van Zouteveen and René Latinne graduated from Vilvoorde in 1894 and 1925 respectively (Goossens personal communication). Van Zouteveen paved the way for Wageningen University to establish landscape architecture education in the Netherlands (Debie in this volume); Latinne became an educator in Vilvoorde and a founding member of the IFLA education committee.

In Portugal, at the Casa Pia in Lisbon, Francisco Simões Margiochi founded the first course of gardening in 1859 and the Froebel school hired women gardeners as educators starting in 1882. Interrupted by political upheavals, educational programmes resumed in 1941 when the *Instituto Superior de Agronomia* opened a programme in landscape architecture education that Francisco Cabral (1908–1992), a graduate from Berlin University, would lead to success (Rodrigues in this volume). In France, during the 1870s, formal gardener education began at the *Jardin des Plantes* in Paris, also involving several gardens in Versailles. Developing from a municipal school for gardeners in 1874 the *École Nationale d'Horticulture* established at the *Potager du Roi* which became the *École Nationale Supérieur d'Horticulture* in 1961 (Blanchon et al. in this volume). During the 1880s, three German states established design courses at institutes of higher learning: Hessen founded the Geisenheim Academy in 1871, Bavaria the Weihenstephan Academy in 1883 and Thuringia the Köstritz Academy in 1887 (Homann and Spithöver 2006, p. 36).

From 1887, the Norwegian gardening pioneer Abel Bergstrøm (1834–1920) became the first head gardener to offer a 550-hour specialisation in design at the *Landbrukshøgskole*, NLH (Norges Landbrukshøgskole 1959, pp. 204–205) (Table 12.1). In 1900, Hans Mikal (Michael) Misvær (1864–1938) continued the development of garden design teaching at NLH, laying the foundations for the 1919 university programme with Moen as professor (Jørgensen in this volume). The Austro-Hungarian monarchy founded the *Höhere Obst- und Gartenbauschule* (Higher Fruit and Horticulture School) in 1895, located in Eisgrub, today Lednice, Czech Republic (Krippner and Lička in this volume). In 1896, the Netherlands had founded the *Rijkstuinbouwschool* (National Gardening School) at the *Landbouwhogeschool* (Higher Agricultural School) in Wageningen, which developed the country's first course in Garden Art (van Buuren 2018). From 1897, Leonard Anthony Springer (1855–1940) led that course; he was a garden designer, always keen to distinguish himself from being a nurseryman. His successor from 1900 was the aforementioned Van Zouteveen (1870–1945). Jan Bijhouwer, one of his students, was the first landscape architecture professor of Wageningen (Debie in this volume). In Britain, in 1881, Edward Milner (1819–1884) became principle of the newly formed British *Crystal Palace School of Gardening* in Sydenham. He was, in 1858, the founder of his own design firm (Craddock 2012).

In North America, similar to developments in Europe, during the latter part of the nineteenth century, ambitious people played important educational roles. Key individuals such as Andrew Jackson Downing (1815–1852) and Frederick Law Olmsted (1822–1903) were instrumental in establishing landscape architecture courses. The earliest school to do so was Harvard University in 1900, and the land-grant agricultural colleges Cornell and Illinois, which established professional degree programmes in 1904 and 1907 respectively. In Illinois, the university programme was founded on a earlier course from 1867 called 'Landscape Gardening' that became 'Garden

Table 12.1 The first six landscape architecture programmes and educators at European universities (after Birli 2016, amended).

Year	University, location	Programme	Educators
1919	Norges Landbrukshøgskole, Ås, Norway	Hagearkitektur, Linje for hagekunst	Olav L. Moen
1929	Landwirtschaftliche Hochschule, Berlin, Germany	Gartengestaltung	Erwin Barth
1930–32	University of Reading, United Kingdom	Landscape Architecture (Diploma course)	Arthur J. Cobb Geoffrey Jellicoe
1934	Manchester School of Art and University of Manchester, United Kingdom	Landscape Architecture	E. P. Mawson Thomas Adams Geoffrey Jellicoe and others
1941	Instituto Superior de Agronomia, Lisbon, Portugal	Cursol livre de arquitectura paisagista	Caldeira Cabral
1946–48	Wageningen Universiteit, The Netherlands	Tuin- en Landschapsarchitectuur	Jan Bijhouwer

Architecture' in 1871 and 'Landscape Architecture' in 1912 (Robinette 1973, vol. 1, p. 87; Steiner and Brooks 1986, p. 27). Several universities followed suit and established professional programmes during the first decades of the twentieth century (Dümpelmann in this volume).

Up to the turn of the century, institutions of higher education did usually not accept women. Schools for women opened instead to give female students the chance to get a landscape design degree. The first of the series of private gardening schools that include design courses in their curricula opened 1889 in Charlottenburg (Berlin), the next was founded by Dr Elvira Castner 1894 in Friedenau/Marienfelde, also Berlin. By 1908, five gardening schools for women operated in Germany (Homann and Spithöver 2006, p. 41). In the USA, women had the opportunity to attend the Lowthorpe School of Landscape Architecture for Women founded in 1901 in Groton, Massachusetts; the Pennsylvania School of Horticulture for Women begun in 1910; and the Cambridge School of Architecture and Landscape Architecture, founded by Henry Atherton Frost in Cambridge, Massachusetts, in 1915 (Dümpelmann in this volume). In Britain, Swanley Horticultural College became a women-only institution in 1903. By 1918, Madeline Agar (1874–1967) was offering the first courses in garden design; in 1911, she published *Garden Design in Theory and Practice* (Fitzsimon 2018). In Vienna, Austria, Yella Hertzka founded, in 1913, the *Höhere Gartenbauschule für Frauen*, an advanced horticultural school for women (Krippner and Lička in this volume).

Early European university education

The first European universities that instituted landscape architecture degree programmes from agricultural and horticultural roots are the Norwegian School of Higher Agricultural Education in Ås, south of Oslo (1919), the School of Higher Agricultural Education in Berlin (1929), the Institute of Higher Agricultural Education in Lisbon (1941), and Wageningen University (in 1946, formal approval 1948). The earliest landscape architectural education in Britain is affiliated with architecture and planning,[1] the first founded in Reading (1930, formal approval 1932), others in Manchester (1934) (Birli 2016, p. 154), in Newcastle-upon-Tyne (1949/50) and (1949) at University College, London (for details see Holden 2016, p. 16).

In 1919 and 1929, Olav Leif Moen and Erwin Barth took the lead in creating university courses. As young men, both got accepted at the *Gärtnerlehranstalt*, earning their professional degrees in 'Garden Art', Barth in 1902 and Moen in 1921. They excelled in design studies and draughtsmanship, acquired knowledge in sciences, particularly botany, and in planting design. In Norway, it was determined from the start that no other but the Agricultural School would be offering the nation's first garden architecture programme and that Olav Leif Moen (1887–1951) would lead it. The school had given him a scholarship to qualify as university professor. From 1918 to 1920, Moen taught landscape gardening at the *Statens Gartnerskole* (National Gardening School). NHL appointed him as docent in 1921 to teach garden art in the professional programme *Hagearkitektur* (Garden Architecture) instituted in 1919. In 1939, Moen was officially appointed as 'Professor of Garden Art' (Jørgensen 2011; Jørgensen in this volume).

In 1929, the Prussian parliament selected the Higher Agricultural School (1934 included into Berlin University, 1949 renamed Humboldt University) that hired Erwin Albert Barth (1880–1933) as professor for *Gartenkunst* (Garden Art) to lead the new programme of *Gartengestaltung* (Garden Design) (see Hopstock in this volume). Barth, after passing the head gardener examination in 1906, had gained prominence among the growing number of directors of city garden departments who, as a group, became a formidable political voice that claimed all urban open space as areas of their professional responsibility. Through their organisation these department heads defined and maintained what today is known as open space planning (Land and

Wenzel 2005). Before starting his full professorship in 1929, Barth's first teaching appointment was in 1920. He lectured two hours per week at the Berlin Technical College. Because of his preference for the affiliation with architecture and engineering, he continued this lectureship, offering additional weekend field trips for the "most promising students" (ibid., p. 426).

During the 1940s, Francisco Caldeira Cabral and Jan Tijs Pieter Bijhouwer developed landscape architecture education in Portugal and the Netherlands. Each had earned a university degree in garden architecture with garden architects as teachers and mentors. In Portugal, Cabral is the first in Europe who was appointed landscape architecture professor with that title (Andresen 2001). From 1931 to 1936, he had studied agricultural engineering at the *Instituto Superior de Agronomia* (Higher Agronomy Institute) in Lisbon. He then enrolled in the programme of Berlin University earning his professional degree in 1939. In 1941, Cabral started the *Cursol livre de arquitectura paisagista* at the *Instituto Superior de Agronomia*, ISA. Building upon his university education and resources of ISA, he not only created a full professional programme; he also began to develop research in landscape architecture as early as 1953 (Rodrigues in this volume). Together with Latinne, he was one of the first members of the IFLA Education Committee (discussed later in the chapter) and, from 1962 to 1966, he served as the fifth IFLA president.

In the Netherlands, Wageningen was given university status in 1918 and, in 1930, started to offer 'Garden Art, Garden Architecture and Horticultural Drawing', which became official exam subjects in 1937. In 1936, Dr Jan Tijs Pieter Bijhouwer (1898–1974) had succeeded his former teacher Van Zouteveen as instructor and generated, in 1946 (formal approval 1948), the country's first full degree programme called *Tuin- en Landschapsarchitectuur* (Garden and Landscape Architecture) (Debie in this volume). For his education, Bijhouwer had worked at several nurseries before he enrolled in horticulture in Wageningen where he became interested in garden art. Insisting, as Moen, Barth and Cabral had done, that the garden and landscape architect should refrain from developing horticultural ambitions, he went one step further and specialised in vegetation studies (Bijhouwer 1926), thus laying the foundations for a science-based approach to design. From 1927 to 1929, he worked as a researcher and instructor of garden design at the Rhode Island School of Design in Providence, USA.[2] From 1948 to 1968, he was, in addition to his position as full professor at Wageningen, extraordinary professor of Landscape Science at the urban development department of the Delft University of Technology lecturing to students of architecture and town planning (Andela 2011, p. 245).

In the United Kingdom, in 1929, the year Erwin Barth took his position in Berlin, the British Association of Garden Architects was founded, changing the name to Institute of Landscape Architects, ILA, in 1930.[3] Three years later, the Institute presented a first draft of a course of study and examination (Downing 1992, p. 72). Building on this draft, an Education Committee that ILA formed in 1938 produced a 'Tentative Syllabus of Education' in 1941, of a five-year course leading to a degree in landscape architecture (ILA 1973 [1945]).[4] Seeing the "immediate need" to form a "school of landscape", and "in order to ideally fulfil the syllabus", the Committee recommends this school to be the combined effort of schools of architecture and town planning, and of horticulture, agriculture, and forestry, with civil engineering being "advantageous" (ILA 1973 [1945], p. 207).[5] In the early 1930s, the University of Reading had started a three-year Diploma course. The emphasis was on horticulture and botany, the latter considered the "basic discipline" (Jacques and Woudstra 2009, p. 245). Initially, the senior lecturer was Arthur J. Cobb, a horticulturist and author of the popular volumes of *Modern Garden Craft – A Guide to the Best Horticultural Practice* (1936). The course was not considered suitable to rank for a university degree and, in addition, Cobb, leaning towards horticulture and aiming to train 'garden designers', did not meet with ILA objectives of broadening the professional palette.

Matters changed in 1934 when Geoffrey Jellicoe (1900–1996), one of the Institute's Education Committee members, became sessional lecturer, and courses on the history of garden art began (Holden 2015, p. 1). In the same year, the Manchester School of Art launched, at its School of Architecture and together with the University of Manchester, a landscape architecture course at undergraduate level. That programme ILA recognizes in 1944. To show their support, ILA members, including E.P. Mawson, T. Adams, J. Jellicoe and others, gave lectures (Csepely-Knorr in this volume).

Mirroring the boundary-crossing ambitions of Barth and Bijhouwer, who, as garden and landscape architects, were getting increasingly interested in architecture and planning, Sir Geoffrey Allan Jellicoe was an architect and architectural historian who became an expert in garden and landscape architecture (Harvey 1998). As a young man he had worked with Parker and Unwin, who were involved in the design of gardens and garden cities (Roe 2007). From 1919, he studied to become an architect at the Architectural Association in London and, in 1923, began research for his first book *Italian Gardens of the Renaissance* (published with John C. Shepherd in 1926). Jellicoe continued to become one of the most influential thinkers and writers of the post-war period in Britain. With his wife Susan, he published *The Landscape of Man: Shaping the Environment from Prehistory to the Present Day*, a textbook classic (Jellicoe and Jellicoe 1975). As an architect, he was not knowledgeable about plants – Susan had the planting design expertise (Fitzsimon 2018, p. 53). Jellicoe continued teaching, including at the University of Greenwich where, from 1979 to 1989, he came as a lecturer and visiting critic. In 1929, he was a founding member of ILA (from 1939 to 1949 it's president) and, in 1948, founding president of IFLA (until 1954 it's President of Honour).

In summary, the first European universities that instituted landscape architecture programmes built them on nineteenth-century foundations, agricultural and horticultural ones, and art and architectural ones. Seven examples illustrate how innovative individuals were the first in Europe to develop curricula and a teaching practice that succeeded in meeting the broadening professional capabilities expected of the garden and landscape designer. The challenge for the profession was to define a common identity, setting it apart from horticulture, architecture and town planning, and for educators to find a balance between art and science.

The beginnings of programme recognition

Research into the travel history of garden artists show how they frequently undertook educational trips (Fischer and Wolschke-Bulmahn 2016). To facilitate the formal exchange of ideas across borders, for strengthening professional identity and discussing issues of common concern, including educational standards, a group of dedicated people started to organise international fora. Conferences were held first in Brussels (1935), then in Paris (1937), Berlin (1938) and Zurich (1939) (Imbert 2007, pp. 6 and 7). The legacy of the interbellum year's activities remains the International Federation of Landscape Architects (IFLA) that representatives from 15 national professional organizations founded in 1948.[6] The formation of IFLA took place in the context of a London conference tour to Cambridge (Anagnostopoulos et al. 2000, p. 6). Lady Allen of Hurtwood, a graduate from the University College, Reading, had proposed the conference in 1946 (Dümpelmann 2015, p. 15).

Educational standards were on the agenda of all conferences since 1935. In 1959, IFLA established the Committee on Education (Robinette 1973, Vol. 2, pp. 211, 212). Next to Zwi Miller from Israel and Hubert B. Owens from the USA, the committee consisted of three European delegates, the Lisbon professor Francisco Cabral, Brenda Colvin from the UK and René Latinne from Vilvoorde representing Belgium. Like Cabral, Latinne and Colvin made important contributions to elevating the application of educational standards to reach international visibility. Through times spent abroad, they had considerable international exposure. Both worked in centres (Brussels, London) that had become the nexus of ideas: drawing on concepts created in Europe and North America, forging together professional politics and education, looking ahead to broadening and at the same time defining their disciplinary field.

For more than 15 years René Latinne (1907–2003) represented Belgium in IFLA and participated in many congresses and meetings worldwide. He studied garden design in Gent and Vilvoorde where he earned his *Diploma van Tuinbouwkundige* in 1925 (mentioned earlier). Latinne was appointed in 1957 to teach landscape architecture at his alma mater Vilvoorde and continued to work in this capacity until 1974. In practice, he designed more than 800 gardens and parks. As a member of the Committee on Education, Latinne will have stressed how professional designers form an understanding of the build environment as landscape where local setting, natural conditions and regional features play equally important roles (Wieser Benedetti and De Gryse, personal communication 2021). As much as Latinne's suggestions where aesthetically motivated, they will have fallen on fertile ground at a time when landscape architects pioneered design approaches that strove to integrate human comfort and natural environments.

Brenda Colvin studied garden design at Swanley Horticultural College, 1919–1921; set up her own practice in 1922; designed many gardens and estates, school grounds, university campus, cemeteries, also industrial landscapes, e.g. around power stations; and she published several books, including *Land and Landscape* in 1947 (Gibson 2011). When Colvin, together with Agar, Jellicoe and others, co-founded the Institute of Landscape Architects, she was a driving force in defining educational requirements. It seems, often Colvin first came up with the ideas (Annabel Downs, personal communication 2021). Colvin had travelled in the USA in 1931 and what she saw hugely influenced her later thinking (Gibson 2011, p. 35).[7]

When presenting their syllabus during the 1930s and 40s (mentioned earlier), ILA did so not only for the purpose of representing professional interest but also to oversee the legitimate practice of the occupation, thus building and maintaining a powerful position as a controlling body. A procedure was established that Colvin and others suggested for IFLA to adopt, namely to take on involving itself in recognizing schools, degrees and defining requirements for examining the skills and competencies necessary for a person to practice. The procedure resembled that of the American Society of Landscape Architects (ASLA) that their Committee on Education (appointed in 1909) had started to use 1938 in school recognition (Robinette 1973, vol. 2, pp. 17,18).

IFLA's Commission on Education recommended the adoption by IFLA of the ASLA process, and that, following North American practice, relevant data should be collected from all schools on a voluntary basis (Anagnostopoulos et al. 2000, p. 22; Williams 2013, p. 461). Soon thereafter, IFLA became engaged with the registration, inspection and recognition of schools, their curricula and their degrees. IFLA regularly publishes reports on the status of education in schools of all member countries. Little did they know that, half a century later, there would be a European Community directive on the recognition of professional qualifications requiring universities and colleges to organize education and award degrees that are compatible in level and content (Vroom 1994, p. 113).

Curricula

By the end of the nineteenth century, a growing number of schools and colleges offered specialist training and, through exchanges of teachers and ideas, their curricula had become similar, beginning to form educational standards (Homann and Spithöver 2006, p. 36). Nineteenth-century curricula included several subjects that would become pillars of landscape architectural education: design training, design history, construction and maintenance, draughtsmanship and painting, botany, horticulture, and plant material, surveying, ground levelling and grading. As teaching staff was small in number, each instructor took on several different subjects and classes.

Curricular structure, content and character expressed the conception of landscape architecture as a discipline that brings together arts and sciences. Curricula and syllabus of the first six university programmes feature teaching units that include, in addition to (studio) training in landscape design, a set of formal courses (lectures, seminars, excursions) that cover several areas of knowledge (Birli 2016, p. 158 and table 5). The professors instructed students on design, and on the history of landscape and of art, particularly of garden art, architecture and landscape painting. In addition, landscape schools would collaborate with art and architecture institutes, or hire specialists as teachers, to offer classes in drawing, sketching, sculpture and model making (Gieseke 2006; Nothhelfer 2008; Jørgensen 2011). Knowledge of plants is fostered at every stage of the programme. For students to excel in planting design and learn about 'plant material' for design was considered the main distinguishing quality of the profession. Making drawings (art) and ecological studies (science), during visits to a wide range of different gardens and landscapes, would help students understand how to use plants in designs and gain knowledge of a plant's aesthetic and environmental properties.[8] For science, particularly at schools of agriculture, taking courses in natural sciences and introductions to horticulture, forestry and horticulture was mandatory. Where landscape architecture is affiliated with technical and architectural schools these would align themselves with horticultural departments where landscape students went to attend classes in subjects such as geology, hydrology, soil sciences, and chemistry, meteorology, geography, and botany and vegetation studies. In the areas of engineering and technology, students would acquire knowledge of site and landscape analysis, learn about topography and ground modelling, prepare working drawings for landscape construction, draining and water management, and develop proficiency in technical specifications and estimating, including costs, and in landscape maintenance and management (ILA 1973 [1945], p. 207).

There were differences, of course, in both structure and content. At Wageningen and Manchester, for example, town planning had a particularly strong part. At Wageningen the curriculum included sociology (Birli 2016, p. 110), a field that ILA introduces for the third year of their syllabus. In Reading, bookkeeping was compulsory (Downing 1992). In Lisbon, students took general mathematics and microbiology (Andresen 2001). Design and landscape history teachers would acquaint students not only with the canonical textbook knowledge of styles from antiquity to the present but also discuss how landscapes are regionally specific (Deming in this volume). There are not only design styles to know, such as 'Italian', 'French' and 'Dutch', but also 'Nordic', 'Mediterranean' and other vernacular cultural landscapes. According to the ILA syllabus, for example, students study "the landscape of England, in so far as it is the result of human activity" (ILA 1973 [1945], p. 207).

For the purpose of school recognition, the ASLA education committee created a formal structure called Minimum Standards for Landscape Architectural Education Programs (Robinette 1973, Vol 2, p. 18). The Standards were first published in 1928 and updated and amended several times thereafter. By the time IFLA began to engage in school recognition, the Standards served as guidance. They define eight "Groups of Studies", each with several subjects: (1) Landscape Architecture, including Landscape Design, Theory, Professional Practice, City and Regional Planning, and History of Landscape Architecture; (2) Architecture, including Architectural Design, History of Architecture and Elements of Architecture; (3) Engineering and Ground Forms, including Mathematics, Surveying, Mechanics of Physics and Landscape Construction; (4) Plant Materials, including Botany, Plant Identification, Ecology and Planting

Design; (5) General Arts, including History; (6) Graphic Expression, including Freehand Drawing, Engineering Drawing, Painting and Basic Design; (7) Verbal Expression, including English, Public Speaking and Modern Language; and (8) Social Science (Robinette 1973, Vol 2, p. 18, 43, 67).

In reviewing examples presented in this chapter, and from correspondence with educators (see acknowledgements and personal communications), the studio and project format is becoming the standard medium to facilitating landscape architectural thinking and for students to become self-regulated learners. Students work on a series of design assignments under the encouragement and critical supervision of their professors and invited critics. To demonstrate how gardens and landscapes be designed to meet aesthetic aims and functional needs, instructors who were also practising landscape architects used their own projects as examples in addition to information gained from reading books and journals. Assignment complexity increases with each term and year of study, starting, for example, with small private and public gardens in year one, playground or estate grounds in year two, a recreation area and town planning scheme in year three and so forth. After successfully completing all courses and assignments, students deliver a major design project and pass oral and written exams. During excursions, students would be visiting design projects and studying plant communities, landscape character and the ways people use and engage with landscape.

Conclusions

Working hard through trying times of nation building and wars, the Spanish Flue Pandemic and the World Economic Crisis, a handful of energetic people succeeded in becoming the first tutors, lecturers and teachers in landscape architecture in several nations. Within a span of about a hundred years, between the 1840s and 1940s, starting with a dual system of pupillage, apprenticeship and schooling and arriving at the formulation of internationally recognised standards and procedures of programme recognition, education pioneers created curricula and courses that met the needs of a dynamically expanding disciplinary field. Where differences in opinion remained, for example between representatives of architectural and horticultural traditions, on the type of education that landscape design practice requires, disputes about the question of specialization "would occupy members of professional organizations for decades" (Imbert 2009, p. 21). A balance between specialist training and generalist education was found where graduates could choose, for their own professional direction, between several areas of specialisation, such as the planning and supervision of work in regard to all kinds of open space, of city and region, of infrastructure projects and state development planning, and as teacher and researcher of landscape architecture.

Since no pedagogic education was mandatory or even offered, the newly appointed professors had only their own experience to develop teaching approaches. The novice teachers also had little guidance when universities asked them to define requirements that a student must complete in order to be awarded a degree. In all of the tasks given to them, the first landscape architecture professors were explorers and pathfinders, including when they set out to design curricula, when they made decisions on learning goals, on types and duration of interaction between students and instructors, on course content, materials and resources, and on processes for evaluating the attainment of educational objectives.

To date, we know little about the day-to-day adventures of "learning how to teach", of building up courses for a "new" profession, about the challenges the first professors encountered at "old" horticultural schools and at institutes where an established elite was guarding architectural realms and traditions of the Art Academy. Archival material and published records do not usually cover realities of the power struggle between politicians, administrators and educators, the rivalry (and fighting) between teachers of different courses, the teacher-student interaction. Interpreting biographical dates provides some hints that, creating new curricula, syllabus and pedagogic approach was the work of visionary thinkers who opened up new areas of thought, forging ahead of practice in recognizing opportunities for societal relevance, preparing graduates to think beyond limited roles of the garden and estate planner (Robinette 1973, vol. 1, p. xi).

Notes

1 Landscape architecture in the UK continued to be affiliated primarily with architecture, in some cases with design and environmental sciences. Only Sheffield has an independent landscape architecture department (Woudstra 2010).
2 Eighteen years before that school's architecture division officially included landscape architecture education.
3 The Landscape Institute (formerly Institute of Landscape Architects) archives are held at MERL (Museum of English Rural Life), University of Reading. www.reading.ac.uk/merl/collections/Archives_A_to_Z/merl-SR_LI.aspx
4 Committee members: Gilbert H. Jenkins (chair), Joseph Addison, Madeline Agar, Brenda Colvin, W. G. Holford, Thomas Sharp and Geoffrey Jellicoe. Ideas for developing the ILA syllabus stem from committee members' own experience at home and from studying curricula at universities in the USA, including Harvard, Berkeley and Pennsylvania.
5 The 1946 syllabus for examination are available from Landscape Institute Archives, archival reference for this document: MERL LI SR LI AD 1/3/1 Papers of the Education Committee (kindly provided by Luca Csepely-Knorr).
6 National representatives who attended two or more of these meetings are Ulla Bodorff (Sweden), Jean Canneel-Claes and René Pechère (both Belgium), Achille Duchêne and Ferdinand Duprat (both France), Walter Leder (Switzerland), Hermann Mattern, Alwin Seifert and Otto Valentien (all three Germany), Maria Teresa Parpagliolo, née Shephard, and Pietro Porcinai (both Italy), and C. Th. Sørensen (Denmark). After the war, German representatives were not invited.
7 Madeline Agar, Geoffrey Jellicoe and several of Colvin's colleagues such as Jaqueline Tyrwhitt and Sylvia Crowe all had connections to the profession in North America (Personal Communication, email 23.07.2021 Karen Fitzsimon).
8 While Moen and Barth used plants to create geometrical and stylized natural forms, Bijhouwer (and his mentor Zouteveen) developed strong natural planting schemes. (The art of natural plant use continued to flourish and evolved to become the ecological planting schemes approach appreciated today for their high degree of sustainability.)

References

Anagnostopoulos, G. L., Dorn, H., Downing, M. F. and Rodel, H. (2000) *IFLA: Past, Present, Future.* Versailles: International Federation of Landscape Architects [Online]. Available at: https://issuu.com/ifla_publications/docs/ifla_green_book (Accessed 8 March 2022).
Andela, G. (2011) *J.T.P. Bijhouwer. Grensverleggend Landschapsarchitect.* Rotterdam: Uitgeverij.
Andresen, T. (2001) *Francisco Caldeira Cabral.* LDT monograph 3. Reigate: Landscape Design Trust.
Bijhouwer, J. T. P. (1926) *Geobotanische studie van de Berger duinen.* Deventer: Ijsel.
Birli, B. (2016) *From Professional Training to Academic Discipline. The Role of International Cooperation in the Development of Landscape Architecture at Higher Education Institutions in Europe.* Unpublished Doctoral dissertation, Technische Universität Wien, Vienna. https://doi.org/10.34726/hss.2016.38887.
Birli, B. and Fetzer, E. (2019) '100 Years of Landscape Architecture Education in Europe: Lessons from the Early Pioneers for Our Visions of the Future', *4D Journal of Landscape Architecture and Garten Art*, 53, pp. 22–27.
Birli, B. and Vugule, K. (2007) *"Rare Knowledge" from the Modernist Period of Landscape Architecture Education.* Final Report LE:NOTRE 2 Project. Vienna/Jelgava.
Brüsch, B. (2010) *Genealogie einer Lehranstalt: Von der gartenmäßigen Nutzung des Landes zur Gründung der Königlichen Gärtnerlehranstalt.* Berlin: Peter Lang Verlag.

Craddock, J. P. (2012) *Paxton's Protege, the Milner White Landscape Gardening Dynasty.* Stevenage: York Publishing Service.

Downing, M. (1992) 'Entry "United Kingdom" in European Foundation for Landscape', in Architecture (ed.) *Teaching Landscape Architecture in Europe.* Brussels: ELFA and Erasmus Bureau, pp. 71–76 [Online]. Available at: https://issuu.com/ifla_publications/docs/efla_blue_book_1992 (Accessed 8 March 2022).

Dümpelmann, S. (2015) *Women, Modernity, and Landscape Architecture.* London: Routledge.

Duquenne, X. (2005) 'Entry, Fuchs, Ludwig', in Beyer, A., Savoy, B. and Tegethoff, W. (eds.) *Allgemeines Künstlerlexikon: Die bildenden Künstler aller Zeiten und Völker*, vol. 46. München: De Gruyter, pp. 61–62.

Fischer, H. and Wolschke-Bulmahn, J. (2016) *Travels and Gardens*, symposium report. Hannover: Druckerei Hartmann.

Fitzsimon, K. (2018) 'Nine Decades, Nine Inspiring Women in Landscape Architecture', *Landscape Journal*, 6(1), pp. 51–56 [Online]. Available at: https://illmanyoungcom.files.wordpress.com/2018/06/landscape-journal-issue-1.pdf (Accessed 8 March 2022).

Gibson, T. (2011) *Brenda Colvin: A Career in Landscape.* London: Frances Lincoln Ltd.

Girot, C. (2016) *The Course of Landscape Architecture: A History of Our Designs on the Natural World, from Prehistory to the Present.* London: Thames and Hudson.

Gieseke, U. (2006) *Wieck, K. Perspektive Landschaft.* Berlin: Institut für Landschaftsarchitektur und Umweltplanung, Technische Universität Berlin, pp. 54, 55, 64.

Harvey, S. (1998) *Geoffrey Jellicoe.* Reigate: LDT Monographs.

Hinz, G. (1980) 'Peter Joseph Lenné, Landscape Architect and Urban Planner', *Landscape Planning*, 7(1), pp. 57–73.

Holden, R. (2015) 'The Rise, Decline and Fall of Landscape Architecture Education in the United Kingdom: A Historical Outline', *Tartu: Proceedings of ECLAS Conference.*

Holden, R. (2016) 'UK Landscape Architecture Education: A Terminal Case?' *EAR Edinburgh Architectural Research Journal*, 34, pp. 7–30.

Homann, K. and Spithöver, M. (2006) *Bedeutung und Arbeitsfelder von Freiraum- und Landschaftsplanerinnen. Von der Professionalisierung seit der Jahrhundertwende bis 1970.* Kassel: Infosystem Planung, Universität Kassel.

ILA Education Committee. (1973) [1945] 'Training in England for Landscape Architecture', in Robinette, G. O. (ed.) *Landscape Architectural Education*, vol. 2. Dubuque, IA: Kendall/Hunt Publishing Company, pp. 206–208.

Imbert, D. (2007) 'Landscape Architects of the World Unite! Professional Organizations, Practice, and Politics, 1935–1948', *JoLA Journal of Landscape Architecture*, 2(1), pp. 6–19.

Imbert, D. (2009) *Between Garden and City. Jean Canneel-Claes and Landscape Modernism.* Pittsburgh: University of Pittsburgh Press.

Jacques, D. and Woudstra, J. (2009) *Landscape Modernism Renounced: The Career of Christopher Tunnard (1910–1979).* Abington: Routledge.

Jellicoe, G. and Jellicoe, S. (1975) *The Landscape of Man: Shaping the Environment from Prehistory to the Present Day.* New York: Viking Press.

Jørgensen, K. (2005) 'Two Centuries of International Influence on Norwegian Landscape Architecture', *Byggekunst, the Norwegian Review of Architecture*, 42–43.

Jørgensen, K. (2011) 'Landscape Architecture in Norway: A Playful Adaption to a Sturdy Nature', *Landscape Architecture China*, 4, pp. 34–43.

Land, D. and Wenzel, J. (ed.) (2005) *Heimat, Natur und Weltstadt. Leben und Werk des Gartenarchitekten Erwin Barth.* Leipzig: Koehler & Amelang.

Norges Landbrukshøgskole (1959) *1859–1959*, Oslo: Grøndahl & Søns Boktrykkeri, *(204–205)* [Online]. Available at: https://nbl.snl.no/Abel_Bergstr%C3%B8m (Accessed 8 March 2022).

Nothhelfer, U. (2008) *Landschaftsarchitekturausbildung – zwischen Topos und topologischem Denken.* Doctoral Dissertation, University of Kaiserslautern, Der Andere Verlag, Lübeck, pp. 46–47.

Robinette, G. O. (1973) *Landscape Architectural Education*, Vol. 1 and 2. Dubuque, IA: Kendall/Hunt Publishing Company.

Roe, M. (2007) 'British Landscape Architecture: History and Education'. *Urban Space Design*, 20(5), pp. 109–117.

Steiner, F. R. and Brooks, K. R. (1986) 'Agricultural Education and Landscape Architecture', *Landscape Journal*, 5(1), pp. 19–32.

Van Buuren, M. (2018) '100 Years of Landscape Architecture in Wageningen?', *TOPOS Online* [Online], Wageningen University. Available at: www.toposonline.nl/2018/100-years-of-landscape-architecture-in-wageningen-pt-1/ (Accessed 8 March 2022).

Vroom, M. (1994) 'Landscape Architecture and Landscape Planning in Europe: Developments in Education and the Need for a Theoretical Basis', *Landscape and Urban Planning*, 30, pp. 113–120.

Williams, T. (2013) 'Some Thoughts on the Education and Training of Landscape Architects', in Newman, C., Nussaume, Y. and Pedroli, B. (eds.) *Proceedings of the UNISCAPE Conference 'Landscape and Imagination: To-wards a New Baseline for Education in a Changing World'*. Ponteder: Bandecchi & Vivaldi, pp. 521–528.

Woudstra, J. (2010) 'The "Sheffield Method" and the First Department of Landscape Architecture in Great Britain', *Garden History*, 38(2), pp. 242–266.

Personal Communications

Annabel Downs, chair of the Friends of the Landscape Archive at Reading, Email communication 23 July 2021.

Jef De Gryse, landscape architect, Brussels, Email communication 8 September – 28 September 2021.

Karen Fitzsimon, School of Architecture and Cities, University of Westminster, London, Email communication 23 July 2021.

Ursula Wieser Benedetti, Belgian Landscape Architecture Archives, Brussels, Email communication 23 July – 8 September 2021.

13
LANDSCAPE GARDENING, OUTDOOR ART, AND LANDSCAPE ARCHITECTURE

The beginning of landscape architecture education in the United States, 1862–1920

Sonja Dümpelmann

This chapter surveys the early years of landscape architecture education in the United States of America, focusing on three characterizing themes. First, it discusses the evolution of courses in landscape gardening and their association with nineteenth-century land-grant colleges versus the development of courses in landscape architecture in architecture schools. Second, it shows how around the turn of the century the formation of the first professional landscape architecture programme at Harvard University's architecture school not only gave expression to the nation's increasing urbanization but also went hand in hand with the foundation of the American Society of Landscape Architects in 1899 and many founding members' preference for "landscape architecture" as the new profession's name. Third, the chapter addresses African and Native American, as well as early female education in horticulture, landscape gardening, design, and architecture that largely occurred in separate schools founded for this explicit purpose after the Civil War and beginning in the early twentieth century.

Education for designing the outdoors

Today's landscape architecture programmes at universities in the United States have both rural and urban roots. European settler colonialism led to the foundation of the first agricultural and horticultural societies in the late eighteenth century with the objective of providing fora for the exchange of experiences and knowledge, that is for the education of farmers who could sustain the settlers' livelihoods. The first professorship in agriculture was established in 1792 at Columbia University in New York City, and when the first educational institution devoted to agriculture opened in Maine in 1822 (Gardiner Lyceum), it "offered courses in surveying and architecture in addition to those in farming and the sciences" (Steiner and Brooks 1986, p. 20). Frederick Law Olmsted who would by the end of the nineteenth century become known as the country's foremost landscape architect, began his self-studies in agriculture in 1845 and for a short while in the late 1840s ran a farm on Staten Island (Olmsted and Kimball 1922, pp. 72–83; McLaughlin and Beveridge 1977, pp. 282–335). In 1890 he reflected that "although for forty years I have had no time to give to agricultural affairs, I still feel myself to belong to the farming community and that all else that I am has grown from the agricultural trunk" (Olmsted cit. in Olmsted and Kimball 1922, pp. 72–73). Olmsted was not alone in this feeling and career path. Landscape architect Horace William Shaler Cleveland had similarly begun his career as a farmer, and their revered acquaintance Andrew Jackson Downing had argued for education in agriculture and was as known for his work in horticulture and the nursery business as for his designs as a landscape gardener.

While the first agricultural colleges were founded during the antebellum years, after the American Civil War in the 1860s, the first courses in what at the time was still called landscape gardening, were taught at the newly founded land-grant colleges. Considered a part of horticulture and therefore of agriculture, landscape gardening was, for example, taught as early as 1868 at the Universities of Illinois and Massachusetts (Steiner and Brooks 1986, p. 27). Like many others of their kind, these institutions were founded upon the Morrill Act of 1862 that determined and facilitated the establishment of agricultural and engineering colleges in each state geared towards educating large parts of the population in scientific and practical engineering skills. The land, or, the funds used to establish the land-grant colleges derived from the federal governments' violent taking of Native American land.

Throughout the nineteenth century, the USA had continued to expand on Native American territory, driving Indigenous peoples onto the verge of extinction. After 1865, all Native Americans were forced to assimilate or live on reservations. Racism, sanctioned in the South by the 1877 Jim Crow laws requiring the separation of white people from "persons of colour" in public transportation and schools, also led most land-grant colleges to exclude not only Native Americans but also African Americans. In response, in 1890, Congress passed the Second Morrill Act, also known as the Agricultural College Act of 1890, requiring states to establish a separate land-grant college for African Americans if they were being excluded from the existing ones. Although claiming to diminish inequality, the Act sanctioned racism and the purported "separate-but-equal" policy. By 1909, sixteen historically Black land-grant institutions had been established (Humphries 1992).

The aesthetics and economics of shaping the land of homesteads and farmsteads were considered integral to agricultural curricula at schools for both white and Black students. At Iowa State Agricultural College (today Iowa State University), Adonijah Strong Welch, the first college president taught a course in landscape gardening himself in 1871. At Michigan Agricultural College (today Michigan State University), a required half-year course in landscape gardening was taught as early as 1863, and in 1885, Liberty Hyde Bailey established the Department of Horticulture and Landscape Gardening there. Trained in agriculture, Bailey would go on to a prolific career at Cornell University, continuing to write about landscape gardening. Inspired by one student's design interests and Olmsted's advice, Bailey became instrumental in establishing Cornell's "outdoor art" programme that began to train future landscape architects in 1904. Administered first by the Department of Outdoor Art, the addition of a Master of Landscape Design degree in 1912 caused a name change to Department of Landscape Art and finally to Landscape Architecture in 1922 (Steiner and Brooks 1986, p. 29).

As one of Cornell's first Black students, Bailey's disciple David A. Williston graduated in 1898 with an agriculture degree and went on to become one of the first professional African American horticulturalists and campus designers. For more than twenty years, Williston worked at the private Tuskegee Normal and Industrial Institute (today Tuskegee University) as Superintendent of Buildings and Grounds and as Director of the Department of Landscape Gardening, passing on to future generations of Black students, knowledge in landscape gardening and design (Weiss 2001; Muckle and Dreck 1982; Tuskegee Institute 1912, pp. 5, 10, 1917, p. 10). Horticulture, including landscape gardening, became a subject taught at Tuskegee in 1895 (*Tuskegee to Date*, 1909,

p. 10). Considered a "landscape handicraft" (Tuskegee Institute 1912, p. 70), courses in landscape gardening and home ornamentation were of a practical nature and taught within the Department of Mechanical Industries (Tuskegee Institute 1903, pp. 101–102, 1904, pp. 109–111). During the school years 1905–06 and 1906–07, students were also able to sign up for a practical course on the Improvement of Grounds, participating in laying out, grading, constructing, and planting their own campus besides receiving basic training in industrial and mechanical drawing (Tuskegee Institute 1906, pp. 79–80, 1907, p. 81, 1913, p. 70). By 1908, landscape gardening courses finally also included small design tasks (Tuskegee Institute 1908, pp. 72–74). Once an independent Department of Landscape Gardening had been created in 1913, students were offered two courses: a one-year practical course in ground and plant care and maintenance, transplanting, and the construction of lawns, drives, walks, and drains; and a two-year course for advanced and post-graduate students that included landscape design encompassing preliminary and detailed grading and planting plans, and "the theory of Landscape Art" taught on the basis of British landscape gardener Edward Kemp's textbook on landscape gardening.[1] A separate course in home ornamentation for senior and postgraduate students in agriculture was extended to include "roadside improvement" as well as "parks, public squares, school grounds, etc." (Tuskegee Institute 1914, p. 88–89). By the early twentieth century, girls and young women in lower-level classes were taught vegetable and ornamental gardening as well as floriculture in separate horticulture and agriculture courses (Tuskegee Institute 1903, pp. 98–99, 1904, p. 109, 1905, p. 107, 1912, p. 80, 1916, pp. 115–117, 1917, p. 110). Tuskegee's educators believed that female students trained in these subjects could contribute to their current and future homes' improvement.

African American students found similar opportunities at the Hampton Normal and Agricultural Institute (today Hampton University) founded in 1868 by white and Black leaders of the American Missionary Association to educate and provide vocational as well as teacher training to freedmen and -women. During the Reconstruction Era (1865–1877), white benevolent societies, Black and white religious denominational organizations (especially the American Missionary Association, and the American Baptiste Home Mission Society), and large philanthropic corporate foundations were instrumental in the development of Black colleges and universities. However, Black leaders, patronizing reformers, missionaries, and philanthropists disagreed on the approaches to African American education. The missionary societies promoted the classical liberal philosophy of education, a stance steadfastly defended beyond the Reconstruction Era by W.E.B. Du Bois. He demanded equality and aspired for African Americans to become leaders transcending white supremacy. In contrast, once the opportunities for African American higher education had diminished with the end of Reconstruction, Booker T. Washington supported colleges' general tendency to turn away from the classical model towards the industrial training model. He argued that African Americans should concentrate on the practical arts of manual labor as this was the work that was available to them. According to Washington, accommodating white supremacy and accepting white leadership could be a way to raise African Americans' status (Brooks and Starks 2011; Devore 1989; Dennis 1989, pp. 101–102).

Washington's alma mater, the Hampton Normal and Agricultural Institute was legally chartered as land-grant school under the Morrill Act in 1870. Horticulture and gardening classes were foundational for both female and male education at the Institute (see Figure 13.1). In landscape gardening classes students learned how to design outdoor yards and grounds surrounding buildings. As later at Tuskegee, teacher training included the establishment of school gardens (Aery 1923, p. 13; Hampton Institute 1903, pp. 9–14, 1909; Shaw 1900). Hampton's emancipatory yet paternalistic and patronizing industrial education that was meant to "civilize" without providing a higher, more academic education, firmly kept African Americans in a socially and

Figure 13.1 Two pages illustrating the horticultural and gardening training of Native American and African American students at Hampton Normal and Agricultural College, ca. 1909.

Source: *Every-day Life at the Hampton Normal and Agricultural Institute* (Hampton, VA: The Hampton Normal and Agricultural Institute, 1909), pp. 13–14.

politically subordinate position upholding white supremacy. It also disregarded many African American vernacular practices of horticulture, and home and yard improvement.[2] Hampton Institute extended its education to Native Americans in 1877. Beginning with a group of former military hostages, it forcefully sought to educate Native Americans so that they would assimilate. Although unique in its biracial student body and in its collection of artefacts related to Native and Black heritage, the Institute was part of a system of schools for Native Americans established on and off the reservations after the Civil War through which the United States sought to oppress and destroy Native American culture, including its agriculture and manifold practices for the cultivation, harvest, stewardship, and gathering of plants.

The Institute's assimilation and acculturation practices also become apparent in the prototypical house garden design illustrated for educational purposes in one of its leaflets (see Figure 13.2; Hampton Institute 1917, pp. 8–14; Dümpelmann 2022). The house is centrally located on the plot, which is organized symmetrically with a privet hedge, box bushes, climbing roses, magnolia, American holly, pink crepe myrtle, dogwood, lavender, and rosemary, all typical of antebellum estate gardens (Ware 1996; Cothran 2003). Symmetry, a characteristic of white plantation owners' gardens, is enforced in the Hampton illustration by two pecan trees

Landscape gardening

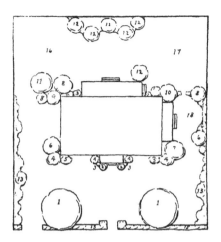

1 Pecan trees 2 Box bushes or red cedars cut low, or lavender 3 Climbing rose 4 Rosemary 5 Lavender 6 Pink crepe myrtle 7 Red cedar allowed to grow tall 8 American holly 9 Dogwood 10 Cherokee rose 11 Magnolia 12 Figs 13 Climbing roses, flowering shrubs and hollyhocks 15 Privet hedge 16 Chicken yard 17 Clothes yard and vegetable garden 18 Rose garden with rose hedge, and climbing roses on the fence.

The garden itself, including the rose garden, should be so related to the house as to seem a part of it, located usually at the sides or back of the house, with high plants or vines connecting and relating the garden to the building.

1 Foundation planting serves to relate the building to the landscape.
2 Trees planted on either side of a house front act as a picture frame as one approaches.
3 By the intelligent use of plant forms the outline of the building is partially concealed, thus securing the charm of mystery.

Figure 13.2   Two pages of a leaflet issued by Hampton Institute illustrating landscape planting and design guidelines, 1917.

Source: The Hampton Normal and Agricultural Institute (1917), *Home Decoration: The Hampton Leaflets* 8(1): pp. 9, 13. Courtesy of Hampton University Archives.

that frame the site's main entrance. (Although pecan trees had long been known to both Indigenous and settler societies, only in the mid-19th century, had they been successfully grafted and cultivated in Louisiana through the skill of an enslaved man known as Antoine ['The Pecan Tree' 1886; Wells 2017, pp. 13–14, 31–60; Perry 2008].) Chicken yards, clothes yards, and vegetable gardens were common for rural homesteads. But Hampton's layout sanctioned a particular spatial order, irrespective of Indigenous and African-American gardening traditions. Women of some Indigenous tribes cultivated corn, beans, and squash in both garden patches and large fields (Wilson, 2008, p. 75; Nelson, 2008); enslaved fieldworkers and later tenant farmers cultivated "provision gardens", where following West African planting practices, they mixed plants with different characteristics to keep insects at bay, shade out weeds, and conserve water and soil nutrients (Wilson 2008, p. 76). And whereas the Cherokee rose, also known as the Carolina rose, was a powerful symbol of suffering for Native Americans[3] – it symbolized the Trail of Tears – it appears in the Hampton leaflet only in the guise in which it was commonly used by white Southerners, as a hedge plant and a mundane ornamental (Cothran 2003, pp. 72–73).

Discrimination also occurred in the appropriation of federal funds for the use by state land-grant colleges to establish agricultural experiment stations provided through the 1887 Hatch Act, and in the cooperative extension service that was created through the 1914 Smith-Lever Act. Instituted to train rural populations unable to receive higher education in agriculture and home economics it only provided funding to the white colleges (Harris 2008). However, in an effort of self-help, Black institutions had already been educating farmers throughout the South since the early twentieth century. For example, at Tuskegee Institute in 1906, Booker T. Washington and George Washington Carver launched a prototypical mobile cooperative extension service. Soon, many women working as extension agents throughout the South were training other women not only how to grow vegetable gardens but how to improve homes and yards, for example by planting flower gardens (Glave 2005, pp. 43–44).

Being among the first to offer instruction in landscape gardening and design, the land-grant colleges and African American agricultural institutes also sought to become successful design examples themselves. For this purpose, they often hired well-known landscape architects, especially Olmsted and his firm, to consult on their campus design. It was also common for educators in landscape gardening to be charged with campus design and maintenance. For example, David Williston worked as a landscape gardener on Tuskegee's campus before also teaching there (Weiss 2001; Muckle and Wilson 1982).[4]

From landscape gardening to landscape architecture

Besides Cornell, the Massachusetts Agricultural College was among the first to establish a Department of Landscape Gardening and to begin a programme in 1902. Its founder, landscape gardener Frank A. Waugh who authored the 1907 textbook *Landscape Gardening: Treatise on the General Principles Governing Outdoor Art* (N.Y.: Orange Judd Co. 1907) was also responsible for the College's campus design (Waugh 1911). As at Cornell in 1922, the Department at Massachusetts Agricultural College was later renamed Landscape Architecture. Whereas in 1930, the Bachelor of Science in Landscape Architecture replaced the original Bachelor of Science in Landscape Gardening, and a five-year Bachelor of Landscape Architecture was added, the 1915 master's program in Landscape Gardening already changed its name to Landscape Architecture three years later. The name changes at Massachusetts Agricultural College and elsewhere reflected two developments that occurred around the turn of the century. First, during the shift from a rural to a more urban nation the tasks for landscape gardeners became broader. Instead of concentrating merely on the design of yards, grounds of farms and homesteads, and of estate gardens, they now also encompassed entire cities and their existing and new infrastructures (Steiner and Brooks 1986, pp. 29–30). Second, colleges and schools of architecture established new courses in landscape architecture. These developments went hand in hand with the foundation of the American Society of Landscape Architects in 1899. They were further inspired by Olmsted's use of the term "landscape architecture" to describe his profession.

Olmsted attributed the title "landscape architect" to French origin (also see introduction to this volume) and although he repeatedly questioned his own title throughout his career, by the late 1880s he had finally come to terms with it as it "better carries the professional idea. It makes more important also the idea of design." He went on to suggest how "'Gardener' includes service corresponding to that of carpenter and mason. Architect does not. Hence it is more discriminating, and prepares the minds of clients for dealing with us on professional principles" (Olmsted 1886, 1921, p. 189). Besides signalling designers' increased confidence and the move from craft to profession, "landscape architecture" also appeared to many designers to better capture the increasing breadth of spatial planning they were carrying out. The design, preservation, and conservation of open space could direct urban and regional development and shape urban form. Many early professionals besides Olmsted also considered themselves both public servants and active agents in the urbanization of the North American continent. Rather than *re*acting to more or less random settlement by embellishing its immediate surroundings as the early landscape gardeners before them, they were *active* in directing urban development in the first place (Dümpelmann 2021). Landscape architects in the USA would, therefore, also come to constitute the first city planners for whom a master's degree was established at Harvard University in 1923 within the School of

Landscape Architecture.[5] Frederick Law Olmsted, Jr. and his assistant Arthur A. Shurcliff had established an undergraduate landscape architecture degree programme already in 1900 within the university's department of architecture. It was turned into a graduate programme in 1906–07 and marking the profession's increasing confidence and independence, in 1908 it was housed in a new Department of Landscape Architecture (Pray [1911] 1973, pp. 39–44; Simo 2000).[6]

When, at the end of the nineteenth century, discussions about the affiliation of landscape architectural education arose, the young Charles Eliot made it clear that landscape architecture belonged to the school of design and not to the school of horticulture. For want of a proper landscape architecture curriculum, Eliot himself, in the early 1880s, had taken courses at the Bussey Institute, Harvard's Department of Agriculture and Horticulture, and then apprenticed in Olmsted's office. Considering landscape architecture an "art of design" that covered "agriculture, forestry, gardening, engineering and even architecture (as ordinarily defined) itself," he had in 1896 advised Harvard's overseers to connect any courses in landscape architecture "with instruction in the other arts, and particularly in connection with Architecture" (Eliot 1896, pp. 630–631).

Architectural education was at the time led by men trained at the École des Beaux-Arts in Paris. It was therefore not only characterized by the study of descriptive geometry, architectural history, and the art of drawing with pencil, ink, charcoal, and washes, but it also heavily focused upon principles of neoclassicism. Translated into landscape architecture this meant that landscapes surrounding buildings – predominantly country house estates – were often characterized by symmetry and geometrical patterns, features inspired by sixteenth- and seventeenth-century French and Italian gardens. Given that the first instructors in Harvard's landscape architecture programme came out of Olmsted's office, beaux-arts interests were paired with the naturalistic style of the senior Olmsted's park landscapes. In 1927, this ultimately caused landscape architect Elbert Peets to criticize the landscape architecture taught at Harvard as characterized by an "incredibly narrow taste" (Peets 1927, p. 100; Pearlman 2007, p. 64).

However, regardless of the degree programmes' name and affiliation with agriculture or architecture and regardless of their varying foci on horticulture versus design, the basic scope and formats of instruction in the early landscape gardening and landscape architecture programmes were quite similar. They included surveying, mapping, grading, plant materials and their use, planting design; as well as education in civic art; city, village and rural improvement; and the design of subdivisions. Even if many early landscape architects disagreed about their titles – some calling themselves landscape architect, others landscape gardener, or landscape artist (Dümpelmann 2014, 2021; Disponzio 2014; Eigen 2014) – they generally agreed upon their vast domain of expertise that became increasingly urban as time progressed.

An opportunity for white women

As a new profession in formation during the time of the first wave of feminism and its suffrage movement, and as a profession with origins that are closely connected to domestic pursuits of gardening and horticulture, landscape architecture soon started to attract women. It provided them with both a chance and a challenge. Although landscape architecture offered women independence and opportunities that many other professions were not yet able to provide, it did not protect them against the discrimination by male colleagues. Female professional training in landscape architecture began around the turn of the century, yet many pioneering female landscape architects were largely self-taught and mentored by male professionals. They came from wealthy families who provided them with the financial means and a supportive intellectual and social environment that often also offered them their first commissions. This was, for example, the case of Beatrix Jones Farrand, the only woman among the eleven founding members of the

American Society of Landscape Architects in 1899. As many contemporary male and female colleagues who designed gardens and grounds for wealthy clients during the Gilded Age and thereafter, Farrand went on a study tour to Europe and much of her work was inspired by French and Italian gardens of the seventeenth and eighteenth centuries.

Left standing in front of closed doors at the Ivy Leagues, women who had the means to seek professional training also entered the programmes at the Massachusetts Institute of Technology, Cornell University, the University of Illinois, and the University of California, Berkeley. Furthermore, they attended the Lowthorpe School of Landscape Architecture for Women founded in 1901 in Groton, MA, by Judith Motley Low; the Pennsylvania School of Horticulture for Women (see Figure 13.3) begun by Jane Haines in Ambler, Pennsylvania, in 1910 and modelled on the British Studley Horticultural and Agricultural College and Swanley Horticultural College as well as German gardening schools for women; and the Cambridge School of Architecture and Landscape Architecture, founded by Henry Atherton Frost in Cambridge, Massachusetts, in 1915. Of these three opportunities, the latter was the only graduate programme that required a previous college degree and the only programme in the entire United States that integrated architecture and landscape architecture (Anderson 1980; *Pennsylvania School of Horticulture for Women* 1911; Lowthorpe School n.d.; Way 2009, pp. 99–127; Libby 2002, 2012; Brown and Maddox 1982; Exley 1914, p. 2).[7] Instruction at these women's schools was delivered by both

Figure 13.3 Photograph by Frances Benjamin Johnston showing the Pennsylvania demonstration kitchen and flower garden at the Pennsylvania School of Horticulture for Women in Ambler, PA in 1919.

Source: Retrieved from the Library of Congress, www.loc.gov/item/2008676032/.

female and male teachers. Reflecting the typical gender bias of the times, men, often practicing landscape architects, taught the technical subjects like surveying and engineering (see Figure 13.4), as well as design. Women predominantly taught plant-related courses like botany, dendrology, greenhouse work and gardening, subjects that were considered a "feminine science" and extensions of the domestic sphere (Kohlstedt 1978; Gianquitto 2007; Schiebinger 1993, p. 36; Shteir 1997). The two-year Lowthorpe programme taught at an estate in Groton, MA, offered two tracks, one in landscape architecture and one in horticulture. Both Lowthorpe's and the Cambridge School's comprehensive curricula profited from the availability of Harvard faculty members and practitioners in the Boston area for instruction. Students were taught in drafting, planting design, elementary architectural drawing, freehand drawing, "Vignola's Orders," and principles of construction. Their education also included the history of architecture and landscape architecture, surveying and engineering, soils, trees and shrubs, hardy herbaceous perennials and annuals, elementary forestry, botany, entomology, greenhouse work and gardening, pomology, and agriculture (Tripp 1912; "Women's School of Landscape Architecture" 1913; Lowthorpe School n.d.). While the Lowthorpe and Pennsylvania Schools prepared women to pursue careers in the nursery business, estate management, and gardening instruction (besides landscape architecture in the case of the Lowthorpe School), the Cambridge School had a strong focus on design education. Yet, in all cases, the expectation was for women's work to remain within the domestic sphere. With a few exceptions, it therefore also took women longer to obtain public and civic design commissions.[8]

Figure 13.4 Surveying class at the Lowthorpe School in Groton, MA.
Source: Lowthorpe School, n.d., without page.

However, for decades already, women had been volunteering their work in civic associations and in the progressive reform and improvement movements, raising funds and rallying for the implementation of public landscapes. Work in civic gardening, especially in the context of the school garden movement between 1890 and 1920, finally provided women with an occupation bridging social reformist volunteer work and landscape design (Lawson 2012; Szcygiel 2012). By offering training for this purpose, in the early twentieth century the Lowthorpe and Pennsylvania Schools of Horticulture complemented the normal schools and African American schools like the Hampton and Tuskegee Institutes.

When the Harvard landscape architecture programme admitted female students for the first time in 1942, they were taking the seats of young men who had been drafted into the United States army. What was initially thought to be a provisional situation to bolster the university's budget would be there to stay. By way of the Second World War, women finally gained admission to the nation's first master's degree programme in landscape architecture (Alofsin 2002, pp. 175–176; Dümpelmann and Beardsley 2015).

Outlook

The beginnings and the early development of professional education in landscape gardening and landscape architecture still reverberate today in academic curricula, in the nature of the discipline's academic homes, and in the composition of today's student bodies and faculty. In contrast to the genesis of many landscape *gardening* courses in co-educational agricultural land-grant colleges and agricultural institutes, and in women's horticultural schools, degree courses in landscape *architecture* developed in colleges and schools of architecture following the Harvard model. Based upon this history, still today, the discipline's education occurs in both agricultural colleges and architecture schools. However, urbanization and the subsequent expansion of the discipline's activities that were increasingly "concerned with promoting the comfort, convenience, and health of urban populations" (Eliot 1910, p. 40), led landscape gardening curricula to be revised and renamed to reflect these changes so that the field of study has uniformly come to be called landscape architecture.

Since the late nineteenth century but with a break from the end of the Second World War until around 1970, the discipline has continued to attract women in increasing numbers (Brown and Maddox 1982, p. 69). Although not reflected in the leadership of professional practice, most student bodies in landscape architecture are today constituted by more female than male students (Pritchard and Martinez Gonzalez 2019). The establishment of women's schools and women's early entry into a fledgling profession was facilitated by, among other things, the field's initially close ties to the domestic sphere and its suitability for self-training. However, it has not prevented gender discrimination.

Whereas the occupation's association with the domestic sphere contributed to paving a way for white women into the profession, landscape gardening's rural roots and landscape architecture's urban bias have exacerbated the challenges faced by African Americans and Native Americans regardless of their gender. The low number of BIPOC (Black, Indigenous, People of Colour) students in today's landscape architecture programmes is related to landscape architecture's historic urban bias and to the association of its practice with wealthy clients and conservative social reform. It is also a result of the systemic racism that has governed education in the United States throughout its history more broadly. Although landscape gardening and design was an integral component of Black college education following the Hampton-Tuskegee model, that is teacher and vocational training in agriculture and industry, it was this same model that hindered African American and Native American access to higher college and university education in landscape architecture. Established and supported by white supremacists who furthered an ideology that rejected African and Native American political power yet recognized the need especially for Black agricultural workers to sustain the South's economy, the Hampton-Tuskegee model embraced students' rural origins and discouraged the acquisition of skills necessary to urban survival in a rapidly modernizing nation (Anderson 1988, pp. 33–78; Lindsey 1995, pp. 176–182). For example, Hampton Institute's white president Hollis B. Frissell explained in 1900 that "The Indian should learn to farm and till the soil. . . is to live in the country, and he should find his comfort and happiness in the flowers, the trees, the rivers, and all nature" (Frissell 1899, pp. 58–59). Courses in landscape gardening and design were therefore not only integral but also central to and complicit in the rural orientation of the Institute's largely manual training that befitted white supremacy. While courses in landscape gardening and design were used quite literally to acculturate African Americans and Native Americans, depriving them of their own vernacular garden cultures, practices, and land, the courses could at best contribute to facilitating the minorities' survival and self-sufficiency in a rapidly modernizing white nation. No matter how related their core subjects were, the courses at agricultural institutes were divorced from what was considered the high art of landscape design taught at Harvard beginning in 1900.

Since then, the breadth of landscape architectural practice has increased. Growing attention is being paid to issues of social and environmental justice in landscape architectural education. The academy today can and must play a role in educating and training a diverse, inclusive student body that as future professionals, educators, and scholars will directly confront racism, discrimination, and inequality.

Notes

1 The two-year course also included the care of woody plants based upon Bernhard Eduard Fernow's book *The Care of Trees* (1911), practical lessons in tree surgery, and nursery work. A textbook by Liberty Hyde Bailey is listed as the basis for the practical landscape gardening course. It could have been L.H. Bailey (1898) *Garden-Making: Suggestions for the Utilizing of Home-Grounds*. New York: Macmillan. For the discrimination against African and Native Americans in landscape gardening and landscape gardening. Also see Dümpelmann 2022.
2 For African American vernacular yards in the Southern United States, see Westmacott 1998; Glave 2005.
3 The Cherokee rose became Georgia's State Flower in 1916, yet its name derived from the Cherokee Nation's wide distribution of the plant based upon the Nation's legend of the plant symbolizing the Trail of Tears, the 1838–1839 forced relocation and often deadly journey of Native Indians from the southeastern United States to a specially designated "Indian Territory" west of the Mississippi River in present-day Oklahoma. Also see Dümpelmann 2022.
4 For Olmsted's consultation on and design of campuses, see, for example Turner 1984, pp. 140–158.
5 A first course in city planning was offered within the School of Landscape Architecture in 1909. The university opened the nation's first school of city planning in 1929.
6 For a first compilation of documents relating to the education and curricula of selected landscape architecture programmes in the United States, see Robinette 1973a, 1973b.
7 The New England Women's Club established an early and short-lived horticultural school for women in 1870. See Kohlstedt 1978, p. 93.
8 For some exceptions, see Way 2009, pp. 16–20.

References

Aery, W. A. (1923) *Hampton Institute: Aims, Methods, and Results*. Hampton: Hampton Institute.
Alofsin, A. (2002) *The Struggle for Modernism: Architecture, Landscape Architecture, and City Planning at Harvard*. New York and London: W. W. Norton.
Anderson, D. M. (1980) *Women, Design, and the Cambridge School*. West Lafayette: PDA Publishers.

Anderson, J. D. (1988) *The Education of Blacks in the South, 1860–1935*. Chapel Hill and London: The University of Carolina Press.

Brooks, F. E. and Starks, G. L. (2011) *Historically Black Colleges and Universities: An Encyclopedia*. Santa Barbara, Denver and Oxford: Greenwood.

Brown, C. R. and Maddox, C. N. (1982) 'Women and the Land: "A Suitable Profession"', *Landscape Architecture*, 72(3), pp. 64–69.

Cothran, J. R. (2003) *Gardens and Historic Plants of the Antebellum South*. Columbia: University of South Carolina Press.

Dennis, M. (1989) 'Progressivism and Higher Education in the New South', in Mohr, C. L. (ed.) *The New Encyclopaedia of Southern Culture, vol. 17: Education*. Chapel Hill: The University of North Carolina Press, pp. 97–102.

Devore, D. E. (1989) 'Black Public Colleges', in Mohr, C. L. (ed.) *The New Encyclopaedia of Southern Culture, vol. 17: Education*. Chapel Hill: The University of North Carolina Press, pp. 50–52.

Disponzio, J. (2014) 'Landscape Architecture: A Brief Account of Origins', *Studies in the History of Gardens and Designed Landscapes*, 34(3), pp. 192–200.

Dümpelmann, S. (2014) 'What's in a Word: On the Politics of Language in Landscape Architecture', *Studies in the History of Gardens and Designed Landscapes*, 34(3), pp. 207–225.

Dümpelmann, S. (2021) '"Landscape Architect Better Carries the Professional Idea": On the Politics of Words in the Professionalization of Landscape Architecture in the United States', in Wolschke-Bulmahn, J. and Clark, R. (eds.) *From Garden Art to Landscape Architecture: Traditions, Re-Evaliations, and Future Perspectives*. Munich: AVM Edition, pp. 55–69.

Dümpelmann, S. (2022) 'Let All Be Educated Alike Up to a Certain Point', *Places Journal*, June. Available at: https://placesjournal.org/article/olmsted-booker-t-washington-landscape-architecture-education/ (Accessed 6 July 2022).

Dümpelmann, S. and Beardsley, J. (2015) 'Introduction: Women, Modernity, and Landscape Architecture', in Dümpelmann, S. and Beardsley, J. (eds.) *Women, Modernity, and Landscape Architecture*. Abingdon and New York: Routledge, pp. 1–14.

Eigen, E. (2014) 'Claiming Landscape as Architecture', *Studies in the History of Gardens and Designed Landscapes*, 34(3), pp. 226–247.

Eliot, C. W. (1896) 'Letter Addressed to Charles Francis Adams, December 12', in Eliot, C. W. (ed.) *Charles Eliot, Landscape Architect*. Boston and New York: Houghton Mifflin, 1902, pp. 630–631.

Eliot, C. W. (1910) 'Letter to the Editors of Landscape Architecture', *Landscape Architecture*, 1(1), p. 40.

Exley, E. (1914) 'Editorial', *Wise-Acres*, 1(1), pp. 1–2.

Frissell, H. B. (1899) 'What Is the Relation of the Indian of the Present Decade to the Indian of the Future?', in *Report of the Superintendent of Indian Schools*. Washington: Government Printing Office, pp. 58–59.

Gianquitto, T. (2007) '"Good Observers of Nature": American Women and the Science of the Natural World, 1820–1885*. Athens: University of Georgia Press.

Glave, D. D. (2005) 'Rural African American Women, Gardening, and Progressive Reform in the South', in Glave, D. D. and Stoll, M. (eds.) *To Love the Wind and the Rain: African Americans and Environmental History*. Pittsburgh: University of Pittsburgh Press, pp. 37–50.

Hampton Institute. (1903) *Hampton Summer Normal Institute 1903*. Hampton: Hampton Institute.

Hampton Institute. (1909) *Every-day Life at the Hampton Normal and Agricultural Institute*. Hampton: The Press of the Hampton Normal and Agricultural Institute.

Hampton Institute. (1917) *The Hampton Leaflets 8, no. 1: Home Decoration*. Hampton: The Press of the Hampton Normal and Agricultural Institute.

Harris, C. V. (2008) '"The Extension Service Is Not an Integration Agency": The Idea of Race in the Cooperative Extension Service', *Agricultural History*, 82(2), pp. 193–219.

Humphries, F. (1992) 'Land-Grant Institutions: Their Struggle for Survival and Equality', in Christy, R. D. and Williamson, L. (eds.) *A Century of Service: Land-Grant Colleges and Universities 1890–1990*. New Brunswick and London: Transaction Publishers, pp. 3–12.

Kohlstedt, S. G. (1978) 'In from the Periphery: American Women in Science, 1830–1880', *Signs: Journal of Women and Culture in Society*, 4(1), pp. 81–96.

Lawson, L. J. (2012) 'Women and the Civic Garden Campaigns of the Progressive Era: "A Woman Has a Feeling about Dirt Which Men Only Pretend to Have . . ."', in Mozingo, L. A. and Jewell, L. (eds.) *Women in Landscape Architecture: Essays on History and Practice*. Jefferson and London: McFarland, pp. 55–68.

Libby, V. (2002) 'Jane Haines' Vision: The Pennsylvania School of Horticulture for Women', *Journal of the New England Garden History Society*, 10, pp. 44–52.

Libby, V. (2012) 'Cultivating Mind, Body and Spirit: Educating the 'New Woman' for Careers in Landscape Architecture', in Mozingo, L. A. and Jewell, L. (eds.) *Women in Landscape Architecture: Essays on History and Practice*. Jefferson and London: McFarland, pp. 69–75.

Lindsey, D. F. (1995) *Indians at Hampton Institute, 1877–1923*. Urbana and Chicago: University of Illinois Press.

Lowthorpe School. (n.d.) *Lowthorpe School of Landscape Architecture, Gardening, and Horticulture for Women: Founded by Mrs. Edward Gilchrist Low*. Groton: Lowthorpe School.

McLaughlin, C. C. and Beveridge, C. E. (eds.) (1977) *The Papers of Frederick Law Olmsted, vol. 1: The Formative Years 1822–1852*. Baltimore and London: The Johns Hopkins University Press.

Muckle, K. and Wilson, D. (1982) 'David Augustus Williston: Pioneering Black Professional', *Landscape Architecture*, 72(1), pp. 82–85.

Nelson, L. A. (2008) 'Native American Agriculture', in Walker, M. and Cobb, J. C. (eds.) *The New Encyclopedia of Southern Culture, vol. 11: Agriculture and Industry*. Chapel Hill: The University of North Carolina Press, pp. 88–92.

Olmsted, F. L. (1886) *Letter to Charles Eliot, 28 October. Frederick Law Olmsted Papers* (Reel 20, frames 680–686). Washington, DC: Library of Congress.

Olmsted, F. L. Sr. (1921) 'A Letter Relating to Professional Practice from F. L. Olmsted Sr. to Charles Eliot', *Landscape Architecture*, 11(4), pp. 189–190.

Olmsted, F. L., Jr. and Kimball, T. (eds.) (1922) *Frederick Law Olmsted Landscape Architect 1822–1903: Early Years and Experiences*. New York and London: G.P. Putnam's.

Pearlman, J. (2007) *Inventing American Modernism: Joseph Hudnut, Walter Gropius, and the Bauhaus Legacy at Harvard*. Charlottesville and London: University of Virginia Press.

Peets, E. (1927) 'The Landscape Priesthood', *The American Mercury*, 10(37), pp. 94–100.

Perry, P. (2008) 'Pecans', in Walker, M. and Cobb, J. C. (eds.) *The New Encyclopedia of Southern Culture, vol. 11: Agriculture and Industry*. Chapel Hill: The University of North Carolina Press, pp. 185–186.

Pray, J. S. (1973) [1911] 'The Department of Landscape Architecture in Harvard University', in Robinette, G. O (ed.) *Landscape Architectural Education, vol 1*. Dubuque: Hunt Pub. Co.

Pritchard, K. D. and Martinez Gonzalez, N. (2019) *Summary of 2019 Annual Reports Submitted to the Landscape Architectural Accreditation Board by Accredited Academic Programs* [Online]. Available at: www.asla.org/uploadedFiles/CMS/Education/Accreditation/2019AnnualReportsSummary.pdf (Accessed 4 November 2020).

Robinette, G. O. (1973a) *Landscape Architectural Education, vol 1*. Dubuque: Hunt Pub. Co.

Robinette, G. O. (1973b) *Landscape Architectural Education, vol 2*. Dubuque: Hunt Pub. Co.

Schiebinger, L. (1993) *Nature's Body: Gender in the Making of Modern Science*. New Brunswick: Rutgers University Press.

Shaw, A. (1900) '"Learning by Doing" at Hampton', *The American Monthly Review of Reviews*, pp. 417–432.

Shteir, A. B. (1997) 'Gender and "Modern" Botany in Victorian England', *Osiris*, 12, pp. 29–38.

Simo, M. (2000) *The Coalescing of Different Forces and Ideas: A History of Landscape Architecture at Harvard 1900–1999*. Cambridge: Harvard University Graduate School of Design.

Steiner, F. R. and Brooks, K. R. (1986) 'Agricultural Education and Landscape Architecture', *Landscape Journal*, 5(1), pp. 19–32.

Szcygiel, B. (2012) '"City Beautiful" Revisited: An Analysis of 19th Century Civic Improvement Efforts', in Mozingo, L. A. and Jewell, L. (eds.) *Women in Landscape Architecture: Essays on History and Practice*. Jefferson and London: McFarland, pp. 95–112.

The Hampton Normal and Agricultural Institute. (1917) *Home Decoration: The Hampton Leaflets*, 8(1).

'The Pecan Tree'. (1886) *Scientific American*, 55(13), p. 202.

The Pennsylvania School of Horticulture for Women. (1911) *Pennsylvania School of Horticulture for Women: Prospectus*. Ambler: The Pennsylvania School of Horticulture for Women.

Tripp, A. F. (1912) 'Lowthorpe School of Landscape Architecture, Gardening and Horticulture for Women,' *Landscape Architecture*, 3(1), pp. 14–18.

Turner, P. V. (1984) *Campus: An American Planning Tradition*. Cambridge and London: The MIT Press.

Tuskegee Institute. (1903) *Twenty-Second Annual Catalog The Tuskegee Normal and Industrial Institute 1902–1903*. Tuskegee: Institute Press.

Tuskegee Institute. (1904) *Twenty-Third Annual Catalog The Tuskegee Normal and Industrial Institute 1903–1904*. Tuskegee: Institute Press.

Tuskegee Institute. (1905) *Twenty-Fourth Annual Catalog The Tuskegee Normal and Industrial Institute 1904–1905*. Tuskegee: Institute Press.
Tuskegee Institute. (1906) *Twenty-Fifth Annual Catalog The Tuskegee Normal and Industrial Institute 1905–1906*. Tuskegee: Institute Press.
Tuskegee Institute. (1907) *Twenty-Sixth Annual Catalog The Tuskegee Normal and Industrial Institute 1906–1907*. Tuskegee: Institute Press.
Tuskegee Institute. (1908) *Twenty-Seventh Annual Catalog The Tuskegee Normal and Industrial Institute 1907–1908*. Tuskegee: Institute Press.
Tuskegee Institute. (1912) *Thirty-First Annual Catalog The Tuskegee Normal and Industrial Institute 1911–1912*. Tuskegee: Institute Press.
Tuskegee Institute. (1913) *Thirty-Second Annual Catalog The Tuskegee Normal and Industrial Institute 1912–1913*. Tuskegee: Institute Press.
Tuskegee Institute. (1914) *Thirty-Third Annual Catalog The Tuskegee Normal and Industrial Institute 1913–1914*. Tuskegee: Institute Press.
Tuskegee Institute. (1916) *Thirty-Fifth Annual Catalog The Tuskegee Normal and Industrial Institute 1915–1916*. Tuskegee: Institute Press.
Tuskegee Institute. (1917) *Thirty-Sixth Annual Catalog The Tuskegee Normal and Industrial Institute 1916–1917*. Tuskegee: Institute Press.
Tuskegee to Date. (1909) Tuskegee: Press of Tuskegee Institute Printing Division.
Ware, E. (1996) 'Formal Ornamental Gardens in the Ante-Bellum South', *Studies in Popular Culture*, 19(2), pp. 49–66.
Waugh, F. A. (1907) *Landscape Gardening: Treatise on the General Principles Governing Outdoor Art*. New York: Orange Judd Co.
Waugh, F. A. (1911) *Studies for the Improvement of the Grounds of the Massachusetts Agricultural College from 1864 to 1911*. Amherst, MA: Massachusetts Agricultural College.
Way, T. (2009) *Unbounded Practice: Women and Landscape Architecture in the Early Twentieth Century*. Charlottesville and London: University of Virginia Press.
Weiss, E. (2001) 'Tuskegee: Landscape in Black and White', *Winterthur Portfolio*, 36(1), pp. 19–37.
Wells, L. (2017) *Pecan: America's Native Nut Tree*. Tuscaloosa: The University of Alabama Press.
Westmacott, R. (1998) *African-American Gardens and Yards in the Rural South*. Knoxville: University of Tennessee Press.
Wilson, C. R. (2008) 'Garden Patches', in Walker, M. and Cobb, J. C. (eds.) *The New Encyclopedia of Southern Culture, vol. 11: Agriculture and Industry*. Chapel Hill: The University of North Carolina Press, pp. 75–79.
'Women's School of Landscape Architecture' (1913) *Park and Cemetery and Landscape Gardening*, 23(9), p. 174.

14
A HUNDRED YEARS OF LANDSCAPE ARCHITECTURE EDUCATION IN ÅS, NORWAY

The pioneering work of Olav Leif Moen

Karsten Jørgensen (†)[1]

In 1919, the Norwegian University of Agriculture at Ås (*Norges landbrukshøgskole*, NLH, currently NMBU) established landscape architecture education at the university level. This was the first academic programme of its kind in Europe (Birli 2016; Jørgensen 2019). This early establishment of a university programme in landscape architecture seems counterintuitive: Norway had a weak tradition in the fields of horticulture and garden art, as compared to, for example Sweden and Denmark, not to mention France and Germany (NLH 1959, p. 396).

Before the university programme in Norway was established in 1919, a large proportion of practicing landscape architects had received their educations at the department of horticulture of the Agricultural University in Ås. The establishment of the university programme seems to have been a result of an interplay between the Agricultural University, which gained its official status as a scientific institution in 1897, and an active supporting group of professional landscape architects. One peculiar detail is that the early academic founding of the discipline seems to be linked to the pronounced need to raise Norway's cultural level in a Nordic and European context in the early decades of the twentieth century. This motivation resonated with the nation-building ethos of the time.

The establishment of the landscape architecture programme in Norway

In 1911, parliament put forward a bill for changes at NLH, which, among other things, proposed greater specialization of subjects via a division into various tracts within the individual departments. Because of World War I, the bill was not passed until 1919, but in the intervening years, the university had the opportunity to prepare for the changes that were to come. In 1917, the university board approved the setting up of a committee to evaluate the future organization of the curricula of the horticultural department. In October of the same year, the committee proposed a division into three programmes: one for fruit and one for growing vegetables, and one for garden architecture. The reasons to establish a chair in garden architecture were:

> Garden architecture is a subject that involves large-scale assignments and it has great development potential. It is an important field for all levels of society. In the teaching

DOI: 10.4324/9781003212645-16

at the college, it has exceeded its formerly modest framework and from now on can only be kept abreast of the times and present developments by having its own chair (. . .). An understanding of the importance of this discipline is now beginning to be realized, which is why garden architecture, sooner or later, will inevitably gain a place at a college in Norway. The committee is convinced that the natural place for such a discipline is at NLH and that it would irreparably damage horticulture and garden architecture itself if it were removed from the department of horticulture and placed in some other educational institution.[2]

The committee may have been inspired by events that took place on the occasion of the 100-years jubilee of the Norwegian Constitution in 1914. Landscape architects Marius Røhne (1883–1966) and Iosef Oscar Nickelsen (1869–1959) designed the jubilee exhibition grounds in Oslo. The exhibition received positive reviews in the press. Carl W. Schnitler (1879–1926), a leading cultural figure and professor of art history at Oslo University who pointed to the need for a high level within garden art if Norway was to be perceived as a cultural nation in Europe, commented on *The Gardens at the Jubilee Exhibition* as follows:

> For the first time in our country we have been able to see truly modern garden complexes in a 'European style.' (. . .) the garden complex at the exhibition is the first significant work that has been implemented here in Norway. (. . .) A lack of architectural attitude is the worst defect in our entire artistic culture. The capacity shown here for subordination and collaboration between architect and gardener[3] is one of the most gratifying things the Jubilee Exhibition has given us.
>
> (Schnitler 1916, translated by author)

The periodical *Kunst og Kultur* (Art and Culture) published a themed issue on landscape architecture on the occasion of the Jubilee. The editor, Harry Fett (1875–1962), who was also the Director of National Antiquities at that time, published an article about American playgrounds in which, by clearly addressing Norwegian urban planners, he analyzes the work of Frederick Law Olmsted (1822–1903) and his efforts to create good places for children in cities such as Chicago (Fett 1914, pp. 199–200). In the same issue, landscape architect I. O. Nickelsen published an article where he mentions the need for a better education within garden art, with a clear 'national' undertone:

> a growing understanding of the enormous development of garden art in recent years will doubtlessly make it crystal clear that, consistently and artistically defensible, gardens can only be created by someone who is able to unite the ability of the architect to design and construct with the artist's sense of composition and the gardener's intimate knowledge of the life-conditions of plants and their effect in the landscape – by the modern landscape gardener or garden architect. Let us hope for a new era also in **Norwegian** [bold in the original] garden art. It will and must come if we want to affirm our position in cultural society in general. Particularly for a tourist country like ours, it is important that our public parks present themselves in the most attractive form possible. Our visitors also assess our level of culture by the state in which our city gardens find themselves.
>
> (Nickelsen 1914, translated by author)

We do not know whether this discourse influenced any conclusion made at the Agricultural University, but the board unanimously decided to recommend for the university to establish a new programme called 'Horticulture II – Garden Architecture', and it approved the announcement of a professorship in Garden Art. In the course of 1918 and 1919, the professorship position was publicly announced three times without any qualified applicants seeking the post.[4] Nevertheless, with the blessing of the ministry, the first students enrolled in 1919. They could apply to the three-year programme after having completed a three-year education in gardening. During the first two years of the university programme, the curriculum included much of the same subjects for garden architecture students as for the other horticulture students: plant and soil sciences, botany, floriculture, freehand drawing and so on. Specialization in what later would be called landscape architecture mainly took place in the third year, in subjects like landscape gardening, building and landscape engineering. The teaching was a combination of lectures, studio assignments and excursions. The university campus, designed by Hans Misvær in 1900, who graduated in horticulture at NLH in 1889, was a prime example of landscape gardening used by the students during the first years.

In 1922, the first students graduated in garden architecture. Classes were small. During the first eight years, from 1921 to 1929, about 14 candidates in total graduated, and in the four following years none at all (Blichner 1989, p. 12). From the mid-1930s, numbers rose gradually. In 1919, Olav Leif Moen, a candidate in horticulture from NLH 1918, was awarded a scholarship to qualify for the professorship position. In 1921, the NLH appointed Moen as docent in garden art. In 1939, Moen's position was upgraded to professorship. Times were hard for the new programme during the first few decades, with only a handful of students graduating each year. On several occasions there were confrontations between Moen and NLH. In 1948, for example, the department of horticulture proposed abolishing the division into lines, so that garden architecture would be one of several main-subject areas under horticulture. Moen protested, gaining increasing support from the profession in these conflicts, but conditions for the study programme were extremely bad. The national association for garden architects *Norsk Hagearkitektlag* drew up a report on the inadequate teaching conditions at NLH. One of the proposals was to transfer the entire study programme to *Norges Tekniske Høyskole* in Trondheim.[5] The department of horticulture, which saw no reason to distinguish between garden architecture and horticulture, rejected the report. When Moen died in 1951, however, the association was asked for advice about the future of the educational programme. The proposal to abolish the programme was dropped and teaching was substantially strengthened. More part-time teachers were employed, and Olav Aspesæter was appointed the new professor in garden art in 1953.

Olav L. Moen as designer and provider of examples for education

Olav L. Moen (1887–1951) grew up in relative poverty in Surnadal, in the Nord-Møre district on the west coast of Norway. His mother died of tuberculosis when he was six years old, and he lost a brother from the disease. In 1905, he studied gardening at *Hylla Hagebruksskole*, and in 1906 at the college in Volda. In 1908, he came to NLH for the first time, and worked as an apprentice at the Department of Horticulture. From 1909 to 1912, he worked as a gardener. From 1912 to 1916, he worked for landscape gardeners and garden architects in Denmark, Germany and Switzerland, including the *Atelier für Gartenkunst* in Zürich. From 1916 to 1918, he studied horticulture at NLH. He was known for his outstanding drawing skills (Blichner 1989, p. 12). After graduating from NLH in 1918, he worked as a teacher at the national gardening school in Oslo, practiced as a garden architect and won a design competition in Oslo.

In 1920, after having received the scholarship from NLH to qualify for the professorship in garden architecture, he attended supplementary education in Berlin and undertook extensive study trips in Germany, Italy, France and England. In 1921, he completed studies at the

Gärtner-Lehranstalt in Berlin, established by the Prussian landscape gardener Peter Joseph Lenné almost a hundred years earlier, and in the same year, he won a competition in Berlin for the design of an allotment garden area. In autumn that same year, he started teaching garden architecture at NLH. He also became head of the development and management of the *NLH Park* (now the *NMBU Park*), an assignment to which he devoted considerable energy. By 1938, he had been responsible for the garden architecture programme for 16 years, under extremely difficult conditions. The grants for both the study programme as well as for the park were kept to a minimum. Nor had he seen anything of the professorship that had been talked about when he was awarded the scholarship in 1920. In short, he was tired of what he regarded as intolerable working conditions. He applied for and was offered the post as principal at The National School of Gardening in Oslo. After pressure from the students, the national associations of garden architects and gardeners, who sent letters to the Ministry of Agriculture that NLH was about to lose a key figure, Moen was finally, in 1939, appointed 'Professor of Garden Art' at NLH.

Moen had reached his goal of becoming a professor of garden architecture, the first in any Nordic country and among the first anywhere in Europe. The next person in Scandinavia to gain such a title was C. Th. Sørensen, who became professor of garden art at the Royal Danish Academy of Fine Arts in Copenhagen in 1954, after having been a lecturer in the discipline there since 1940. In the field, C. Th. Sørensen is perhaps best known as a practicing landscape architect, but he was also a respected teacher. He initiated the Nordic 'summer schools' for professionals in the 1930s – in Copenhagen 1933, Oslo 1935, Helsinki 1936 and Gothenburg 1939 (Lund 2002, p. 15). Moen attended several of these events.

Moen was a popular and committed teacher, and he was often seen among the students working in the *tegnesal*, literally 'drawing studio' (Moen in his 'studio jacket' (see Figure 14.1). Staff and students considered Moen an excellent lecturer (Blichner 1989). Besides his teaching and his work maintaining the NLH Park, Moen was active as a practicing landscape architect. He referred to his own projects in his lectures, and especially the NLH park design and construction was utilized in teaching, for example during field visits and hands-on construction. Didactically, Moen used the power of the good example; students should be able to get good ideas by visiting the different parts of the park. And students did take part in some construction

Figure 14.1 The original caption of this picture is: "Course for horticultural officials in garden art and garden architecture" at NLH in January 1931. Among the books on the table is *Die Praxis der Friedhofsgärtnerei*. (Docent Olav L. Moen is on the right.)

Source: NMBU Historical Archive of Norwegian Landscape Architecture.

Figure 14.2 Students working in the NLH park during the 1940s.

Source: NMBU Historical Archive of Norwegian Landscape Architecture.

work, particularly for the sports fields and park around the students' association building from 1930. Here almost everything was built as *dugnad* (voluntary work), mainly during summer holidays and other breaks (see Figure 14.2). This was a strong tradition within the university; when it was founded as an agricultural college in 1859, the students were a considerable proportion of the workforce.

According to student's and Moen's notebooks,[6] the design programme was very important in Moen's teaching; it included information on future park uses, as well as information on soil, planting and construction solidity. However, design style was always a topic, and drawing skills highly valued. Early on in his career Moen was strongly influenced by the neo-classicist style, and for the most part he maintained this style in his own projects, even though he argued in favour of functionalism in some of his writing after 1930 (Blichner 1989). In pre-1930 articles he mentions the Renaissance and Baroque gardens as sources of inspiration. He later adopts a different position and promotes the landscape style. In 1935, he writes that gardens have become simpler and more objective, and that they are now designed more rationally and genuinely than was the case throughout the 19th century. His unpublished lectures show that he was also inspired by Reginald Blomfield (1856–1942) and Thomas Mawson (1861–1933) from England, and Leberecht Migge (1881–1935) from Germany. He argues against the so-called natural garden advanced by Willy Lange (1864–1941), whom he had met in Berlin. The two approaches, the neoclassical and the landscape style, can be seen, even in combination, in some of his public parks, for example in Drøbak (1931), Mysen People's Park (1938), in Horten (1945) and Harstad

(1947). In 1938, he was awarded first prize in all three sections of a landscape architecture competition for farm gardens to suit small, medium-sized and large farms. In 1941, he won both first and second prize (along with O. Reisæter) in an urban development competition in Namsos. Most of these projects were predominantly variants of the neoclassicist style, with the landscape style appearing in some sections. Towards the end of his career, Moen increasingly also designed buildings in addition to gardens. In some of these projects we can detect a clear functionalist influence. Typical examples are rows of funkis houses and gardens in Ås and in Moss.

His major work, however, is the park at NLH, mainly realized in the period between 1924 and 1935. This project exhibits many of Moen's most important ideals and motifs (see Figure 14.3).

The park has different parts that are clearly defined. Each space is designed to allow for certain activities. For example, the auditorium serves as outdoor lecture hall, a 'mock cemetery' as a demonstration area for students. The park also includes traditional activities like sports, play, and so on. The clearly defined areas are linked to axes and views, and proportions that are closely linked to the architecture of the buildings. When Moen was appointed professor at NLH, the construction of Tårnbygningen (the Tower Building) was starting up. This building was designed by the architect Ole Sverre (1865–1952) who had also designed the other main buildings on campus 20 years earlier. The earlier buildings are characterized as a blend of *Jugendstil* and 'dragon style', with animal and plant motifs and clear national references, whereas the Tower Building is neoclassicist. Moen's park plan was based on the neoclassicist style, but he would keep some of the old parts of the park in the landscape style, with gently rolling terrain and winding paths, as well as some elements in a *Jugend* or Arts and Crafts style. The neoclassicist idea is clearly exposed in the way the park harmonizes with the buildings. The new central park space Moen created was called *Storeplenen* ('the great lawn') where 'the Clock Building' from 1900 was emphasized as the 'main building'. The centre of the whole composition is *Storeplenen*, and via a system of axes and vistas, the buildings and the old and new parts of the park are integrated into a unified composition. In addition to the central park space, *Storeplenen*, a number of adjacent park sections have been designed. A characteristic example is *Staudehagen* (garden of perennials), which lies north of Tårnbygningen.

This is clearly inspired by Arts and Crafts. Here there is a small, rectangular pool, a small fountain with a lion's head and a terrace with a place to sit that is framed by six symmetrically placed columnar oak trees. Directly northeast of the main courtyard lies the other experiential point of

Figure 14.3   Olav L. Moen: the 1924 park plan in perspective from ca. 1935. Perspectives like these mainly served teaching purposes and as examples.

Source: NMBU Historical Archive of Norwegian Landscape Architecture.

gravity, *Svanedammen* (the Swan Pond). This section contrasts strongly with the former one – *Storeplenen*. At the Swan Pond the style is romantic and picturesque, particularly in the so-called Niagara Falls, a small water garden at the mouth of the pond. As NLH gradually expanded its sphere of activities, new areas were regularly incorporated into the park. Around 1930, Moen was one of the keenest advocates of establishing the new student society building, *Studentsamfunnet*. The architect Thorleif Jensen (1903–1972) won the competition and designed the building characterized as a 'gem of functionalist architecture' in Norway. Moen's plan for the park at *Studentsamfunnet* is kept in the same style. A triangular reflecting pond 'Dimple', the sports ground *Storbrand* and the dance floor at *Skogsdammen* reveal the functionalist influence in Moen's production.

Moen involved himself in the public debate of both subject-related and social issues. He was keenly interested in children's environments during their formative years, believing that good play areas for children was one of the most important things to establish in connection with urban development. His pioneer efforts and commitment in this field have helped landscape architecture gain the breadth and influence it now has in Norway. He was strongly involved in debates in the students' society at NLH, and his engagement would also show in his lectures in the classroom. 'Happy children make good people' was a favourite saying[7] (From notebooks, university archives). Moen would bring his social engagement into lecture hall and studio. An example of his involvement in the public debate is the discussion about Vigelandsparken in Oslo. Moen sharply criticised the sculptor Vigeland's plan for the design of the landscape. In 1924, the city of Oslo gave Gustav Vigeland a studio and left the responsibility for the designing of the park with the sculptor himself. Olav L. Moen found this to be a scandalous decision. At a meeting of the Oslo Gardener Association when Vigeland's plans were to be implemented in February 1932, Moen gave a lecture in which he presented his criticism of the plans, including the lack of appreciation of natural precondition of the site, and he showed a sketch for an alternative park design. The criticism was printed and followed up in several newspapers as well as in the Norwegian Gardener Association's periodical.

The legacy of Olav Moen: landscape architecture education in Norway

Over a period of three decades, Olav L. Moen acted as the leading figure in Norwegian landscape architecture, until he died in 1951. His main legacy is the way he developed the first higher education programme in landscape architecture in Europe, and thus laid the ground for the strong position that the subject has in Norway today. Several times under his leadership, the study programme faced serious economic problems, but survived, thanks to Moen's strong position. Towards the end of the 20th century, the Ås school experienced strong growth. Student numbers quadrupled between the 1960s and the 1990s.

The profession, as well as the education, enjoys great respect in academia and in society in general. Today, the School of Landscape Architecture at the Norwegian University of Life Sciences, NMBU, is host to three programmes: a five-year master's in landscape architecture, a recently established two-year master's in landscape architecture for global sustainability, and a three-year bachelor's degree programme in landscape engineering. The two landscape architecture programmes have an annual total intake of 60 students and are among the most popular programmes at the university. In addition to the landscape architecture school at NMBU, the Oslo School of Architecture and Design, AHO, have since 2004 offered a two-year master's programme in landscape architecture, since 2013, in collaboration with the University of Tromsø. In 2016, the programme was extended to a full five-year master's programme. The students will be based in Oslo the first three years and in Tromsø the final two years. The total intake is 30 students per year, bringing the national number of student intake in landscape architecture to 90.

The two schools have a collaboration agreement, optimizing the special qualities in each school, and the results from student work as well as research and development projects are frequently acknowledged both nationally and internationally. The strategies seem to be increased collaboration between the different branches of architecture, landscape architecture, urbanism, environment and design. All this would hardly have been possible without the strong foundation laid by Olav Moen in the first half of the twentieth century.

Notes

1 We thank Annegret Dietze-Schirdewahn for carefully checking sources in this manuscript.
2 NLH Archive, from the senate's protocol 'Skolerådsprotokoll 1917–18', Archive number: RA/S-1572; translation by author. It is not clear to which chair the committee is possibly referring, but it is presumably at Norwegian University of Technology (NTH) in Trondheim, where the first study programme in architecture in the country was established in 1910.
3 *Aftenposten* was the leading national newspaper in 1914 (and it still is in terms of number of subscribers). The titles 'gardener', 'landscape gardener' and 'garden architect' are used fairly synonymously during this period. In Schnitler's article, Nickelsen and Røhne are sometimes described as 'gardeners', at other times as 'landscape gardeners', while *Aftenposten* uses 'garden architect' (Jørgensen and Torbjörn 1999).
4 On 3 October 1919, however, the Danish garden architect P. Wad from Odense was recommended for the post by a unanimous committee and by the board, but in the minutes of the meeting held by the latter on 20 November of the same year, it is noted that the ministry was unwilling to submit him for the professorship. NLH asked the ministry to reconsider its decision, but to no effect (NLH Archive, folder/section «NLH»).
5 The executive committee of NHL recommended that the study of the discipline should continue to be at NLH, but that it ought to be separated from the department of horticulture and that a new professorship ought to be established in the discipline. One member, Torborg Zimmer (Frølich), dissented from this recommendation suggesting the programme was transferred to the Norwegian University of Technology in Trondheim, NTH (Bruun 2004, p. 88).
6 Both notebooks are stored in the Historical Archive of Norwegian Landscape Architecture. Collection Olav Moen: https://blogg.nmbu.no/ila-samling/2018/02/olav-leif-moen-pioneer-in-norwegian-landscape-architecture/
7 From Moen's notebook, see endnote 6.

References

Birli, B. (2016) *From Professional Training to Academic Discipline*. Dissertation, Fakultät für Architektur und Raumplanung Fachbereich Landschaftsplanung und Gartenkunst, Technische Universität Wien.

Blichner, B. C. (1989) *Olav Leif Moen (1887–1951): en landskapsarkitekt i brytningen mellom nyklassisisme og funksjonalisme*. Master Thesis, NMBU.

Bruun, M. (2004) *75 år for landskap og utemiljø Norske landskapsarkitekters forening 1929–2004*. Oslo: NLA.

Fett, H. (1914) 'Amerikanske lekepladser', *Kunst og Kultur*, 4(3).

Jørgensen, K. (2019) 'An Important Step in Our Artistic Regeneration', in Osuldsen, J. (ed.) *Outdoor Voices. The Pioneer Era of Norwegian Landscape Architecture*. Oslo: Orfeus Publishing.

Jørgensen, K. and Torbjörn, S. (1999) 'Om etableringen av landskapsarkitektutdanningen i Norge og Sverige', in Eggen, M., Geelmuyden, A. K. and Jørgensen, K. (eds.) *Landskapet vi lever i*. Oslo: Norsk Arkitekturforlag, pp. 251–267.

Lund, A. (2002) *Danmarks Havekunst III*. Copenhagen: Arkitektens Forlag.

Nickelsen, I. O. (1914) 'Byer med parker, trær, lekepladser og blomster: Moderne havekunst', *Kunst og Kultur*, 4(3).

NLH (1959) *NLH, 100 år: Norges Landbrukshøgskole 1859–1959. Festskrift*. Oslo: Grøndahl & Søns Boktrykkeri.

Schnitler, C. W. (1916) *Norske haver i gammel og ny tid. Norsk havekunsts historie med oversigter over de europæiske havers utvikling*. Kristiania: Alb. Cammermeyers Forlag.

15
GERMAN GARDEN DESIGN EDUCATION IN THE EARLY 20TH CENTURY

Lars Hopstock

Introduction

On 15 October 1930, the *Landwirtschaftliche Hochschule* (Agricultural College) Berlin, Germany, was the second European university after the one in Norway (in 1919, see Jørgensen in this volume) to start offering a professional landscape architecture graduate programme. The decision for an agricultural college to install the programme had been anything but straightforward. Initially, during the first three decades of the 20th century, a variety of different institutions had been under consideration. A central figure in the discourses accompanying this process was Erwin Barth (1880–1933) who, in 1929, became the first professor to lead the new programme.[1] He had been head of the municipal garden departments in the cities of Lübeck (1908–1911) and Charlottenburg (1912–1926), and, from 1926, garden director general for the German capital more than five years after the formation of the new municipality of Greater Berlin. This chapter explains how Barth embodies several strands of the early 20th century garden design education debate. During the 19th century, in German-speaking countries the field was generally referred to as *Gartenkunst* (literally 'Garden Art'). After 1900 progressives started to refer to themselves as *Gartenarchitekt* ('Garden Architect'). In this chapter, for consistency, the terms 'Garden Designer' and 'Garden Design' are used. This also corresponds with the focus on *design* education as championed by the protagonists presented later in the chapter. The artistic qualification of practicing garden designers had been subject to criticism ever since Peter Josef Lenné (1789–1866) had first lobbied for formal design education, and beyond. When in 1823 at Potsdam the *Königliche Gärtner-Lehranstalt* was established as the first higher gardener academy in Europe, this happened not least to improve the artistic skills of future designers (P. J. Lenné quoted in Jühlke 1872, p. 27).

The aim of this chapter is to expose the diversity of developments and the wide spectrum of discourses that characterise pre-landscape architecture education in Germany. Looking at today little-known developments, this exploration offers a glimpse into perspectives that presented themselves at art colleges and institutes of technology before professional landscape architecture education became tied, for many years, to one agricultural university in Berlin. Sources employed include documents located in different archives, articles that appeared in German language journals and secondary literature. Essential are two published doctoral theses, one on Erwin Barth (Land 2005) and one on Walter von Engelhardt (Grützner 1998).

The 'academisation' debate

The antecedents of landscape architectural education in Germany are the 19th-century institutions called *Höhere Gärtnerlehranstalten* and *Höhere Lehranstalten für Gartenbau*, here referred to as 'horticultural institutes'.[2] They offered higher education and vocational training for gardeners, including courses in design. Many of these institutes originate from older institutions such as courtly forestry offices and pomology schools. They were the only institutes of higher education that offered state-approved educational programmes in garden design and that were certified to officially confer the highest educational diplomas (cf. 'Denkschrift. . .' 1930, pp. 20–21, Lange 1921, p. 63) (see Figures 15.1 and 15.2).

Many things that changed around the turn of the century had a notable impact on garden design education. The work domain expanded and increasingly included public commissions as strategic and socially oriented open space planning had become a key issue in urban development, leading to growing demands for garden design specialists in administrative positions (Gröning 2012, pp. 158–62). Accordingly, one incentive for attaining higher education and respected titles was for municipal garden intendants to be able to compete at eye-level with architects working in civil service. A second strong incentive for improving professional education had come with the momentous garden reform that dominated the garden design profession at the turn of the 20th century. All professional groups working in the field of garden design faced the superiority of the architects' academic or administrative titles. One would expect them to have shared a common interest and to be joining forces. Instead, the momentum generated towards establishing a new kind of higher horticultural education was impeded by vanities and

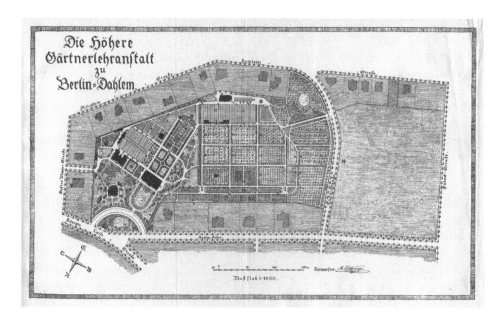

Figure 15.1 Plan for the grounds of the *Höhere Gärtnerlehranstalt* in Dahlem, designed by Theodor Echtermeyer, director of the institute; in north–south direction, in the middle axis of the main college building, lies the geometric rose and perennial gardens. The director's villa is situated in the centre of the garden in the landscape style in the north-eastern corner.

Source: Höhere Gärtnerlehranstalt Berlin-Dahlem (ed) (1924) *Denkschrift zum 100-jährigen Bestehen der Höheren Gärtnerlehranstalt Berlin-Dahlem*. Frankfurt (Oder): Trowitzsch & Sohn [no pagination/last page].

Figure 15.2 Drawing hall of the *Höhere Gärtnerlehranstalt* in Dahlem.

Source: Höhere Gärtnerlehranstalt Berlin-Dahlem (ed) (1924) *Denkschrift zum 100-jährigen Bestehen der Höheren Gärtnerlehranstalt Berlin-Dahlem*. Frankfurt (Oder): Trowitzsch & Sohn, p. 120.

contrasting notions of where such an education should be established. The disputes also reflect a diversity of professional identities defined by different traditions of plant breeding, engineering and technology, architecture and art. Anti-educational views also existed. The influential Hermann Koenig (1883–1961), for example, expressed an opinion that was wide-spread amongst independent garden designers, repeatedly warning of over-scientification and of *Akademisierung* (academisation) (Cf. Koenig 1912, pp. 335–340, 1929, pp. 15–17).

After World War I and the foundation of the first German republic, discussions focussed on four alternative educational perspectives: (i) for the most renowned of the existing horticultural institutes, the one at Dahlem (founded at Potsdam in 1823), to be upgraded as an independent college with the right to award academic degrees in horticulture and in garden design; (ii) the incorporation of the Dahlem institute as a new department into the Berlin Agricultural College (in case the independent college idea was unattainable); (iii) the introduction of a curriculum for garden design at an art college in order to reach the modern ideal of exchange amongst the arts; and (iv) a special department of garden design at a *Technische Hochschule* (institute of technology) (Högg 1920, pp. 111–113, cf. Land and Wenzel 2005, p. 432). Debate amongst proponents of these four options continued for many years. The diversity of arguments was also represented in a committee that the *Deutsche Gesellschaft für Gartenkunst* (DGfG, German Society for Garden Art) established in 1919 with the aim of finding agreement.[3] One member of the committee was Erwin Barth. Virtually until the day of Barth's appointment as professor in 1929, the debate remained unresolved.

Erwin Barth and the first full professional programme in Germany

Erwin Barth's qualification to lead a garden design programme was uncontested. After a long career that began at the firm of the independent garden designer and nursery owner Reinhold Hoemann (1870–1961) he had finally been appointed garden director general of the capital city. What came as a surprise was the fact that, after decades of debate and experimentation, it was an agricultural college that would offer the first professional landscape architecture study programme.

The decision of the Prussian parliament for the state Agricultural College to offer garden design education was spurred to action by a motion of a fraction within the profession that had failed to consult with any of the relevant professional organisations. Most of the progressive designers disagreed with the decision as they had hoped for a solution involving an art academy or an institute of technology. The *Bund Deutscher Gartenarchitekten* (BDGA) (German garden architect association), headed by Hermann Koenig, protested most stridently. The young association of independent garden designers, founded in 1913, condemned the idea of 'their art', *Gartenkunst*, being taught in the context of an agricultural institution (Milchert 1983, pp. 428–429). Barth himself was not convinced; being engaged as lecturer at the *Gärtnerlehranstalt* in Dahlem since 1920 and at the *Technische Hochschule* in Berlin since 1921, he had publicly made the case for an artistically focussed course to be offered at an art college (Barth 1920). Behind the scenes, he had also lobbied for a full professional programme in garden design to be installed at the *Technische Hochschule* (Land and Wenzel 2005, pp. 433–435).

On 1 October 1929, the *Landwirtschaftliche Hochschule* appointed two new professors, one for *Gartenkunst* (Garden Design) and one for *Gärtnerischer Pflanzenbau* (Horticultural Plant Cultivation).[4] The former was Erwin Barth, the latter Erich Maurer (1884–1981), specialist for pomiculture and at that time managing director at one of the largest tree nurseries in the world, the *Spät'sche Baumschulen* (ibid.; Escher 2010, pp. 611–612). Together, Barth and Maurer formed the new *Abteilung Gartenbau und Gartengestaltung* (Division of Horticulture and Garden Design) that had the right to confer a special doctoral degree in horticulture, the "*Dr. hort.*" (Barth 1930). In conjunction with the new division, two new curricula, or tracks (*Fachrichtungen*), were introduced: *Gartengestaltung* (Garden Design) and *Gartenbau* (Horticulture) (ibid.). The tense economic situation following the stock market crash the same month entailed that both Barth and Maurer had to fill a double position; they also served as professors at the Horticultural Institute at Dahlem, where all teaching took place. In April 1930, the division grew and Erich Kemmer (1895–1976), another Dahlem teacher, was appointed professor for *Gewerblicher Obstbau* (commercial fruit growing). An officially approved syllabus was still lacking and students of the new programmes were allowed to attend classes at three institutions, the *Höhere Gärtnerlehranstalt* in Dahlem, the *Landwirtschaftliche Hochschule* and the *Technische Hochschule*. For the track of garden design, some credits related to sciences were substituted with credits related to architecture, and these had to be earned at the *Technische Hochschule* (Ibid.; Land and Wenzel 2005, pp. 444–445). Barth involved all three institutions when developing the new syllabus and study plan and stipulating entry requirements such as proof of internships.

The garden design curriculum became effective on 15 October 1930 (Land and Wenzel 2005, p. 439). Only a few students had the opportunity to study with Barth before his suicide on 8 July 1933, shortly after the Nazis' rise to power (see Gröning in this volume).

Garden design at art colleges

The main path taken at the Agricultural College is generally considered a breakthrough. However, there were several significant side tracks leading towards a stronger representation of garden design in the architectural education system in Germany. Most of these are little known today. For the following, one has to go about two decades back in time. In July 1908, the DGfG, at its general assembly in Potsdam, by a narrow majority voted for establishing a commission consisting of Walter von Engelhardt, Fritz Encke and Reinhold Hoemann to explore possibilities of introducing a professional garden design programme at a *Kunstgewerbeschule* (College of Applied Arts) (Grützner 1998, pp. 135–137). The group contacted the Ministry of Commerce that had the regulatory supervision over trade and business schools. In this ministry, the architect Hermann Muthesius (1861–1927) was charged with exactly the matter of modernising art colleges. By that time, it had already been decided to open a new architecture department at the Düsseldorf *Kunstgewerbeschule*, under the direction of the monumentalist architect Wilhelm Kreis (1873–1955), successor of Peter Behrens as head of that school. Attaching a garden design curriculum to the new department – on a pilot basis – made much sense. Muthesius knew von Engelhardt; years before, the garden designer had contacted him to express his appreciation of Muthesius' significant contribution to the garden reform movement (Grützner 1998, pp. 80–81). Muthesius then officially suggested Kreis and von Engelhardt as teachers in the new study programme.[5] Headwind came from people who disagreed with an artistically oriented form of garden design education. There were those who claimed a prior kinship with horticulture and agriculture, and those who dreamt of a true academic elevation of garden design that only an institute of higher education could offer.[6] In order to appease the opponents of the plan and to get the Ministry of Agriculture to agree, the commission reduced technical horticultural subjects to a minimum and made a certificate from the horticultural institutes an entry requirement (Grützner 1998, p. 138). The plan of joining garden design with the other arts was realised in 1909. Two decades before Barth's appointment in Berlin, Walter Baron von Engelhardt became the first in Germany to offer courses to future garden designers at an institution ranking higher than the traditional horticultural institutes.

The teaching he offered was in the form of a 'class' called *Fachklasse für Gartenkunst* (Hoemann 1910), and it was supposed to take two to three semesters. Von Engelhardt had the rare distinction of being an academic *and* a practicing garden designer. After graduating in botany from the University of Dorpat (today Tartu, Estonia), he had started a career as botanist with the Imperial Academy of Sciences at St. Petersburg; but he became weary of it and had decided to attend the *Höhere Gärtnerlehranstalt* at Potsdam between 1891 and 1893 where the widely respected Fritz Encke (1861–1931) was his teacher. Von Engelhardt had then established his own garden design company and tree nursery in his home region Livonia. It was destroyed in the course of the Russianisation, which prompted him to move west (Grützner 1998, pp. 19–36).

At Düsseldorf, in the winter term of 1909–1910, the first six students were admitted to the newly established garden design course (Grützner 1998, pp. 141–142). In the coming years, student numbers rose (ibid.). Teaching was studio-based. Students were given functionally oriented design tasks, and von Engelhard would ask them to consider imaginary clients and their requirements. In the assignments, he would incorporate irregularly located existing trees in order to avoid schematic and stiffly symmetric designs (Hoemann 1910). In 1919, in the course of a post-war national reform of the art college system, the Düsseldorf college was dissolved and its architecture school merged with the *Kunstakademie* (Art Academy), which entailed the introduction of an artistic admission examination (Jensen 1919, p. 304). Von Engelhardt's contract was maintained; he held his successful lecture once a week and critiqued one of three atelier classes for two and a half hours, under the general supervision of the head of the department of architecture, Wilhelm Kreis (Grützner 1998, p. 187 (fn. 895), 221). However, some years after the National Socialists rose to power in 1933, the garden atelier class seems to have faded out.[7] For a while, the emancipation of garden design amongst the arts was cut short. It gained momentum again in 1947, when Hermann Mattern (1902–1971) co-founded a new college of art based on Bauhaus ideas in the city of Kassel under the name *Werkakademie*, complete with a landscape department under his guidance (Hopstock 2012, pp. 22–27).

Apart from Düsseldorf, at least one other art college had also experimented with a garden design class. It was the *Unterrichtsanstalt des Kunstgewerbemuseums*, the teaching department of the Berlin Museum of Applied Art, that introduced a department for garden design in 1919 under the guidance of the architect Franz Seeck (1874–1944).[8] The initially low demand for this class increased over the years, despite the teachers at the *Höhere Gärtnerlehranstalt* at Dahlem regarding it as unwelcomed competition (Land and Wenzel 2005, p. 431). It continued to exist when in 1924 the institution was joined with the art academy at Charlottenburg, but a programme with a full curriculum was never established.[9] Some prominent students of the Dahlem institute benefitted from the opportunity to attend Seeck's garden design studio class, including Herta Hammerbacher (1900–1984) and Ulrich Wolf (1902–1967) (Hammerbacher 1982, p. 21). Hammerbacher's fiancé Hermann Mattern mentions the lectures on garden history that the art historian Wolfgang Sörrensen (1882–1965) had given at the same institution.[10]

Researchers have given little attention to the two cases presented earlier, the garden design courses at Düsseldorf and Berlin art colleges. There are two main reasons. Firstly, both courses were short-lived. Secondly, in 1910, the general assembly of the DGfG had more or less pulled the plug on these projects when it passed a resolution that declared the implementation of a possibility for studying garden design at an institute of technology their main goal (Wieler 1911, p. 138). This resolution won the vote of the assembly only by a hair's breadth, but many delegates were unhappy with the views that von Engelhardt, Barth, Hoemann and others presented mainly because, with respect to certificates and the degrees they offered, art colleges were not a truly academic option of higher education.

Garden design education at *Technische Hochschulen*

Garden design started to appear as a subject in study programmes not only at art colleges but also at institutes of technology, where it was included in architecture curricula. One reason is that urban planning was evolving as a discipline. Influential academic teachers in this field like Cornelius Gurlitt (1850–1938) at Dresden and Karl Henrici (1842–1927) at Aachen, who can be associated with Camillo Sitte's artistic, perception-oriented approach, also addressed the questions of modern urban green (cf. Hopstock 2022). Another reason for architecture schools to include garden design is the garden reform around 1900 that considerably raised architects' interest in the design of geometrically structured green spaces. The subject then enriched the architectural palette and became part of architectural education. Professional garden designers benefitted indirectly as their field grew in importance as did their esteem amongst architects. At the same time, many garden designers feared an increasing competition by architects. Three examples serve to illustrate how broad the spectrum of architecture schools involving garden design was. At the *Technische Hochschule* in Hanover, the city garden director Julius Trip (1857–1907) lectured about garden history (Laible 1904, p. 183). As a young man, Erwin Barth had worked with the progressive Trip, who was chairing the DGfG from 1905 until his death in 1907. A different case was the eminent art historian and architect Cornelius Gurlitt at the *Technische Hochschule* in Dresden. In the lectures he started giving in 1904, he delivered, spiced with sarcasm, volleys of critique on traditionalist green spaces, at the same time contrasting these with classic and progressive examples (Laible 1904, p. 183). Barth lectured at the *Technische Hochschule* in Berlin starting in 1921. In 1927 he was awarded with an honorary professorship.[11] He explicitly included no designing exercises into his lectures in order to "[. . .] protect the non-expert from amateurism in this field" ("Persönliche Nachrichten" 1921, p. 340, author's translation). Architects should merely be enlightened as to the high social relevance of contemporary garden culture that existed in German cities.

Another interesting development took place at Aachen. In 1911, the botanist Arved Ludwig Wieler (1858–1943) disclosed plans for a two-semester curriculum for garden designers with an artistic focus to be offered at the *Technische Hochschule* in Aachen (cf.: Erkes 1912). His reasoning was that graduates from the horticultural institutes had good knowledge of gardening and plants but lacked design skills and artistic taste, both of which the new course was to focus on. The faculty members at Aachen comprised some well-respected personalities such as the Impressionist painter August von Brandis (1859–1947) and Karl Henrici (Wieler 1911, p. 140). The programme most probably did not conclude with a formal graduation certificate.[12]

Perhaps the most colourful figure in the context of garden design education at the institutes of technology was Max Laeuger (1864–1952), a multitalented architect-artist, influential ceramicist and Werkbund co-founder. In early 20th-century garden history, he is famous for having produced *the* quintessential reference work for the reformist ('architectonic') garden at Mannheim, the temporary *Internationale Kunst- und große Gartenbau-Ausstellung* ('International Art Exhibition and Great Gadten Exhibition') laid out to celebrate the 300th town anniversary in 1907 (see Figure 15.3).

Laeuger was professor of drawing at the *Kunstgewerbeschule* in Karlsruhe since 1894. He also lectured at the at the *Technische Hochschule* in Karlsruhe, where he became full professor for *Figuratives Zeichnen* (figurative drawing) in 1898 and broadened his subject to include interior design and garden design.[13] Thus, in 1898, the *Technische Hochschule* at Karlsruhe was probably the only university-level institution in Germany where the field of garden design was represented beyond mere lectures on garden history.[14] From documents in his estate, it appears that

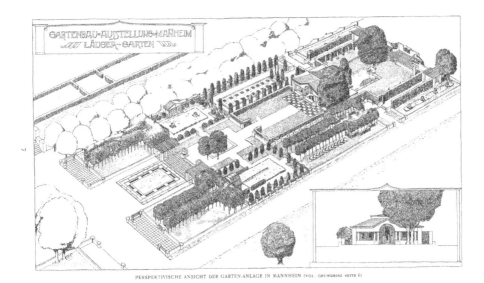

Figure 15.3 Jubilee Exhibition at Mannheim in 1907. This compartmentalised garden by Max Laeuger with bathing house at the centre represents only a part of his overall design that also included a series of exhibition buildings.

Source: N.N. (1907) *Gärten von M. Laeuger: Ausstellung Mannheim*. München: Bruckmann, p. 7.

considering gardens was but one of several means for Laeuger to discuss general questions of spatial perception and colour contrast. The complete embeddedness of the subject of the garden into the general design propaedeutic may originate in Laeuger's ideal of the *Gesamtkunstwerk*, which culminated in his work on the gigantic Bunge estate in Aardenhout, Netherlands (1908–1911) (Cf. Schumann 2014, pp. 182–183). Looking at his respective publications and at student work kept at the archive at Karlsruhe, he clearly took plants and vegetation as serious as buildings when discussing the general foundations of spatial design. Laeuger's *Kunsthandbücher* ("Art Handbooks", 1937–39; see Figure 15.4) presented 'good' versus 'bad' examples in a bold and simple way.

Laeugner's pedagogic concept seems to resonate in a seminal garden reform polemic, the volume on gardens in the series *Kulturarbeiten* (1902) by Paul Schultze Naumburg (1869–1949), who had studied at Karlsruhe. More evidence for Laeuger's popularity one can find in form of dedications in best-selling books on architecture and garden design (cf. Hopstock 2022).

Conclusions

Starting to shed light on the multi-branched nature of educational histories in the field that became landscape architecture in Germany, one may appreciate the significance of personal networks and the connectedness of professional identities, both of which reveal how individuals and groups acting in concert governed developments. The dynamics within the professional associations are a topic worth exploring further,[15] including the role that garden designers played within the Werkbund – almost all names appearing in this chapter were members.[16] Contacts were sometimes actively sought, but non-intentional ones appear to be equally significant as biographical connections between Barth, von Engelhardt, Encke and Hoemann illustrate. It seems that those devoted to furthering the profession through education inspired each other across generations. The key players were highly educated and experienced professionals. In the case of von Engelhardt, for example, it can be assumed that his literacy and scientific research experience will have contributed to his critical mind and his teaching competence. As city garden directors, both von Engelhardt and Barth held the highest administrative-political position a garden designer could reach at the time.

In deciding what institution of higher education landscape architecture education would be affiliated with, pragmatism prevailed. Apparently, the group that voted for the agricultural college had commanded the better means and networks. It was able to push through with their plan for starting a combined university education of garden designers and horticulturists in 1930. Barth, again in a pragmatic way, felt the need to defend this development in a detailed account of his conceptualising of the new curriculum (Barth 1930).

The partly short-lived examples from Berlin, Dresden, Aachen and Karlsruhe clearly document how the institutes of technology were relevant areas of innovation leading the garden design discipline gradually towards higher academic levels. More research is needed to learn about the involved institutions' role in the progress made in this field. More knowledge would become available about teaching contents as practiced and about student biographies, as well as for quantitative analyses with regard to the development of curricula, syllabi, certificates and student numbers.

A central aspect of studies in the field of pre-landscape architecture design education is that of the continuities and discontinuities with regard to different institutions. In the decades after 1945, many of the old *Gärtnerlehranstalten* became universities of applied sciences, for example the ones in Dahlem (Berlin), Freising-Weihenstephan and Geisenheim. The professional

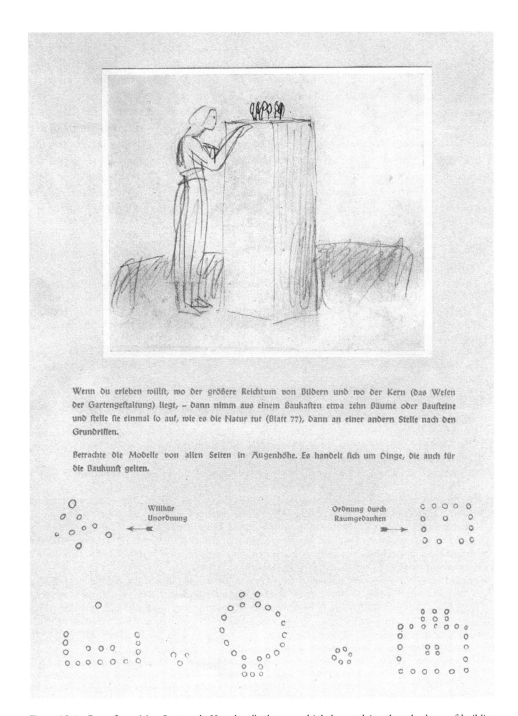

Figure 15.4 Page from Max Laeuger's *Kunsthandbücher* on which he explains that the laws of building are also valid for arranging trees; the annotations with the arrows read (left) "Arbitrariness, disorder" and (right) "Order through spatial ideas".

Source: Laeuger, Max (1938) *Kunsthandbücher*. Vol 2 ('Grundsätzliches über Malerei, Städtebau, Gartenkunst und Reklame'). Pinneberg bei Hamburg: Beig, p. 78.

programmes today offered at the Technical Universities of Berlin and Dresden both have roots in the Agricultural College of Berlin where education started in 1929–1930 (Wimmer 2009, pp. 34–35). Other continuities may still come to light in future studies. At the same time, there are many discontinuities, and these seem to link with economic depressions such as those occurring during the Weimar Republic. Under the circumstances, even the most successful horticultural institute might be terminated, such as, for example the one at Proskau, Silesia, that fell victim to austerity measures around 1922 (Duthweiler 2007, p. 138). Efforts to establish new curricula faded into oblivion soon after they had been initiated. In the light of the stories told in this chapter, it is worth exploring how landscape architecture programmes developed that German universities started to offer after 1945 and to what extent they were building on efforts of earlier years. This is a matter of doing basic research and of synthesising pieces of information contained in countless studies that have already appeared over a period of many decades.

Notes

1 For all biographic information on Barth in this chapter, see Land and Wenzel 2005.
2 These institutes, in the literature also called 'gardener academies', were located in Geisenheim (Prussia), Köstritz (Thuringia), Pillnitz (Saxony), Potsdam (Prussia, in 1903 transferred to Dahlem), Proskau (Prussia, closed 1922/23) and Weihenstephan (Bavaria) (cf. 'Denkschrift. . .' 1930, p. 21). Until the early 1920s, the Prussian institutes seem to have had higher entry requirements than others, which was important for graduates striving for civil service employment (Lange 1921, p. 63).
3 The DGfG had been established as a progressive alternative to the traditionalist *Verein deutscher Gartenkünstler* (Association of German Garden Artists, VdG), which in turn was dissolved into the DGfG in 1910. Confusingly, in 1914, another association with the same acronym was founded, the *Verein der Gartenarchitekten e. V.* (VdG) (Hennebo 1973, p. 10). Members of the mentioned committee were Erwin Barth, Carl Heicke, Hermann Kube, Christian Roselius, Friedrich Scherer, Wolfgang Singer, Heinrich Wiepking and Fritz Zahn (Heicke 1919, pp. 144–152 (150)).
4 A first step had actually been taken in 1924, when the *Landwirtschaftliche Hochschule* appointed Theodor Echtermeyer, head of the *Höhere Gärtenerlehranstalt* in Dahlem, as honorary professor (Höhere Gärtnerlehranstalt Berlin-Dahlem 1924, p. 30).
5 He informed von Engelhardt about this step; see typescript of letter from H. Muthesius to W. von Engelhardt, 16/01/1909, Werkbund-Archiv Berlin, file D102–178.
6 The more traditionalist VdG – Muthesius suspected Willy Lange to be behind it – was also interfering with the DGfG's plan (Grützner 1998, p. 137). Lange preferred an association with agriculture. As example for this position, expressed some years later, see: Lange 1923, pp. 197–19.
7 In a report in the context of the *Gleichschaltung* (quoted in Grützner 1998, p. 221) von Engelhardt was evaluated as "outstanding educator", deemed superfluous and suspended from office. The decision was reversed one year later.
8 Bruno Paul (1874–1968), the influential reformer who headed the institution, published a brochure in 1919 that lists *Gartenkunst* alongside architecture and interior design in the section *Baukunst* (Paul 1919, p. 10). The so-called *Abteilung für Gartenkunst* ('Department of Garden Design', probably existing merely in the person of Franz Seeck) was joined with the Department of Architecture in 1921; see Archive of the University of the Arts, fonds no. 7/2 ('Organisation der Unterrichtsanstalt').
9 The new college was named *Vereinigte Staatsschulen für freie und angewandte Kunst*.
10 Letter from Mattern to Hammerbacher (without no.), 10 February 1927, Herta Hammerbacher collection at the University Archive of the Technische Universität Berlin.
11 See the entry in the *Catalogus professorum* on the website of the University Archive of the Technical University of Berlin, https://cp.tu-berlin.de/person/376 (Accessed 03 December 2021).
12 Evidence of the programme exists in contemporary journals, but not in the university archive at Aachen.
13 The syllabus programme documented at the archive of the Karlsruhe Institute of Technology for the first time lists Laeuger in winter term 1895/6 with "exercises in decoration" (*Übungen im Dekorieren*).
14 The university never introduced a full professional programme for landscape architecture. However, with hindsight, Laeuger's teaching was considered so successful that the appointment in 1965 of the Swede

Gunnar Martinsson (1924–2012) as the new chair for Landscape and Garden happened with reference to Laeuger (personal information, Dr. Gerhard Kabierske of the Archiv für Architektur und Ingenieurbau at the Karlsruhe Institute of Technology, 06 October 2020). Martinsson's successors were the similarly distinguished Dieter Kienast (1945–1998) and, as of today, the French landscape architect Henri Bava (*1957).

15 Some studies already exist, e.g. Gert Gröning and Joachim Wolschke-Bulmahn, *DGGL. Deutsche Gesellschaft für Gartenkunst und Landschaftspflege e.V., 1887–1987: Ein Rückblick auf 100 Jahre DGGL*, series Schriftenreihe der DGGL, 10 (Berlin: Boskett, 1987).

16 Von Engelhardt and Encke joined by invitation already the year after the association's foundation on 05 October 1907. Founding members were twelve artists – Peter Behrens, Theodor Fischer, Josef Hoffmann, Wilhelm Kreis, Max Laeuger, Adelbert Niemeyer, Josef Olbrich, Bruno Paul, Richard Riemerschmid, J.J. Scharvogel, Paul Schultze-Naumburg, Fritz Schumacher – and twelve firms. See the chronology of Deutscher Werkbund on the association's web site. Available at: www.deutscher-werkbund.de/wir-im-dwb/werkbund-geschichte/chronik-des-deutschen-werkbundes-1907-bis-1932/ (Accessed 3 December 2021).

References

Barth, E. (1920) 'Ausbildung der Gartenkünstler an Kunsthochschulen', *Die Gartenkunst*, 22(9), pp. 136–137.

Barth, E. (1930) 'Die hochschulmäßige Ausbildung des Gartenarchitekten', *Der Deutsche Gartenarchitekt*, 7(9), pp. 101–103.

'Denkschrift zum gesetzlichen Schutz der Berufsbezeichnung "Gartenarchitekt"' (1930) [original document written in 1928]. *Der Deutsche Gartenarchitekt*, 7(2), pp. 20–21.

Duthweiler, S. (2007) 'Die Königlich Preußische Gärtnerlehranstalt zu Proskau. Ein Baustein in der Geschichte moderner Gartenarchitektur', *Die Gartenkunst*, 19(1), pp. 127–142.

Erkes, A. (1912) 'Studium der Gartenarchitekten an der technischen Hochschule zu Aachen', *Die Gartenkunst*, 17(11), p. 180.

Escher, F. (2010) 'Späth, Franz Ludwig', *Neue Deutsche Biographie*, 24, pp. 611–612 [Online]. Available at: www.deutsche-biographie.de/pnd117648663.html#ndbcontent (Accessed 30 July 2021).

Gröning, G. (2012) 'Die Institutionalisierung der Gartenkunst in der kommunalen Verwaltung und in der Ausbildung im 19. und 20. Jahrhundert', in Schweizer, S. and Winter, S. (eds.) *Gartenkunst in Deutschland: von der Frühen Neuzeit bis zur Gegenwart. Geschichte, Themen, Perspektiven*. Regensburg: Schnell+Steiner, pp. 158–180.

Grützner, F. (1998) *Gartenkunst zwischen Tradition und Fortschritt: Walter Baron von Engelhardt (1864–1940)*. Bonn: Lemmens.

Hammerbacher, H. (1982) 'Frühe Arbeitsjahre Hermann Matterns von 1926 bis ca. 1939', in Akademie der Künste, A. (ed.), *Hermann Mattern 1902–1971: Gärten, Gartenlandschaften, Häuser*. Series Akademie-Katalog, 135. Berlin: Akademie der Künste & Technische Universität Berlin, pp. 21–23.

Heicke, C. (1919) 'Die Tagung in Weimar: Bericht über die XXXII. Hauptversammlung der Deutschen Gesellschaft für Gartenkunst, Weimar 24.-29. September 1919', *Die Gartenkunst*, 32(11), pp. 144–152.

Hennebo, D. (1973) 'Gartenkünstler – Gartenarchitekt – Landschaftsarchitekt, Versuch einer Übersicht über die Entwicklung des Berufes und Berufstandes in Deutschland von den Anfängen bis zur Neugründung des BDGA im Jahre 1948', in BDLA (ed.) *Der Landschafts-Architekt. Das Berufsbild des Garten- und Landschafts-Architekten*. München: Callwey, pp. 7–21.

Hoemann, R. (1910) 'Von der Fachklasse für Gartenkunst an der Düsseldorfer Kunst- und Gewerbeschule', *Die Gartenkunst*, 12(7), p.123.

Hopstock, L. (2012) 'Vom Bauhaus zum Studium generale: Der Landschaftsarchitekt Hermann Mattern (1902–1971) als Lehrer', *Stadt+Grün*, 61(7), pp. 22–27.

Hopstock, L. (2022) 'The garden as *Raumkunstwerk*: The Role of Early 20th-Century Architecture Schools for the Modernisation of Landscape Architecture Education | Le jardin comme *Raumkunstwerk* : le rôle des écoles d'architecture dans la modernisation de l'enseignement de l'architecture du paysage au début du XXe siècle', *Projets de paysage*, special issue 'Devenir paysagiste' [Online]. Available at: https://doi.org/10.4000/paysage.27504.

Högg, E. (1920) 'Gedanken zur Neugestaltung des Studiums der Gartenkunst', *Die Gartenkunst*, 22(8), pp. 111–114.

Jensen, H. (1919) 'Unser Bildungsunglück und die staatliche Kunstakademie zu Düsseldorf', *Die Gartenwelt*, 23(38), pp. 303–304.

Jühlke, F. (1872) *Die Königliche Landesbaumschule und Gärtner-Lehranstalt zu Potsdam: Geschichtliche Darstellung ihrer Gründung, Wirksamkeit und Resultate, nebst Culturbeiträgen*. Berlin: Wiegandt & Hempel.

Koenig, H. (1912) 'Non scholae, sed vitae discimus: Ein Beitrag zur Ausbildung des Gartenarchitekten (Deutschland)', *Die Gartenkunst*, 14(22), pp. 335–340.

Koenig, H. (1929) 'Zur Hochschulfrage (Deutschland)', *Der Deutsche Gartenarchitekt*, 6(2), pp. 15–17.

Laible, P. (1904) 'Gartenkunst an der Dresdener Technischen Hochschule', *Die Gartenkunst*, 6(10), pp. 183–184.

Land, D. (2005), *Erwin Barth (1880–1933). Leben und Werk eines Gartenarchitekten im zeitgenössischen Kontext*, doctoral dissertation, Fakultät VII – Architektur Umwelt Gesellschaft, Technische Universität Berlin, Berlin.

Lange, T. (1921) *Der Gärtnerberuf. Ein Führer und Berater von der Lehrzeit bis zur Selbständigkeit*. 2nd ed. Berlin: P. Parey.

Lange, W. (1923) 'Betrachtungen am Weg zur Hochschule', *Die Gartenwelt*, 27(22), pp. 197–199.

Milchert, J. (1983) 'Die Entstehung des Hochschulstudiums für Gartenarchitekten an der Landwirtschaftlichen Hochschule in Berlin im Jahre 1929', *Das Gartenamt*, 32(7), pp. 428–437.

Paul, B. (1919) *Erziehung der Künstler an staatlichen Schulen*. Berlin: Unterrichtsanstalt des Kunstgewerbemuseums.

'Persönliche Nachrichten' [section title] (1921) *Die Gartenwelt*, 25(34), p. 340.

Schumann, U. M. (2014) 'Arkadische Mauern, unauffällige Erfolge – Max Laeuger als Architekt', in Badisches Landesmuseum (ed.), *Max Laeuger. Garten Kunst Werk*. Karlsruhe: Badisches Landesmuseum, pp. 177–89.

Wieler (1911) 'Studium für Gartenarchitekten an der Technischen Hochschule zu Aachen', *Die Gartenkunst*, 13(8), pp. 138–142.

Wimmer, C. A. (2009) 'Die Bibliothek des Berliner Instituts für Landschafts- und Freiraumplanung und seiner Vorgänger seit 1929. Ein Beitrag zur Geschichte der Berliner Lehre und Forschung im Fach Garten- und Landschaftsarchitektur', in Heinrich, F. (ed.), *Zwölf Aufsätze für Vroni Heinrich zu Gartenkunst und Landschaftsplanung*. Berlin: Universitätsverlag der TU Berlin, pp. 29–46.

16
LANDSCAPE ARCHITECTURE UNIVERSITY EDUCATION UNDER NATIONAL SOCIALISM IN GERMANY

Gert Gröning[1]

Using relevant publications and documents held in archives, this chapter presents evidence that shows how the decline of a supposedly superior Western civilization, a distinctive feature of National Socialist ideology, reflects in landscape architectural education at German universities. Compared with the age and number of studies in philosophy, medicine, law and so on at German universities, studies in landscape architecture are more recent and modest in number. The penetration of National Socialist ideology, not through official sources of propaganda, but rather through the efforts of zealous professionals as documented later for landscape and garden design, had its equivalent in other areas of thought and in other professions (Etlin 2002).[2]

Beginnings of landscape architecture education in the German university

In the German principalities and states, there were various hopes of and aspirations to establish a university education in the area of professional garden and landscape design, especially in the second half of the 19th century. For example, the *Verein deutscher Gartenkünstler* (German Garden Artists Association), founded in 1887, specifies in paragraph 3 of its statutes the aim of "introducing an extensive training of garden artists in a suitable teaching establishment and ultimately the establishment of an institute of higher learning for garden art" (Gröning and Wolschke-Bulmahn 1987a, p. 22), but this did not happen until 1929 during the Weimar Republic when a course of studies for landscape architecture was introduced in the *Landwirtschaftliche Hochschule* (Agricultural College) Berlin.[3]

The 1930 plan of Erwin Barth (1880–1933) for the study programme of landscape architecture in Berlin (Land and Wenzel 2005)[4] was far more comprehensive than the subjects addressed in Gustav Meyer's 19th-century *Lehrbuch der schönen Gartenkunst* (Teaching Manual of Fine Gardening) (Meyer 1860). Barth's elaborate plan for a six-semester course aimed at training students for professional practice. He drew attention to his proposal that "the Department of Horticulture and Garden Design at the Agricultural College will enable students to study for a doctorate degree" (Barth 1930, p. 102; this and all following quotes are translated from German to English by the author) and emphasized the academic nature of the course. But Barth and his assistant, the garden designer Wilhelm Feldmann (1903–?), who worked in Hanover, had no more than six semesters to train students, according to the curriculum (Anon 1929). For inexplicable

DOI: 10.4324/9781003212645-18

reasons, a few months after the NSDAP (*Nationalsozialistische Deutsche Arbeiter Partei*, National Socialist German Workers' Party) had won the elections and Adolf Hitler (1889–1945) had become *Reichskanzler*, Barth committed suicide in July 1933 (Land 2004).

The National Socialist takeover of landscape architecture studies

In the initial stages of National Socialism the way was open, as Hermann Koenig (1883–1961) (Gröning and Wolschke-Bulmahn 1997, pp. 193–195), the Hamburg garden architect and editor of the magazine *Der Deutsche Gartenarchitekt* wrote in 1934, to follow "the genius of Adolf Hitler, who gave us a new worldview (Weltanschauung)" (Koenig 1934, p. 347) and appoint a new holder of the Chair at the *Institut für Gartengestaltung an der Landwirtschaftlichen Hochschule Berlin* (Institute for Garden Design at the Agricultural College in Berlin). As the garden architect Hertha von Oven noted in her report of the meeting of the Society for Garden Art on 29 July 1933, "Hitler had done the appropriate 'groundwork' for the seed to grow" (von Oven 1933, p. 405).

National Socialism was supposedly meant to replace "in the new state, at the same pace as the progress made in Western civilization, what had become unusable with fixed laws of a social order that was increasingly c u l t u r e – d o m i n a t e d [spaced in the original, GG]" (Saathoff 1933, pp. 225–226). The National Socialists wanted to be the only ones to determine what was true and what was not true. On 1 July 1934, the Barth's position, vacant after his suicide, was filled by Heinrich Friedrich Wiepking (1891–1973) (Gröning and Wolschke-Bulmahn 1987b; Gröning 2001, 2006), whose only professional qualification was a certificate of apprenticeship (Kellner 1998). To qualify for holding a chair at an academic institute of higher education under the National Socialists apparently required no proof of study, let alone an academic degree. It was sufficient to be able to manage an office successfully and perform as a National Socialist loudspeaker, which other professional colleagues could do as well, besides Wiepking. The opportunity to do research in this field at an institute of higher education, envisaged by Barth, was deliberately ignored.

In 1933, Wiepking maintained somewhat grandiloquently:

> Should one lift the green mantle of love with which nature cloaks ignorance, that fashionable and fallacious Romanticism, one would discover a gruesome mass murder of millions of plants, whose false and unhorticultural planting has consumed astronomical amounts of money and made some former garden-lovers cynical and sceptical.
> (Wiepking-Jürgensmann 1933, p. 193)

Wiepking's lectures in the winter semester of 1934–1935 were the starting point of the National Socialist transformation of landscape and garden studies at the university (Goetze 1934). Lecturing was a special challenge for him. Without any supporting proof, he claimed there was a shift from garden to landscape in "hands-on practice" (Wiepking-Jürgensmann 1937a, p. 3).

The concept of university education in garden and landscape studies under National Socialism

As early as 1920 Wiepking had proved that he was unqualified to work as a university teacher in a democratic society like the Weimar Republic. In an article described as a "pre-study" on Frederick the Great (1712–1786), he advocated renewed settlement of the East "to counteract

Sarmatian and Hunnish non-culture (*Unkultur*)" (Wiepking 1920, p. 77).[5] In the same article, his appreciation of King Frederick is positively dripping with superlatives:

> He, Frederick, the frequently despised despot, became through his deeds the most ideal creator, the greatest, the truest Socialist of our blood and our country. May this be recognized by all 'fairweather' social revolutionaries and democrats and may lunacy and 'party-muddle' (*Parteisuppe*) be left to one side. We have to leave megalomania to one side and together with Frederick, acknowledge that we are, at best, merely 'insignificant marionettes' operated by divine hands.
> (Wiepking 1920, p. 71)

Wiepking also expressed the unsubstantiated belief that gardeners were inept: "We gardeners lack depth and strength and ability" (Wiepking 1920, p. 75). In direct opposition to what he actually did, he maintained that his concern was reconciliation: "Our aim is not to judge but to reconcile and construct from the ground upwards" (Wiepking 1920, p. 78).

As newly appointed professor of Garden Design, he gave an "inaugural lecture" published in 1935, in which he clambered from one cliché to the next, on the subject of *The profession and tasks of a garden designer* (Wiepking 1935a). The term 'inaugural lecture' is a misnomer, because it normally follows a successful post-doctoral process, marking the beginning of entitlement to deliver university lectures (*venia legendi*) and is tied to the title of *Privatdozent* (unsalaried lecturer paid by student fees), none of which applies in this case.

Without even attempting to understand the social development of the previous decades, Wiepking began to construct a fairy-tale image of ruralization, which he wanted to support with "landscape study" (*Landschaftskunde*) and "new village communities" (Wiepking 1935a, p. 43). In the course of his professorial activity as a National Socialist sympathizer who used "blood and soil" (*Blut und Boden*) language, Wiepking, who believed that he was following a good "gardening bloodline" (Wiepking 1935a, p. 43) from Lenné to Barth, wanted to shift the focus from the town to the village landscape.

Because Wiepking had nothing to lecture about, he came up with the idea of a "competition for the acquisition of teaching material" in 1935 (Wiepking-Jürgensmann 1935b, p. 33). The competition was open to "all blameless German citizens"; and the Institute for Garden Design (*Institut für Gartengestaltung*) at the University of Berlin should have the "right to use any of the submitted texts for lectures" (Wiepking-Jürgensmann 1935b, p. 33). The key to success in the competition was tied to the supposedly inherent superiority of respectable Germanness. Once again, a pre-eminent right claimed by National Socialists was meant to remove the connection with the hated "liberalist" (Dörr 1939, p. 200) concept of universal human rights that characterized the Weimar Republic.

Wiepking clearly found it difficult to switch from a view suitable for a private office to a perspective appropriate to a university and hence to academic work. Moreover, his complete lack of knowledge of military technology led him to a concept of "defence landscape," and as early as 1937 he integrated military considerations into the university course (Wiepking-Jürgensmann 1937a, 1937b).

In May 1936, Wiepking applied to the German Research Association (*Deutsche Forschungsgemeinschaft* (DFG)) and boasted that he was researching, with his students who could not have spent more than three semesters "studying" with him, the "German cultural landscapes inside and beyond the German borders on the basis of ethnic history and settlement techniques" (Wiepking 1936, letter). The DFG did not respond. Wiepking tried again in early 1937 and enlisted the support of his friend Konrad Meyer (1901–1973) (Gröning and Wolschke-Bulmahn 1987b;

Figure 16.1 Certificates of Honour for especially good achievements in the first Reich achievement competition of German universities and colleges (*Ehrenurkunden für besonders gute Leistungen im 1. Reichsleistungskampf der deutschen Hoch- und Fachschulen*) for Gert Kragh, Werner Lendholt, Gerhard Neef, Gerhard Prasser, Dietrich Roosinck, Ulrich Schmidt.

Source: *Die Gartenkunst*, 1937, 50 (1), p. 1.

Jung 2020), the Deputy Director of the DFG and Director of the Department of Agricultural Science and General Biology (*Fachgliederung Landbauwissenschaft und Allgemeine Biologie*) of the National Socialist Research Council founded in 1937. Meyer, who joined the NSDAP in 1932 and was the incumbent chairholder of Agriculture and Agrarian Politics at the Friedrich-Wilhelm-Universität Berlin, was an SS officer and in 1942 drew up the genocidal National Socialist *Generalplan Ost* (Pospieszalski 1958;[6] Heiber 1958;[7] Madajczyk 1988; Rössler and Schleiermacher 1993).[8] Wiepking wrote that in December 1936 "he had allowed Meyer to look at three diploma degree theses" and that Meyer had assured him that "he was ready to do his best to support this work" (Wiepking 1937, letter).

The theses in question were probably those of Werner Lendholt (1912–1980) (Gröning and Wolschke-Bulmahn 1997, pp. 223–224), Hans-Ulrich Schmidt (1912–2006)[9] and Fritz Rose (1908–1942), whom Wiepking in his sloppy way regarded as "pointing in some ways to the right path" (Wiepking-Jürgensmann 1937a, p. 3). Their task was not only to demonstrate the landscape significance of Germanic bronze-age prehistory but also the "total connection between land and people" as well as to clarify "the first obligation of the landscape designer" (Wiepking-Jürgensmann 1937a, p. 3). In the "First National Competition" (*Reichsleistungskampf*) in 1936 of German higher education establishments (*Hoch- und Fachhochschulen*) two of the authors received a distinction for their work (see Figure 16.1).

Drawings and explanations of student plans show that, at least in the case of the younger students in the first years, it was all about planning of houses and allotment gardens (Wiesner 1937). University education in garden and landscape studies in National Socialist Germany (Ignatius 1939)[10] continued to be an area that did not require academic work.

Consequences of war for National Socialist landscape designers

In 1939, Wiepking renamed the Institute for Garden Design (*Institut für Gartengestaltung*) into the Institute for Landscape and Garden Design of the University of Berlin (*Institut für Landschafts- und Gartengestaltung der Universität Berlin*) (Anon 1940, p. 4). The name change was intended to express in the Institute's title the shift of professional activity, of which he himself had no experience, from garden to landscape.

After Germany invaded Poland in September 1939, apparently opening up new planning opportunities for landscape designers (Gröning 1989; Gröning and Wolschke-Bulmahn 1989), the army conscripted many students. In the territories that from 1939 onward were regarded as the German East (*Deutscher Osten*) (Mindt and Hansen 1940; Hoffmann 1942), Wiepking saw "an urgent task for our students" (Wiepking-Jürgensmann 1939, p. 193). Referring to his 1920 article about King Frederick as a great colonizer, he pointed out that he could at that time "have had no idea that in such a short time Hitler would be able to realize Socialism and to create a German territorial empire of a size never before seen" (Wiepking-Jürgensmann 1939, p. 193).

Wiepking now saw undreamt-of possibilities of employment for the landscape designers being trained by him. "But today I believe that, after the Reich is finally secured, there will be a period of florescence for German landscape and garden designers that will exceed what even the most fervent of us dared to dream of". These are words of a man who was sufficiently deluded to think, like so many others did, that he knew what was really happening (Wiepking-Jürgensmann 1939, p. 193). Those who had studied under him were to see the start of "a complete re-Germanification of the large territory and find a final solution which would result in this seminal country's being one of the most beautiful in the crown of blooming German territories. Only then will the future forever be secure!" (Wiepking-Jürgensmann 1939, p. 193).

According to the ideas developed in the *Generalplan Ost* by Konrad Meyer, "German people were to put down roots in a completely foreign environment and to secure for themselves a continued biological existence (Meyer 1942, p. 18). The inhuman attitude cultivated in the training of the new generation of landscape designers can be seen in their contributions to the "National Professional Competition of German Institutes of Higher Education 1939" (*Reichsberufswettkampf der deutschen Hoch- und Fachschulen 1939*). In this event, several of Wiepking's students were mentioned with distinction in the category of Space and Settlement, for "care paid to the landscape of an endangered area in the border area of Posen-West Prussia" (Barth et al. 1939, p. 194). The names of the students were Jürgen Barth (1911–2001)[11] (Heinrich 2004), Hans W. Schmidt,[12] Fritz Müller,[13] Walter Hollmann, Willy Reinardy, Helmut Löhmer,[14] Hans-Otto Sachs, Peter-Fritz Gabriel, Erich Paul, Willi Koberg and Peter Krause. As new landscape planners, Wiepking's students, in keeping with National Socialist ideology, considered "completely new blood in the border area. . . very important. . . the best farming families of high achievement. . . in the endangered areas in isolation from others" (ibid., p. 196). "Everything uneconomical, squalid or foreign [was] contrary to their planning and [had] to be removed sooner or later" (ibid., p. 197).

The social-military reality began to be displaced by boastful claims of the National Socialist propagandists, and these were also heard by students of landscape design in Berlin. At the beginning of 1940, in the vilifying words of SS member Josef Pertl (1899–1989), the director of the urban gardens in Berlin "history has taken a course that is not surprising in view of the arrogance of the English and the bloodlust of World Jewry behind them (. . . *Blutgier des dahinter stehenden Weltjudentums* . . .)" (Pertl 1940, p. 1). Boastfully, Pertl asserted: "whereas we cannot know precisely what forms and dimensions this war will assume, about one thing we can be certain, that victory will be ours" (Pertl 1940, p. 1).

Garden design, as advocated by Wiepking's predecessor Barth as part of the university degree course, had disappeared after the invasion of Poland. Wiepking believed that "Aesthetic thoughts", as may have been present in the old garden designer, "play no role in the current landscape policy" (Wiepking-Jürgensmann 1940b, p. 92).

Wiepking confidently affirmed: "We have to construct such a total defence landscape [*Wehrlandschaft*] that farmers and soldiers can enjoy the untrammelled and enduring pleasure [of security]" (Wiepking-Jürgensmann 1940a, p. 116). How these German landscapes were to look in the East, he hoped to be able to explain by means of a plan that showed "the backbone of a new German defence landscape" (Wiepking-Jürgensmann 1940b, p. 96). The idea of defence landscape Wiepking set out in detail in his 1942 book *Die Landschaftsfibel* (Landscape Primer) and made use of the 1940 glacial valley plan of Posen. This work presented "a visible, meaningful, and relief-enhancing landscape of high value from a military and natural economic perspective" (Wiepking-Jürgensmann 1942, p. 322; see also Gröning and Wolschke-Bulmahn 2021).

On the assumption that National Socialist leaders would believe him unreservedly, Wiepking was bold enough to maintain what could not be substantiated by close scrutiny or was not convincing with the planning suggestions derived from his work. Emphasizing the importance of his idea of a defence landscape, pretending to be a moral instance and vaguely blaming others, Wiepking believed that it would be "an inexcusable omission if the landscape designer were not seriously and responsibly to point out and endeavour to ensure that future landscape design had to be regarded as a clear defence measure that was to be implemented" (Wiepking-Jürgensmann 1942, pp. 321, 323).

Of the approximately 100 students who had completed their course and of the 70 in the middle of it, there would be at most 40 students available, because of other duties, for landscape maintenance in the East, "a number that is by no means sufficient, based on current knowledge of the situation, to be able to perform the tasks in a few years" (Wiepking 1940, letter).

The continuing indifference to working out an academic basis for university education in landscape design was expressed again at a meeting of the Advisory Councils and State Group Leaders of the German Society for Garden Art (*Beiräte und Landesgruppenleiter der Deutschen Gesellschaft für Gartenkunst e.V.*) with the representatives of higher education institutes on 15–16 November 1940, at which Wiepking thought that a second and "perhaps even a third higher education institute for landscape and garden design" (Mappes 1941, p. 24) might be required.

Many gaps reveal how threadbare the National Socialist landscape education was. In 1940 the Institute was moribund.[15] In the following years the possibilities of attracting "young academically interested people. . . to acquire the title of 'Diplomgärtner' (graduate gardener) by means of a university study course, which would enable them to rise to leading administrative and research posts" got worse because of the war (Genthe 1943, pp. 1–2). On 22 June 1941 Germany attacked the Soviet Union, thereby opening a new front to the war, the end of which was "not the successful invasion of Moscow in autumn 1941 hoped for by the German leaders" (Rürup 1991, p. 8).

Regular study was no longer possible for many. In August 1943 Wiepking unscrupulously gave a student in the institute of landscape and garden design, Max Fischer (1902–1979),[16] the thesis project of "being responsible for landscaping the new town of Auschwitz in Upper Silesia" which would reflect "the green political trains of thought. . . of his own origin" (Wiepking-Jürgensmann 1943). The student in charge of the planned submission date of 1 December 1943 was unable to attend "because of serious bomb damage to his place of residence", and the appointment was postponed until 7 February 1944. The concentration camp at Auschwitz had been set up in the Polish city of Oświęcim in 1940 (Rudorff 2018).

Wiepking left Berlin on 14 February 1945 before the Soviet army, which had been advancing rapidly from the beginning of January 1945 to the Oder River only 60 km east of Berlin, entered the city and captured it on 2 May 1945 (Kellner 1998, p. 280).

Thus, the National Socialist-influenced training of landscape designers in the only university in the country with such a programme came to an inglorious end. To understand and explain how the National Socialist character of university courses made itself felt in the period of landscape architecture after the collapse of National Socialism has yet to be undertaken.

Notes

1 I am grateful to Anthony Alcock and Richard A. Etlin for their help with the translation.
2 For studies of National Socialist zealotry in the various professions and arts, as well considerations of official propaganda, see Richard A. Etlin, ed., *Art, Culture, and Media under the Third Reich* (Chicago:

University of Chicago Press, 2002), especially the "Introduction" and Chapters 1–8, 11–12. This anthology includes Joachim Wolschke-Bulmahn and Gert Gröning (1988), "The National Socialist Garden and Landscape Ideal: *Bodenständigkeit* (Rootedness in the Soil)," pp. 73–97.

3 See Gröning and Joachim Wolschke-Bulmahn 1987, chapter "Zum Beitrag der DGGL zur Hochschulausbildung" (About the DGGL contribution to university education), pp. 85–86; see also Wolschke-Bulmahn and Gröning 1988, chapter "Die Stellung der Berufsverbände zum Hochschulstudium" (The positions of professional associations to university education), pp. 52–58.

4 See also Gröning and Wolschke-Bulmahn 1997, pp. 26–27.

5 On the concept of "Sarmatian landscapes" see Pollack 2005.

6 For a reprint of "Stellungnahme und Gedanken zum Generalplan Ost des Reichsführers SS" by Erhard Wetzel (1903–1975) of 27 April 1942 see pages 347 to 369.

7 The text of this, with comments, is printed on pp. 297–324. [Online]. Available at www.ifz-muenchen.de/heftarchiv/1958_3_5_heiber.pdf.

8 For the *Generalplan Ost* (General Plan East) Meyer envisioned a 25- to 30-year span for the Germanization of the "*Marken und Stützpunkte*" (provinces and strongholds) (1942, p. 73), with 3,345,805 German people needed (1942, p. 76). The first *Generalplan Ost* version dates from 15 July 1941. The June 1942 version is available online at: www.1000dokumente.de/index.html?c=dokument_de&dokument=0138_gpo&st=GENERALPLAN%20OST&l=de, accessed 2021-01-22.

9 Schmidt was Director of the Office of Gardens, Cemeteries and Forestry in Bielefeld 1947–1976. As a former colleague of Wiepking's in the late 1930s, he wrote the article "Krankheiten und Entwicklungsschäden des Stadtmenschen und ihre Abhilfe durch öffentliche Grünflächen" (Diseases and developmental damage in the urban population and mitigation of them through public green spaces) (pp. 113–118) in the *Festschrift für Heinrich Friedrich Wiepking* eds. Konrad Buchwald, Werner Lendholt and Konrad Meyer, Beiträge zur Landespflege, vol. 1, Stuttgart 1963.

10 Ignatius gives a contemporary overview of academic training in Australia, Denmark, Germany, Austrian Germany, England, France, Estonia, Finland, Italy, Norway, Poland, Switzerland, Czechoslovakia, Hungary and the USA.

11 Son of Erwin Barth and from 1971 to 1976 professor at TU Berlin.

12 After World War II he served as municipal garden director at Wilhelmshaven.

13 In the late 1930s he worked as *Gartengestalter* in Guben.

14 1943 Löhmer had become an official (*Generalreferent*) responsible for landscape maintenance in the district of Danzig-Westpreußen (Gröning und Wolschke-Bulmahn 1987, p. 69, 70, 110, 187, 189). A *Generalreferent* was put in office by the Reich Governor (*Reichsstatthalter*). His job was to develop the superior planning of landscape design, to co-ordinate specific planning of special administrations and steer its implementation on the basis of the General Order of the Reich Commissoner for the Strengthening of German Volkishness (*Reichskommissar für die Festigung deutschen Volkstums*). In the 1960s and 1970s Löhmer was the director of urban gardens office (*Stadtgartenamt*) in Braunschweig.

15 According to the incomplete information in the "Sammlung Diplomarbeiten 1935–1944 bei Wiepking, Institut für Landschaftsarchitektur, Universität Hannover" in 1938 nine and in 1939 six diploma (master) theses were written. No such theses appear to have been written in 1940 and 1941; for 1942, one, and for 1943 and 1944, there are two each on record, and again none for 1945.

16 From 1936 to 1967 Fischer was head of *Gartenwesen* (Horticultural Department) at Badische Anilin und Soda Fabriken (BASF). 1952, with Wiepking's support, he received a Dr title at the Faculty of Land Maintenance and Horticulture (*Landespflege und Gartenbau*) at the TH Hanover with the thesis *Stadtvolk und Stadtgrün* (Urban people and urban green), in which he introduced his "green planning for Auschwitz. . . with later made unrecognizable toponym" (Gutschow, N. (2001) *Ordnungswahn. Architekten planen im "eingedeutschten Osten" 1933–1945*, Bauwelt Fundamente, vol. 115. Basel: Birkhäuser, p. 12).

References

Anon. (1929) 'Standesnachrichten', *Die Gartenkunst*, 42(12), p. 3.
Anon. (1940) 'Institut für Gartengestaltung der Universität Berlin', *Die Gartenkunst*, 53(1), p. 4.
Barth, E. (1930) 'Die hochschulmäßige Ausbildung des Gartenarchitekten', *Der Deutsche Gartenarchitekt*, 7(8), pp. 101–103.

Barth, J., Gabriel, P.-F., Hollmann, W., Löhmer, H., Müller, F. and Schmidt, H. W. (1939) 'Die Gestaltung des oberen Dobrinkatales und der Stadt Preußisch-Friedland', *Die Gartenkunst*, 52(10), pp. 194–198.
Buchwald, K., Lendholt, W. and Meyer, K. (eds.) (1963) *Beiträge zur Landespflege*, vol. 1. Stuttgart: Eugen Ulmer.
Dörr, H. (1939) 'Landschaftsgestaltung und Raumordnung', *Die Gartenkunst*, 52(10), pp. 199–208.
Etlin, R. A. (ed.) (2002) *Art, Culture, and Media Under the Third Reich*. Chicago: University of Chicago Press.
Genthe, I. (1943) 'Ist der Gärtnerberuf lohnend?', *Die Gartenkunst*, 56(3), pp. 1–2.
Goetze, P. (1934) 'Berufung', *Die Gartenkunst*, 47(9), p. 148.
Gröning, G. (ed.) (1989) *Planung in Polen im Nationalsozialismus*. Berlin: Universität der Künste.
Gröning, G. (2001) 'Wiepking-Jürgensmann, Heinrich Friedrich 1891–1973, German Landscape Architect', in Shoemaker, C. A. (ed.) *Chicago Botanic Garden Encyclopedia of Gardens, History and Design*, vol. 3. Chicago: Dearborn, pp. 1429–1431.
Gröning, G. (2006) 'Wiepking-Jürgensmann, Heinrich Friedrich', in Taylor, P. (ed.) *The Oxford Companion to the Garden*. Oxford: Oxford University Press, p. 511.
Gröning, G. and Wolschke-Bulmahn, J. (1987a) *1887–1987 DGGL Deutsche Gesellschaft für Gartenkunst und Landschaftspflege e.V. Ein Rückblick auf 100 Jahre DGGL*. Berlin: Boskett Verlag.
Gröning, G. and Wolschke-Bulmahn, J. (1987b) *Die Liebe zur Landschaft, Teil III: Der Drang nach Osten. Zur Entwicklung der Landespflege im Nationalsozialismus und während des Zweiten Weltkrieges in den "eingegliederten Ostgebieten", Arbeiten zur sozialwissenschaftlich orientierten Freiraumplanung*, vol. 9. München: Minerva Publikation.
Gröning, G. and Wolschke-Bulmahn, J. (1989) '1. September 1939, Der Überfall auf Polen als Ausgangspunkt "totaler" Landespflege', *RaumPlanung*, (46/47), pp. 149–153.
Gröning, G. and Wolschke-Bulmahn, J. (1997) *Grüne Biographien. Biographisches Handbuch zur Landschaftsarchitektur des 20. Jahrhunderts in Deutschland*. Berlin: Patzer Verlag.
Gröning, G. and Wolschke-Bulmahn, J. (2021) 'The Concept of "Defense Landscape" (Wehrlandschaft) in National Socialist Landscape Planning', in Tchikine, A. and Davis, J. D. (eds.) *Military Landscapes. Dumbarton Oaks Colloquium on the History of Landscape Architecture*. Washington, DC: Trustees for Harvard University, pp. 201–220.
Gutschow, N. (2001) *Ordnungswahn. Architekten planen im "eingedeutschten Osten" 1933–1945. Bauwelt Fundamente*, vol. 115. Basel: Birkhäuser.
Heiber, H. (1958) 'Der Generalplan Ost', *Vierteljahrshefte für Zeitgeschichte*, 6(3), pp. 281–325.
Heinrich, V. (2004) 'Im Schatten. Erinnerungen an Jürgen Barth', in Schöbel, S. (ed.) *Aufhebungen, Urbane Landschaftsarchitektur als Aufgabe*. Berlin: Wissenschaftlicher Verlag, pp. 95–99.
Hoffmann, H. (ed.) (1942) *Deutscher Osten. Land der Zukunft*. München: Heinrich Hoffmann Verlag.
Ignatius, I. G. W. (1939) 'Sektion 15: Ausbildungswesen, Generalbericht: Die Entwicklung des gartenbaulichen Schulwesens und seine Bindung an die gartenbauliche Entwicklung der Länder', in Guenther, F. (ed.) *12. Internationaler Gartenbau Kongress Berlin 1938*, vol. II. Berlin: Der Reichsminister für Ernährung und Landwirtschaft, pp. 1071–1168.
Jung, M. (2020) *Eine neue Zeit. Ein neuer Geist? Eine Untersuchung über die NS-Belastung der nach 1945 an der Technischen Hochschule Hannover tätigen Professoren unter besonderer Berücksichtigung der Rektoren und Senatsmitglieder*. Petersberg: Michael Imhof Verlag.
Kellner, U. (1998) *Heinrich Friedrich Wiepking (1891–1973) Leben Lehre Werk*. (Dr. Ing.). Fachbereich Landschaftsarchitektur und Umweltentwicklung. Doctoral dissertation, Universität Hannover, Hannover.
Koenig, H. (1934) 'Zur Grün-Politik der Kleinstadt', *Der Deutsche Gartenarchitekt*, 11, pp. 346–348.
Land, D. (2004) 'Erwin Barth (1880–1933). Fragen zum Freitod eines Gartenarchitekten', in Schöbel, S. (ed.) *Aufhebungen – Urbane Landschaftsarchitektur als Aufgabe. Eine Anthologie für Jürgen Wenzel*. Berlin: Wissenschaftlicher Verlag, pp. 107–117.
Land, D. and Wenzel, J. (2005). *Heimat, Natur und Weltstadt: Leben und Werk des Gartenarchitekten Erwin Barth*. Leipzig: Verlag Koehler & Amelang.
Madajczyk, C. (1988) *Die Okkupationspolitik Nazideutschlands in Polen 1939–1945*. Köln: Pahl-Rugenstein.
Mappes, M. (1941) 'Hochschulfragen – Fachschulfragen Nachwuchsplanung. Bericht über die Tagung der Beiräte und Landesgruppenleiter der Deutschen Gesellschaft für Gartenkunst e.V. mit den Vertretern der Hochschule und der Fachschulen am 15. und 16. November 1940 im "Haus des deutschen Gartenbaues" Berlin', *Die Gartenkunst*, 54(2), pp. 23–33.
Meyer, G. (1860) *Lehrbuch der schönen Gartenkunst*. Berlin: Riegels Verlagsbuchhandlung.
Meyer, K. (1942) *Der Generalplan Ost*. Berlin: Machine-Typed Manuscript.

Mindt, E. and Hansen, W. (1940) *Was weisst du vom deutschen Osten? Geschichte und Kultur des Deutschen Ostraumes*. Berlin: Verlagsgemeinschaft Ebner & Ebner.

Pertl, J. (1940) 'An alle Mitglieder unserer Gesellschaft und Leser der "Gartenkunst"', *Die Gartenkunst*, 53(1), pp. 1–2.

Pollack, M. (ed.) (2005) *Sarmatische Landschaften. Nachrichten aus Litauen, Beloruss, der Ukraine, Polen und Deutschland*. Frankfurt am Main: S. Fischer Verlag.

Pospieszalski, K. M. (1958) 'Hitlerowska Polemika z "Generalplan Ost" Reichsführera SS', *Przegląd Zachodni*, 2, pp. 347–369.

Rössler, M. and Schleiermacher, S. (eds.) (1993) *Der "Generalplan Ost". Hauptlinien der nationalsozialistischen Planungs- und Vernichtungspolitik*. Berlin: Akademie Verlag.

Rudorff, A. (Bearb.) (2018) *Die Verfolgung und Ermordung der europäischen Juden durch das nationalsozialistische Deutschland 1933–1945, vol. 16, Das KZ Auschwitz 1942–1945 und die Zeit der Todesmärsche 1944/45*. Berlin: De Gruyter/Oldenbourg.

Rürup, R. (ed.) (1991) *Der Krieg gegen die Sowjetunion 1941–1945. Eine Dokumentation*. Berlin: Argon Verlag.

Saathoff, J. (1933) 'Die gärtnerischen Belange im neuen Staat', *Die Gartenwelt*, 37(19), pp. 225–226.

von Oven, H. (1933) 'Tagung der "Gesellschaft für Gartenkunst"', *Die Gartenwelt*, 37(33), pp. 404–406.

Wiepking, H. F. (1920) 'Friedrich der Große und Wir', *Die Gartenkunst*, 33(5), pp. 69–78.

Wiepking, H., Letter of 19.5.1936 to Deutsche Forschungsgemeinschaft. Bundesarchiv Koblenz R73/15698.

Wiepking, H., Letter of 6.1.1937 to Deutsche Forschungsgemeinschaft. Bundesarchiv Koblenz R73/15698.

Wiepking, H., Letter of 15.10.1940 to Konrad Meyer. Bundesarchiv Koblenz, R 49/2064.

Wiepking-Jürgensmann, H. Fr. (1933) 'Gartengedanken und Gartenbilder', *Monatshefte für Baukunst und Städtebau*, 17, pp. 193–200.

Wiepking-Jürgensmann, H. Fr. (1935a) 'Der Beruf und die Aufgaben des Gartengestalters', *Die Gartenkunst*, 48(3), pp. 41–46.

Wiepking-Jürgensmann, H. Fr. (1935b) 'Wettbewerb', *Die Gartenschönheit*, 16(4), p. 33.

Wiepking-Jürgensmann, H. Fr. (1937a) 'Gedanken über Ausbildungsfragen', *Die Gartenkunst*, 50(1), pp. 1–3.

Wiepking-Jürgensmann, H. Fr. (1937b) 'Um die Erhaltung der Kulturlandschaft', *Raumforschung und Raumordnung*, 2(3), pp. 119–122.

Wiepking-Jürgensmann, H. Fr. (1939) 'Der Deutsche Osten. Eine vordringliche Aufgabe für unsere Studierenden', *Die Gartenkunst*, 52(10), p. 193.

Wiepking-Jürgensmann, H. Fr. (1940a) 'Die Landschaftspflege in Schlesien', in Oberpräsident, Verwaltung des schlesischen Provinzialverbandes (ed.) *Almanach*. Breslau: Schlesische Verlagsanstalt und Druckerei Karl Klossok KG, pp. 103–118.

Wiepking-Jürgensmann, H. Fr. (1940b) 'Aufgaben und Ziele deutscher Landschaftspolitik', *Die Gartenkunst*, 53(6), pp. 81–96.

Wiepking-Jürgensmann, H. Fr. (1942) *Die Landschaftsfibel*. Berlin: Deutsche Landbuchhandlung Sohnrey & Co.

Wiepking-Jürgensmann, H. Fr. (1943) *Diplom-Hausaufgabe für den Kandidaten der Landschafts- und Gartengestaltung Max Fischer*. Berlin: Institut für Landschafts- und Gartengestaltung, Universität Berlin, Machine-Typed.

Wiesner, K. (1937) 'Grundsätzliches über die Einführung der jungen Studierenden in die Aufgaben der Gestaltung', *Die Gartenkunst*, 50(1), pp. 7–16.

Wolschke-Bulmahn, J. and Gröning, G. (1988) *1913–1988, 75 Jahre Bund Deutscher Landschafts-Architekten BDLA. Teil 1 Zur Entwicklung der Interessenverbände der Gartenarchitekten in der Weimarer Republik und im Nationalsozialismus*. Bonn: Köllen Druck + Verlag.

17
'TO BROADEN THE OUTLOOK OF TRAINING'

The first landscape course in Manchester

Luca Csepely-Knorr

Introduction

Education in landscape architecture in Britain strongly interlinks with the questions of professional standards and regulations, set by the Landscape Institute. The Institute, previously called the Institute of Landscape Architects (ILA), has played a key role in the development of landscape architecture curricula and examination standards since its foundation in 1929. This chapter looks into the first few decades of ILA's work in arguing and campaigning for professional training, and discusses one of the earliest landscape architecture courses in Britain, at the Manchester School of Architecture, the history of the first few decades of which has only recently been rediscovered. These new insights help to create a more nuanced understanding of the history of British landscape architecture education, and its links to the broader questions of institutional and professional development.

Sources and methods

By using the early history of landscape architecture education in Manchester as a purposive case study, this chapter highlights the intricate links that exist between the history of education and professionalization of landscape architecture in the UK (Swaffield 2016). It underlines the changing understanding of landscape architecture, how the focus switched from garden design to public spaces and ultimately to landscape planning, and how this affected higher education curricula and practice.

This research uses primary text-based archival material from two main sources. To explain the professional setting, the papers of ILA's Educational Committee are analyzed; they are held in the Special Collections at the Museum of English Rural Life in Reading (SR LI AD 1/1/2, SR LI AD 2/1/2/11, SR LI AD 1/3/1). This set of data helps to scrutinize the Institute's approach to professionalization and education, and will be compared with the changes in the course structure in Manchester.

The analysis of the evolving curriculum in Manchester is examined through university catalogues and committee minutes held in the Special Collections at the Sir Kenneth Green Library in Manchester. The chapter follows the history of the course until the early 1950s, when landscape architecture education was on pause in Manchester, before its re-launch in a new format in the 1960s. To better understand the course and its students in the wider professional and socio-political context, the research analyzes registration cards – documenting the application and journey of students – also held in the Special Collections archives. These cards give information about students entering the course, including their previous work experience.

Calls for specialist landscape architecture education

In June 1924, the prominent landscape architect and civic designer T. H. Mawson wrote a letter to the editor of the *Manchester Guardian*, C. P. Scott, offering an article about the urge to create a university course to train landscape architects (A/M57/3–4). The editor found the idea interesting and asked Mawson to submit his argument as a letter to the editor. On 11 July 1924 the letter, titled 'Landscape Architecture. The Need for a Training Centre' appeared in the 'Correspondence' section of the paper (Mawson 1924). Mawson, who signed the letter as 'Past President of the Town Planning Institute', had long been a keen advocate for the formation of specialist training centres for landscape architects. Being a self-trained designer, who started his career as a nurseryman, Mawson was early to identify the importance of specialist education for landscape architecture and the contribution this would make towards its recognition among the built environment professions. Most importantly, as Mawson argued, there was an immense need for well-trained designers to shape public spaces, rather than only focusing on private gardens. This emphasis on private spaces derived from the existing education system: up until the twentieth century, the practice of landscape architecture was taught primarily through apprenticeships. In the 1880s, the 'School of the Art of Landscape Gardening and Improvement of Estates' at Crystal Palace was established. Then there were garden design courses also at colleges such as Swanley or Glynde for ladies, the former being the alma mater of the first two female presidents of ILA, Brenda Colvin and Sylvia Crowe (Hextable Heritage Society n.d.; Meredith 2003; Woudstra 2010).

Mawson tirelessly argued how public parks and other open spaces in cities were overlooked from a professional point of view and how landscape architects should design these instead of them being 'almost entirely the work of amateurs' (Mawson 1911a). Specialist education was therefore a key to the betterment of cities. The first-time landscape architecture was integrated into a university course was also linked to Mawson: he was the first lecturer of 'Landscape Design' at the Department of Civic Design in Liverpool between 1909 and 1920. His lecture notes are proof of his commitment to integrate landscape design into town planning: his course dealt with a variety of aspects of the public realm: from questions of design to the details of planting and construction (Mawson 1915–1916). As the most successful designer of the period, Mawson also contributed to teaching elsewhere by giving lectures, including at courses such as Swanley Horticultural College where Madeleine Agar taught Garden Design (Colvin 1979).

Mawson had first-hand experience of landscape architectural education in an international context: just before World War I, his growing international reputation led him to be invited to prestigious universities in the USA, at Harvard, Berkeley, and the University of Pennsylvania. Here, he carefully studied the curriculum, and integrated these experiences into his own writings and lectures. He was a champion of creating and maintaining links with the American Society of Landscape Architects (ASLA). He published regularly in their journal *Landscape Architecture*, about the state of the profession in the UK, its history and contemporary issues (Mawson 1911b, 1911c, 1917). In 1912, Mawson, assisted by his son Edward Prentice Mawson, published a pamphlet describing a School of Landscape Architecture in Caton Hall, near Lancaster (Mawson 1921). The two-year-long curriculum mirrored Mawson's understanding of landscape architecture encompassing both garden design and the urban scale: from public parks to park systems. While the first year's focus was on the basics of garden design, drawing and surveying skills and tree identification, the second year broadened the knowledge base to public parks, playgrounds, boulevards and green systems. As well as the classroom education, the students were expected to work on the grounds of the school to gain practical experience.

In 1923, Mawson became president of the Town Planning Institute (TPI – today Royal Town Planning Institute RTPI) in the UK, and was a founding member of the Royal Fine Arts Commission in 1924. He used his presidential address at the TPI to champion the role and importance of landscape architecture in town planning and concluded that 'I have urged the claims of landscape architecture as a delightful profession whose contribution to the elevation of human life and happiness is beneficially rich in opportunity' (Mawson 1923, p. 44).

The realization of Mawson's vision for a profession that is equal to other built environment professions and has its own educational system was in strong connection with the birth of another organisation: the Institute of Landscape Architects (ILA), today the Landscape Institute (LI).

The Institute of Landscape Architects and the first degree course in landscape architecture in Britain

The Institute of Landscape Architects was originally established as the British Association of Garden Architects in 1929, following a meeting at the Chelsea Flower Show in 1928. The meeting to discuss the creation of a professional institute for Garden Designers was called by Richard Sudell. Interestingly, Mawson, the champion of professional education did not agree with the creation of a new institute. As Brenda Colvin, founding member and first female president of the Institute, remembers: 'Thomas Mawson had told us at Swanley that there was not enough work to justify founding an Institute; that his firm and that of Milner White would absorb all there was in this country' (Colvin 1979).

The name of the organization was changed to Institute of Landscape Architects (ILA) within a year on the recommendation of Thomas Adams (Powers 2020, p. 103). As Brenda Colvin wrote, this was an important step in defining the scope of the profession: 'It would have taken us much longer to arrive at the full scope the profession has today – if we had arrived at it at all' because (Thompson 2014, p. 40)

> the wider application of landscape architecture are those of regional planning. While the main work of members may be at present laying out medium sized gardens, we had to remember that our most important contribution to the life of the nation would be in the wider field.
>
> (Rettig 1983, p. 3)

Colvin also convinced the two 'great names' Mawson and White to be the first and second presidents of the Institute.

When the ILA was established, it was small, and to be able to gain recognition, they invited well-established professionals from related fields, such as town planning and architecture. As there was no accredited training where people could earn a degree in landscape architecture, ILA members came from very different backgrounds. The intake from architecture and town planning was prevailing. As Sylvia Crowe, a very prominent landscape architect, Swanley graduate and second female president of the ILA, remembered:

> We had influential people from allied professions, like Patrick Abercrombie, who was a great supporter of the Institute, and Thomas Sharp, Lord Holford and Lord Reith, who also gave great support – people from allied professions who realized we had a contribution to make, something to marry in to their work. There was a missionary

optimism mingled with anxiety that you could not overcome the destruction and invasion of the English countryside in time.

(Harvey 1987, p. 34)

In 1930, the University of Reading became the first to launch a diploma course in landscape architecture. Although the course itself is listed in the calendars of the university, the catalogues do not give much detail about the structure of the course other than its three years' long duration – most probably due to it being a diploma course rather than a traditional degree programme. Contemporary press cuttings stored in the ILA archives do give more insight into the course, but there remains much to be uncovered (SR LI AD 2/1/2/11). The press release emphasizes that the course will cover 'every necessary phase of the subject', such as art, botany, bookkeeping, physics, chemistry, surveying and levelling, horticulture, building construction. The course was organized jointly by the Fine Art and Horticulture departments, and the Hinfield Horticultural Station gave opportunity to the students to gain practical knowledge. Arthur J. Cobb, funding member of the ILA and successful horticulturalist, was the first member of the institute to contribute to the course (Holden 2015). Edward Prentice Mawson became external examiner in 1933, and, in 1934, Geoffrey Jellicoe, then lecturer at the prestigious Architectural Association in London was invited as sessional lecturer. Russell Page succeeded Jellicoe in 1938.

The first landscape architecture course in Manchester

In 1934, ten years after Mawson's letter appeared in the Manchester Guardian and a few years after the first diploma course in landscape architecture at the University of Reading had started, the School of Architecture at Manchester's Municipal School of Art launched a new course in landscape architecture, established by the Manchester Education Committee (Prospectus 1939–1940, p. 1). The course was a pioneering collaboration between the School of Architecture and the University of Manchester. The artistic and architectural elements of the course were taught through lectures and studio units, under the direction of the staff of the School of Architecture. The science element of the discipline, through the study of horticulture, was facilitated via the University's 'Botany Department' (Prospectus 1939–1940, p. 1). Being part of the Municipal School of Art gave students more opportunities to discover links between the fine arts, architecture, landscape and town planning, while also being part of the university's scientific life.

To show their support and – most probably – as an attempt to direct the public's attention to the profession, prominent ILA members gave public lectures as part of the curriculum. In the academic year of 1939–40, for example Edward Prentice Mawson, Thomas Adams, Geoffrey Jellicoe, Richard Sudell, Edward White, John T. Jeffrey and William G. Holford were all involved in the lecture programme as well as other guest lecturers, such as Brenda Colvin (Prospectus 1939–1940, p. 3).

According to the prospectuses of the school, one reason to found the course was to 'meet the call for an Art training' in landscape architecture; because the 'architectural aspect of landscape work [was] becoming more important, and there [was] a desire to broaden the outlook and training of those likely to be in charge of public parks and open spaces in the future' (Prospectus 1935–1936, p. 23). The statement and the emphasis on the importance of well-designed public spaces clearly referred to Mawson's aims to educate a generation of professionals for the improvement of the public domain. The syllabus was created in consultation with prominent professionals, such as Edward Prentice Mawson (then President of the ILA), who had trained as an architect in the École des Beaux Arts in Paris, the Manchester Parks Superintendent J. Richardson and Manchester City Architect G. Noel Hill. The courses were open for architectural students, 'wishing to add a knowledge of landscape architecture to their qualifications' and, – as landscape architecture as a profession as well at the time – was open to students with no previous training, creating a way of social mobility through education (Prospectus 1935–1936, p. 23).

Students

An analysis of the registration cards for students who enrolled to study at the Manchester School of Art and its predecessors reveals some intricate details about the student cohorts. For the purpose of this chapter, I have analyzed the records from 1934 to 1952. During this period, 662 students registered to architecture courses and 92 to landscape architecture. Interestingly, during the first year landscape architecture was on offer it attracted more students than architecture (23 vs 13). However, the numbers were declining on the landscape course over the years, while they were growing on the architecture course (see Table 17.1).

In terms of gender, ten female students enrolled to the landscape course as opposed to the 82 male students, but this was still a much higher proportion than the architecture course, where we see the registration of only 35 women out of 662 students. The background, training and place of work at the time of the application at the landscape course gives a very interesting overview of the recognition and direction of the training. While most male applicants already in employment were gardeners, working for local park corporations in the Manchester area, women came from a variety of jobs, such as teacher, secretary, typist or cadet nurse. Only one was a gardener. This shows, firstly, how the course's appeal was local and within the area that is known today as Greater Manchester. Secondly, the course appears to meet Mawson's idea to train people already working with and in public parks to gain design skills. The stark difference in the background and training between the

Table 17.1 Student numbers registered to the architecture and landscape architecture courses in Manchester, 1934–1952.

Year	Landscape architecture total	Landscape architecture male	Landscape architecture female	Architecture total	Architecture male	Architecture female
1934	23	22	1	13	13	0
1935	6	5	1	25	23	1
1936	8	8	0	24	24	0
1937	14	14	0	21	20	1
1938	24	23	1	13	13	0
1939	5	5	0	22	21	1
1940	1	1	0	17	15	2
1941	2	1	1	46	37	9
1942	0	0	0	27	25	2
1943	2	1	1	28	26	2
1944	0	0	0	46	37	9
1945	1	0	1	40	35	5
1946	2	2	0	69	66	3
1947	0	0	0	73	64	9
1948	8	8	0	64	61	3
1949	0	0	0	43	42	1
1950	1	1	0	48	43	5
1951	2	0	2	33	30	3
1952	4	1	3	23	22	1

Source: Luca Csepely-Knorr.

different genders is something worthy of attention, and of more research. It is not unique to Manchester – the research of applications to become 'probational' members to ILA shows a very similar pattern: women came from all different background, from law to library studies.

The curriculum

The Manchester curriculum continued developing through the 1930s. Until 1936, all students attended the same three-year-long programme. During their first year, they enrolled to study Studio Practice in Architecture, Landscape Construction, History of Styles of Architecture and Colour and Ornament. The second year consisted of Garden Lay-out and Design, History of Garden Design and Landscape Estimating. The third year focussed on Park and Garden Design, Landscape Surveying and Technique of Landscape Draughtmanship (see Figure 17.1).

From 1937, two different courses were created, to accommodate previous education and experience. Juniors and apprentices entered a two-year course in Preliminary Drawing, seniors entered another two-year programme ancillary to, or following, their horticultural education (see Figure 17.2). Throughout the first few decades, the University of Manchester provided the botanical education, as it was set out at the establishment of the course.

The course for juniors and apprentices in the first year consisted of students studying the basics of Geometrical Drawing, Colours, Patterns, Scale drawings and Plans (the structure of any preliminary courses at the School of Art). During their second year, students took part in sessions for Free Drawing, Plants, Sketching Plans, Colour and Simple Design. The course for seniors was also delivered in two years. In the first year, they attended the unit on Garden Layout with Free and Constructional Drawing and Colour and Simple Design as well as studies in Landscape Construction. During the second year, students extended their knowledge through the studies of Park and Garden Layout II where they analyzed various types and styles of designs, including their historic antecedents. These were complemented with studies on Surveying and Estimating (Prospectus 1937–1938, pp. 2–3).

The courses were designed to enable prospective students to engage in university studies whilst also working: lectures were organized in the evenings, however, students of the 'senior' course had the opportunity to take certain architectural units as well as the ones from the landscape curriculum. The analysis of course contents shows a very strong orientation towards garden and park design, while questions of urban or regional design do not seem to appear in the curriculum. While this was in line with the main scope of the profession in the 1930s, where garden design prevailed, the profession was heading towards new directions, and new scales of designed landscapes, especially after the war.

Professional recognition and training – the developing ILA strategy and the closure of the course in Manchester

While, in the beginning, only the Institute of Park Administration accredited the Manchester course, by 1944 the studies also led to Associateship of the Institute of Landscape Architects (Prospectus 1944, p. 7). In the context of the development of professional education and professionalization of landscape architecture, entry to the Institute after graduation was an important step. However, the set-up of professional standards was equally important for the Institute itself. If they aimed to be considered equal to other professional bodies, they needed vigorous examination systems, such as the Royal Institute of British Architects (RIBA)'s or the Town Planning Institute's.

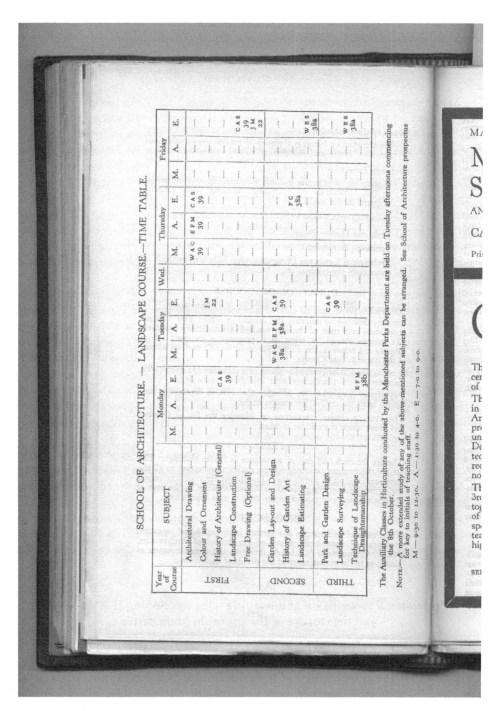

Figure 17.1 Course structure in the 1935–1936 School Catalogue.

Source: Manchester School of Art archive at Manchester Metropolitan University Special Collections Museum.

'To broaden the outlook of training'

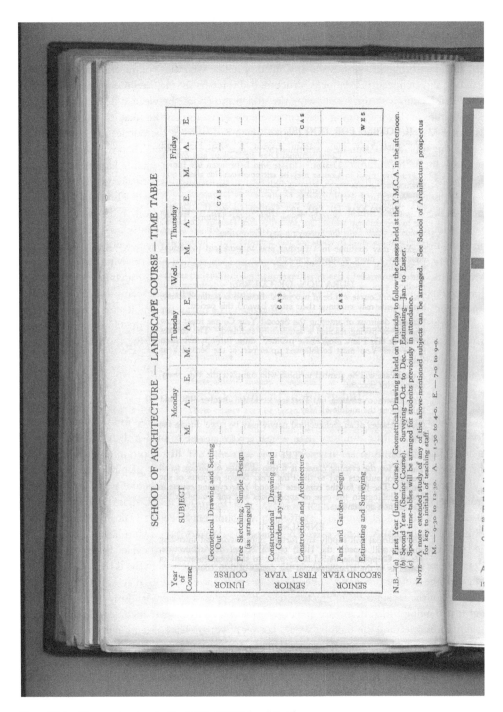

Figure 17.2 Course structure in the 1937–1938 School Catalogue.

Source: Manchester School of Art archive at Manchester Metropolitan University Special Collections Museum.

In 1931, Giffard Woolley was working on a proposed syllabus for the examination of applicants aiming to become members of the Institute. In a letter to Robert Mattocks, Edward Prentice Mawson criticized the focus of these exam plans being focussed on gardens, rather than larger-scale spaces (SR LI AD 2/1/2/12). In his response, Mattocks drew up a syllabus that aimed to 'bring the syllabus and examination in line with those of the existing Institutions', planning and architecture. Mattocks' structure had three levels: the 'Preliminary', 'Intermediate' and 'Final Exam', with the last allowing the students to choose between and specialize either in Horticulture or Town Planning. Students were expected to pass the final exam before getting Associate Member status of the ILA. Despite these early efforts, the publication of the syllabus took a long time. In 1938 an Educational Committee was set up, which was working on creating proposals for both educational opportunities and the examination system. The work continued during the war, and a first 'Provisional Syllabus of Education' was finally published September 1941 in the Wartime Journal of the Institute. As Jellicoe mentioned: 'This remarkable document dealt with the Preliminary, Intermediate and Final Examination, and was accepted as a basis of future education' (Jellicoe 1985, p. 14). The Manchester course's accreditation by the Institute in 1944 meant that they were in line with the aims and objectives of the profession.

In 1949, the course was renamed to Landscape Design. In 1951, the Municipal College of Art became the Regional College of Art, and beyond the 1952–53 academic year, the course does not appear in the school catalogue any longer. Although there seems to be some student enrolment in the late 1950s, these students seemed to enrol in the architecture course and had some landscape architecture tuition within that. The reasons behind the closure of the course have yet to be revealed. However, the professional development and the Institute's decision to remain independent might shed some light on these.

Independent Institute and education

As the Institute of Landscape Architects was relatively young compared to its counterparts, the Town Planning Institute (founded in 1914), the Royal Institute of British Architects (founded in 1837) and the Institute of Surveyors (founded in 1868), it invited prominent members from the related professions. As Sylvia Crowe remembered 'most of our members were architects and/or town planners and to get them realise that landscape architecture was a third different profession was not always easy' (Harvey 1987, p. 34). Already in 1934, at the Town Planning Institute's (TPI) annual conference and summer school at St Peter's Hall in Oxford, Gilbert Jenkins, himself an ILA member, recommended the merger of the two (Gibson 2011, p. 124). In 1941, the proposed association with the TPI was on the agenda again. Between 1943 and 1945, the next question whether the Institute should assimilate within the RIBA was dominant. Through the merger the RIBA aimed to create landscape education, that was 'an extension of the work of the recognized architecture schools' as opposed to new, landscape training centres, and that training in landscape architecture should be offered for trained architects after their qualification (meaning the training would last seven years). They also argued that the Institute should call themselves the Institute for Landscape Design, rather than architecture (Gibson 2011, p. 124). Although the ILA leadership was sympathetic at first, there were two key members who opposed it, Brenda Colvin and Sylvia Crowe. Both women were trained at Swanley Horticultural College as opposed to being architects and planners. One of the main reasons behind their opposition to the new educational system (beyond the exceptionally lengthy training process) was that they doubted that qualified architects would wish to take on 'purely rural work such as forestry'. Both of these they considered as key to the profession, and worried that the interest of trained architects 'would be primarily geometrical rather than biological' (SR LI

AD1/1/2). The Institute voted for independence and their efforts to create their own training centres became crucial in the next few years. However, the shift towards the questions of the 'rural' and large-scale landscape planning, infrastructure and forestry became stronger yet.

In 1946, the Institute published its new examination syllabus. The topics included Design, History of Urban and Rural Landscapes, Surveying, Park and Garden Construction, Geology and Soil and Ecology – a much stronger emphasis on the natural sciences (SR LI AD 1/3/1). A few years later, in 1948, the Institute raised enough money to create two permanent lectureships. Brian Hackett was appointed as first full-time senior lecturer at Durham University, and Peter Youngmann took up his post at the University College London. They were both trained planners and based at Departments of Town and Country Planning.

The later history of landscape architecture in Manchester

Throughout the next two decades more courses were to come both at certificate, diploma, undergraduate and post-graduate levels at different universities, throughout the United Kingdom. The Manchester course did not appear in the ILA papers and was all but forgotten, until now. Whether the ILA did not support it anymore or the college felt that the establishment of new courses created too much competition remains unknown. Nevertheless, the new courses were more in line with Colvin's vision about landscape architecture's focus on landscape and regional planning, rather than Mawson's vision about the artistic and skilful design of public parks, something the Manchester course was set up to address.

Manchester relaunched the landscape course in the 1960s, parallel to other universities, such as Leeds and Sheffield. The 1965 Prospectus advertised a 'certificate course' in landscape architecture for students who have already passed final examination at another course and a follow up 'diploma course' preparing them for the intermediate and final examinations of the Institute. When, in 1968–69, the university minutes discussed the creation of a full-time undergraduate landscape architecture course, and both part-time and full-time post-graduate courses, the situation was completely different from the 1950s. The College launched a new department, the Department of Environmental Design, in 1968–69, and within this department, three independent schools existed: the School of Architecture, School of Landscape Architecture and School of Interior Design. The new, four-year course in landscape architecture led to a Faculty Diploma and Associate Membership to the ILA. The College became Manchester Polytechnic in 1970 and Manchester Metropolitan University in 1992. Its landscape education has been continuous. The University of Manchester – the partner in the joint landscape course launched its own course in 1967. It was part of the Department of Planning and Landscape and was closed in 1992 (Holden 2015; Roe 2007; Woudstra 2010). However, since September 2021 the Master of Landscape Architecture courses in Manchester are validated by both universities (Manchester Metropolitan University and the University of Manchester).

In the early-to-mid 1940s, the prospectuses of the Municipal College of Art had a complex diagram as a cover image that illustrates the intricate links and overlaps that exist between the disciplines of Art, Architecture and Design (Figure 17.3). The inclusion of landscape architecture is an expression of a holistic understanding of the built environment professions. Although the Manchester course was discontinued in the 1950s, it had a strong effect not just on students enrolled in this course but also on architecture students. The Special Collections Gallery in MMU's Sir Kenneth Green library has recently acquired the student work of the architect Gordon Hodgkinson. His careful plant studies, drawings and briefs are proof of the holistic and deep understanding the student developed of outdoor space, and its links to buildings. Hodgkinson's portfolio is illustration of and testament to the success of landscape education in the late 1940s.

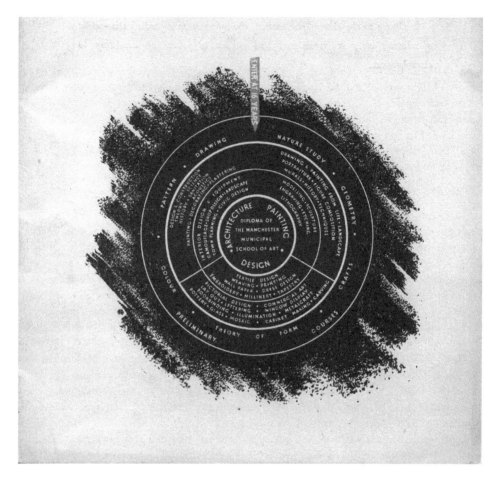

Figure 17.3 Diagram of the courses taught at the Manchester Municipal School of Art on the front page of the school's 1946–1947 Prospectus.

Source: Manchester School of Art archive at Manchester Metropolitan University Special Collections Museum.

References

Colvin, B. (1979) 'Beginnings', *Landscape Design*, 125, p. 8.
Gibson, T. (2011) *Brenda Colvin. A Career in Landscape.* London: Frances Lincoln Limited.
Harvey, S. (ed.) (1987) *Reflections on Landscape. The Lives and Work of Six British Landscape Architects.* London: Gower Technical Press. Interview with Sylvia Crowe, p. 34.
Hextable Heritage Society (n.d.) *Remarkable Women of Swanley Horticultural College* (sa) [Online]. Available at: www.hextable-heritage.co.uk/Remarkable%20women.pdf (Accessed 7 January 2020).
Holden, R. (2015) *The Rise, Decline and Fall of Landscape Architecture Education in the United Kingdom: A Historical Outline.* Paper presented at the ECLAS Conference in Tartu [Online]. Available at: www.academia.edu/16232741/The_rise_decline_and_fall_of_landscape_architecture_education_in_the_United_Kingdom_a_historical_outline (Accessed 29 December 2020).
Jellicoe, G. (1985) 'The Wartime Journal of the Institute of Landscape Architects', in Harvey, S. and Rettig, S. (eds.) *Fifty Years of Landscape Design.* London: The Landscape Press, pp. 9–26.
Mawson, T. H. (1911a) *Civic Art.* London: B. T. Batsford.
Mawson, T. H. (1911b) 'The School of Civic Design at the Liverpool University', *Landscape Architecture. A Quarterly Magazine: Official Organ of the American Society of Landscape Architects*, 1(3), p. 109.

Mawson, T. H. (1911c) 'Landscape Architecture in England', *Landscape Architecture. A Quarterly Magazine: Official Organ of the American Society of Landscape Architects*, 1(3), pp. 110–114.

Mawson, T. H. (1915–1916) *Landscape Design*. Lectures Given by Thomas H Mawson Honorary ARIBA Lecturer on Landscape Design at the University of Liverpool. Manuscript.

Mawson, T. H. (1924) 'Correspondence. Landscape Architecture. The Need for a Training Centre. To the Editor of the Manchester Guardian', *The Manchester Guardian*, 11 July, p. 16.

Mawson, T. H. (1917) 'The Retrospect and Prospect of Landscape Architecture in Britain', *Landscape Architecture. A Quarterly Magazine: Official Organ of the American Society of Landscape Architects*, 7(3), pp. 109–115.

Mawson, T. H. (1923) 'Presidential Address', *Journal of the Town Planning Institute*, 10(2), pp. 33–47.

Mawson, T. H. and Mawson, E. P. (1921) *School of Landscape Architecture Caton Hall, Near Lancashire*. Caton Hall: Lancashire.

Meredith, A. (2003) 'Horticultural Education in England, 1900–40: Middle-Class Women and Private Gardening Schools', *Garden History*, 31(1), pp. 67–79.

Powers, A. (2020) 'Geoffrey Jellicoe and the Landscape Profession', in Charlton, S. and Harwood, E. (eds.) *100 20th Century Gardens and Landscapes*. London: The 20th Century Society, pp. 100–107.

Prospectus. (1935–1936) Prospectus of the Municipal School of Architecture Manchester, City of Manchester Education Committee.

Prospectus. (1937–1938) Manchester Municipal School of Art and School of Trade Craft Training School of Architecture Prospectus Manchester, City of Manchester Education Committee.

Prospectus. (1939–1940) Manchester Municipal School of Art and School of Trade Craft Training, School of Architecture Prospectus Manchester, City of Manchester Education Committee.

Prospectus. (1944) Manchester Municipal School of Art. Architecture Prospectus.

Rettig, S. (1983) The Creation of Professional Status: The Institute of Landscape Architects between 1929 and 1955. Unpublished manuscript p. 3 MERL SR LI AD 2/1/1/28.

Roe, M. (2007) 'British Landscape Architecture: History and Education', *Urban Space Design*, 20(5), pp. 109–117.

Swaffield, S (2016) 'Case studies', in van den Brink, A. et al. (eds.) *Research in Landscape Architecture Methods and Methodology*. London: Routledge, pp. 105–119.

Thompson, I. (2014) *Landscape Architecture: A Very Short Introduction*. Oxford: Oxford University Press.

Woudstra, J. (2010) 'The "Sheffield Method" and the First Department of Landscape Architecture in Great Britain', *Garden History*, 38(2), pp. 242–266.

Unpublished Archival Materials

SR LI AD 1/1/2 Landscape Institute Council Minutes Museum of English Rural Life, Landscape Institute Archives.

SR LI AD 1/3/1 Landscape Institute Education Committee Papers Museum of English Rural Life, Landscape Institute Archives.

SR LI AD 2/1/2/11 Press cuttings related to the Landscape Course in Reading. Museum of English Rural Life, Landscape Institute Archives.

A/M57/3–4. University of Manchester Special Collections. Mawson's 1924 letter to C. P. Scott Editor, Manchester Guardian.

18
LANDSCAPE ARCHITECTURE EDUCATION HISTORY IN PORTUGAL

The pioneering roles of Francisco Caldeira Cabral and Francisco Simões Margiochi

Ana Duarte Rodrigues

Introduction

In 1941, the University of Lisbon launched a landscape architecture programme. It was one of the first programmes of its kind in Europe. The curriculum developed based on the pioneering work of Francisco Caldeira Cabral (1908–1992) and on nineteenth-century studies of gardening and horticulture that anchor in agricultural studies as a multidisciplinary field gathering knowledge on agronomic sciences, botany, hydraulics and art.

Focusing on landscape architecture education in Portugal and its beginnings at the Institute of Agronomy in 1941, this chapter traces the roots of this education back into the nineteenth century. It does so not only because the passion and knowledge propagated by gardeners and horticulturists became the basis for the development of landscape architecture as a profession (Marques 2009; Raxworthy 2018), but also on the grounds that the curricula of the earliest study programmes on gardening echoes in the first landscape architecture degree programme in Portugal. This chapter also demonstrates the central role played by Lisbon City Council in promoting practical gardening education and professionalization.

This chapter brings out the prominent roles played by the landscape architect Francisco Caldeira Cabral in the twentieth century (Andresen 2003; Antunes 2019) and by the agronomist Francisco Simões Margiochi (1848–1904) in the nineteenth century (Rodrigues and Simões 2017; Rodrigues 2017a, pp. 124–130, 2020, pp. 107–110).

Research methods include sampling and analyzing primary sources, such as nineteenth-century contracts for education assistants who were gardeners, photographs of children playing with gardening utensils, printed sources of the nineteenth-century study programmes, drawings of the practical school for gardeners at Casa Pia, the minutes of the Lisbon City Council and several articles in periodicals. For the research on the twentieth century I relied on the work of Andresen and on Cabral's publications. Documents and publications were hand-searched and qualitative textual analysis performed, including content analysis and thematic analysis.

Sampling includes materials held by the Municipal Archive of Lisbon, the National Archives, the National Library of Portugal and the Archive of Casa Pia, an important orphanage in Lisbon.

The first landscape architecture degree programme in Lisbon

In 1941, the agricultural school *Instituto Superior de Agronomia* (ISA), in Lisbon launched the first landscape architecture degree programme in Portugal. It did so on the initiative of Francisco Caldeira Cabral, the first Portuguese to graduate from a university programme in landscape architecture. Cabral earned his professional degree in 1939 at the Friedrich-Wilhelms-Universität in Berlin, which had started to run its programme in 1930 (see Gröning in this volume).

Cabral was born into a bourgeois milieu. His father was a physician. During the troubled political and economic times of the 1920s, when the Republic of Portugal was about to become a dictatorship, the academic environment was anything but desirable and Cabral was sent abroad, first to a Jesuit college in northern Spain for high school. In 1925, Cabral enrolled as a chemistry student at the *Technische Hochschule* in Berlin. For health reasons, he was obliged to return home and, back in Portugal, he studied agricultural engineering at ISA where he completed his degree in 1936 (Andresen 2003, p. 21). When Cabral concluded his BA as agronomist, his title was *engenheiro agrónomo* (agronomic engineer, Cabral 1935, front cover).

However, his first work experience brought him close to landscape architecture. While attending his last year at ISA, he was invited by the Lisbon City Council to replace the horticulturist Vieira da Silva as head of gardens and cemeteries. The city council had hoped to hire an expert specialized in horticulture but recognized there was no one who had the required profile (Andresen 2003, pp. 21–23) because, by then, there was no department dedicated to gardens (Tostões 1992). As explained later, the once-flourishing period of horticulture and garden education had ended in the late nineteenth century (Rodrigues 2020). Cabral, wishing to educate himself as the expert the city council was seeking, applied for a grant to study landscape architecture. In a letter addressed to the Ministry of Education, Cabral acknowledges the difference between two distinct fields that find themselves under the larger umbrella of gardening. The first is gardening as a special branch of horticulture, with an education that, at the time in Portugal, was in an incipient stage. The second is landscape architecture, a field that did not yet exist at all in the country (Andresen 2003, p. 22).

In Germany, Cabral had encountered state-of-the-art landscape architecture education. The national socialist regime had an impact on every aspect of life in Germany and, after enrolling in landscape architecture in Berlin, Cabral's studies began by addressing issues of "eminently national character." Students acquired practical knowledge in gardening, horticulture, arboriculture for public and private parks and gardens and sports fields, as well as nurseries. Cabral sent both quarterly reports and letters to his professors and colleagues at the ISA providing detailed information about his ongoing studies and experiences. He praised the practical component of his study programme, which included an internship with one of the best nursery and park designers, comparable in his opinion to the most famous Portuguese horticulturist of that time, Moreira da Silva.[1] In parallel, he produced drawings, watercolours, perspectives, designs and plans for gardens and parks, and he detailed the history of art of gardens and architecture as well as their means of construction. Finally, he went on field trips around Berlin and on an excursion to Italy (Andresen 2003, p. 30).

Around the time Cabral was finalizing his degree programme, he collaborated on the project for the Jamor National Stadium, one of the Estado Novo promises that served to launch Cabral as an expert in urban planning, a field of growing importance.

It is in the year of 1943 that we have full awareness of what Cabral was advocating. He makes some conferences in which he explains what landscape architecture is and what a landscape architect is. He does not quote German sources, but Elliot Brandt, from Harvard (Cabral 1993, p. 25). In one of this series of conferences he assumes he is a landscape architect:

> Many will be certainly disappointed with my words, as they would expect that I as 'Arquitecto Paisagista', derived from the French *architecte paysager*, would come here to praise the tree and the forest or give recipes to embellish landscape, however, you will listen to scientific and technical concepts.
>
> (Cabral 1943a, p. 53)

In 1940, Caldeira Cabral was hired by the ISA as coordinator of the discipline of Organographic Design and, in October of that same year, he started to coordinate the discipline of Rural Constructions. At the same time, he was authorized to start an 'experimental course' on Landscape Architecture, which began in the academic year 1941–1942. In 1942, the Ministry of National Education approved the study programme, as well as its regulation, with the status of *Curso Livre de Arquitectura Paisagista* (Free Course on Landscape Architecture) in 1942 (Nunes 2011, p. 109). It was open to students of Agronomy or Silviculture study programmes, strengthening from the beginning their curricula with disciplines of Landscape Architecture, which by then was considered a specialization.

Two years after earning his university degree Cabral started working as instructor of landscape architecture. Initially, he had around ten students. The programme launched at the ISA ran from Monday to Friday between 6 pm and 8 pm over a four-year period of study. The subjects required underpin the agronomic sciences core of the new degree. Landscape architecture students attended botany, mathematics, chemistry, microbiology, topography, agriculture, forestry, horticulture, viniculture, rural construction and agricultural hydraulics. In addition, in the first year, landscape architecture students were expected to attend classes in garden construction, geometry, drawing and perspective. In the second year, they studied garden plants, architecture and construction, before taking classes in landscape architecture and history of art in the third and fourth years. In the final year, the programme included urban planning and colonization (Andresen 2003, p. 44). The first students with a university degree appeared in 1946.[2] As this *Curso Livre* represented an addition, a specialization, to the study programme of agronomy engineering or forest engineering, most academic certificates were *Diploma de engenheiro agrónomo e arquiteto paisagista* (diploma of agronomic engineer and landscape architect) (Nunes 2011, p. 110). Then, practitioners would use the one they preferred. For example, Ilídio Araújo, author of the book *Arte Paisagista e Arte dos Jardins em Portugal* (1962), always signed as landscape architect and agronomic engineer (so, he decided that he was first an architect and then an engineer, although his diploma was written the other way around).

However, Cabral was unable to gain the support he needed inside the ISA to promote landscape architecture in a realm of agronomists. He even lacked the means to build a specialist landscape architecture library (Araújo 2009, p. 19). Only in the 1950s were more effective bridges established between the technical disciplines and their artistic and humanistic counterparts.

In 1956, the Lisbon Society of Geography organized a cycle of conferences dedicated to the protection of nature. As the opening speaker, Cabral focused on nature conservation and the role landscape architecture would play. Cabral stated that ever since Le Nôtre, exceptional gardeners had been working as landscape architects integrating horticultural knowledge with technical ability and artistic sensibility. In the future, Cabral stressed, landscape architects should turn their attention also to the protection of nature (Cabral 1956, p. 152). Essentially, he said, the aims of landscape architecture aligned with the objectives of nature conservation. They were both born in the nineteenth century as a reaction to the damaging effects of the Industrial

Revolution. Recognizing how landscape architecture was tied more to ecology and the environmental sciences to its agricultural beginnings was a path Cabral pursued early in his career.

While the first landscape architecture curriculum in Portugal was based on a German model (Antunes 2019), it also may be understood as a palimpsest of nineteenth-century gardening and horticultural studies launched by Francisco Simões Margiochi.

Gardening and horticultural studies in the late nineteenth century

In the nineteenth century, horticulturists and gardeners in several countries considered professional studies essential, for example Belgium, France, Germany, the United Kingdom and the USA (Woudstra 2010; Limido 2018, p. 78). Following the 1834 civil war victory of the Liberals, the Lisbon City Council started to deliver on long-awaited improvements, including grey and green infrastructure (Depietri and McPhearson 2017). According to the new Liberal agenda, improvements included constructing not only sewage, hydraulic systems and public lightening; they also led the city council to establish green open space. In 1840, before its French counterpart in Paris followed in 1855, Lisbon City Council set up its first Department of Gardens and Green Grounds (Rodrigues 2020, pp. 49–61, 2022b; Santini 2018, pp. 34–38).

New public gardens and parks included the city's most famous, the Estrela Garden and the Park of Liberty, now called Park Edward VII, the Campo Grande, the boulevard-like Avenue of Liberty, and also tree-lining avenues in keeping with the city's expansion northwards (Le Cunff 2000; Rodrigues 2017b, 2022a, Rodrigues and Simões 2022).

The Estrela Garden, inaugurated in 1852, was nominated as the headquarters of the Department of Gardens and Green Grounds in 1859. Henceforth, the department gardener-in-chief was located there as were the municipal nurseries alongside channelling the bulk of the budget and the working efforts of the department into this garden. The Estrela Garden also served as a school for practitioners.[3] A gardener might begin working as an 'assistant' and develop to become a 'municipal gardener' while rising to the position of 'gardener-in-chief'. Such was the case of João Francisco da Silva (Rodrigues 2020, pp. 115–132). As he designed and prepared plans for new gardens and parks, the Lisbon chief gardener provided an example that younger gardeners were to follow. Theoretical grounds could be pursued by these practitioners at the specialized library of the Department, created in 1872 (Rodrigues 2017b, 2020, pp. 141–205).

In addition, the Estrela Garden facilities actively promoted horticultural and gardening education to the public, including children. In 1882, an important educational institute opened on the site – the Froebel School. It was the first kindergarten ever established in Portugal. According to the theories of the German psychologist Friedrich Froebel (1782–1852), children under the age of six should attend an educational institution; several countries followed (Allen 1988a, 1988b; Wollons 2000; Albisetti 2009; Murcia and Ruiz-Funes 2019). In Froebel's words, children are "human plants" and should be educated accordingly (Ferreira 1882, p. 3). Gardens and gardening played important roles. Children should have opportunities to grow up in contact with nature. In their education, alongside with physical education gym and games, gardening and horticulture should be part of the curriculum. Period photographs depict children playing with gardening utensils in the Estrela Garden (Figures 18.1 through 18.3). Gardening education for children in Portugal followed a path similar to that established in the USA (Latter 1906; Kohlstedt 2008). Evidence exists that female gardeners were hired as educational auxiliaries for the Lisboan Froebel school (Rodrigues 2017a, pp. 38, 39, 70), offering an employment opportunity for women incorporating the chance to simultaneously work on gardening and education, similar to what happened in other countries (Mozingo and Jewell 2012; Opitz 2013; Dümpelmann and Beardsley 2015). Although there were no schools for women gardeners, it is

Figure 18.1 Children playing at the Estrela Garden with gardening utensils, 1927.
Source: ANTT, EPJS/SF/001–001/0005/0706B.

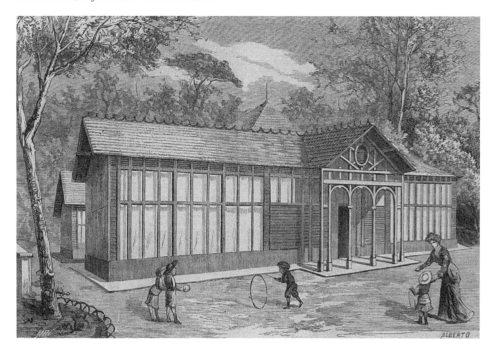

Figure 18.2 Engraving of the Froebel School at the Estrela Garden in 1883.
Source: Published in *Occidente*, n° 146, January 1883: 12.

Figure 18.3 The building of the former Froebel School at the Estrela Garden in Lisbon, in 2015.
Source: Photo by Ana Duarte Rodrigues.

a fact that women gardeners were hired by the city council to act as educational assistants at the Froebel School of the Estrela Garden. Not only this fact was divulged in magazines, as contracts hiring these women as *monitora jardineira* (assistant gardeners) stand as an evidence of that reality (Rodrigues 2017a, p. 39).

The key actor bringing together the rise in importance of horticulture and gardening on the municipality's agenda was Francisco Simões Margiochi. He was the councillor of the Department of Gardens and Green Grounds between 1872 and 1875 and its specialized library was one of his initiatives (Silva 1899; Rodrigues 2017a, pp. 113–166, 2020, pp. 107–110; 141–205). After leaving that position, Margiochi became the director of the Casa Pia charitable orphanage in Lisbon. Aware of the need for professional gardeners to implement the new works under development by the city council, in 1895, he established an agriculture and gardening study programme for orphaned children aged 10–12 (*Portaria* n° 205 in Silva 1903, pp. 10–32). The Casa Pia's study programme was inspired by the *École Spéciale d'Arboriculture et de Jardinage de Bastia in Corsega*,[4] as well as by French orphanages' study programmes such as the ones in Salvert, Sedière and Saint Martin des Donets.[5] The course was to provide theoretical and practical preparation for *jardineiros paizagistas* (landscape gardeners) (Silva 1903, pp. 38–39). Students attended theoretical classes and acquired practical knowledge by working on pilot-gardens, groves and vegetable gardens, for example in the monastic enclosure of the sixteenth-century Monastery of Jerónimos in Belém, an area of Lisbon (Figure 18.4). In addition, gardener-horticulture students also attended classes on growing special breeds and crops, horticulture and gardening, hygiene, garden and park management, practices for growing plants in greenhouses as well as agricultural

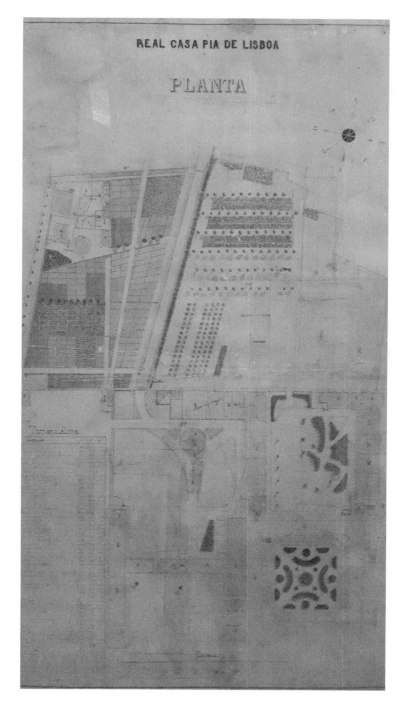

Figure 18.4 Plan of the monastic enclosure of the Monastery of Jerónimos, Lisbon, in which the premises for the School of Gardening was developed. Colour drawing made by Alphonse Jules Picard, between 1867 and 1879.

Source: Casa Pia de Lisboa – Centro Cultural Casapiano CPL CCC _planta_picard_1867_1879.

accounting and domestic animal breeding (Silva 1903, pp. 26–33). Students went on fieldtrips to visit some famous gardens in the Lisbon region, such as the Palace of Fronteira with its rose collection and recently introduced palm trees (Rodrigues and Luna 2015).

The Casa Pia study programme lasted no longer than two years. Margiochi and the Marquis of Fronteira then pooled their efforts to promote the field and, in 1898, founded the Royal Society of Horticulture. Margiochi became its president while the Marquis of Fronteira headed the gardening section. In general terms, horticultural societies fostered the development of gardening education as they almost inevitably ran their own publications, whether in the form of journals, bulletins, almanacs, organized exhibitions and were often interlinked with launching and running study programmes and courses for gardeners and horticulturists (Fletcher 1968; Elliot 2004), as it also occurred in Portugal.

Under the Royal Society of Horticulture umbrella, the fashion for botany and gardening and a true passion for horticulture were nurtured with the seeds for gardening education planted in Portugal through horticultural exhibitions, amateur collections and periodical publications. Margiochi again proposed a plan for a Practical School of Horticulture that would offer gardener-horticulture courses formerly provided at Casa Pia (*Boletim* v. 5, 1899, p. 78). Unable to find a site for its location, the project was never implemented (*Boletim* v. 7, 1900, p. 98). Moreover, the society itself did not survive Margiochi's very enthusiastic entrepreneurship and, following his dismissal, the society closed in 1906 after eight years packed with activities and initiatives (Rodrigues and Simões 2017; Rodrigues 2017a, pp. 121–166).

Following an epoch of flourishing horticultural activities in the second half of the nineteenth century, the Department of Gardens and Green Grounds of the Lisbon City Council, established in 1840, had disappeared by the end of the late nineteenth century. The *Jornal de Horticultura Pratica* (*Journal of Practical Horticulture*) edited by the famous horticulturist José Marques Loureiro ended in 1892 after 22 years of continual publication. Projects for the Park of Liberty, designed by Henri Lusseau (in 1889) and for Campo Grande, envisioned by Ressano Garcia (in 1903) were never implemented. The first study programme for gardeners at Casa Pia did not find a successor and the plans for such education under the auspices of the Royal Society never came to fruition.

A gap of four decades passed before new projects for establishing garden and landscape architecture education started. The gap was primarily due to national and international political and economic turmoil of the first half of the twentieth century. Afterwards, landscape architecture education initiatives resumed during the *Estado Novo* dictatorship (Andresen 2003; Antunes 2019).

Conclusion

Focusing on the establishing of landscape architecture education in Portugal, this chapter explains how the first courses on gardening started up under the Liberal regime. The educational programme displayed curricula similar to the ones that universities developed decades later. However, dramatic historical circumstances such as the regicide, followed by the declaration of the First Republic in 1910, Portugal's participation in World War I, subsequent political and financial turmoil and, finally, bankruptcy and the shift from a democratic republican regime to a dictatorship delayed what one would have expected to seamlessly emerge from the nineteenth-century developments in gardening education.

Lisbon's gardens and vegetation, considered in the nineteenth century as 'semi-exotic', and the winner of the International Green Capital 2020 Prize, owes part of its success to the work carried out by the Lisbon City Council Department of Gardens and Green

Grounds established in 1840. The Liberal agenda for the city's modernization paved the ground for a school of practitioners where municipal gardeners gained their education and performed the same kind of activities that professional landscape architects would some decades later.

During the 1940s, partly due to the need for embellishing the city of Lisbon for the Portuguese World Exhibition in 1940, attention returned to gardening and horticulture. The Lisbon City Council Department of Gardens and Green Grounds reopened in 1938. It was in this context of renewed interest that the Portuguese landscape architect Francisco Caldeira Cabral, after returning from Germany with a university degree, was able to establish the first experimental degree in landscape architecture in Portugal.

Notes

1 The horticulturist Moreira da Silva stems from a nineteenth-century family of horticulturists. In the first decades of the twentieth century, the family owned the most famous nursery and had reached an important status and wealth (Rodrigues 2020, pp. 445–447). This family still holds remarkable nurseries in Lisbon.
2 Among them were some of the most famous landscape architects in Portugal: Manuel Azevedo Coutinho, Edgar Fontes, Gonçalo Ribeiro Telles, Ilídio de Araújo, António Campello, Viana Barreto, Álvaro Dentinho, and Manuel Sousa da Câmara.
3 As councillor Ricardo Teixeira Duarte states in his 1859 report, *Archivo Municipal de Lisboa*, n° 8, February 1860, p. 61.
4 Historical Archive of Casa Pia of Lisbon, Registo de Ofícios Expedidos, Administração, n° 2, cota 419, f. 22.
5 Historical Archive of Casa Pia of Lisbon, Registo de Ofícios Expedidos, Administração, n° 2, cota 419, f. 47 v.-48.

References

Albisetti, J. (2009) 'Froebel Crosses the Alps: Introducing the Kindergarten in Italy', *History of Education Quarterly*, 49(2), pp. 159–169.
Allen, A. T. (1988a) 'Let Us Live with Our Children: Kindergarten Movements in Germany and the United States, 1840–1914', *History of Education Quarterly*, 28(1), pp. 23–48.
Allen, A. T. (1988b) 'Spiritual Motherhood: German Feminists and the Kindergarten Movement, 1848–1911', *History of Education Quarterly. Special Issue: Educational Policy and Reform in Modern Germany*, 22(3), pp. 319–339.
Andresen, T. (ed.) (2003) *Do Estádio Nacional ao Jardim Gulbenkian. Francisco Caldeira Cabral e a primeira geração de arquitetos paisagistas (1940–1970)*. Lisboa: Fundação Calouste Gulbenkian.
Antunes, A. C. D. S. (2019) *A influência alemã na génese da Arquitetura Paisagista em Portugal*. PhD, Faculdade de Ciências da Universidade do Porto.
Araújo, I. (2009) 'Ao professor Francisco Caldeira Cabral', in de Abreu, A. C. (ed.) *Francisco Caldeira Cabral: Memórias do Mestre no Centenário do seu Nascimento*. Lisboa: Associação Portuguesa dos Arquitectos Paisagistas, pp. 20–24.
Boletim da Real Sociedade Nacional de Horticultura de Portugal, vol. 5. (1899) Lisboa: Real Sociedade Nacional de Horticultura de Portugal.
Boletim da Real Sociedade Nacional de Horticultura de Portugal, vol. 7. (1900) Lisboa: Real Sociedade Nacional de Horticultura de Portugal.
Cabral, F. C. (1935). *Relatório de Tirocinio do Curso de Engenheiro Agrónomo de Francisco Caldeira Cabral*. Lisboa: Instituto Superior de Agronomia.
Cabral, F. C. (1943a) *Lições proferidas no Instituto Superior de Agronomia*. Lisboa: Universidade Técnica de Lisboa.
Cabral, F. C. (1943b) 'Zonagem sob o ponto de vista paisagista', in *Anais do Instituto Superior de Agronomia*. Lisboa: Instituto Superior de Agronomia, pp. 53–63.
Cabral, F. C. (1956) 'Protecção à Natureza e Arquitectura Paisagista', in *Sep. Boletim da Sociedade de Geografia de Lisboa*, pp. 151–162.

Cabral, F. C. (1993) *Fundamentos da Arquitectura Paisagista*. Lisboa: Instituto da Conservação da Natureza.
Depietri, Y. and McPhearson, T. (2017) 'Integrating the Grey, Green, and Blue in Cities: Nature-Based Solutions for Climate Change Adaptation and Risk Reduction', in Kabisch, N., Korn, H., Stadler, J. and Bonn, A. (eds.) *Nature-Based Solutions to Climate Change Adaptation in Urban Areas. Theory and Practice of Urban Sustainability Transitions*. Cham: Springer, pp. 91–109.
Dümpelmann, S. and Beardsley, J. (eds). (2015) *Women, Modernity and Landscape Architecture*. London and New York: Routledge.
Elliot, B. (2004) *The Royal Horticultural Society: A History, 1804–2004*. Chichester: Phillimore and Company Limited.
Ferreira, T. (1882) 'Escolas infantis ou jardins de Froebel: apontamentos para a sua história em Portugal', *Froebel: Revista de Instrucção Primária*, 1, pp. 2–3.
Fletcher, H. (1968) *The Story of the Royal Horticultural Society: 1804–1968*. Oxford: Oxford University Press for the Royal Horticultural Society.
Kohlstedt, S. G. (2008) 'A Better Crop of Boys and Girls: The School Gardening Movement, 1890–1920', *History of Education Quarterly*, 8(1), pp. 58–93.
Latter, L. (1906) *School Gardening for Little Children*. London: Swan Sonnenschein & Co.
Le Cunff, F. (2000) *Parques e jardins de Lisboa, 1764–1932: do Passeio Público ao parque Eduardo VII*. Master thesis on Art History, Faculdade de Ciências Sociais e Humanas/NOVA.
Limido, L. (2018) 'La formation des architectes-paysagistes depuis Jean-Pierre Barillet-Deschamps', in Audouy, M., Le Dantec, J.-P., Nussaume, Y. and Santini, C. (eds.) *Le grand Pari(s) d'Alphand. Création et transmission d'un paysage urbain*. Paris: Éditions de la Villette, pp. 75–89.
Marques, T. D. P. (2009) *Dos jardineiros paisagistas e horticultores do Porto de oitocentos ao modernismo na arquitetura paisagista em Portugal*. PhD, Faculdade de Ciências da Universidade do Porto.
Mozingo, L. A. and Jewell, L. (eds.) (2012) *Women in Landscape Architecture: Essays on History and Practice*. Jefferson and London: McFarland & Company, Inc.
Murcia, J. P. M. and Ruiz-Funes, J. M. (2019) 'Froebel and the Teaching of Botany: The Garden in the Kindergarten Model School of Madrid', *Paedagogica Historica. International Journal of the History of Education*, 56(1–2), pp. 200–216.
Nunes, O. P. (2011) *O Arquitecto paisagista em Portugal: a construção do grupo profissional e o seu regime justificativo de acção perante a legislação que o "regula"*. Master's thesis, Faculdade de Ciências Sociais e Humanas da Universidade Nova de Lisboa.
Opitz, D. L. (2013) 'A Triumph of Brains Over Brute'. Women and Science at the Horticultural College, Swanley, 1890–1910', *Isis*, 104, pp. 30–62.
Portaria da administração da Real Casa Pia, n° 205, de 23 de Julho de 1896, published in da Silva, J. E. D. da (1903), *A Escola de Agricultura Pratica da Real Casa Pia de Lisboa*. Lisboa: A Liberal – Officina Typographica, pp. 10–32.
Raxworthy, J. (2018) *Overgrown: Practices Between Landscape Architecture and Gardening*. Cambridge, MA: MIT Press.
Rodrigues, A. D. (2017a) *Horticultura para todos*. Lisboa: Biblioteca Nacional de Portugal.
Rodrigues, A. D. (2017b) 'Greening the City of Lisbon under the French Influence of the Second Half of the Nineteenth Century', *Garden History*, 45(2), pp. 224–250.
Rodrigues, A. D. (2020) *O Triunfo dos Jardins. O pelouro dos Passeios e Arvoredos (1840–1900)*. Lisboa: Biblioteca Nacional de Portugal/Lisboa Capital Verde Europeia.
Rodrigues, A. D. (2022a) 'Trees, Nurseries, Tree-Lined Streets and the Making of Modern Lisbon (1840–1886)', in Simões, A. and Diogo, M. P. (eds.) *Science, Technology and Medicine in the construction of Lisbon*. Leiden: Brill.
Rodrigues, A. D. (2022b) 'Building Green Urban Expertise: Politicians, Agronomists, Gardeners and Engineers at Lisbon City Council (1840–1900)', *Urban History*, 2022(1), pp. 1–20.
Rodrigues, A. D. and de Luna, R. C. (2015) 'The 8th Marquis of Fronteira's Taste of Gardening in its English Cultural Context', *Gardens & Landscapes of Portugal*, 2015(3), pp. 30–59.
Rodrigues, A. D. and Simões, A. (2017) 'Horticulture in Portugal 1850–1900: The Role of Science and Public Utility in Shaping Knowledge', *Annals of Science*, 74(3), pp. 192–213.
Rodrigues, A. D. and Simões, A. (2022) 'A Liberal Garden. The Estrela Garden and the Meaning of Being Public', in Simões, A. and Diogo, M. P. (eds.), *Science, Technology and Medicine in the Construction of Lisbon*. Leiden: Brill.

Santini, C. (2018) 'Construire le paysage de Paris. Alphand et ses équipes (1855–1891)', in Audouy, M., Le Dantec, J.-P., Nussaume, Y. and Santini, C. (eds.) *Le grand Pari(s) d'Alphand. Création et transmission d'un paysage urbain*. Paris: Éditions de la Villette, pp. 75–89.

Silva, J. E. D. da (1899) *Francisco Simões Margiochi, par do Reino e presidente da Real Sociedade Nacional de Horticultura de Portugal*. Lisboa: A Liberal – Officina Typographica.

Silva, J. E. D. da (1903) *A Escola de Agricultura Pratica da Real Casa Pia de Lisboa*. Lisboa: A Liberal – Officina Typographica.

Tostões, A. (1992) *Monsanto, Parque Eduardo VII, Campo Grande: Keil do Amaral, Arquitecto dos espaços verdes de Lisboa*. Lisboa: Salamandra.

Wollons, R. (2000). *Kindergartens and Cultures: The Global Diffusion of an Idea*. New Haven: Yale University Press.

Woudstra, J. (2010) 'The 'Sheffield Method' and the First Department of Landscape Architecture in Great Britain', *Garden History*, 38(2), pp. 242–266.

19
DUTCH LANDSCAPE ARCHITECTURE EDUCATION IN THE FIRST HALF OF THE TWENTIETH CENTURY

The pioneering roles of Hartogh Heys Van Zouteveen and Bijhouwer

Patricia Debie

Introduction

This chapter provides insight into the beginnings of Dutch landscape architecture education. At the end of the nineteenth century, the Netherlands had become an internationally renowned horticultural centre and prosperity among citizens led to growing demands for gardens. Many gardens where realized by horticulturist who, in pre-landscape architecture times, dominated in number over 'garden architects'. The type and quality of design that commercial horticulturalist-designers offered caused an unstoppable flow of criticism. Architects and members of the public were mocking them as diminutive parks in the "vermicelli style" (Hartogh Heys van Zouteveen 1915, p. 647, 1911, p. 6, 1918, p. 85). Critics found it scandalous how poorly educated horticulturalist even used the title 'garden architect' (see Figure 19.1).

Aiming for the betterment of the design field, horticultural organizations started to pay more attention to garden architecture and an education where composition and drawing combines with planting design and scientific knowledge. Common secondary horticultural schools such as the ones at Boskoop and Frederiksoord offered a good general practical surface but little science and theory (Springer 1895, pp. 49–54). Evening schools for craftsmen did offer individual garden design lessons and drawing courses, however without including science-based education. Outlining how garden and landscape architecture education in Wageningen developed, this chapter explains which content, domains and scientific approach became important and it provides insight into how Wageningen students began to obtain the skills required to work as garden and landscape architects.

Methods include sampling and analyzing primary sources, such as private letters, letters of recommendation, official documents such as subscription lists, meeting reports, school curricula and programmes, learning material and inventory. Secondary sources include journal articles and other kinds of publications. Documents and publications were hand-searched and qualitative textual analysis performed, including content analysis and thematic analysis.

DOI: 10.4324/9781003212645-21

Figure 19.1 Advertisement at the plant nursery Moerheim Ruys at Dedemsvaart.
Source: *Onze tuinen; geïllustreerd weekblad voor amateur tuiniers* (14), no. 36, 1920 559 in: www.delpher.nl/.

From practical to scientific education – the example of Wageningen

In 1890, the agricultural school at Wageningen developed a new two-year course called Garden Art (de Vos 1887, p. 245, 1890, pp. 341–343) and then, in 1896, established a Horticulture Department that offered an educational programme combining horticultural knowledge with individual design education, managerial skills and drawing courses. Aiming to counterbalance the emerging influence of architects and urban planners in the field of garden design, the garden art course had a broad science base. A sound education and a distinguishing title would set 'garden architects' apart also from commercial horticultural-designers (Hartogh Heys van Zouteveen 1918, p. 2). During the first two years, students would get an overview of the whole field. They would take sciences for the following two years. Eventually, a four-year broad scientific course offered at the Community School was connected to the university education (van der Haar 1993, p. 82). As there was little appreciation for the field of garden architecture, the course was initially only optional (Springer 1895, pp. 49–54).

The educational programme had two strands. One was garden art, garden architecture and composition where students would learn the art of designing plans for gardens, the history of garden art and of garden architecture (Landbouwhoogeschool Wageningen 1927, p. 39). These domains give a clear indication how specific knowledge and skills are needed to be employed as a garden architect, namely first the aesthetics and design of gardens and parks in which the planning and space-mass ratios were treated. Secondly, the education of hand- (or natural) drawing, as fundamental toolbox for the garden and landscape architect was important to develop a sense of line, form and colour (Hartogh Heys van Zouteveen 1916, p. 29).

The Wageningen lector Hartogh Heys van Zouteveen was aware of the term 'landscape architecture' used in the USA and the UK. As early as 1903, he had organized international excursions to learn more about the profession of landscape architecture (Hartogh Heys van Zouteveen 1915, p. 648). In 1911, he had attended a conference about the architectural elements in protecting the beauty of the Netherlands and pleaded for the profession and title *Tuinarchitect* (garden architect):

> And... if we now establish, that the architectural engineer wants to adapt the architectural element into the landscape and that the Garden architect has the task of adapting the landscape onto the buildings. (...) Garden art, ... Garden architect... I did not hear those words!
> (Hartogh Heys van Zouteveen 1913) p. 187; translation by author)

Around 1913, inspired by international examples (see Birli et al. in this volume), Van Zouteveen proposed educational reforms aimed at developing garden architecture as 'landscape architecture', and to integrate it with urban design and architecture (Hartogh Heys van Zouteveen 1913, p. 648). He suggested that not only the Agricultural University in Wageningen, but also the Technical University in Delft and the Academy of Fine Arts in Amsterdam needed a course in garden and landscape architecture (Hartogh Heys van Zouteveen 1913, pp. 2–3, 1918, pp. 7, 37–38 and 77–78). Only in this way, the field of garden and landscape architecture education could stand out and be separate from the common pragmatic horticultural training that Wageningen offered so far.

In 1930, the first year of landscape architecture education, initially called *Hogeschoolopleiding voor tuin- en landschapsarchitecten* was set up as a broad academic programme that included distinctively more aesthetics, composition and design compared to the horticultural training of earlier courses. The programme was accessible not only to all students of the Agricultural University (including forestry and agriculture departments) but also to everyone who worked in practice, such as architects and architecture students of Delft or Amsterdam (Hartogh Heys van Zouteveen 1913, 1918, pp. 77–78, 1928, pp. 4–7). The main subjects considered necessary for landscape architecture education included Art History, Nature Drawing, Garden and Landscape Architecture, Architecture and Urban Planning, Public Art, Landscape Painting, Perspective and Surveying. Additional courses were subjects that existed in earlier programmes, such as Plant Systematics and Geography, Plant Breeding, Chemistry and Fertilization, Plant Diseases, Agronomy, Horticulture Technology, Genetics, Botany, Microbiology, and Colonial Agronomy Dutch-Indian Agrarian Law.[1]

In addition to combining compositional design skills and science-based knowledge, working on a larger scale also played a role in the broadening of education. Over the years, insights into spatial planning in tune with social developments grew substantially, as did the need for training in the design of larger areas such as city landscapes. In the initial programme, a course in the direction of spatial planning as a combination of research, planning and design of space on a universal basis was lacking. To meet growing needs for designers, during the 1930s and 1940s, the university prepared to establish a department for garden and landscape architecture.[2] Finally, the continuous efforts and pleas of Van Zouteveen were recognized.

Educational pioneers

In Wageningen, from 1867 to 1900, the garden architect Leonard Anthony Springer was the first teacher of the course called Garden Art. Between 1900 and 1935, the garden and landscape architect Van Zouteveen was his successor. The profession title of garden architect was official since the establishment of the *Vereniging van Nederlandsche Tuinkunstenaars*, also known as the *Bond van Nederlandse Tuinarchitecten* (B.N.T.) in 1922. Van Zouteveen is considered as the founder of the Dutch garden and landscape architecture education programme in Wageningen (Debie 2011). In 1936, his successor was his former student Jan Tijs Pieter Bijhouwer, a landscape architect and urban planner who, in 1946, became the first professor of garden and landscape architecture in Wageningen and two years later was appointed professor of garden and landscape architecture at the Technical University in Delft. With him, respect for the discipline and the training of garden and landscape architects in Wageningen grew.

In addition to landscape architects, teachers in other fields are important in developing the skills required for garden and landscape architectural education. Springer pointed out how by developing their sense of aesthetics and architectural composition, garden artists were able to distinguish themselves from the limited skills of craftsmen (Moes 2002, p. 20). Students needed to train as independent and critical designers who did not simply follow the French or German garden fashion, but who would find their own styles (Springer 1895, pp. 49–54).

Van Zouteveen stated quite firmly that he was "being discouraged by the abuses in the trade prevailing in the Netherlands" (Hartogh Heys van Zouteveen 1915, p. 647). Firstly, he thought it was necessary to realize a well-founded education on scientific grounds. The main goal was "to cultivate in- and outside the school an awareness and appreciation of garden art as a high art form".[3] This concerned not only the public opinion, but also that of fellow construction artists, such as architects and urban planners. Secondly, Van Zouteveen referred to the great importance of garden art as a link for three institutions, namely the Agricultural College, the Academy of Visual Arts and the Technical College (Hartogh Heys van Zouteveen 1918, pp. 37–38). Only in that way, garden art could be an independent field additional to contemporary architecture. This would help shaping more natural gardens and parks, but also create well-functioning urban spaces, preserving natural beauty and bringing it into the living areas of cities and villages for people's health and well-being, were history provided an important source of inspiration (see Figure 19.2). Planting should serve as building blocks in such a way that it enriches existing plant communities also called 'plant societies' (Hartogh Heys van Zouteveen 1913, pp. 3, 17, 1927, p. 2). Thirdly, Van Zouteveen promotes garden art as an expression of social life and the important role that plant communities play in it (see Figure 19.3).

As a 'cultural philosopher', Van Zouteveen describes his ideology and ideas in three publications: *Boomen en heesters in parken en tuinen* (Trees and shrubs in parks and gardens, 1908), *De Siertuin* (The ornamental garden, 1920), and *De ideeële en reeële beteekenis van bosschen en natuurmonumenten voor de menschelijke samenleving* (The ideal and real significance of forests and natural monuments for human society, 1931). Three of his students, Mien Ruys, Hein Otto and Jan Pieter Bijhouwer, considered plant knowledge and 'plant sociology' to be a fundamental part of garden and landscape architecture, in contrast to the predominantly materialistic and technical Wageningen culture. Mien Ruys and Hein Otto translate their teacher's ideas into a quest to develop a new independent Dutch garden art, with strong natural planting

Figure 19.2 An example of a detailed garden design (1899) for the insane hospital Grave, containing a floor plan, a plan of construction and profiles.

Source: Special Collections, University Wageningen inv.no. 01.222.01, 01.222.02, 01.222.03.

Dutch landscape architecture education

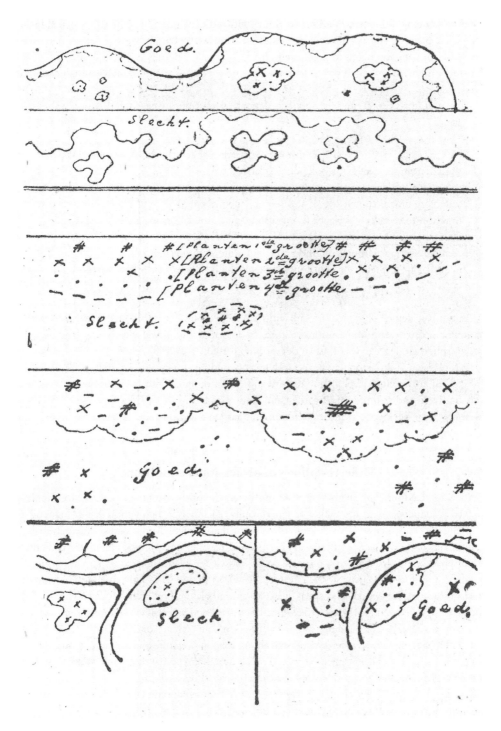

Figure 19.3 Drawing instructions with good (*goed*) and wrong (*slecht*) examples of plant schemes to create natural plant communities.

Source: Hartogh Heys van Zouteveen, H.F., *de Siertuin: Geïllustreerd Handboek ten dienste van vakman en liefhebber en van inrichtingen voor tuinbouwonderwijs*, De Bilt 1920, 6.

schemes (Andela 2011, p. 196). This art flourished and evolved to become the sustainable and ecological planting approach categorized and well-known as Dutch 'wave gardens'. The starting point for applying plants to meet their environmental properties and the question of what a national garden art intended to be their teacher and predecessor Van Zouteveen had explicitly advocated.

The profile of the landscape architect, according to Bijhouwer, should strongly focus on practice.

> (A landscape architect is) a craftsman who approaches the task; searches for what the environment tells, for coherence, clarity and readability, for a human scale and the essential characteristic of the use, in which the starting point for the design is found. A broad knowledge of soil science, geomorphology and plant science, with insight into the effect of space in relation to mass, the landscape architect make a legible design with a recognizable articulation. Technical knowledge of urban planning, architecture, agricultural engineering, hydraulic engineering and wood culture are needed, not to practice it, but to understand the insights and working methods.
>
> (Andela 2011, p. 178)[4]

Bijhouwer describes garden art as having a beneficial modesty that was simple, but with a naturalness, clarity and unintentional nature. In addition, Bijhouwer clearly saw landscape, or geomorphology, as a carrier of urban development (ibid., 197–198).

It took until 1946 before landscape architects would be appreciated for their skills and to be recognized at the same level as architecture and urban design. With his positions at Wageningen and Delft and as professor of garden and landscape architecture, Bijhouwer could promote these specific skills needed for the self-confident garden and landscape architect to preserve the traditional images of the Dutch countryside landscapes.

Different pillars of professional education

Garden and landscape architecture evolved, in the course of the twentieth century, from initially viewed as an independent and critical designer mainly of gardens to a profession that includes urban planning and encompasses a wide array of technical knowledge. To distinguish themselves from plantsmen, the garden and landscape architect should have no horticultural ambitions (Bijhouwer 1945, pp. 44–45). To work as a 'creative artist' knowledge of building styles and art principles formed a tool for recognizing aesthetic values. Towards the middle of the century, the identity of the landscape architect had become that of an "all-around planner" or "stedebouwer" who develops the planning of city and landscape (Cleyndert 1944). As "building master", the landscape architect was seen as "designer of building, soil and vegetation" but also as "organizer and designer".[5] In line with changing views, there came a need for training in design in and for larger spatial dimensions, such as park and landscape projects and urban development that dealt with a combination of research, planning and design of space on a more universal basis to gain a well-respected position (Cleyndert 1944, p. 1). The changing views also echoed in the educational programme. Garden and landscape architecture became the important link between building art, visual arts and urban planning for three existing courses: the Agricultural University in Wageningen, the Academy of Art in Amsterdam and the Technical University in Delft (Hartogh Heys van Zouteveen 1918, pp. 37–38).

As teachers in Wageningen, Van Zouteveen and Bijhouwer developed and applied their own teaching methods. In their opinion, the four years of academic training contained four essential domains that a garden and landscape architect had to master: Art History, Plant Sociology, Architecture and Urban Planning, and Folk Art such as drawing and related arts.

First, Art History fed through the lines of the curriculum like a thread. With Springer, in the course *Kunststijl*, students learned to study the artistic "national past" to develop their own aesthetic sense (Moes 2002, pp. 20, 64). Van Zouteveen and Bijhouwer employed the study of history as a way to recognize garden art as an art form (Hartogh Heys Van Zouteveen 1918, p. 7). They saw the combination of knowledge of architecture and visual art as a means to become a good landscape architect, necessary for the available positions for planners and landscape architects at government, provincial and municipal services. With their special knowledge, engineers from Wageningen could perform these specific activities, and compete with planners from Delft.[6]

Secondly, knowledge in plant systematics was important (see Figure 19.3). Van Zouteveen was looking for a way to develop a new independent Dutch garden art with new plant systematics similar to the ideas of Édouard François André and William Robinson in the nineteenth century (Debie 2011, pp. 45–46). The application of planting in so-called plant communities attuned to their surroundings he strongly advocated in his publications, as described earlier.

A third pillar in the education was the study of relationships between architecture and urban planning with garden and landscape architecture (Hartogh Heys van Zouteveen 1911). With his personal education and urban planning activities, Van Zouteveen discovered quite early the unlimited scope of garden and landscape architecture. He was educated at the École d'Horticulture in Vilvoorde where Louis Fuchs was his teacher, one of the most important garden architects of Belgium at the time who had urban planning experience.[7] In 1895, Van Zouteveen had worked in his practice in Brussel (Debie 2011, p. 10; Hartogh Heys van Zouteveen 1928, p. 2.). He participated in several urban planning lectures, which he incorporated into his publications (Debie 2011, pp. 83–87; Hartogh Heys van Zouteveen 1913, 1927). In addition, in 1921, he was a member of the Urban Planning Department of the Dutch Institute for Housing and Urban Planning (Andela 2011, pp. 64, 67). His library collection contained forty modern book and plate works on architecture, urban planning and maps illustrating the development of the Dutch road systems.[8] Particularly noteworthy is his participation in the international competition for the urban expansion of Gross-Berlin in 1909. The subsequent designation as a member of the committee for the construction of Belgium, together with the successful Belgian urban planner Louis van der Swaelmen to create an urban encyclopaedia, shows his great affinity with urban planning.[9] According to Bijhouwer, the careful attention of the Wageningen engineer is necessary between landscape architecture and urban planning (Bijhouwer 1947, p. 9).

The fourth and most important pillar in the education as a garden and landscape architect was drawing skills based on Folk Arts. Initially, drawing and sketching was not included as a compulsory subject in the exam package of the garden and landscape architecture training (Hartogh Heys van Zouteveen 1915, p. 617). Drawing and sketching needed to be developed as part of a science-based garden and landscape architecture education, not only an instrumental and practical skill, but a competence needed to analyze site and landscape. The remarkably innovative ideas about drawing methods Van Zouteveen demonstrated in his plans and his building regulations map of his entry for the 'Wettbewerb Gross-Berlin', the urban design contest in 1909. They held various data on transportation, industry, greenery and construction classes that use a colour-coding system.[10] His ideas were much appreciated and drawn up in 1935 as standardization of garden drawings for the Netherlands participating in "Commission 8" (Debie 2011, p. 58).

In 1937, Landscape Architecture and Garden Art became an official exam subject. Despite an attempt to link up with the scientific level of Architecture in Delft, drawing practice was not part of the compulsory exam subject and major.[11] Although the teachers themselves developed specific drawing education, the main criticism about the active horticultural landscapers was that they were unable to draw by nature and created poor spatial design solutions. In 1942, a contest organized by the Dutch Union of Garden architects (B.N.T.) yielded disappointing results. The poor quality was blamed on "a serious gap in education" (Andela 2011, p. 176). "The experienced landscaper prefers to draw on the site instead of on the paper" (Cleyndert 1944, p. 6). Eventually in 1947 for one year and then for the long term in 1951 a teacher was hired and hand drawing became a required subject.

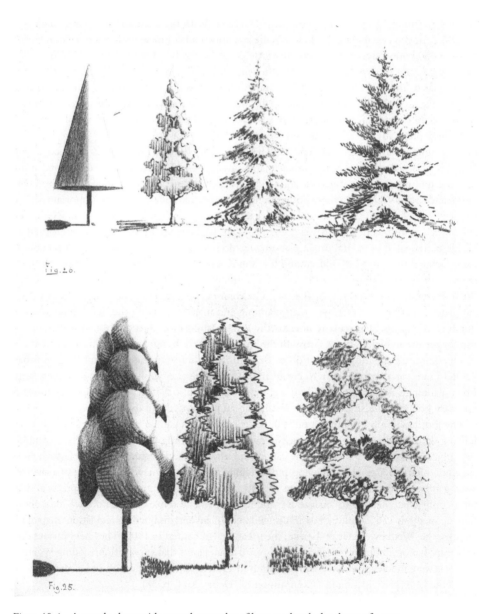

Figure 19.4 A sample sheet with several examples of how to sketch the shape of a tree.

Source: J. Landwehr, *Tekenen van bomen en heesters, Potloodstudies voor tuintekenaars, leerlingen van tuintekencursussen en bouwkundigen*, Amsterdam 19, z.p.

Drawing became an essential visualization tool and an important tool to analyze and develop design concepts (see Figure 19.4). Drawing was the medium practiced in several subjects, including Composition/Aesthetics, Botany, Urban Design and Art History.

In the curriculum, drawing courses were called Nature Drawing, Hand Drawing, Line Drawing and Perspective Drawing. Because most landscape architecture students came from third grade elementary schools, they had never had a scribing tool or a brush in their hands before entering the university. The first university years were allocated to line drawing, to provide a good basis for further education (Hartogh Heys van Zouteveen 1915, p. 617).

To develop drawing skills and a sense of composition, the first of Springer's students copied the perspective drawings of architectural elements from example books in the course Garden Art (Bijhouwer 1950). Van Zouteveen did not encourage stencilling. He stimulated students to create and design "all talking and measuring" (Hartogh Heys van Zouteveen 1911, p. 2). In a combined course, design and sketching of simple compositions of parks and gardens were practised to design a small garden (Tilborgh and Hoogenboom 1982, pp. 43–44). Students also practiced drawing parts of plants, landscape motifs and ornaments from memory (Landbouwhoogeschool Wageningen 1927, p. 39). The drawing and colouring of simple plots and leaf shapes, twigs and flowers were part of the finger skills[12] (see Figure 19.4).

Bijhouwer encouraged students during outdoor excursions to improve their drawing skills with hand and perspective sketches.[13] His sample books *Perspectief-constructie zonder vertekening voor tuinontwerpers* (1954) and *Waarnemen en Ontwerpen* (1954) were presented next to *Pencil Pictures* and *Pencil Broadsides* of Theodore Kautsky and *Tekenen van Bomen en Heesters* (1950) from J. Landwehr, *The artistic Anatomy of Trees* by Rex Vicat Cole and *Sketching* and *Rendering in Pencil and Drawing with Pen and Ink* were appointed as textbooks, but should certainly not be copied (Bijhouwer 1950).

With all the efforts to improve drawing skills, the garden and landscape teachers stressed how a professional could only develop self-esteem and be accepted in public if a proper exam course Drawing was part of the garden and landscape architecture education.

Conclusion

The development of garden architecture, as the profession was known during pre-landscape architecture times of the first half of the twentieth century, is closely tied to the establishment and development of its education. For thirty-five years, the lector Hartogh Heys van Zouteveen pleaded for an independent department of garden architecture at the university level. As one of the first of its kind in Europe, such a department was officially set up in 1946 in Wageningen. For many years, its predecessor course had existed as a minor, mainly aimed at educating horticulturists, keeping design and fields such as urban planning out of sight. Only after World War II the inspection of the horticultural education realized that regional planning issues in the Netherlands were not possible without input from skilled garden and landscape architects.

Aesthetic skills were developed by educating drawing skills and by studying urban planning, architecture and art history. These were considered necessary for a skilful garden and landscape architect, focusing on, apart from drawing plants, the spatial ordering of planting and space-mass ratios of the design of gardens and parks. However, it was the professionalization of the art and drawing education that led garden and landscape architects to a better connection with the scientific level of planners.

The importance of broadly, interdisciplinary educated garden and landscape architects has since been embraced. Current practice shows that garden and landscape architects today have a close affinity with planning at all levels and scales including urban or regional planning. Exercising this trade requires, as the first professor of garden and landscape architecture Bijhouwer states, "a pronounced ability and in addition, both technical competence as a developed social capacity". Garden and landscape architects should work as generalists to oversee the coherence of the whole and, where necessary, to restore it (Andela 2011, p. 193). However, their metier mainly depends on the convincing handling of the drawing pen.

Notes

1 Letter no. 1743, Wageningen, 16 juni 1942, to the college of Rector Magnificus and Assessors of the Agricultural College, from the Committee and Letter 19 mei 1937, and the Professors of the Agricultural College at Wageningen, from the lecturer in hand drawing H. Ramaer. (GA, toegang 0740, Landbouwhogeschool Wageningen, inv.no. 1558 and 1144 Letter *mbt Ramaer /Bijhouwer studierichting tuinkunst en natuurteekenen*).
2 Note regarding the education as a landscape architect and urban planner, from H. Cleyndert Azn, 6. (GA, toegang 0740, Landbouwhogeschool Wageningen, inv.no. 1558, Tuin – en landschapsarchitectuur instelling studierichting korrespondentie 1945–1950).
3 Hartogh Heys van Zouteveen, *Moet de opleiding van den landschapsarchitect plaats vinden in Delft of in Wageningen*, Wageningen 1928, 2.
4 J.T.P. Bijhouwer succeeds Van Zouteveen as teacher of Garden Art (1936), and became professor of Garden Art, Garden Architecture and Horticultural Drawing at the Agricultural College (1939). He was appointed professor of Garden and Landscape Architecture (1946–1966) and teacher of Garden Art at the Agricultural College in Wageningen, and became extraordinary professor of Landscape Science at the urban development department of the Delft University of Technology (1948–1968), and was a guest lecturer at various foreign universities (1951–1952/1953/1958–1963). Andela 2011, p. 245.
5 Letter no. 1830, Apeldoorn, 5 mei 1944, to the President-Curator of the Agricultural College at Wageningen, of the Department of education, science and culture protection. H. Cleyndert, 3. with postscript of the Director of the Bureau of the National Office for the Nationale Plan dr. ir. F. Bakker Schut. (GA, toegang 0740, Landbouwhogeschool Wageningen, inv.no. 1558).
6 Letter no. 1830, Apeldoorn, 5 mei 1944, to the President-Curator of the Agricultural College at Wageningen, of the Department of education, science and culture protection, from H. Cleyndert with the Director of the Bureau of the National Office for the Nationale Plan. (GA, toegang 0740, Landbouwhogeschool Wageningen, inv.no. 1558, *Tuin – en landschapsarchitectuur instelling studierichting korrespondentie 1945–1950*).
7 Fuchs participated on the city development of Brussel with an urban proposal in 1858 to extend the plans of Jourdan and De Joncker. E. Smellinckx. *Urbanisme in Brussel, 1830–1860* (S.A.B., B.C. Séance de 23 août 1856, 82–86. *Rapport van de sectie Openbare Werken*.) www.ethesis.net/urbanisme/urbanisme_hfst_4.htm <consulted on 19 September 2021>. In 1861 Fuchs became the first urban landscape architect of Brussels and designed a national exposition (1880) and a world exposition (1885).
8 Overview of submitted pieces by Van Zouteveen for an exhibition at the agricultural college in Wageningen mentioned in 'Tuinkunst', *Nieuweieuwe Rotterdamsche Courant*, Rotterdam 5 may 1929. In 1911 his salary was raised up to ƒ 1.600, a year. He used part of his salary for setting up the library. In 1913 he expanded this library to include the next collection: I. *Gardenarchitecture*: 69 general works, 19 books of history, 52 pictures and descriptions, total of 140 objects (in 1915 extended up to 160 pieces). II *Architecture*: 21 pieces general works, 13 urban planning books, total 34 books (in 1915 about 40 books), III *Botany*: 11 general works, 34 biology and ecological geographic works, 24 stuks systematic (Flora and dendrology), total 69 works, IV. *Praktic handbooks*: 70 pieces. Hartogh Heys van Zouteveen [1915], pp. 648–649.
9 In 1924, Van Zouteveen's entry at Gross-Berlin is shown at an exhibition. Van Zouteveen made an impressive number for his entry. (34 Maps in 4 wallets, FORUM Spec. Coll. inv.no. RPk.IV-G, collection Wettbewerb Gross-Berlin 236445, Overview maps sheet 2–5, schale 1:25000 and WUR Spec. Coll. inv.no. R346A11, *collection Wettbewerb Gross-Berlin* 236445).
10 WUR, Spec. Coll. Toegang Springer Collection, map A31, inv. no. 01.856. 02 and 01.856. 01.
11 Letter of Ramaer, Wageningen, 24 februari 1937, to the Senat of the Agricultural Collegd at Wageningen, 1. (GA, toegang 0740, Landbouwhogeschool Wageningen, inv.no. 1144, *brieven examenvak natuurteekenen Ramaer-Bijhouwer 1935–1938*).
12 Letter 7 januari 1923 to the college from Ramaer, (GA, toegang 0740, Landbouwhogeschool Wageningen, inv.no. 2538, *pleidooi cursus handtekenen Ramaer 1922–1927*).

13 Letter no. 326, 7 febr 1950, of the professors J.T.P. Bijhouwer and H.J. Venema, to the College of Curators at LHS at Wageningen. (GA, toegang 0740, Landbouwhogeschool Wageningen, inv.no. 1639, *aanstelling leraar handtekenen 1950–1955*).

References

Andela, G. (2011) *J.T.P. Bijhouwer, Grensverleggend landschapsarchitect*. Rotterdam: 010 Uitgeverij.
Anonymous. (1929) 'Tuinkunst', *Nieuweieuwe Rotterdamsche Courant*, 5 februari 1929, 83(35), p. 9.
Bijhouwer, J. T. P. (1945) 'Een tuinarchitect moet beslist geen eigen kweekerij hebben', *De Boomkwekerij*, 7(7), pp. 44–45.
Bijhouwer, J. T. P. (1947) 'Inaugural speech', in Bijhouwer, J. T. P. (ed.) *De wijkgedachte: Rede uitgesproken bij de aanvaarding van het ambt van hoogleeraar van de Landbouwhoogeschool te Wageningen, op 18 maart 1947*. p. 9. [Online]. Wageningen: Wageningen University & Research, [viewed 21 juli 2020]. Available at: https://library.wur.nl/WebQuery/groenekennis/html
Bijhouwer, J. T. P. (1950) 'Voorwoord', in Landwehr, J. (ed.) *Potloodstudies voor tuintekenaars, leerlingen van tuintekencursussen en bouwkundigen*. Amsterdam: N.V. Wed. J. Ahrend & Zoon.
Cleyndert Azn, H. [F. Bakker Schut] (1944) *Nota betreffende een opleiding tot landschapsarchitect en planoloog*, 1. Letter no. 1830, Apeldoorn, 5 mei 1944.
Debie, P. H. M. (2011) *'Ziehier een kleinen grondslag, waarop gij verder zult kunnen voortarbeiden', De nalatenschap van tuinkunstenaar H.F. Hartogh Heys van Zouteveen (Delft, 13 juli 1870 – Wageningen, 23 maart 1943)*. MA thesis, Universiteit Utrecht, Zeist.
de Vos, C. (1887) 'Eene Tuinbouwschool', *Het Nederlandsche Tuinbouwblad*, 3(42), p. 245.
de Vos, C. (1890) 'Eene Tuinbouwschool', *Het Nederlandsche Tuinbouwblad*, 3(42), pp. 341–343.
Hartogh Heys van Zouteveen, H. F. (ca. 1911a) *Grundlegende Ideeën auf dem Gebiete des Staedtebau's: festgelegt in Erlaeuterungsberichte des unter Motto Parkstadt Grosse Zuege in 1910 eingesandtes Entwurf von Gross – Berlin*. Wageningen.
Hartogh Heys van Zouteveen, H. F. (ca. 1911) *Tuinaanleg, Het een en ander over den aanleg van kleine tuintjes*. Wageningen.
Hartogh Heys van Zouteveen, H. F. (1913a) *Het een en ander over de verhouding tusschen architect en tuinarchitect en over het verband tusschen tuinarchitect[uur], bouwkunst en de aanverwante kunsten, overdruk uit het Bouwkundig Weekblad*. Wageningen.
Hartogh Heys van Zouteveen, H. F. (1913b) 'Het een en ander over de verhouding tusschen architect en tuinarchitect en over het verband tusschen tuinarchitect[uur], bouwkunst en de aanverwante kunsten', *Bouwkundig Weekblad*, (33), pp. 187–189.
Hartogh Heys van Zouteveen, H. F. (ca. 1915) *Het onderwijs in tuinarchitectuur aan de Rijks Hoogere Land-, Tuin- en Boschbouwschool te Wageningen*. z.pl. Wageningen.
Hartogh Heys van Zouteveen, H. F. (1916) *'De geschiedenis der Tuinkunst als vak van Studie', overdruk uit De Tuinbouw*. z.pl. Wageningen.
Hartogh Heys van Zouteveen, H. F. (1918) *Het onderwijs in tuinarchitectuur en tuinkunst aan de landbouwhoogeschool te Wageningen*. 'Mededeelingen van de landbouwhoogeschool, deel XIV', Wageningen.
Hartogh Heys van Zouteveen, H. F. (1920) *De Siertuin: Geïllustreerd Handboek ten dienste van vakman en liefhebber en van inrichtingen voor tuinbouwonderwijs*. Zutphen: P. van Belkum Az.
Hartogh Heys van Zouteveen, H. F. (1927) *Het verband tusschen landschappelijke tuinkunst en stedebouw*. s'Gravenhage.
Hartogh Heys van Zouteveen, H. F. (1928) *Moet de opleiding van den landschapsarchitect plaats vinden in Delft of in Wageningen*. Wageningen.
Landbouwhoogeschool Wageningen. (1927) *Programma voor het studiejaar 1927–1928*. Wageningen.
Moes, C. D. H. (2002) *L.A. Springer 1855–1940, Tuinarchitect, Dendroloog*. Rotterdam: Hef Publishers.
Springer, L. (1895) 'Tuinarchitectuur en tuinbouwonderwijs', *Tijdschrift voor Tuinbouw*, 1, pp. 49–54.
van der Haar, J. (1993) *Van School naar hogeschool, 1873–1945. De geschiedenis van de landbouwuniversiteit Wageningen*. Wageningen.
van Nes, K. C., Poortman, H. A. C. and Cleyndert Azn., H. (ca. 1926) *Rapport commissie inzake vorming landschap-architecten*, z.pl. Wageningen.
van Tilborgh, L. and Hoogenboom, A. (1982) *Tekenen destijds. Utrechts tekenonderwijs in de 18^e en 19^e eeuw*. Utrecht: Het Spectrum.

20
EARLY HISTORY OF LANDSCAPE ARCHITECTURE EDUCATION INITIATIVES IN ROMANIA

The pioneering work of Friedrich Rebhuhn

Alexandru Mexi

Introduction

Landscape architecture is a young profession in Romania. The first university programme was established in 1998 and the profession was formally acknowledged in 2014, when it was included in the Romanian Code of Occupations. However, landscape architects from Central and Western Europe have been working in Romania for a long time. Initiatives to create a school to educate landscape gardeners dates back to the 19th century, and early official records of gardening, garden design and landscape architecture courses date back to 1913. In particular, during the early and mid-20th century, Friedrich Rebhuhn, a German professional who had spent most of his time working in Romania, became engaged in personal endeavours to establish a school for landscape gardeners and landscape architects. Even though his efforts came to no avail at first, those of his successors did. This chapter presents Rebhuhn as a pioneer of landscape architecture education. At the same time, it is the first report on the early history of landscape architecture education in Romania. It also discusses the significance and effects that some intentions and initiatives had to start and design a landscape architecture school.

Research and publications on landscape architecture in Romania in general, and on the history of landscape architecture education in particular are scarce. Few publications that mention fragments of the history of landscape architecture education in Romania exists. This chapter reports on research that uses archival documents located both in public and in private collections. Included are correspondence, official requests, legislation, and historic school and university curricula. In addition to analysing primary sources, a number of publications and other secondary sources such as newspaper articles, legislation and autobiographies have been hand-searched and analysed. All findings from relevant documentation were studied and synthesised to find answers to the question about "how landscape architecture education developed in Romania?" Comparisons between different approaches to landscape architecture education, as well as between different historic curricula were made in order to observe the evolution of how landscape architecture education was seen in different periods of time.

A need for gardeners, apprentices and skilled workers

Until the mid-19th century, most private gardens from the Romanian historical provinces of Wallachia and Moldavia, south and east of the Carpathians, were not designed by trained gardeners or architects but, instead, simply planted with a "natural and cultivated mix of flowers, weeds, roses and fruit trees" (Vintilă-Ghitulescu 2015, this and all following quotes are translated by author). Things changed after 1829, when the two principalities became Tsarist protectorates. Russia issued "Organic Regulations", directives similar to a constitution, one for Wallachia (1831) and one for Moldavia (1832). Russia considered the "Organic Regulations" as "necessary for the well-being of some regions neighbouring our kingdom and as a measure that will strengthen the foundation of our political influence on the East" (Regulamentele Organice ale Valahiei și Moldovei 1944); the regulations were also an instrument of reorganisation of the two Romanian provinces according to modern principles (Moldovan 2013). To help implement the "Organic Regulations", sub-regulations on economy, public administration, sanitation, education and so on were issued. Some of them related to the beautification of cities, especially of the two capitals, Bucharest in Wallachia and Iași, in Moldavia. Regulations included urban development guidelines and referred to the obligation of the creation of walking grounds and of municipal public parks (Regulamentele Organice ale Valahiei și Moldovei 1944).

As there were no local specialists in garden design; the newly created public administrations called for gardeners and landscape gardeners from Central and Western Europe. With their support, they created more than 50 public parks during the years between about 1830 and 1916 – the year when Romania became involved in World War I (Mexi 2020). The foreign gardeners who worked on public projects were also called to design parks and gardens on private estates. Since they were working at the same time on several projects, both for public administrations and for private owners, they were in constant need of apprentices and skilled workers able to supervise and implement the projects they were designing and to properly maintain the grounds (see ANR, selection and Mexi 2020) (Figure 20.1). Solicitations asking local authorities to finance and develop training courses for workers were to no avail and, since there were no trained gardeners in Romania, foreign specialists employed apprentices from Western Europe and untrained local workers (ANR, selection).

During the 19th century, schooling was a general problem in the two Romanian provinces. At the beginning of that century "the population was little educated and not at all involved in the practice of a craft" (Vintilă-Ghitulescu 2015). Craftsmen, such as carpenters, stonemasons, blacksmiths, masons and so on existed in small numbers and they were poorly prepared. However, towards the end of the century, due to a large number of youth who decided to study in different European countries, things changed. Schools opened across the country and the numbers of publications grew in fields such as arts and architecture, literature, music, mathematics, sciences, medicine and others (Popa 2018; Vintilă-Ghitulescu 2015; Djuvara 2015). However, concerning gardening and garden design, no dedicated schools opened and neither did any publications on these topics become available (Mexi 2020). Places where people could find garden related education were garden projects and nurseries such as the ones owned of the Faraudo and Leyvraz families. These places offered learning-by-doing about garden design, planting principles and techniques, garden management, tree pruning and so on (Mexi 2020).

The oldest curriculum mentioning garden and landscape design courses dates back to 1913 and belongs to the Grozăvești School of Horticulture (*** 1913). Apart from courses such as tree care, potted plants, greenhouse plants, vegetable and fruit tree growing, accounting, chemistry, botany, geology or zoology, the curriculum also mentioned garden architecture, hand and

Figure 20.1 Romanian workers on the worksite of Bibescu Park in Craiova, ca. 1900.

Source: Redont, É. (1904). *Ville de Craiova – Promenades, parcs, squares, jardins publics et avenue*. Craiova, p. 3.

linear drawing, geometry and scale reduction. The course concerning garden architecture is described as

> The garden besides us and before our house, the garden of a villa, the park, public plantations, alignments and streets. The ground and its transformations according to garden art. Earthworks, terraces, the lake, basins, etc. Water installations, alley constructions, lawns, mosaics, garden furniture. Reports and estimates. General information about garden architecture.
>
> (*** 1913)

However, no reference exists about the length of study periods or about the instructors who would have taught these courses. Since no information about the history of this curriculum exists, it is impossible to know if there was a similar older version of it, if the curriculum changed or even disappeared after 1913.

The Secondary School of Horticulture in Bucharest (not a professional school similar to a college or university) also offered garden design courses. A curriculum from the school year of 1929–1930 includes subjects such as Dendrology and Architecture in both the first and second year of study, and Drawing in the third and fourth years. The curriculum of 1929–1930 is more complex than the one of 1913 and it offers a better understanding of the courses. Apart from garden design techniques, planting patterns, vegetation management and other similar topics, in the fourth year of study, students studied the history of gardens; namely "the history of the regular and geometric gardens of the Orient, Greece, France and of the Netherlands and the irregular and sculptural gardens of China, Japan, England and Germany" (ANIC, fond Fritz Rebhuhn). However, here too, there is no information on the teaching staff or on the history of the curriculum. Based on archival material, no other curricula seem to have existed before 1998. Despite the evidence of curricula on garden design, official interest in establishing a professional school to educate and train gardeners remained low. Schools had either poorly trained instructors or no qualified teaching staff at all. One of the foreign garden designers who came to Romania at that time was Friedrich Rebhuhn. He decided to try to establish a formal school of landscape gardening and landscape architecture.

Friedrich Rebhuhn. A personal endeavour

Friedrich Rebhuhn (1883–1958)[1] began to learn the practice of gardening in 1897 with his uncle, E. Schreiter, in Schlanz, Breslau district. From 1900 to 1903, he worked in a commercial garden in the town of Mitweida; in a flower and rose greenhouse in Koswig, near Dresden, Saxony; and in a cemetery garden and flower binding company in Bochum, Ruhr district (ANIC, fond Fritz Rebhuhn). After six years in the "specialties of the gardening profession" (ANIC, fond Fritz Rebhuhn), he decided to become "a good landscape architect – a specialty I love the most" (ANIC, fond Fritz Rebhuhn) and, consequently, decided to acquire theoretical knowledge in this field.[2] From 1903 to 1904, he attended the Oranienburg Horticultural Institute near Berlin. Then, after having served in the army for two years, in 1906 he took on the position of deputy director at the Horticultural Institute in Weinheim, near Heidelberg, where he also was professor of practical horticulture, dendrology and topography. It is most probable that this experience as educator in the early part of his career drove him to pursue a personal endeavour of establishing a school in Romania – the country where he would live for most of his life.

In order to enrich his specialised knowledge, he worked as a gardener in Zürich (1908–1909) and in the largest nurseries in France, Crue et fils, Chatné, and in the botanical garden of Paris (ANIC, fond Fritz Rebhuhn). He became member of the Association of German Dendrologists and, in 1910 took part in the annual congress of dendrologists in Brussels. During this event, the head of the horticultural section of Bucharest invited Rebhuhn to take over the position of gardener-architect for the city. He signed a one-year contract, hoping that after the end of the contract period he would travel to Japan "to learn about its ancient horticultural art" (ANIC, fond Fritz Rebhuhn). However, less than three months after being employed in the public administration of Bucharest, the director died, and the vacant post was offered to Rebhuhn. He accepted and remained in Romania until 1957 when he decided to move to Brasil.

Rebhuhn designed and remodelled numerous public parks and private gardens, studied the Romanian landscapes and flora, and he published his research findings in newspapers, horticultural magazines and in a book (ANIC, fond Fritz Rebhuhn). He became director of the House of Gardens in Bucharest and of other public departments responsible with the green spaces of the capital city, as well as director of the Royal Horticultural Department (Mexi and Zaharia 2020).

First attempts to establish a school for landscape gardeners

In May 1923, Friedrich Rebhuhn publishes an article entitled "Peisagistul" ("The Landscape Architect") in the Romanian Horticulture Magazine (Figure 20.2). He develops ideas about educating landscape architects and also articulates his view of the profession:

> By landscape architect we mean the creator of gardens. Creating gardens is the highest stage of horticulture. It includes almost all branches of horticulture and that is why

ANUL I   No. 3   1 MAIU 1923

REVISTA HORTICOLĂ

PUBLICAȚIE LUNARĂ

ORGANUL SOCIETĂȚII DE HORTICULTURĂ DIN ROMÂNIA

ABONAMENTUL 100 LEI ANUAL
PENTRU MEMBRII SOCIETĂȚII ȘI AUTORITĂȚI 60 LEI ANUAL
PENTRU ANUNȚURI ȘI RECLAME:
1|8 DE PAGINĂ PE COPERTĂ . . . . ANUAL 1000 LEI
1|8 " IN TEXT . . . . 600 LEI

REDACȚIA ȘI ADMINISTRAȚIA:
LA SEDIUL SOCIETĂȚII DE HORTICULTURĂ DIN ROMÂNIA
CALEA ȘERBAN-VODĂ No. 221. — BUCUREȘTI

PEISAGISTUL

Una din întrebările cele mai de seamă pentru horticultura românească este de dezvoltarea și activitatea horticultorului peisagist. De aceia voesc să atrag atenția, în linii generale, asupra calităților ce trebue să le posede un grădinar care îmbrățișează această ramură, și ce anume trebue să învețe, pentru ca să devie un adevărat peisagist.

Prin peisagist, înțelegem creatorul de grădini. Crearea grădinilor este treapta cea mai înaltă a horticulturii. Ea cuprinde aproape toate ramurile horticulturii și de aceea, cu drept cuvânt se numește „Artă horticolă". Un perfect horticultor-artist va putea deveni numai acela care'i va cunoaște toate ramurile și va stăpâni cunoștințe întinse în această ramură.

Peisagistul trebue să se intereseze tot atât de mult de legumicultură și pomologie, ca și de floricultură; pe urmă trebue să fie un bun pepinierist și botanist; să aibă noțiuni de științe naturale, vedere clară despre formațiunea pământului, și să fie înzestrat de la natură cu prinderea adevăratelor ei frumuseți. In general, peisagistul trebue să fie capabil, să dea adevărata însemnătate și înțelegere tuturor legilor naturii. Aceasta trebue să se întinză asupra simțimântului, obiceiurile și religii poporului.

Tânărul grădinar care intenționează să îmbrățișeze această artă, trebue să știe să scoată în evidență legile formelor, liniilor și culorilor. Prin conducerea drumurilor prin alcătuirea figurilor și prin împărțirea terenului, el face ca într'o grădină sau parc să se nască impresii la fie care pas.

Figure 20.2 Friedrich Rebhuhn's article in the Romanian Horticulture Magazine, 1913.

Source: ANR – The Romanian National Archives; ANIC – The Romanian National Central Archives, Fond Friedrich Rebhuhn, folder 18.

it is rightly called "Horticultural Art". One can become a perfect horticultural artist only if he will know all branches of horticulture and will master extensive knowledge in garden art. The landscape architect must be as interested in vegetable growing and pomology as in floriculture; then he must be a good nurseryman and botanist; he has to have notions of natural sciences, a clear view of the formation of the earth, and be endowed by nature with the apprehension of its true beauty. In general, the landscape architect must be capable to give true meaning and understanding to all laws of nature.
(Rebhuhn 1923)

However, in this article Rebhuhn does not mention anything about the establishment of a specialised school.

Seven years later Rebhuhn writes another article, this time in the *Universul* newspaper. Under the title "The amateur of gardens and the gardener. The role of the landscape architect", he presents the role of the garden designer and the skills that such a professional should possess, and he subtly points out the need for a school specialised in garden design (Rebhuhn 1930). In 1938, Rebhuhn publishes a paper about "The education of a garden artist", saying, "the cultural and artistical importance of horticulture is still too little known and there is little interest in the importance of the gardener's role in the cultural education of the people" (Rebhuhn 1938). He states that the profession of horticulture and garden design requires theoretical and practical knowledge. Rebhuhn briefly describes how a school of horticulture and landscape gardening should be organised, and continues by discussing how "establishing a horticultural university in Romania will be very difficult these days especially because we still don't have enough trained professors able to teach horticulture and gardening" (Rebhuhn 1938). It was the first time in the history of landscape architecture in Romania when a professional was publicly asking for the establishment of a school for landscape gardeners and landscape architecture respectively.

Less than a month later, another article appears again in the *Universul*, this time the title reads "The landscape management" (Rebhuhn 1938). In it, Rebhuhn thoroughly describes the term "landscape" and all the issues that Romanian landscapes were facing at that time, particularly underlining the improper visual effects roads and railroads have on natural and cultivated landscapes. He then asks for a law for the protection of landscapes, a legal document similar to the ones made for the "National Park" in Bucharest, and that the state authorities should work on educating people about landscapes. Rebhuhn continues by elaborating the idea of establishing a school for landscape architecture, this time focusing on landscape in general and not particularly on gardens alone. His requests will remain unanswered and no school of landscape architecture established for years to come. However, Rebhuhn continued pursuing his mission. "Several times officially being asked to recommend a skilled gardener, I was confronted with unpleasant and ridiculous situations" (Rebhuhn 1940). This is how Friedrich Rebhuhn begins another article in the *Universul* newspaper, entitled "The mistakes of our horticultural education". It appeared in 1940 and is the first written newspaper article in which we find references of an existing, yet weak public school of gardeners (not of landscape gardeners or architects). However, no references about the whereabouts of this school or about its curricula are made. In this article, apart from numerous and profound critics, the German landscape gardener recommends the division of horticulture schools into three sections, one "for vegetable farming", one "for orchards" and one "for aesthetic horticulture and garden management" (Rebhuhn 1940). He adds that the first two schools should be created in the "regions of the country that best fit their purpose" and the school for aesthetic horticulture and garden management should be established near the capital city of Bucharest.

Official requests for a study programme

On the 1st of October 1940, Rebhuhn writes directly to the Minister of Agriculture and Estates. In his letter, he starts by informing the Minister that "for over 30 years, the Bucharest House of Gardens had the opportunity to observe the flows in the education that students received at the School of Horticulture, an institution of the Ministry of Agriculture and Estates" (ANIC, fond Fritz Rebhuhn, translated by author). He continues by saying that the current study programme is similar to programmes used in countries such as England, France and Germany 70 to 80 years ago (ANIC, fond Fritz Rebhuhn). Apart from bemoaning the outdated curricula, Friedrich Rebhuhn also mentions few qualified professors exist and continues by offering the Minister a solution similar to the one "applied 20 years ago in Germany" (ANIC, fond Fritz Rebhuhn). He says

> The management of horticultural institutions must be represented only by gardeners with practical and theoretical education, and the rest of the school leaders, at least 90%, must have the same horticultural studies. Only students, who, depending on the education institutions, will be graduates of the lower or upper high school cycle, with a mandatory 3-year horticultural practice, presenting good certificates, must be enrolled in horticulture schools. Horticulture schools must be divided into three main groups: schools for vegetable farming, schools for orchards and schools for aesthetic horticulture and garden management.
>
> (ANIC, fond Fritz Rebhuhn)

Until the end of the Second World War and the abdication of the king in December 1947, no progress was made to install a programme for landscape architecture education. Rebhuhn waited until the mid-1950s when he, yet again, despite close professional relationships with the former Royal Family, kept his position as head of public institutions responsible for the design and management of urban green spaces even during communist times and continued to make requests for the establishment of a landscape architecture school. Towards the end of his career, in 1955–1956, when he was working at the Department of Green Spaces within the *Institutul Proiect București* – IPB (The Bucharest Institute for Planning), he sent several letters to institutions such as the Council of Ministers, the Central Committee of the Romanian Worker's Party and the Bucharest People's Council. He "pleaded [. . .] for the importance of garden construction in Romania and its economic, sanitary, social and aesthetic significance" (ANIC, Fritz Rebhuhn), as well as for the establishment of specialised educational institutions in order to prepare future specialists to deal with garden design and maintenance. Rebhuhn even repeatedly showed his intention and willingness to guide young architects "who have sensitivity and interest in garden architecture and green space design" (ANIC, fond Fritz Rebhuhn) in the department he was coordinating. If his requests would have been taken into consideration, as Rebhuhn personally stated in his memories, "these young people would have learned, among other, about plants and the specific requirements of different species, about the aesthetics of green space design and about landscaping techniques" (ANIC, fond Fritz Rebhuhn).

These were not the only proposals Rebhuhn officially made. He also proposed that each People's Council should have a horticultural service and be legally obliged to train apprentices "in proportion to the importance of the service" (ANIC, fond Fritz Rebhuhn). The chief gardener or "most educated" person in this department should teach five hours of theoretical lessons on the horticultural practice they usually carry out in the field (ANIC, fond Fritz

Rebhuhn). Apprentices or students who were to attend these courses would have obtained a "gardener's apprentice" diploma after completing three years of study and after passing an exam (ANIC, fond Fritz Rebhuhn). Rebhuhn's proposal continued with the following recommendations:

> In schools of horticulture and of landscape architecture (!), only those possessing a "gardener's apprentice" certificate should be accepted and only trained specialists, especially horticulturists, should teach. A "Horticultural Institute" should be established and its purpose and obligations should be analogous to the Agronomic Institute and the Forestry Institute.
>
> (ANIC, fond Fritz Rebhuhn)

In a document from 1957, referring to "the forest engineer's manual of 1955" (ANIC, fond Fritz Rebhuhn; Bălănică and Stinghe 1955), Rebhuhn is pointing out how "standards, norms and regulations regarding different specializations, including landscape architecture, are written by forest engineers and these are nothing but copies of old foreign texts and they are not relevant to the particular climate, soil and weather conditions of Romania" (ANIC, fond Fritz Rebhuhn). He adds that because of this there are no well-trained specialists that can design and maintain parks and gardens and thus the state and Bucharest in particular are "loosing hundreds of millions and will continue to loose until things will change" (ANIC, fond Fritz Rebhuhn).

Neither during monarchy, nor during communism did Rebhuhn succeed in convincing the public authorities to establish a school dedicated to landscape architecture or to garden design. However, his desire to create such an institution we can assume will partially have come true in the Botanical Gardens in Bucharest.

Theory and practice – the school within the Botanical Gardens

Friedrich Rebhuhn's persistent requests were never successful and neither a school of landscape architecture nor of garden design was opened, nor did the schools of horticulture transform as he suggested. However, Rebhuhn did not rely solely on public administrations to create a new or reorganise an existing school. To achieve educational goals, he took the matter in his own hands and designed, during the 1930s and 1940s, a garden design school within the Botanical Gardens in Bucharest (ANIC, fond Fritz Rebhuhn). Evidence of this project can be found in a book manuscript entitled "Theory, practice and aesthetics of green spaces in urban planning" (Rebhuhn 1942). The book contains 31 chapters on planting designs, vegetation and categories of urban open spaces, green space management and public institutions. In Chapter 20, Rebhuhn goes into detail about school gardens and gardens for outdoor studying. In that chapter he refers to a school designed within the Botanical Gardens. He includes a picture portraying students learning outdoors (probably not a picture taken in the Botanical Gardens) (Figure 20.3) and a plan of the Botanical Gardens, signed and dated 27.VII.1943, indicating the site designed as outdoor school. Lacking further details about this school such as a curriculum, it is uncertain whether Rebhuhn ever fully realised the project. Judging on the basis of a number of sketches and notes made by Rebhuhn on different documents found in archives (personal, city hall of Bucharest, Royal archives, etc.), we may assume that he did teach courses for gardeners and apprentices employed within the institutions responsible for the management of both public and private parks and of the Botanical Gardens.

Figure 20.3 The picture Friedrich Rebhuhn shows in his manuscript "Theory, practice and aesthetics of green spaces in urban planning" (1942), portraying students learning outdoors.

Source: ANR – The Romanian National Archives; ANIC – The Romanian National Central Archives, Fond Friedrich Rebhuhn, folder 4a.

The opening of the first school of landscape architecture in Romania

For the years after Rebhuhn left Romania, in 1957, no other relevant documents regarding the establishment of a landscape architecture school in Romania were found. Some landscape architecture and garden design courses were taught in faculties of horticulture, forestry and architecture, but there was not a school or study programme for landscape architecture. After the fall of communism in 1989, three people, Ana Felicia Iliescu, a horticulturalist, Florin Teodosiu, a forest engineer, and Valentin Donose, an architect, teamed up and started working on plans for establishing a school dedicated to landscape architecture education. These three professionals had been active in garden and park design, urban planning, plant care and so on, and even worked together in practice on a number of different projects.

In 1998, after continuous efforts, they succeeded in establishing a school of landscape architecture. At first, it was a landscape architecture specialisation within the Faculty of Horticulture, University of Agronomic Sciences and Veterinary Medicine in Bucharest. They had effectively argued for their school project by saying:

> The absence of landscape architects is incompletely supplemented by architects, horticultural, forest and even construction engineers whose activity is marked by insufficient professional training, reflecting negatively in the current situation of the green spaces in Romania. Romania's urban and industrial development in this century has demonstrated the stringency of environmental protection issues, the need to enhance

the natural landscape [. . .] to which are added the neglect and even sacrifice of city parks and gardens, with serious consequences of ecological and aesthetic alteration of the environment [. . .] impose in the present times and in the future the competent approach of the problems of environmental protection in our country in which an important role belongs to the landscape specialists. The profession of landscape architect, having a vast field of action, becomes even more necessary in the context of political, social and economic changes in Romania. In the conditions of a judicious economic policy, implications in the territorial landscape planning, in the creation, exploitation and maintenance of the landscaped areas cannot be neglected. The planning and efficiency of investments in both the public and private sectors, as well as the good management of the budgetary effort in the field of green spaces, can be ensured only by practicing this profession with professionalism.

(Iliescu, private archive)

The first school of landscape architecture opened 140 years after the first foreign garden designers were asking public administration to organise garden design courses for Romanians working on the sites of public parks and private gardens. Apart from the three specialists that finally achieved establishing the school, other professionals were engaged in the process as well. They came from both the University of Agronomic Sciences and Veterinary Medicine in Bucharest, as well as from other universities, research and planning centres such as the Technical University of Civil Engineering in Bucharest, "Ion Mincu" Institute of Architecture or Proiect București S.A. Many of these highly engaged people taught different courses at this landscape architecture school (Figure 20.4).

Conclusions/challenges and future development

The first intentions for establishing a school for garden design and management in Romania date back to the 19th century and were supported by many foreign specialists employed by private owners and public administrations to create private gardens and municipal parks in Romania. They all saw the need for properly trained gardeners who would be able to supervise the construction works on the parks they were designing and to properly maintain the grounds after the inauguration. The most vocal of all foreign specialists was the German landscape gardener Friedrich Rebhuhn who settled in Romania in 1910, remained, and worked here for most of his life. He pursued a personal endeavour in establishing a school for landscape architecture/garden design. As director of public institutions and even head of the Royal Horticultural Department, he used his influence to pursue the government, and especially the Ministry of Agriculture and Estates either to establish such a school or to reorganise the existing schools of horticulture. Despite persistent efforts, publications, official letters and complaints, Rebhuhn was not able to achieve his goal during his lifetime, neither during monarchy, nor communism. However, in 1998, 41 years after his departure from Romania, the first school of landscape architecture opened in Bucharest. It did so after much hard work and continuous efforts by a team of three Romanian specialists, a horticulturalist, a forest engineer and an architect. Currently, universities of Timișoara, Iași, Cluj, Craiova, Oradea and Bucharest are offering landscape architecture education programmes. In 2003, the first students graduated with a professional degree. The profession was officially recognised in 2014 when it was introduced into the Romanian Code of Occupations.

Landscape architecture continues to evolve as a profession. It is receiving more and more interest, particularly from private landowners, but also from public, local and governmental

UNIVERSITÉ DES SCIENCES AGRONOMIQUES
ET MEDICINE VETERINAIRE – BUCAREST
FACULTÉ D'HORTICULTURE
Spécialisation: ARCHITECTURE DU PAYSAGE
                    Ingénieur paysagiste
Durée: 5 années (10 semestres à 14 semaines)

PLAN D'ENSEIGNEMENT

1-ère
Disciplines obligatoires

Géometrie descriptive	56
Perspective	56
Dessin et techniques graphiques	112
Architecture du paysage	112
Etude de projets	84
Histoire des arts visuels	28
Topographie	84
Géographie des paysages	42
Botanique	112
Economie politique/Politologie	42
Langues modernes	56
Education physique	56
Stage	120

Disciplines facultatives
Relation publiques
Langues modernes (2-ème options)

2-ème
Disciplines obligatoires

Pédologie	56
Météorologie	56
Dendrologie	112
Physiologie végétale	56
Informatique	56
Mathématique-Statistique	56
Architecture du paysage	42
Etude de projets	98
Dessin et techniques graphiques	88
Utilisation des logiciels d'architecture paysagère	42
Langues modernes	56
Education physique	56
Stage	120

Disciplines facultatives
Psychologie
Langues modernes (2-ème options)

3-ème
Disciplines obligatoires

Ecologie et protéction de l'environement	42
Etude de projets	196
Design	84
Planification du paysage	56
Matériaux et construction	56
Aménagement du territoire	28
Urbanisme	28
Floriculture et gazon	42
Silviculture	28
Production du mteriel à planter	56
Utilisation des logiciels d'architecture paysagère	56
Stage	120

Disciplines facultatives
Pédagogie
Langues modernes

4-ème
Disciplines obligatoires

Améliorations foncières	56
Plante d'interieur et art floral	56
Réseaux	56
Photogrammétrie	70
Planification du paysage	182
Etude de projets	140
Chemins et terrassements	56
Techniques d'aménagement	56
Machines	56
Protection des végétaux	56
Stage	120

Disciplines facultatives
Pédagogie
Langues modernes

5-ème
Disciplines obligatoires

Etude de projets	84
Planification du paysage	98
Géstion des espaces verts	42
Législation et économie des travaux paysagers	56
Sociologie	28
Management	42
Marketing	42

Disciplines facultatives
Birotiques
Législation et droit international
Langues modernes
   Tout le 2-ème semestre: travaux de fin d'études

Figure 20.4 One of the final drafts of the curriculum of 1998, translated in French.
Source: Ana Felicia Iliescu, private archive.

administrations. Professionals are now engaged in projects for the design and even for the management of some parks and gardens and are engaged in projects related to urban planning, environmental protection, cultural heritage, legislation amendments and so on. Some professionals are teaching university courses in Romania and elsewhere, while others are working in public administrations. In addition, professional associations such as AsoP (the Romanian Association of Landscape Architects) are promoting the profession on a national scale and are arguing for its recognition and for the involvement of landscape architects in projects related to urban planning, heritage protection, environment protection, safeguarding of natural landscapes and so on.

Notes

1 1957 or 1965 according to different sources.
2 Extract from the original text in Romanian reads as following:

> După o activitate practică de 6 ani în specialitatățile meseriei de grădinar mi-a venit clar că pentru a deveni un bun peisagist – specialitate pe care o iubesc cel mai mult – trebuie să-mi însușesc și cunoștințele teoretice care țin de aceasta.

Bibliography

Bălănică, T. and Stinghe, V. (1955) *Manualul inginerului forestier*. Bucharest: EdituraTehnică.
Cărămăzin, V. (1957) *Arhitectura peisajelor. Principiile compoziției estetico-sanitare ale spațiilor verzi*. Stalin (Brașov): Litografia Învățământului.
Ciubotaru, M. (2014) 'Grădinile publice din Iași în secolul al XIX-lea', Grădina Publică din Copou. *Monumentul*, XVI, pp. 368–409.
Damé, F. (1907) *Bucarest en 1906*. Bucharest: Socec and Cie Éditeurs.
Djuvara, N. (2015) *O scurtă istorie ilustrată a românilor*. Bucharest: Humanitas.
Faraudo, M., Faraudo, C. and Monroy, D. (n.y.) *Marcel Faraudo, navigator pe mapamond (consemnări și autobiografie despre proiectele și realizările sale)*. n.p.: n.e.
Iliescu, A. F. (2008) *Arhitectură peisageră*. Bucharest: Ceres.
Marcus, R. (1958) *Parcuri și grădini din România*. Bucharest: Editura Tehnică.
Marinache, O. (2016) *Grădini bucureștene și peisagiști din spațiul cultural german*. Bucharest: Editura Istoria Artei.
Mexi, A. (2018) 'Friedrich Rebhuhn și Grădina Cișmigiu în prima jumătate a secolului XX', *Revista Bibliotecii Academiei Române*, No. 2(year III), pp. 41–52.
Mexi, A., Bogdan, C., Burcuș, A., Chiriac, A., Petrică, M, Toma, A. and Vaideș, A. (2018) *Prin parcuri publice din sudul României*. Bucharest: Simetria.
Mexi, A. and Zaharia, R. (2020) *Friedrich Rebhuhn și grădinile României*. Bucharest: Arché.
Moldovan, H. (2013) *Johann Schlatter. Cultura occidentală și arhitectura românească (1831–1866)*. Bucharest: Simetria.
Popa, O. P, Bulborea, C. G., Dragoș, M. and Popa, L. O. (2018) *Kiseleff nr. 1. De 110 ani orașul crește în jurul lui!*. Bucharest: Editura Muzeului Antipa.
Preda, M. and Palade, L. (1973) *Arhitectura Peisageră*. Bucharest: Ceres.
Rebhuhn, F. (1923) 'Peisagistul', *Revista Horticolă*, 1 April, pp. 41–43.
Rebhuhn, F. (1930) 'Amatorul de grădini și grădinarul', Rolul arhitectului peisagist. *Universul*, 15 March, p. 3.
Rebhuhn, F. (1938a) 'Educația unui grădinar artist', *Universul*, 17 February, p. 13.
Rebhuhn, F. (1938b) 'Ingrijirea peisajului', *Universul*, 9 March, p. 7.
Rebhuhn, F. (1940) 'Greșelile învățământului nostru horticol', *Universul*, 2 September, p. 2.
Rebhuhn, F. (1942) *Îngrijirea și bogăția peisagiului românesc*. Bucharest: Universul.
Redont, É. (1904) *Ville de Craiova – Promenades, parcs, squares, jardins publics et avenue*. Craiova.
Tudora, I. (2011) 'Teaching Landscape Architecture: Tuning Programs in Europe for a Common Policy. The Romanian Case', *Agriculture and Environment Supplement*, pp. 283–297.

Vintilă-Ghitulescu, C. (2015) *Patimă și desfătare, despre lucruriile mărunte ale vieții cotidiene în societatea românească*. Bucharest: Humanitas.

Vîrtosu, E., Vîrtosu, I. and Oprescu, H. (1936) *Începuturi edilitare 1830–1832. Documente Istoria Bucureștilor*. Bucharest: Tipografia de Artă și Editură Leopold Geller.

***. (1913) *Programul învățământului practic și teoretic în cei patru ani în Școala de Horticultură dela Grozăvești*. Bucharest: Atelierele Socec & Co.

***. (1944) *Regulamentele Organice ale Valahiei și Moldovei*, vol. I. Bucharest: Întreprinderile "Eminescu" S.A.

Public Archives

ANIC – The Romanian National Central Archives.
ANIC, fond Casa Regală – Castele și Palate, selection.
ANIC, fond Friedrich Rebhuhn, folders 1 to 27.
ANIC, fond REAZ, selection.
ANR, Dolj county, fond Primăria orașului Craiova, selection.
Public archives, including groups of records ('fond'), ANR – The Romanian National Archives.
Tudora, I. (2015) 'Arhitectura peisagistică în București', *Igloo* [Online]. Available at: www.igloo.ro/arhitectura-peisagistica-in-bucuresti/ (Accessed June 2021).

Private archives and unpublished papers

Ana Felicia Iliescu, private archive.
Curriculum Vitae, Valentin Donose, unpublished.

21
LANDSCAPE ARCHITECTURAL EDUCATION IN CROATIA

Petra Pereković and Monika Kamenečki

Introduction

This chapter provides an overview of landscape architectural education in Croatia. Setting the stage, it first describes how the Horticultural Society of Zagreb contributed to generating resources and ideas for professional education. At the beginning of the 20th century, gardening courses were forerunners of first, secondary and then higher landscape architecture education. In 1968, the University of Zagreb established a formal degree programme called Landscape Design, followed, in 1996 and 2005, by undergraduate and graduate programmes in landscape architecture. This chapter takes a historian's look at the development of landscape architecture education, including the interdisciplinary approach specific to education in Croatia, and the pioneering role that a handful of personalities have played. The leading figures in educational development were, in the first half of the 20th century, Vale Vouk, Zdravko Arnold and Ciril Jeglič, and in the second half of that century, Elza Polak, Dušan Ogrin and Branka Aničić.

In addition to analysing primary sources, such as printed materials of different periods and archival documents such as official documents, this chapter reviews a number of publications and other secondary sources such as newsletters and professional magazines. It synthesises findings from relevant documentation to describe how landscape architecture education developed in Croatia.

Development of the garden architecture in Croatia

In pre-landscape architecture times in Croatia, gardening and ornamental gardening were part of the agricultural professions in the wider biotechnical field. In step with general developments of specialisation during the late 19th century, professional designers called themselves ornamental gardeners, a term that includes open space design and growing plants for that purpose.[1] In contrast, the term 'gardening' generally included the cultivation of plants for food such as vegetables and fruit production. Developments of Ornamental Gardening continued in the socio-political context of the Austro-Hungarian monarchy until it dissolved in 1918 (see also Krippner/Lička in this volume). At that time, no gardening schools existed in the former Kingdom of Serbs, Croats and Slovenes, that became the Kingdom of Yugoslavia in 1929.

DOI: 10.4324/9781003212645-23

During that period, societal interest in public parks, squares, streets and private ornamental gardens grew as new cultural trends poured into Croatia from different parts of the world. The first modern planning documents of the new country, dating from the second half of the 19th century, are indicators of a growing demand for cities to establish public parks and squares dedicated for sociality, leisure and entertainment (Knežević 1996). The need arises for professionals capable of designing, constructing and maintaining green urban areas. During these pre-landscape architecture times, the profession begins to develop education and recognition. First, it is the organisation of professional education including secondary and higher education. Second, it is the general promotion of the field including an embedding of the profession into the legislative system and a growing public awareness for open space design.

At the beginning of the 20th century, ornamental gardening gradually became garden architecture. In many cases, Croatian cities recruited gardeners and garden architects from outside of the country to plan, design and maintain gardens, parks, promenades and other open spaces. In addition, they often imported plants and equipment from outside of the country. For example, Zagreb hired the well-known garden architect Rudolph Siebeck, director of public city parks and open spaces of Vienna, to design Zrinjevac Park (Knežević 1996). For the construction of the Zrinjevac Park, the City Council ordered iron benches from Vienna and procured 300 plane trees (*Platanus acerifolia*) from Trieste, which in 1872 arrived from Udine (Knežević 1996). The great demand for gardeners is visible in professional magazines[2] that published numerous employment advertisements for foreign gardeners, and foreign nurseries published ranges of their stock and price lists. At the time, city authorities, being aware of the lack of professional staff, started encouraging reforms in gardening. For example, at the beginning of the 20th century, the Zagreb administration conducted a survey of the city's gardening service. Findings indicated the need to reorganise the gardening service (Vouk 1934a). In 1890, the 'committee for examining the basics for designing the park in Tuškanac and Josipovac' started to work, and in 1881, the Zagreb city administration established a gardening committee that monitored and directed gardening projects and related professional issues (Knežević 1996). City gardeners were encouraged to acquire additional education. For example, 1884, the Zagreb city administration sends its gardener Josip Peklar on a trip to Austria and Germany to study progress made in ornamental gardening (Knežević 1996). At the same time, some of the first city parks and gardens are created with implementation of nursery plants and production of urban equipment that were made in Croatia.[3]

In the first half of the 20th century, the gardener profession in Croatia changed significantly, particularly in the field of garden architecture. Most notably, the city services and general professional activities were taken over by educated garden architects. Prominent examples include Angela Rotkvić (graduate of secondary gardening school in Božjakovina), Silvana Seissel (completed a gardening course in Zagreb) and the first formally educated garden architects Ciril Jeglič, Pavao Ungar and Smiljan Klaić (who completed university studies in garden architecture or additional education in Prague, Vienna, Paris and Berlin). With their engagement, the first modern parks and children's playgrounds in Croatia were created.[4] Just before World War II, the first park design competition in the Kingdom of Yugoslavia was announced, that of Krešimir's park in 1937–38 in Zagreb, which is considered the first modern park in the country (Barišić 2002). In addition to the first garden architects in the country, professionals from natural and biotechnical fields also contributed to the development of the profession and education in the field of garden architecture, including botanists and agronomists (Pereković and Kamenečki 2019). Among them, the highest contributions came from Vale Vouk (first president of Horticultural Society in Zagreb) and Zdravko Arnold (first assistant professor in the field of horticulture in the country) (Figure 21.1).

Figure 21.1 Many pioneers of education in the field of garden architecture have been educated and/or have continued their educations abroad. For example, Zdravko Arnold was educated in Vienna (Hochschule for Bodenkultur), Zagreb (Faculty of Agriculture and Forestry) and in Paris (specialist study of gardening and garden art). Z. Arnold's book of records during his schooling in Vienna.

Source: Archive of the Department of ornamental plants, landscape architecture and garden art.

Horticultural Society in Zagreb

In the first half of the 20th century, the Horticultural Society in Zagreb,[5] founded in 1932, contributed significantly to the development of the profession and to professional education. After returning home from studying abroad, young engineers[6] brought new and advanced ideas about horticulture and garden architecture to Croatia. The main task of the Society was to promote those branches of gardening that were considered the least developed: horticulture and garden architecture (Jeglič 1934; Vouk 1934a, 1934b; Pirnat, 1935; Jellachic, 1934). The Society established publications of different professional magazines, organised public lectures and professional excursions. To improve education, members held monthly lectures on the profession at the Zagreb Radio Station and in other local societies.[7] The Society also directs envoys to take part in international exhibitions and symposia dedicated to horticulture or garden architecture. For example, the Society sent representatives to the following events: XI. International Horticultural Congress in Rome in 1935 (envoy Z. Arnold); Congress of Garden Architects in Paris in 1937 (envoy C. Jeglič; see more in Imbert 2007); XII. International Congress of Horticulture in Berlin in1939 (envoy C. Jeglič). From these and some earlier events (e.g. VIII. International Horticultural Congress in Vienna, 1927), Croatian envoys took messages that they conveyed to Society members, such as the need to organise and improve higher education and diploma in the field of garden architecture, the protection of professional interest of garden architects and their education and the important role of garden architects in public service (Arnold 1927, 1938a).

In 1934, the Society started publishing the professional magazine *Naš vrt* (Our Garden). The Society's postulates find expression in the subtitle of the magazine that reads *Review of Garden Culture*, 'the garden as part of cultural life and hygienic, psychological and aesthetic sides of the garden' (Board of the Horticultural Society 1934). The magazine published articles on the subject of garden architecture, including debates that discuss the state of local education and the analysis of education abroad.[8] The first and most difficult task the Society took on during the first years of operation was the organisation and fostering of professional education. In 1932, the Society sent a request to the City of Zagreb for gardening courses and a proposal for the establishment of a higher technical gardening school (State Archives in Zagreb 1932). The Society also participated in the preparation of laws, such as for advancement of horticulture (in 1935).

Gardening courses as forerunners of secondary and higher education

Starting in 1932, the Horticultural Society in Zagreb began to offer gardening courses. Held twice a year (winter and autumn courses), they had different programmes for amateurs and professional staff.

Before that time, the first known courses on gardening were organised after 1903, on the governor's estate in Božjakovina near Zagreb (before the establishment of the secondary gardening school in the same place). After 1915, Vale Vouk offered gardening courses at the Botanical Garden in Zagreb, such as a multi-month gardening course for war invalids with practical and theoretical training, and evening gardening courses for gardeners employed at the Botanical Garden. The first gardening course that the Society offered was the autumn (amateur) Gardening Course for Ladies in 1933 (see Figure 21.2). Forty-four students enrolled and instructors were, among others, garden architects C. Jeglič and P. Ungar (Arnold 1934). The courses included fieldwork and demonstration of practical exercises. Study programmes

Figure 21.2 Participants and teachers of a gardening course held at the Botanical Garden in 1932, Zagreb.
Source: Archive of the Department of ornamental plants, landscape architecture and garden art.

printed in the magazine *Naš vrt* (years 1934–1938) included subjects of all branches of gardening. Learning outcomes of amateur courses were the knowledge of general gardening, while the professional courses also covered subjects of garden architecture, including Garden Design, Garden Construction, Practical Geometry and Drawing Plans, Public Open Spaces and Parks, Review of the History of Garden Art, Small Family Gardens, Public Gardens and Their Meaning.

Vocational courses were classified as 'lower gardening school' that offered training for participants to qualify as assistant gardener. In order to advance and become gardener, an apprenticeship of many years of practical work in gardening was required in addition to completing courses. Candidates had to pass an exam taken in front of a professional commission (Vouk, 1934c). Professionals who aspired to upgrade their education even further had to attend higher gardening schools outside of the country. Only then would they be able to carry the title 'Gardening Technician' and to become independent planners for gardens and parks or to lead plant nurseries and gardening companies (Vouk 1934c). Lacking institutes of higher education of their own, the Society considered the courses it had established only as a temporary solution. The aim for Croatia remained, to have secondary vocational schools and institutes of higher education in the future.

Secondary education in gardening and garden architecture

The first gardening school in Croatia established secondary education in 1932 in Božjakovina near Zagreb on the foundations of an earlier agricultural school. Garden architect C. Jeglič founded and led the gardening school and the school operated until 1943, when it was closed due to the war. This school initially operated as a two-year and later a three-year programme. In the first two years, students took theoretical subjects, while the third year was mostly dedicated to practical work. The focus of the school's work was on garden design and included a course dedicated to the design of parks (Polak 1980). Several generations of students earned their degree at this school. After World War II, these graduates acted as experts in horticulture, and in 'landscape architecture' as the profession was soon to be called.

In 1932, the Society made proposals for the first complete programme of gardening education in Croatia. The idea was to establish three gardening schools on the territory of the Kingdom of Yugoslavia, two of them in Croatia. The first school proposal was the Secondary Gardening Technical School in Zagreb for growing flowers, ornamental trees and shrubs, and for garden art and garden architecture, while the other was the Secondary Gardening School for Mediterranean Cultures located in Split or in Dubrovnik. For preparing their proposal, the Society analysed schools and educational programmes in France, Germany, the Netherlands and Czechoslovakia. They prepared a report describing programmes of many higher, secondary and lower schools and colleges as well as the professional occupations achieved through education (Arnold et al. 1934). The authors of the report considered a number of institutes of higher education in the field of gardening and garden architecture that differed in educational structure. Each of the institutions offers one or two specialisations, some in vegetable growing, fruit growing, ornamental trees, floriculture and general production of horticultural plants and commercial gardening, others dedicated to garden art education, design and the construction of gardens that acquire the title 'garden architect' (Arnold et al. 1934). The authors of the report note that gardening education appears most developed in the Netherlands and Germany and in terms of modern garden construction and social garden art, neither Wageningen in the Netherlands nor Reading in England can 'compete with Berlin'. The report suggests to upgrade domestic education 'especially in the aesthetic and artistic direction' and that gardening schools must teach 'specialists in natural sciences, agriculture and architecture' (Vouk 1934d, 1934e, translation by authors).

The proposed programme for the gardening school in Zagreb includes general subjects in horticulture in the first three years and in the fourth year a specialisation in one of the branches of horticulture: either vegetable or floristry, dendrology and nurseries, and the design and construction of gardens. According to the draft statue of the school, the curriculum includes the following garden architecture subjects: Garden Design and Construction, Garden Technique, Gardening Architecture and Engineering and History of Garden Art (Horticultural Society 1934, translation by authors). Due to economic circumstances and pre-war events, gardening schools became reality only after World War II. The first opened in Brezovica in 1946 (1946–1949). In 1958 the gardening school *Arboretum Opeka* in Vinica followed. Secondary education in these schools and in other gardening schools became available in Croatia during this period. However, the programmes did not include a specialised orientation towards the profession of garden and landscape architecture, but a wider range of knowledge for the titles of 'Gardening Technician' and 'General Gardener'.

Higher education in garden and landscape architecture

The beginning of higher education in horticulture and garden architecture in Croatia is the year 1937, when the University of Zagreb, Faculty of Agriculture, established the Department of Horticulture (decision no. 21819, 8 December 1937). This was the first department of its kind in the Kingdom of Yugoslavia. The aims of the Department activities were improving national horticulture and promoting all its branches, such as floriculture, vegetable growing, fruit growing, garden dendrology and garden architecture (Arnold 1938b). The department was well funded, however, pre-war and war-related circumstances delayed the development of education in garden architecture.

After World War II, there were two major incentives to develop the profession. First, during the war, almost all green areas had been neglected and were in need of renewal. Second, plans for new urban development foresaw a large share of open and green areas. A 'Landscape Project' was needed to address each of these challenges. During the same time, the Horticultural Society was re-formed and, as of 1955, operated under the name Horticultural Society of the People's Republic of Croatia. The Society established sections for its main branches of activity: one section for floriculture, one for nursery and one for landscape architecture. It decided to join the international organisation IFLA in 1957 (Jurčić 1957). The Society intensively dealt with the topics of professional development, especially due to the uncoordinated criteria of work in the profession of landscape architecture – diverse structure of experts, inconsistencies in the formation of project documentation, general standards and professional terminology. In that time, the term 'landscape architecture' began to replace 'garden architecture' and the profession broadened its tasks. For the first time, the roles of landscape architects in spatial planning and nature protection are discussed and recognised (Klaić 1987; Marušić 1987; Mirčevska 1987).

In the middle of the 20th century, landscape architects began their educations at either the Faculty of Agriculture, the Faculty of Forestry or the Faculty of Architecture. Their programmes were insufficient to meet standards set for a full landscape architectural education (Ungar 1956; Polak 1968). Within the three faculties of the University of Zagreb, study programmes contained elements that were relevant in landscape architectural education. However, no faculty alone provided students with the entire spectrum of subjects necessary to gain sufficient professional knowledge. The Faculty of Agriculture and Faculty of Forestry provided mainly general natural sciences and biotechnical education, while the Faculty of Architecture focused on art, construction design and spatial planning (Milić 1976). Students, who chose garden architecture after graduating in general agronomy, would intern and specialise in the department according to a special programme (Jurčić 1987). After completing their studies on each of the faculties, professionals mainly established themselves as designers through practical work in the field and by passing the state professional exam. There were separate professional exams in horticultural production and garden and landscape design (Jurčić 1987). Once passed, the exam gave authorisation for planning and design open spaces, under the professional title 'Landscape Architect'.

The first landscape architecture study programme

One of the most significant steps in the direction of educating experts in the field of landscape architecture was taken in 1968, when Zagreb University established the first interfaculty postgraduate programme called Landscape Design. To organise and offer a comprehensive landscape architecture programme, segments of biotechnical, technical and artistic fields needed to be fused and joint studies offered at the Faculty of Agriculture, Faculty of Forestry and Faculty of Architecture. These faculties, in cooperation with the Biotechnical Faculty in Ljubljana (Slovenia),[9] finally formed a landscape architecture programme that started in 1968. It was a two-year programme divided into four semesters. After completion, students earned the degree Master of Landscape Architecture. Graduates from each of the participating faculties could apply to the integrated study programme who had at least one year of practice in the profession of landscape architecture. They had to pass an entrance exam and, depending on previous education, enrol in a certain number of differentiated subjects. The programme (see Table 21.1) was offered from 1968 to 1986 and a number of eminent teachers in the fields of landscape architecture, horticulture, forestry, architecture and urbanism served as instructors. On the occasion of the opening of the postgraduate study, Josip Seissel explains that the programme is not just adding to existing knowledge of architecture, agronomy or forestry, but forming a new way of looking at the complex landscape problems imposed by the intertwining of nature and human action in nature. Seissel insisted that this programme is laying the foundations for a new discipline, the domain of landscape architecture that is established in many parts of the world (Seissel 1968). According to Elza Polak, the first head of this programme, the curriculum was oriented towards landscape planning and design and contained more than 1000 hours of lectures and seminar design work (Polak 1980).

Finally, in 1996, based on the interfaculty postgraduate programme Landscape Design, the faculties of Agriculture, Forestry and Architecture established the first graduate programme of landscape architecture in Croatia (see Table 21.2). The Department of Ornamental Plants, Landscape Architecture and Garden Art of the Faculty of Agriculture was appointed as the holder of the study programme, with participation of different experts who were involved in shaping the curriculum.[10] An analysis of the professional market and landscape architecture profile revealed a lack of educational content and staff. Based on the analysis of European landscape architecture programmes, the new graduate study programme developed further (Aničić 2008). In 1996, a five-year (nine semesters) interfaculty graduate study programme started enrolling students. The faculties of Architecture, Philosophy, Geodesy, Forestry and Science participated along with the Faculty of Agriculture. The largest contribution to the organisation and teaching of this programme was, at the turn of the 21st century, provided by landscape architects. They are Branka Aničić, full professor and long-time head of department and study and Dušan Ogrin, full professor. Both were founders, initiators and the greatest promoters of the first graduate study of landscape architecture in Croatia.

Table 21.1 The curriculum for the first Landscape Design study in Croatia – interfaculty postgraduate study held from 1968 to 1985 at University of Zagreb (Faculty of Agriculture – leading faculty, Faculty of Forestry and Faculty of Architecture). Head of studies: Professor Elza Polak.

Curriculum courses in 1968	Teaching methods: lectures, seminars and exercises	Teacher and academic title (at the time of the establishment of the study)	Profession
History of fine arts	- lectures	Prof. A. Mohorovičić	Architect
Drawing and spatial – model shaping	- lectures, exercises	Prof. K. Tompa Prof. J. Vaništa	Artists
Descriptive geometry	- lectures, exercises	Prof. V. Niče	Civil engineer
Theory and history of landscape design	- lectures	Asst. Prof. D. Ogrin	Landscape architect
Urbanism and public greenery	- project seminar	Prof. J. Seissel	Architect
Garden architecture	- project seminar	Asst. Prof. D. Ogrin	Landscape architect
Landscape design	- project seminar	P. Ungar, BCs Prof. C. Jeglič	Landscape architects
Phytocenology 1 and 2 with the application of climatology	- lectures	Prof. J. Kovačević	Agronomy engineer – herbologist
Selected chapters from pedology	- lectures, exercises	Prof. A. Škorić	Agronomy engineer – pedologist
Selected chapters in botany	- lectures, exercises	Prof. N. Plavšić	Agronomy engineer – agricultural botany
Selected chapters in plant nutrition	- lectures	Prof. M. Anić	Forestry engineer
Park plants 1 – trees and shrubs	- lectures, exercises	Prof. C. Jeglič Asst. Prof. D. Ogrin	Landscape architects
Park plants 2 – perennial and annual plants	- lectures, exercises	Prof. E. Polak	Agronomy engineer – horticulture
Biological bases of forest cultivation	- lectures	J. Šafar, BCs	Forestry engineer
Soil use and landscape morphology	- lectures	Prof. J. Roglić	Geography engineer
Garden and landscape engineering	- lectures, exercises	Z. Kani, BCs Prof. B. Milić	Landscape architect Ruglike – urbanist
Nature preservation	- lectures	Ratko Kevo, BCs	Forestry engineer
Urbanism and communal technology – spatial planning	- lectures, exercises	Prof. B. Milić	Architect – urbanist
Basics of regional planning	- lectures, exercises	Asst. Prof. A. Marinović Uzelac	Architect – urbanist
Urban sociology	- lectures	Asst. Prof. A. Marinović Uzelac	Architect – urbanist

Source: The authors.

Table 21.2 The curriculum for the interfaculty graduate study 'landscape architecture' taught by University of Zagreb in 1996 (Faculty of Agriculture – leading faculty, Faculty of Forestry, Faculty of Architecture, Faculty of Geodesy, Faculty of Science, Faculty of Humanities and Social Sciences). Head of studies: Professor Branka Aničić.

Curriculum courses in 1996	Teaching methods: lectures/exercises	Teacher and academic title (at the time of the establishment of the study)	Profession
Introduction to landscape design	45/60*	Prof. B. Aničić	Landscape architect
Open space design	45/90*	Prof. B. Aničić	Landscape architect
Landscape construction	60/60*	Prof. B. Aničić	Landscape architect
Theory and evolution of landscape design	60/30	Prof. D. Ogrin	Landscape architect
Landscape analysis and evaluation	60/90**	Prof. J. Marušič	Landscape architect
Urban landscape design	60/60*	Prof. B. Aničić	Landscape architect
Landscape planning	60/90**	Prof. J. Marušič	Landscape architect
Art history	60/00	Prof. M. Jurković	Art historian
Drawing	00/60	Prof. A. Vulin	Architectural engineer
Drawing and plastic design	00/150	Prof. A. Vulin	Architectural engineer
Basics of urbanism	60/30	Asst. Prof. N. Lipovac	Architect – urbanist
Basic of urban planning	30/30	Asst. Prof. N. Lipovac	Architect – urbanist
Botany	60/60	Prof. N. Hulina	Agronomy engineer – agricultural botany
Park plants	60/30	Asst. Prof. I. Vršek	Agronomy engineer – horticulture
Agro climatology	45/15	Prof. I. Penzar	Engineer of physics and geophysics
General and landscape ecology	30/15	Asst. Prof. M. Kerovac	Ecology engineer
Pedology	75/60	Prof.dr. Vidaček	Agronomy engineer – pedologist
Geodesy with basic of cartography	30/75	Prof. K. Šimičić	Geodesy engineer
Geomorphology	60/30	Prof. A. Bognar	Geography engineer
Basics of Agriculture	45/30	Asst. Prof. M. Mesić	Agronomy engineer
Landscape elements and dynamics	60/15	Prof. M. Sić	Geography engineer
Phytocenology	30/15	Prof. J. Vukelić	Forestry engineer
Photogrammetry and remote sensing	30/30	Asst. Prof. V. Kušan	Forestry engineer
Introduction to ecological psychology	30/00	Prof. D. Ajduković	Environmental psychologist
Urban sociology	30/15	Prof. O. Čaldarović	Sociologist
Basics of spatial planning	30/30	Prof. O. Grgurević	Urbanist
Protection of soil and water	30/15	Prof. F. Bašić / Asst. Prof. D. Romić	Agronomy engineer / Agronomy engineer

(Continued)

Table 21.2 (Continued)

Curriculum courses in 1996	Teaching methods: lectures/exercises	Teacher and academic title (at the time of the establishment of the study)	Profession
Nature and environment protection	30/15	Asst. Prof. Ž. Španjol	Forestry engineer
Water management	30/30	Prof. D. Petošić	Agronomy engineer
Protected spaces		Prof. J. Borošić	Agronomy engineer
Biometrics and experiment planning in plant breeding	45/60	Prof. Đ. Vasilj	Agronomy engineer
Chemistry	75/60	Asst. Prof. Lj. Đumija	Chemical engineer
Mathematics	45/30	Prof. V. Hitrec	Mathematician
Informatics	45/30	Prof. V. Grbavac	Informatician
Society and the state	30/00	Prof. A. Kolega	–
Agricultural economics	60/30	Prof. T. Žimbrek	Agronomy engineer
English language	00/120	V. Arbanas Markotić, MSc	–
Physical and health culture	–	R. Caput Jogunica	Kinesiology engineer

Source: * landscape design project exercises and seminars.
** landscape planning exercises and seminars.

Conclusions

The period between World War I and II was a key time for the forming of garden architecture education in Croatia. The key actors were engineers of natural and biotechnical professions, botanists and agronomists, and the first educated garden architects who received their education abroad. At that time, education in the field of landscape design, called garden architecture, was exclusively developed. The pioneers of this education, such as V. Vouk, Z. Arnold and C. Jeglič, E. Polak and P. Ungar, clearly delineated the field of ornamental gardening and developed two educational branches: horticulture and garden architecture. On these foundations, they formed secondary education curricula.

In the middle of the 20th century, the term 'garden architecture' was replaced by the term 'landscape architecture' in step with the expansion of the tasks of the profession. Landscape design is being upgraded with the field of landscape planning and thus there is a need to modernise education. The first university study of landscape architecture in Croatia in 1968 was based on these foundations. Interdisciplinarity and interfaculty cooperation became increasingly important in the ensuing success of the curriculum of the first undergraduate and graduate study of landscape architecture. Main protagonists of this curriculum at different time intervals of the 20th century were E. Polak, B. Aničić and D. Ogrin.

The relatively long time the development of landscape architecture education took in Croatia, which stretched for almost a hundred years, has retained some idiosyncrasies that are visible today. This primarily refers to the interdisciplinarity of teaching through the involvement of diverse professions in the process of education of landscape architects, including experts in landscape architecture, horticulture, architecture and urbanism, fine arts, forestry and other professions. In addition, particular characteristics retained throughout the decades are visible in the basic teaching structure of landscape architects, which has represented a model of combining lectures from the original courses until today, interconnected practical and design exercises.

Notes

1. Jellachich defines three types of gardening activities: gardening, production and design (Jellachich 1934). The first two categories include 'luxury' production of plants for aesthetic purposes and 'economic' production of plants for food purposes. The third category is the design of gardens, parks and other open spaces. It includes 'physical work' such as construction and maintenance of open space and 'intellectual work' such as creating ideas and plans for open spaces, also known as 'garden architecture'.
2. *Uzorni vrtlar* and *Uzorni gospodar* (The Exemplary Gardener and The Exemplary Master).
3. The fence around the Academic Square in Zagreb was made by the Zagreb locksmith A. Mesić (Knežević 1996); M. Lenuci and R. Melkus developed the Zagreb model of public stairs around 1883, which are considered to be the first element of urban equipment designed in Zagreb (Knežević 2003); Park Tuškanc, designed by J. Peklar in 1890; "The Green Horseshoe" – "U"-shaped system of parks and squares in centre of Zagreb; and other projects.
4. King Petar Krešimir IV Square in Zagreb (C. Jeglič, 1938), Park I Hrvatske štedionice in Zagreb (S. Klaić I C. Jeglič, 1940).
5. In 1935, the Society changed its name to Horticultural Society of the Kingdom of Yugoslavia and since 1940 it has been operating under the name Croatian Horticultural Society.
6. The main protagonists in the landscape architecture profession in the Society were V. Vouk (botanist, member of a Horticultural society in Vienna), Z. Arnold (agronomist with a specialisation in the field of garden architecture and garden art in Paris) and C. Jeglič (garden architect).
7. For example: "Planning a cemetery as a problem of contemporary garden art" (C. Jeglič, 1932); "Past, present and future of Maksimir Park" (V. Vouk, 1936); "Gardens of Dubrovnik and surroundings" (Z. Arnold, 1936); "Design and maintenance of the home garden" (C. Jeglič, 1936).
8. For example: "Gardening education in Germany and the Netherlands" (C. Jeglič), "Gardening education in France" (Z. Arnold), "Gardening education in Czechoslovakia" (G. Chadraba); "Legal status of gardening in Germany and professional training of gardeners" (J. Jellachich), "What kind of gardening schools do we need" (V. Vouk).
9. Commission for the organization and study programme: Faculty of Forestry: M. Anić; Faculty of Agriculture: E. Polak and P. Ungar; Faculty of Architecture: B. Milić, Biotechnical Faculty in Ljubljana: C. Jeglič and D. Ogrin (Polak 1968).
10. Committee for curriculum development: Faculty of Agriculture in Zagreb – B. Aničić, L. Sošić, J. I. Barčić; Biotechnical Faculty in Ljubljana – D. Ogrin; Faculty of Forestry in Zagreb – B. Prpić; Faculty of Architecture – A. Marinović Uzelac, B. Milić, M. Obad Ščitaroci; Faculty of Philosophy in Zagreb – O. Čaldarović; representative of the Ministry of Construction and Environmental Protection – M. Salaj (Aničić 2008).

References

Aničić, B. (2008) 'Studij krajobrazna arhitektura (1996–2005)', *Jubilarni zbornik 40 godina Studija krajobrazne arhitekture*, Zagreb: Sveučilište u Zagrebu Agronomski fakultet, Zavod za ukrasno bilje, krajobraznu arhitekturu i vrtnu umjetnost, str. 3–8. (In Croatian)

Arnold, Z. (1927) 'VIII. Internacionalni hortikulturni kongres', *Uzorni vrtlar*, br. 12, god. V, str. 122–123. (In Croatian)

Arnold, Z. (1934) 'Društvene vijesti – izvještaj tajnika o radu društva', *Naš vrt*, god I., sv. 1–2, str. 42–43. (In Croatian)

Arnold, Z. (1938a) 'Na kongresu vrtnih arhitekata u Parizu', *Naš vrt*, god. V, mart-april, str. 55–57. (In Croatian)

Arnold, Z. (1938b) 'Zavod za vrtlarstvo na Poljoprivredno šumarskom fakultetu u Zagrebu', *Naš vrt*, god. V, maj-juni, pp. 129–130. (In Croatian)

Arnold, Z., Jeglič, C. and i Chadraba, G. (1934) 'Vrtlarsko školstvo u inozemstvu', *Naš vrt*, god I, jul-august, sv. 7–8, pp. 135–139. (In Croatian)

Barišić, Z. (2002) 'Trg kralja Petra Krešimira IV u Zagrebu – urbanističko – arhitektonska i perivojna geneza', *Prostor*, 1(23), pp. 77–91. (In Croatian)
Board of the Horticultural Society. (1934) 'Što hoćemo', *Naš vrt*, god. I, januar-februar, svezak 1–2, p. 1. (In Croatian)
Gradsko poglavarstvo Zagreb. (1932) 'Kazala općih spisa Gradskog poglavarstva 1932–1935', State Archives in Zagreb. (In Croatian)
Horticultural Society in Zagreb. (1934) 'Nacrt statuta srednje vrtlarske škole u Zagrebu', *Naš vrt*, god. I, juli-august, svezak 7–9, pp. 142–144. (In Croatian)
Imbert, D. (2007) 'Landscape Architects of the World, Unite! Professional Organizations, Practice, and Politics, 1935–1948', *Journal of Landscape Architecture*, 2(1), pp. 6–19.
Jeglič, C. (1934) 'Glavna skupština Hortikulturnog društva za godinu 1933 – izvještaj', *Naš vrt*, god. I, mart-april, svezak 3–4, pp. 82–87. (In Croatian)
Jellachich, J. (1934) 'Odnošaj vrtlarstva prema drugim granama privrede i njegov položaj u našem pravu', *Naš vrt*, god I., May – June, svezak 5–6, pp. 109–112. (In Croatian)
Jurčić, V. (1957) 'Glavna skupština Hortikulturnog udruženja Narodne Republike Hrvatske', *Hortikultura*, god. III, br. 1, pp. 29–33. (In Croatian)
Jurčić, V. (1987) 'Vrtlarstvo i oblikovanje pejzaža u visokoškolskom obrazovanju SR Hrvatske', *Zbornik radova s naučnog skupa Uloga pejzažne arhitekture u razvoju i uređivanje zemlje*, Beograd, 23–25 September, pp. 63–69. (In Croatian)
Klaić, S. (1987) 'Uloga pejzažnog arhitekte u prostornom planiranju i projektiranju', *Zbornik radova s naučnog skupa Uloga pejzažne arhitekture u razvoju i uređivanje zemlje*, Beograd, 23–25 September, pp. 103–108. (In Croatian)
Knežević, S. (1996) *Zagrebačka Zelena potkova*. Zagreb: Školska Knjiga, pp. 14–41. (In Croatian)
Knežević, S. (2003) *Zagreb u središtu*. Zagreb: Barbat, pp. 9–25. (In Croatian)
Marušič, J. (1987) 'Varovalno planiranje kot način urejenja krajine. Zbornik radova s naučnog skupa', *Uloga pejzažne arhitekture u razvoju i uređivanje zemlje*, Beograd, 23–25 September, pp. 119–132. (In Slovenian)
Milić, B. (1976) 'Visoko školsko obrazovanje stručnjaka za područje hortikulture, uređenje i planiranje pejzaža', *Uloga i značaj zelenila za stanovništvo Zagreba i njegove regije*, zbornik savjetovanja, Stablo mladosti, Zagreb, 10–11 June, pp. 156–160. (In Croatian)
Mirčevska, S. (1987) 'Mogućnost uključivanja pejzažne arhitekture u planiranje prostora i zaštiti čovjekove okoline. Zbornik radova s naučnog skupa', *Uloga pejzažne arhitekture u razvoju i uređivanje zemlje*, Beograd, 23–25 September, pp. 133–138. (In Slovenian)
Pereković, P. and Kamenečki, M. (2019) 'A Historical Overview of Landscape Architecture Profession in Croatia from 1900 to 1945', *ACS – Agriculture Conspectus Scientificus*, 84(2), pp. 127–134.
Pirnat, S. (1935) 'Hortikultura na Poljoprivrednom fakultetu Univerziteta u Zagrebu', *Naš vrt*, god. II, juli-august, svezak 7–8, pp. 125–126. (In Croatian)
Polak, E. (1968) 'Postdiplomski studij Pejzažne arhitekture', *Hortikultura*, br. 1, pp. 29–30. (In Croatian)
Polak, E. (1980) '70. Obljetnica prof. dr. Elze Polak', *Hortikultura*, br. 4, pp. 28–31. (In Croatian)
Seissel, J. (1968) 'U povodu otvaranja postdiplomskog studija oblikovanja pejzaža', *Hortikultura*, br. 4, pp. 101–103. (In Croatian)
Ungar, P. (1956) 'Osnovano je Hortikulturno udruženje NR Hrvatske', *Hortikultura*, god. II, br. 1, pp. 1–3. (In Croatian)
Vouk, V. (1934a) 'Nastojanje oko podizanja vrtlarske srednje škole u Zagrebu', *Naš vrt*, sv. 7–8, jul august, pp. 139–144. (In Croatian)
Vouk, V. (1934b) 'Dajte nam vrtlarske škole', *Naš vrt*, god. 1, juli-august, svezak 7–8, pp. 131–132. (In Croatian)
Vouk, V. (1934c) 'Glavna skupština Hortikulturnog društva – govor predsjednika', *Naš vrt*, god. II, januar – februar, str. 1–2, pp. 60–62. (In Croatian)
Vouk, V. (1934d) 'Kakove škole trebamo', *Naš vrt*, god. 1, juli-august, svezak 7–8, str. 133–134. (In Croatian)
Vouk, V. (1934e) 'Glavna godišnja skupštini Hortikulturnog društva: O potrebi osnutka vrtlarskih škola, govor predsjednika društva', *Naš vrt*, god. I, januar-februar, svezak 1–2, pp. 40–41. (In Croatian)

22
LANDSCAPE ARCHITECTURAL EDUCATION IN HUNGARY
The pioneering work of Béla Rerrich, Imre Ormos and Mihály Mőcsényi

Albert Fekete

Introduction

With over 660 students and 55 staff members, the Institute of Landscape Architecture, Urban Planning and Garden Art, in Budapest, at the Hungarian University of Agriculture and Life Sciences (MATE) (the 'Institute' hereafter) is one of the largest institutes of higher education in Central Europe. This chapter provides an overview of the history of landscape architectural education in Hungary and highlights the role of prominent educators. Recognising three periods, the account begins with the pioneering period before 1945, continues with the time of consolidation between 1945 and 1992, and closes with recent educational developments.

In 1894, the Horticultural Institute was founded. In 1908, Béla Rerrich became professor and began to teach garden architecture as an independent discipline. In 1939, the Institute was renamed as the Hungarian Royal Horticultural Academy and the Department of Garden Art was established. During the second period, planning tasks gained importance. Mihály Mőcsényi became professor and, under his guidance, landscape planning education started officially. In 1963, professor Imre Ormos presented a proposal for an independent educational programme of Landscape and Garden Design. In 1969, the university-level college was declared University of Horticulture. It developed Landscape Architecture and Planning programmes. More recently, in 1992, the government established the independent Faculty of Landscape Architecture, Protection and Reclamation. Denominations and affiliations changed several times since. As of 2021, landscape architectural education is part of the Institute of Landscape Architecture, Urban Planning and Garden Art housed in the Hungarian University of Agriculture and Life Sciences.

Several former and current staff members have studied and documented the history of Hungarian landscape architecture education. This chapter is based on archival research and biographies collected by Mihály Mőcsényi (Mőcsényi 1993, 2008), Imre Jámbor (Jámbor 2003, 2012) and by the author. The chapter also draws information from interviews that Sándor Bardóczi conducted with Mihály Mőcsényi from 2012 to 2014 (Bardóczi 2012–2014), and from publications in the official journal of the Hungarian Chamber of Architects and in Conoisseur of the Land-enigmas (Buella et al. 2013). Included is information that was collected for the book *Landscape Architecture in Higher Education – 25th Anniversary of the Faculty of Landscape Architecture and Urbanism* (Csemez et al. 2017). In addition, this chapter refers to historic accounts that are included in published studies and reports (Alföldy 2006; Baloghné 2003; Csemez 2008; Csepely-Knorr 2011; Hajós 2016; Kenyeres 1982).

Garden and landscape architectural education between 1894 and 1945

During the time of the Austro-Hungarian monarchy, horticultural colleges offered design education (see Krippner and Lička in this volume). In Hungary, the private Horticultural School, founded in 1853 by Ferenc Entz, became the Royal Hungarian School of Horticulture in 1894, and higher education courses in garden art were launched (Jámbor 2003). The subject Drawing and Painting was important throughout the entire programme, taking up 20% of student's workload, counting six hours a week in the first year, and four hours a week in both the second and third years. During the 1896/97 school term, the subject of Garden Design was offered. Students would take the course during the third year for four hours a week. At the beginning, István Révész was the instructor of Garden Design, and László Gyulai taught Drawing and Painting until 1910. Later, the university hired a number of professionals. According to a school yearbook from 1919, the design instructors were Dezső Angyal, Dezső Morbitzer (head gardener of Budapest from 1930 to 1940) and Béla Rerrich (Mőcsényi 2008). Educators who influenced the development of the landscape education and profession in Hungary also included Mátyás Mohácsi (1881–1970), László Dalányi (1928–2007) and András Balogh (1919–1992). This chapter puts the focus on three professionals, who each left substantial marks on Hungarian landscape architecture education. They are Béla Rerrich, Imre Ormos and Mihály Mőcsényi.

In 1908, the Royal Hungarian School of Horticulture that had existed for 14 years at the time, hired Béla Rerrich (1881–1932), architect and professor at the Technical University of Budapest, and asked him to teach Garden Art and Garden Construction (Hajós 2016). The main subjects at that time were Horticultural Studies, Garden Technology and Plant Species and Varieties. After accepting the appointment, Rerrich began to shape the subject of Garden Art into a course of its own right, and he prepared summaries of the knowledge base required for courses in Garden Art and Design. The professional sources recommended for students by Rerrich were architectural and horticultural journals published in Hungary at that time. An updated international collection of books was also available for the students in the library of the school. Rerrich extended the existing curriculum first with Garden Design and Drawing, and then with Garden Architecture and Cultivation. In these courses, he was both the leader and lecturer (Hajós 2016). Rerrich gained national reputation as an architect, and for his garden and open space designs. He was the first in Hungary who subscribed to *Arts and Crafts* and *Jugendstil*, and he contributed to the manifestation and development of secessionist movement in the country (Alföldy 2006). He researched and started to teach the positive social and well-being effects of green areas in the city (Rerrich 1919). According to his own professional training, he set up a garden design education based on architectural principles. He recognised the importance of integrating ecological and artistic knowledge in the education of garden and landscape architects. The geometrical garden, the scenic garden, landscape architecture and many other disciplinary terms became defined through Rerrich's contributions to the Hungarian literature and higher education.

Rerrich's international activities manifest themselves in professional cooperation, for example with renowned designers such as Thomas Mawson (Csepely 2011). Rerrich's successor was Imre Ormos (1903–1979). Ormos graduated in 1926 from the Pázmány Péter University of Sciences, with a doctoral dissertation in the field of art history and aesthetics (Mőcsényi 2003). From 1927 to1929, Ormos spent two years in Turkey as 'green chancellor' of Kemal Atatürk. At that time, Ankara was being developed as the capital of the country. Ormos was involved in

Figure 22.1 Eyebird sketch of the Hungarian Royal School of Horticulture, designed by Béla Rerrich, around 1930.

Source: Archive of the Institute of Landscape Architecture, Urban Planning and Garden Design, MATE.

the process and designed several gardens, estates and a cemetery, thus gaining experience that he later used in teaching design (Csemez 2008). As an architect, Rerrich conceived of geometric garden where open space is defined by hedges and lines of trees.

Ormos started his educational career as a monthly employee in 1932. In 1936, he worked officially as an assistant lecturer in the position of an assistant professor (Baloghné 2003).

When the Department of Garden Art was established, in 1939, Imre Ormos (1903–1979) had a leading role. In 1943, Ormos organised a garden design seminar. The seminar was a studio-based course that became firmly installed in the curriculum. This course represents the moment when the first professional degree becomes available in the country. The Landscape and Garden Design Division of the Chamber of Hungarian Architects accepted the course and started to offer membership and accreditation to all who graduated from the Department of Garden Art ever since.

Landscape architecture education between 1945 and 1992

After World War II, the provisional government published the decree '8740/1945 M.E.' and, on that basis, established the University of Agriculture in Budapest. The formerly independent College of Gardening and Viticulture was included in the university and became the Faculty of Horticulture and Viticulture. College and university courses continued to run parallel until 1947. The professional education led by Imre Ormos was provided by the Department and Institute of Garden Design. Classic garden and landscape architecture disciplines developed vigorously and, by 1952, the number of courses, seminars and colloquia in the Department had increased to 12. Every student had to prepare garden plans over the duration of three semesters.

The first garden history research and planning commission that the department took on, in 1947, related to historic research, survey and preparation of reconstruction plans for the castle park of Lengyel in Tolna County. Mihály Mőcsényi, assistant professor with Ormos, completed this assignment. In 1952, Imre Ormos set up frameworks for the research of historic gardens. The fruits of this work was that a sub-specialisation emerges. Within a few years, landscape and garden architects excelled in this specialisation and became renowned professionals who laid the

Figure 22.2 Professor Ormos and his students during a seminar in the 1960s.
Source: Archive of the Institute of Landscape Architecture, Urban Planning and Garden Design, MATE.

foundations of Hungarian garden history research and restoration. In 1956 Ormos published *A kerttervezés története és gyakorlata* (*The History and Practice of Landscape Design*), a book that became famous and, with several new editions, a kind of 'bible' for generations of landscape architects. The next editions of this book is still used in education (Ormos 1967).

During the 1950s, ties between the development of educational programmes and professional and social needs strengthened. In planning, for example, it was mainly due to the launch of the so-called 15-year housing programme for one million new dwellings, that the new and now independent landscape design programme established in 1963–64. According to the housing programme, 60% of the new properties were planned as apartment blocks. Societal needs and education went hand in hand. Along with real-world projects, landscape architecture education focused on the development of large community parks. The landscape design programme produced the professionals needed to get the huge task done (Mőcsényi 1993). The establishment of the independent landscape design programme in Budapest happened when landscape architectural education started to rise worldwide, while numbers of independent education programmes were still small. The launch of the separate programme is linked to the name of Imre Ormos. In Hungary, Ormos is considered the founder of both the curriculum and the school of landscape design, and of landscape architecture education in general.

Ormos' programme-based education is at the foundation of the so-called Ormos-school, built in considerable measure on his planning activity of four decades of practice. He introduced a methodological framework in higher education, a new teaching approach highlighting the importance of including a comprehensive breadth of different scales in landscape architecture, from the design of the villa garden to the development plan of green infrastructure, and of entire recreational landscapes such those of Lake Balaton (Jámbor 2003). His innovative research and educational activities led to the blossoming of Hungarian landscape architecture. Many of his former students became leaders in education and practice, including, for instance, Mihály Mőcsényi, Imre Jámbor, Attila Csemez, Péter Csima, Ilona Ormos and Károly Őrsi (Dalányi 2003). In 1969, the University of Horticulture was established as the successor of the Faculty of Horticulture and Viticulture. The five-year-long curriculum was split into two programmes: Horticultural Engineering and Landscape Architecture. Since that time, the landscape and horticulture programmes have existed side by side. Admission, education and training were based on separate curricula from year one. However, the title of the degree remained 'Kertészmérnök' 'Horticulturist', with the programme indicated as a supplement attached to the official certificate.

Before Ormos retired, in 1969, Mihály Mőcsényi (1919–2017) took on the leadership of the landscape architecture programme. Under his guidance a series of reforms were applied to the curriculum during the 1970s. New landscape and planning related subjects were introduced, such as regional planning. Mőcsényi developed the so-called 2 steps programme, a predecessor of the Bologna system. In 'step 1' the school offered a garden constructor and designer programme of three years and in 'step 2' a landscape architecture programme of two additional years. However, the curriculum developed for this educational system was not approved by the University Council. Mőcsényi, as a conscious leader, took drastic measures to ensure the academic future of the programme. He sent his assistant professors abroad to acquire international research experience, to build professional contacts and finally to achieve a building up of an overall recognised doctoral degree system. In this context, Imre Jámbor and Attila Csemez spent four years at Dresden Technical University (East Germany at that time), receiving their doctoral degrees in 1978 and in 1979 respectively. Both of them became significant personalities in Hungarian higher education, in the profession of landscape architecture and, further on, as deans of the Faculty (Buella 2017).

By the early 1980s, the University of Horticulture had firmly established its institutional structure. It was a ridged one, and attempts to create a new institute out of only two departments would have been challenging. The Department of Landscape and Garden Design decided to split into two and to create the Institute of Environmental Management and Dendrology. Object-based design and landscape planning became distinct areas of practice within the professional and academic field of landscape architecture, as was practice in many countries at that time. In 1982, the school established two departments, marking the two main fields of the profession: the Department of Garden Design (garden design, building and maintenance; urban ecology) and the Department of Landscape Planning (nature and landscape protection, landscape rehabilitation, landscape and regional planning). The aim was to broaden the professional palette and to introduce, into the Hungarian system of higher education, the entire spectrum of landscape architecture (Mőcsényi 1993). In the same year, the educational programme was renamed and became the Programme of Landscape Architecture and Garden Design.

During the 1980s, mainly due to Mihály Mőcsényi's efforts, Hungarian landscape architecture decisively entered the international arena. The world of landscape architecture learned about and began to appreciate Hungarian professional achievements during the time Mőcsényi was vice president of the Central Region (Africa and Europe) of IFLA, the International Federation of Landscape Architects from 1979 to 1983, and then president between 1986 and 1990. The positive response offered internationally was at least partly due to a successful IFLA General Assembly meeting and conference held in Siófok, Hungaria, in 1984, organised by Mőcsényi (1984).

Figure 22.3 Landscape assessment seminar nowadays (2019). Research, planning and design based on fieldwork and landscape survey represents the basis of the current master's-level education.

Source: Archive of the Institute of Landscape Architecture, Urban Planning and Garden Design, MATE.

Landscape architecture education since 1992

Due to ongoing university reorganisation, after 1992, landscape architecture programmes changed denomination and affiliations several times. In 2000, it moved from the University of Horticulture and Food Industry to Szent István University. In 2003, the Faculty of Landscape Architecture moved to the Budapest University of Economic Sciences and Public Administration in 2004 to Corvinus University of Budapest, and in 2016 back to Szent István University. In 2004, the Faculty became the Faculty of Landscape Architecture, FLA. Since 2021, the Institute of Landscape Architecture, Urban Planning and Garden Art is part of the Hungarian University of Agriculture and Life Sciences.

After a brief overview of the general development of the education, in this last paragraph, I will discuss some educational aspects in more detail.

By the beginning of the 1990s, the practice of green space planning and landscape planning grew significantly. Educational programmes developed accordingly and, in 1992, based on Governmental Decree 1059/1992 (X. 27.), the University of Horticulture established the Faculty of Landscape Architecture, Protection and Development. The Rector was Mőcsényi. He became the first Dean of the Faculty and served in this position from 1992 to 1993, when he retired. He insisted on including all three areas of disciplinary practice in the name of the Faculty. The first term 'Tájépítészet' (Landscape Architecture), he said, refers to the field of the profession. 'Tájvédelem' (Landscape Protection) emphasises the professional dedication to approaches that are protective and preventative. 'Fejlesztés' (Development) refers to the importance of professional activity in supporting the concept of a dynamic and investment-oriented market (Csemez 2007). When the university established the independent Faculty, it formed five departments: Garden and Urban Design, Garden Art, Garden Technology and Engineering, Landscape

Protection and Reclamation, Landscape Planning and Regional Development, together representing the breadth of landscape architectural practice at that time. Initially, the number of students was 20–25 per class. Moving forward, the Faculty continued to build educational programmes and the number of students increased considerably. By the turn of the century, 70–80 students were admitted each year. A credit-based study system was introduced from the academic year 1996–97, and this required a more flexible educational programme, for example, some subjects had to be offered in both semesters, which wasn't the case before.

In 2000, Imre Jámbor introduced the term 'szabadtér' (open space), a novelty in Hungarian professional terminology and a complex subject that covers every open space and its function in the urban landscape. The knowledge base of subjects previously taught as Garden Architecture and Garden Design needed to be extended. Open Space Architecture and Open Space Design were added as new sub-disciplines to the educational agenda. In 2000, professionals accepted both terms during the Green Space Management Conference in Debrecen, open space planning officially became part of the professional realm.

A new curriculum was established in the academic year 2001–02, where first and second-year students all took basic landscape architectural classes and then, at the end of the second year, would choose one of four specialisations: Garden Design and Urban Management lead by Professor Imre Jámbor, Landscape Protection and Reclamation lead by Professor Péter Csima, Landscape and Regional Planning lead by Professor Attila Csemez and Garden Heritage Protection lead by Professor Ilona Baloghné Ormos.

Based on the application by Szent István University, the government established the five-year-long chartered Urban Systems Engineer programme, by Government Decree 240/2001 (XII.10.). The new programme started in the academic year 2003–04. This way, the Faculty of Landscape Architecture now offered two university degrees which are closely linked, leading to the strengthening of the Faculty and the diversification of the education it offers. Imre Jámbor, the Dean of the Faculty at that time, played a major role in setting up the new university programme.

In line with European developments in higher education, the Faculty implemented the statutes set by the Bologna process. Since 2006, the undergraduate programme in Landscape Management and Garden Construction was launched with 120–140 students and offering three specialisations: Garden Engineer, Landscape Engineer and Urban Manager. Upon completion of the programme, the university awards a BSc degree. The purpose of the BSc in Landscape Management and Garden Construction programme is to train students so that they are capable of managing and developing landscapes, urban environment, green spaces, gardens and open spaces; to take on duties in construction, management and protection related to environmental quality; are capable to fulfil positions at specific authorities and councils; and deal with planning that does not require a chartered status. The curriculum is composed of six plus one semesters, of which half a year is spent in professional practice. During practice time, students spend 2×6 weeks (plus 2×1 weeks for reporting) as interns with a local and/or international company to learn about planning, plan implementation, landscape maintenance and management, general construction, environmental management, urban management, public services, ecology, landscape protection and policy administration. Those keen to continue their studies can work towards earning their master's (MSc) degree, which is a prerequisite for gaining a chartered planner status. The master's programme in landscape architecture continues specialisations established earlier, although these are not specified in the landscape architecture degree the university awards.

In 2007, the school again broadened its programme structure and introduced the graduate programme in Urban Planning, coupled with the founding of a new Department of Urban

Planning and Design. The name of the Department of Garden and Urban Design was changed to Department of Garden and Open Space Design, and the number of departments within the Faculty increased to six. Urban Planning is a new programme leading to a Master in Urban Planning degree. It is technically separate from landscape architecture education. It admits students mainly with landscape architectural or architectural backgrounds and BSc degrees. In 2015, the title of the Faculty changed to Faculty of Landscape Architecture and Urbanism, reflecting the existence of the graduate programme in Urban Systems Engineering and the related undergraduate specialisation.

With 737 students enrolled and 41 full-time staff members, the Faculty reached an all-time maximum in 2009. In the academic year 2009–10, the academic portfolio was extended with the graduate programme in Landscape Architecture and Garden Art where students earn a master's degree. It was offered in February 2012 for the first time. The curriculum is four semesters. As of 2014–15, the programme is offered in English under the name Master in Landscape Architecture and Garden Art, MLA. The graduate programme is related mainly to the Department of Garden and Open Space Design and the Department of Garden Art and Landscape Techniques.

In addition to these two departments, three more units support the educational and research activity of the Faculty: the Arboretum in Szarvas, the Landscape and Urban Analytics Laboratory and the Liveable Urban Landscape Workshop. Beside the university library, a specialised Faculty library with over 5000 volumes is available for all students to use.

In 2016, the Faculty of Landscape Architecture and Urbanism became part of Szent István University without any changes in the Faculty structure or educational programmes. The Faculty offers training in three scientific areas, with one seven-semester undergraduate programme in Landscape Management and Garden Construction and three graduate programmes of four semesters each. Doctoral studies are now part of the Doctoral School of Landscape Architecture and Landscape Ecology (TTDI) and offer five different programmes with six semesters each.

Along with reorganisations of the Hungarian Higher education system, Szent István University, together with the National Agricultural Innovative Research Centre, became the Hungarian University of Agriculture and Life Sciences in 2021. It has 21 independent Institutions replacing the former Faculties. The Institute of Landscape Architecture, Urban Planning and Garden Art, ILAUGA, is integrated into this new structure. In addition to the five existing departments, ILAUGA has two more units, the Department of Floriculture and Dendrology, and the Ornamental Plants Research Group.

Educational philosophy, professional practice and internationalisation

The Budapest approach focusses on a professional education of evidence-based design, on strong links with practice and on learning and teaching by doing. More than 50% of our lecturers – mostly those teaching design subjects – are professionals working in practice. They build their teaching on experience and raise the standards and authenticity of designing and planning assignments. The representatives of the Student Government have regular informal discussions with the management of the Institute about practical problems and experiences, and a full report is discussed at the beginning and the end of each semester. Promoting the profession in society is one of our goals, with particular regard to the challenges of dynamic societal transformations. Accordingly, building on the foundations of classical professional education, we are developing new educational areas including landscape and democracy, green city and safe city, landscape and energy transition, ecosystem services, participatory planning and community-based design. These and other new themes have been built in to the educational programmes at all levels.

Engaging with professional practice was mandatory for students from the very beginning of garden and landscape architecture education. After implementing the statutes of the Bologna reform in 2007, expectations for practice learning is different at undergraduate and graduate levels. At undergraduate level, practice time is included in the curriculum as an independent, full semester activity (14 weeks during the last semester). At the graduate level, practice time is four weeks long, and the students do their internships during the summer holidays. Professional practice is formally organised through official agreements between the university and the institution or office that offer internships. The practice contact person and the responsible staff member recognise the practice-time reports. Experience made in the context of professional practice offers opportunities to review educational programmes and to assess the status and requirements of the market. Student reports about professional practice provide a direct feedback of the educational quality and expectations of professional partners.

The Faculty's leadership in the field is closely tied to the Hungarian Association of Landscape Architects, the Landscape Architecture and Garden Design Division at the Chamber of Hungarian Architects, and the Hungarian Society for Urban Planning. By inviting input and opinions from practice, we are developing and launching practice-oriented educational programmes. Professionals also facilitate the internships for students, thus strengthening ties between the Faculty and planning offices, public authorities concerned with landscape architecture and urban development. Students can organise their practice time at more than 250 national and foreign partners of the Faculty.

Another form of extra-university activities that includes students are planning-cooperation with municipalities that offer a variety of opportunities for real-world involvement. For example, teachers and students, while participating in the development of international courses under the umbrella of Landscape and Democracy (LED, as part of the Erasmus+ programme), the Faculty cooperated with the municipality of Törökbálint in organising an international student workshop. Students also profit from international knowledge exchange. In 2012, the Faculty supported the Sapientia Transilvanian Hungarian University in Marosvásárhely, Romania, in developing new curricula and undergraduate courses. A cooperation of long standing exists with the University of Massachusets in Amherst, where Hungarian-born Julius Fábos has been a professor for many years. The *Fábos Conference on Landscape and Greenway Planning* is held every three years to bring together experts who are influencing landscape planning, policymaking and greenway planning from the local to the international level. Ilona Baloghné Ormos connected the Faculty with the Entente Floral international, and, starting in 1994, she built a research relationship with the Humboldt-University in Berlin, Germany. These and other cooperations have helped make the Budapest education increasingly international and to implement an English language graduate programme. Since 2012, the Faculty is continuously working towards developing international relations by actively taking part in student mobility programmes through bilateral agreements and multilateral networks. The number of official international partners regarding student and staff mobility in 2020 was 59, and the proportion of international students reached 15%.

Conclusions and outlook

The University of Life Sciences in Budapest offers the only landscape architecture education at university level in Hungary. The programmes are unique in Hungarian higher education and research in several ways. First, they are set in the educational context of three different disciplines: agriculture, engineering and arts. Second, they are connected through cross-border exchanges with landscape architecture programmes in several parts of the World, including those that the Faculty helped getting started. Third, the Budapest school has a leading role among the landscape schools in the Central and Eastern European region.

The efficiency of our programmes is demonstrated in the consistently high average of final exam grades each year. The most important tasks are to respond to expectations to expand educational programmes, to improve research potentials and to react promptly and appropriately to the challenges of the higher education market. In the current competitive era, it is imperative to improve communication aiming at first-rate and authentic media presence that gives value-based publicity and presents the institution of higher learning in a realistic and favourable way. Good education-based communication, in turn, helps to continuously make society aware of landscape architecture and to ensure its public appreciation. In the same vein, we will expand on successful cooperation with municipalities and civil organisations, and take on design, planning, development and research commissions on township, micro-region and county levels.

References

Alföldy, G. (2006) 'Rerrich Béla', in Beke, L., Gazda, I., Szász, Z. and Szörényi, L. (ed.) *Nemzeti évfordulóink*. Budapest: Nemzeti Kulturális Örökség Minisztériuma Nemzeti Évfordulóink Titkársága, p. 36.

Baloghné, O. I. (ed.) (2003) *Ormos Imre 1903–1979. Centenáriumi kiállítás – Centennial exhibition*. Budapest: Ormos Imre Foundation- BKÁE Dept of Garden Art-Hungarian Association of Landscape Architects.

Bardóczi, S. (2012–2014) *Mőcsényi esszék 1–15* [Online]. Available at: https://epiteszforum.hu/dosszie/mocsenyi-esszek (Accessed 2 March 2021).

Buella, M. (2017) 'Dr. Mőcsényi Mihály, tájépítész. Ember nélkül nincs táj', *Országépítő*, 2017(3), pp. 61–67.

Buella, M., Zajti, G. and Zajti, B. (2013) *Tájtitkok tudói* [Online]. Available at: https://nava.hu/id/1635394 (Accessed 8 May 2021).

Csemez, A. (2008) *15 éves a Tájépítészeti Kar – 1992–2007*. Budapest: Budapesti Corvinus Egyetem.

Csemez, A., Csima, P., Fekete, A., Jámbor, I. and Schneller, I. (2017) *Landscape Architecture in Higher Education. 25'th Annyversary of the Faculty of Landscape Architecture and Urbanism*. Budapest: Szent István Egyetem, Tájépítészeti és Településtervezési Kar.

Csepely-Knorr, L. (2011) *Modern Landscape Architecture. The Evolution of Public Park Theory Until the End of the 1930's*. Budapest: Corvinus University of Budapest Doctoral School of Landscape Architecture and Landscape Ecology.

Dalányi, L. (2003) 'Imre Ormos' planning activity', in Baloghné, O. I. (ed.) *Ormos Imre 1903–1979. Centenáriumi kiállítás – Centennial Exhibition*, vol. 110. Budapest: Ormos Imre Foundation- BKÁE Dept of Garden Art-Hungarian Association of Landscape Architects, pp. 73–77.

Hajós, G. (2016) *Rerrich Béla építész és kertművész munkássága*. Budapest: Építésügyi Tájékoztatási Központ Kft.

Jámbor, I. (2003) 'A tájépítészeti oktatás múltja és jelene', in Éva, Z. K., Ormos, B., Glits, I., Sáray, M., Tar Imola, T. G. and Hámori, Z. (eds.) *150 év a kertészettudományi, élelmiszertudományi és tájépítészeti oktatás szolgálatában 1853–2003*. Budapest: Szent István Egyetem, pp. 137–138.

Jámbor, I. (2012) 'Education from Garden Design to Landscape Architecture in Hungary', *4D Journal of Landscape Architecture and Garden Art*, 27, pp. 13–24.

Kenyeres, Á. (ed.) (1982) *Magyar Életrajzi Lexikon 1–3*. 2. vol. 506. [Online]. Available at: https://adtplus.arcanum.hu/hu/view/MagyarEletrajziLexikon_2/?pg=507&layout=s (Accessed 2 March 2021).

Mőcsényi, M. (1984) *IFLA XXII World Congress at lake Balaton, Hungary, 1984 – Program*. Budapest: Magyar Agártudományi Egyesület.

Mőcsényi, M. (1993) *Tájépítészeti munkásságom*. Doktori Disszertáció Tézisei. Magyar Tudományos Akadémia Könyvtára, Budapest.

Mőcsényi, M. (2003) 'The "Ormos school"', in Baloghné, O. I. (ed.) *Ormos Imre 1903–1979. Centenáriumi kiállítás – Centennial exhibition*. Budapest: Ormos Imre Foundation- BKÁE Dept of Garden Art-Hungarian Association of Landscape Architects, pp. 64–72.

Mőcsényi, M. (2008) 'Levél a tájépítészeknek a „tájkertész" képzés honi eredetéről', *4D Tájépítészeti és Kertművészeti Folyóirat*, 9. szám, 2008.2–3. oldal.

Ormos, I. (1967) *A kerttervezés története és gyakorlata*. 2nd reworked ed. Budapest: Mezőgazdasági Kiadó.

Rerrich, B. (1919) 'A modern városépítészet szociális irányú kertművészeti feladatairól I', *Magyar Mérnök- és Építész-Egylet Közlönye*, 53, pp. 14–15. szám 127–130.

23
LANDSCAPE ARCHITECTURE EDUCATION HISTORY IN SLOVAKIA AND THE CZECH REPUBLIC

Attila Tóth, Ján Supuka, Katarína Kristiánová, Jan Vaněk, Alena Salašová, and Vladimír Sitta

Introduction

Covering six decades, this chapter provides an account of landscape architecture education history in Slovakia and the Czech Republic, including more than four decades when landscape architecture developed as an independent field of study. Education in Czechoslovakia built upon the horticultural and garden art tradition of the former Austro-Hungarian Monarchy. With the new republic, new professional associations, secondary schools, higher education facilities, and research institutions were established. University education in landscape architecture evolved after World War II. University studies in landscape architecture have been offered at five universities – three in Czechia and two in Slovakia (currently only one). Comparing the establishment and development of landscape architecture in Czechoslovakia, Slovakia, and the Czech Republic to a tree that roots in agricultural and horticultural traditions of the former Austro-Hungarian Monarchy, its stem grows in former Czechoslovakia where the profession of *Sadovnictví a krajinářství* (original term for Landscape Architecture) starts developing in 1962 as part of Horticulture and since 1979 as an independent field of study in Lednice (Moravia). In 1993, the stem continued growing into the Czech and the Slovak branches, with Lednice, Nitra, Bratislava, and Prague growing in its crown. Highlighting the many pathways that education took, educators from all five schools present this chapter. Our aim is to fill knowledge gaps in the patchwork of European education history through introducing Czechoslovak, Slovak, and Czech landscape schools, education, research, practice, and personalities. The information presented in this chapter are from university archives, personal notes of the authors, and available published sources.

During pre-landscape architecture times, the profession was referred to as *krasosadovníctví* (beauty gardening), *umělecké zahradnictví* (artistic horticulture), and *sadovnictví* (garden architecture). The word *sadovnictví* derives from "sad" or "libosad" meaning a place of joy and beauty (Mareček 2018, p. 8). The Czech term *sadovnictví* is defined as "design, establishment and maintenance of ornamental orchards". The equivalent Slovak term *sadovníctvo* as "design and maintenance of ornamental orchards and parks". The term *sad* (orchard) traditionally refers not only to fruit tree plantations but also to public ornamental gardens and parks for leisure. The term *krajina* means landscape in both languages, while *krajinářství* (Czech) and *krajinárstvo* (Slovak) refer to both landscape painting and landscape design.

The roots: horticultural education in the Austro-Hungarian Monarchy in the 19th and early 20th century

Secondary agricultural and horticultural education in Czechoslovakia was built upon a foundation laid during Austro-Hungarian Monarchy times. In the 18th and 19th century, horticultural education was offered as a one- to two-year programme in schools in Vienna and Budapest. At the end of the 19th century, further schools were established, also in Czech and Moravian lands. For instance, in 1885 a wine- and fruit-growing school was established in Mělník. In 1895, the *Höhere Obst-Gartenbauschule* (Higher Fruit-Growing and Horticultural School) was established in Eisgrub, Lednice na Moravě.

For horticultural education in the 19th-century Austro-Hungarian Monarchy, Tlustý (2017, pp. 11–13) identifies four main types. One is education in traditional horticultural families, usually from father to son. Another is professional training, usually provided by renowned chief court gardeners in manorial and castle gardens. A third is university education; as long as no university education in horticulture was available, some horticulturists took botanical lectures at the university in Prague in a three-year programme that ended with a final exam. Finally, gardeners also went on study travels and took on trainees; it was common for gardeners to go to famous royal gardens and horticultural companies throughout Europe, especially to Germany, the Netherlands, France, and England, to widen their knowledge. The subject of garden and park design was mainly included in apprenticeships in court gardens and parks, as well as in educational journeys, during which young horticulturalists became knowledgeable about design. In the last decades of the 19th century, there were new possibilities of garden art and design education, such as training courses in different cities of the monarchy. Later, around the turn of the century, several horticultural, viticultural, and pomological schools were established, many of which offered garden design courses (Tlustý 2017, pp. 11–13; also see Fekete, Krippner and Lička, Pereković and Kamenecki in this volume).

The stem starts growing: horticulture and garden architecture education in the new republic of Czechoslovakia in the interwar period

The republic established new schools or further developed already existing ones, for instance in Brno-Bohunice (1919), Chrudim (1920), and Mělník (1921)/Czechia and Modra (1884/1922), Malinovo (1923)/Slovakia (Mareček 2018; Supuka 2019; Zámečník 2019).

In the young democratic republic of Czechoslovakia (1918–1939), horticulture, including garden and park design, was part of the agricultural sector covered by the Ministry of Agriculture and the Academy of Agricultural Sciences. In 1919 the Imperial Union of Czechoslovak Horticulture Organizations was founded in Prague (in 1923 renamed to Union of Provincial Units of Czechoslovak Horticulture), the first agricultural university was established in Brno (renamed Mendel University in Brno in 2010). Professionals of this period, including educators, were actively involved in professional organisations, which facilitated exchange. The Dendrological Society was founded in 1922 in Průhonice. Understanding the importance of facilitating research in horticulture to further develop this profession, the government acquired the noble estate in Průhonice from Count Arnošt Emanuel Silva-Tarouca in 1927 and established a horticultural research institute (since 2000 the Silva Tarouca Research Institute of Landscape and Ornamental Gardening). Research in Průhonice focussed on dendrology, nursery, urban green space design, restoration of historical parks, and floriculture. Researchers working at this institute became the first educators in garden art and design in Czechoslovakia. Many cities conceptually implemented green spaces and landscape architecture by commissioning so-called sadové úřady (gardening offices).

Important professional manifestation of the newly developing field of garden architecture were garden design offices of František Josef Thomayer, Josef Vaněk, Josef Kumpán, and Josef Miniberger, who were highly influential personalities of garden architecture in the interwar period (Steinová 2018; Ottomanská and Steinová 2017). Some of these influential garden architects were also involved in education and awareness raising to some extent. Kumpán was examination commissioner for teachers at higher horticultural schools, Vaněk was active in publishing and lecturing on garden design and related fields, Thomayer published in journals and established the horticultural school in Prague.

In the period between 1926 and 1945, two dendrologists mainly represented the garden design field, including education – Jaromír Scholz and Bohumil Kavka. They were the first lecturers of landscape aesthetics and dendrology in Lednice. Scholz was initiator of a creative linkage between aesthetics and ecology, an enthusiastic admirer of arts, establisher of university education in horticulture in Lednice. Kavka also admired arts and aesthetics, supervised the education in *Sadovnictví*, dendrology, and floriculture within the horticultural field at the University of Agriculture in Prague and in Lednice (Mareček 2017; Mareček 2019).

The stem grows stronger: landscape architecture as an independent study programme

After the end of World War II, in the period between 1945 and 1948, four secondary schools with a professional specialisation in Horticulture were established in the reunited Czechoslovakia (Jureková 2005). In the following decades, until 1989, more than 20 secondary agricultural schools were established in Slovakia (Supuka and Tóth 2019).

After World War II, separate fruit-growing and vegetable-growing research institutes were established, and the research centre in Průhonice could fully focus on *krasozahradnictví/okrasné zahradnictví* (ornamental horticulture) and *sadovnictví* (linguistically related to the Russian *sadovodstvo*/садоводство), the field that would become landscape architecture. Based on his research, Kavka published several important works, including study books for high schools and universities (Kavka et al. 1970). Design research and teaching was integrated with dendrology, nursery, and floriculture (Mareček 2017). Besides Průhonice, research on ornamental plants and urban green space design was also conducted in Arborétum Mlyňany of the Slovak Academy of Sciences in Slovakia (founded by Count István Ambrózy-Migazzi in 1892). Its Institute of Dendrobiology (1953–1993) generated three influential founding educators in Nitra – Pavol Vreštiak, Pavel Hrubík, and Ján Supuka. Research at this institute focussed on historical and contemporary parks, as well as protection and design of urban and landscape greenery. Publications of the institute have been implemented in landscape architecture teaching. The most important works include the *Atlas of Extension of Introduced Woody Plants in Slovakia and Zoning of their Cultivation* (Benčať 1982), *Slovak Parks and Gardens* (Steinhübel 1990), *Scientific Foundations of the Bratislava Green Space System* (Tomaško 1967); *Historical Parks and Gardens in Slovakia* (Tomaško 2004), *Park Forest Design* (Supuka and Vreštiak 1984); *Ecological Principles of Green Space Design and Protection* (Supuka 1991), *Analysis and Proposal of Urban Green Space Design and Protection in Slovakia to 1990–2000* (Benčať et al. 1979) and many others.

University education in the field of *sadovnictví* that became landscape architecture started in the context of horticulture in 1947 at the two agricultural universities, one in Brno and the other one in Prague. In the third and fourth year, study programmes included specialisations on *sadovnictví*. Teaching was provided in Průhonice by Bohumil Kavka in the form of practical training, and in Lednice by Jiří Scholz, Bohdan Wagner, and Jaroslav Machovec. Research on landscape design, management, plant resources, and production floriculture were mainly done in the form of applied research and counselling. Its pedagogical importance consisted in the fact that it linked the assortment, technological, and compositional aspects. Researchers authored many important publications, including study books and taught at both agricultural universities. Until 1950, Horticulture as a discipline was part of the study programme Common Agriculture at the Faculty of Agronomy of the Agricultural University in Brno.

In 1951, when horticultural studies moved to Lednice, the horticultural school estate was established and the Department of Horticulture was founded in 1952, initially led by Jiří Scholz. It took until 1962 that *sadovnictví* (garden design) became an independent specialisation within Horticulture and until 1979 that *Sadovnictví a krajinářství* (garden and landscape architecture) became an independent field of study. Since then, Lednice had been the headquarters of garden and landscape architecture education for the whole of Czechoslovakia. In 1985, when the Faculty of Horticulture was founded, "architecture" was not yet literally included in the Czech or Slovak title that the university awarded to its graduates, however the programme was already dedicated to *landscape architecture* by contents.

In 1963, Czechoslovakia started the construction of motorways. This was a great design challenge. Landscape architectural offices were called to choose the most suitable vegetation, and to test and select different meadow types and woody plant assortments. In 1965, the exhibition organisation Flora Olomouc (Czechia) was established, followed by Flóra Bratislava (1966, Slovakia), which since had an important role for professional practice and education as a source of inspiration and exchange. In 1967, the Botanical Garden and arboretum at the Mendel University in Brno was extended; designed by Ivar Otruba, it became an important manifestation of contemporary landscape architecture on one hand, with a significant educational importance on the other hand. Ivar Otruba was an influential personality of Czechoslovak and later Czech landscape architecture education and published a study book on garden architecture for secondary schools and universities (Otruba 2002), as well as other theory books on Italian and French gardens. Further study books on landscape architecture were published by Kavka et al. (1970) and Wagner (1989, 1990).

Meanwhile in Prague, the idea of a new agricultural university was born and in 1961 the Department of Plant Breeding and Gardening was established, led by professor Červenka and Garden Design taught by Jiří Mareček, who is the author of the landscape architectural design of the area of the Czech University of Life Sciences (ČZU) in Suchdol and its special educational part – the *Libosad* (ornamental garden). Mareček taught garden and landscape architecture at both the ČZU in Prague and Mendel University in Brno for three decades (1961–1991).

The stem branches out: the period after 1992

In 1989, communist rule in the country ended. In 1993, Czechoslovakia was dissolved, with its constituent states becoming the independent states of the Czech Republic and Slovakia. Landscape architecture education continued developing in both countries. At the Faculty of Horticulture of the Agricultural University in Brno, the field of study *Sadovnictví a krajinářství* was renamed as *Záhradní a krajinářská architektura* (Garden and Landscape Architecture) in 1995 and the programme was officially accredited by the government in 2001. The most influential educators include Miloš Pejchal, Jiří Damec, Pavel Šimek, Tatiana Kuťková, Dana Wilhelmová, Alena Salašová, Petr Kučera, Milan Rajnoch, Přemysl Krejčiřík, and others.

In 1995, the Faculty of Horticulture and Landscape Engineering (FHLE) was constituted at the Slovak University of Agriculture (SUA) in Nitra (established in 1952). This faculty acquired the accreditation for university education, among others in the field of Garden and Landscape Architecture, in a five-year study programme. In 2003, within the new accreditation, the Ministry of Education approved the right of providing university education at the first, second, and third levels of study, as well as the right of habilitation (associate professor appointment) and inauguration (university professor appointment). Doctoral studies, habilitation, and inauguration in landscape architecture are currently only available at two schools – Lednice and Nitra. In 2012, the Landscape Architecture programme offered by the Slovak University of Agriculture in Nitra was recognised by the International Federation of Landscape Architects (IFLA Europe) for the duration of five years (2012–2017), prolonged for another five years (2018 to 2023). Education and research in landscape architecture in Nitra was conducted by two departments – Department of Garden and Landscape Architecture and Department of Planting Design and Maintenance (in 2021 merged into Institute of Landscape Architecture led by Attila Tóth). The most influential educators include Pavol Vreštiak, Pavel Hrubík, Ján Supuka, Anna Jakábová, Zdenka Rózová, Ľubica Feriancová, Viera Paganová, Roberta Štěpánková, Ľuboš Moravčík, Gabriel Kuczman, Dagmar Hillová, Attila Tóth, Katarína Miklášová, Ján Kollár, and others. One of the most comprehensive study books in landscape architecture was elaborated by a team of 20 co-authors led by Ján Supuka on vegetation structures in built environments, with a particular focus on parks and gardens (Supuka et al. 2008).

In 2003, based on the new Act on Universities and Higher Education in Slovakia, both faculties in Nitra and in Bratislava elaborated a comprehensive curriculum of a new study programme entitled Landscape and Garden Architecture, which was approved by the Accreditation Committee of the Slovak government. The study programme was elaborated by scientific and pedagogical personalities of Nitra (Ján Supuka and Pavel Hrubík) and Bratislava (Peter Gál and Maroš Finka). Following a successful accreditation, both faculties established their study programmes in Landscape Architecture.

Until the mid-20th century, horticultural schools and agricultural universities were the only places to offer garden and landscape design education in Czechoslovakia. This has changed when landscape education started to be integrated to urban design, territorial planning and architecture studies at the Faculty of Architecture of the Slovak University of Technology (STU) Bratislava. This process was initiated by the enlightened educator Emanuel Hruška (1906–1989), a Czech architect and urban designer who understood the importance and mutual relationships between landscape, urban planning, and architecture (Hruška 1945, 1946). He founded the theoretic basis of Slovak (landscape) urbanism and territorial planning, while integrating environmental aspects. The department he led between 1950 and 1961 worked on many design and research assignments in the field of landscape planning. Hruška was followed by his younger colleague Milan Kodoň (1929–2001), who developed a specific profile as the "architect of landscape". Kodoň can be considered as one of the early pathfinders of landscape architecture in Slovakia. In 1964 he started teaching the special subject *Krajina a technické dielo* (Landscape and Technical Work), within which he taught the wider context of harmonising relationships between technical works and landscape (Kodoň 1965). He authored many urbanistic and territorial studies focusing on the protection of natural environments. He was the first president of the Czechoslovak Branch of IFLA – International Federation of Landscape Architects (1969–1976). Thanks to his extraordinary drawing skills with a characteristic drawing style, he considerably enriched the theory and techniques of landscape drawing and landscape teaching (Kristiánová 2015a, 2016), see Figure 23.1.

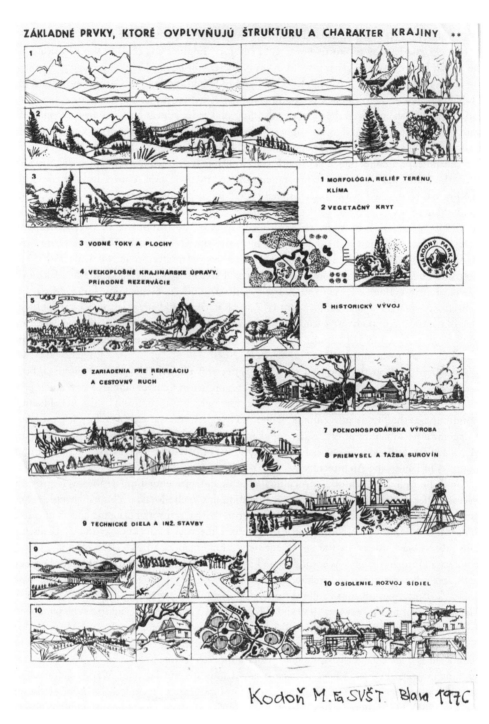

Figure 23.1 Some characteristic hand drawings by Milan Kodoň, whose landscape drawings were instrumental in teaching. He used drawing as an interactive teaching tool.

Source: The archive of ILA SUA Nitra.

The landscape branch of the Department of Urbanism and Territorial Planning was, during the 1970s, developed further by Peter Gál and Karol Kattoš. During the 1980s, research with landscape focus provided the foundation for key study books to be published for the field of landscape architecture (Gál and Kodoň 1981; Kodoň and Gál 1989). Building on Hruška's and Kodoň's foundational work, Peter Gál established the independent Department of Landscape and Park Architecture in 1990. Landscape architecture was first taught as part of Urban Design (1990–2003), later as an independent study programme entitled Landscape Architecture and Landscape Planning (2003–2015). Gál tried to integrate landscape aspects, especially landscape ecology and dendrology, also into the professional profile of architects. In 1996, he initiated the tradition of the colloquium of landscape departments as a platform for presentation of results in teaching and research in Slovakia and Czechia. The establishment of an independent department of landscape architecture was an important achievement in the field of landscape education, which meant the beginning of systematic teaching of landscape subjects at the Faculty of Architecture at STU Bratislava. During the first years of its existence the department provided education for architects on landscape architecture in the following structure: lecture – seminar – design studio, which was finalised in the last year of the master study with the possibility of choosing a study focus on landscape architecture, with a specialised master's thesis (Kristiánová 2015a, 2016; Kristiánová and Stankoci 2015). This model was realised with small adjustments until 2003, when the independent study programme Landscape and Garden Architecture was accredited.

A specificity of the landscape architecture programme at the technical university was a strong share of architectural, urbanistic, planning, and design subjects that developed the design skills of landscape architecture students (Kristiánová 2015b). Professor Gál became dean of the faculty in 2002 and due to necessities of rationalisation and effectivisation of the faculty structure, a decision was made to merge the Department of Urbanistic and Architectural Systems, Department of Settlement Design and Department of Landscape and Park Architecture into a new Institute of Urbanism in 2003, led by Maroš Finka. The new institute included urbanism, territorial development, landscape architecture and landscape planning, spatial planning and management. This institute continued providing education in landscape and garden architecture and it organised the ECLAS Conference in 2006 on Cultural Dimensions of Urban Landscapes (Gál 2015). In 2007 the master study programme landscape architecture and landscape planning lost the accreditation and only the bachelor's degree programme remained. From 2008, the bachelor's programme was provided by the new Institute of Garden and Landscape Architecture led by Ingrid Belčáková, followed by Tamara Reháčková and Katarína Kristiánová. Students who completed their bachelor cycle at the Faculty of Architecture, continued their master's studies either in Nitra or at the Faculty of Civil Engineering in the Landscape Engineering programme. The Faculty of Architecture accredited a new four-year bachelor's programme Landscape and Garden Architecture in 2014, which replaced the three-year programme Landscape Architecture and Landscape Planning. In 2016 the decision was made to discontinue the bachelor's study programme Landscape and Garden Architecture in Bratislava. Most of the students continued their studies at other schools, mainly in Nitra. The main reason for this decision was the continuously low number of students compared to the programme of architecture and urbanism, as well as the missing continuity in the master's cycle. After this, the Institute of Landscape and Garden Architecture was merged again with the Institute of Urban Design and Planning, while the original staff continue research, design, and teaching landscape architecture to students of architecture and urban design (Kristiánová et al. 2017).

Since 2016, the study programme of landscape and garden architecture in Slovakia is officially accredited and offered only by the Slovak University of Agriculture in Nitra, where the

Figure 23.2 Outdoor drawing classes belong to important learning practices, especially within the subjects of Drawing, Architectural Drawing, and Landscape Drawing at SUA Nitra.

Source: Photo by Attila Tóth, September 2017.

bachelor's study programme includes three semesters of hand drawing – basics of drawing, architectural drawing, and landscape drawing; see Figure 23.2.

This study programme, along with the study programmes at the Mendel University in Brno and the Czech University of Life Sciences in Prague have been recognised by the International Federation of Landscape Architects (IFLA Europe). Both Slovak universities and the Mendel University in Brno are part of the European Council of Landscape Architecture Schools.

The field of study Landscape Architecture was later established also at the Faculty of Agrobiology, Food and Natural Resources at the Czech University of Life Sciences (ČZU) (department since 2008, accreditation since 2012, led by Jan Vaněk) and at the Faculty of Architecture of the Czech Technical University in Prague (ČVUT) (department since 2009, accreditation since 2015, led by Vladimír Sitta). At ČZU, the establishment of the programme can be considered as a re-establishment of a former tradition, while at ČVUT, the youngest programme of landscape architecture in former Czechoslovakia is mainly linked to the personality of Vladimír Sitta.

Conclusions

The history of landscape architecture in former Czechoslovakia is strongly linked with agriculture and horticulture. The oldest school of landscape architecture was established in Moravia (Lednice). After the division of Czechoslovakia, the school in Lednice continued and the new school in Nitra was established, similarly at a horticultural faculty of an agricultural university. Landscape Architecture education was also established at the technical university in Bratislava, where it built upon a long-term landscape focus as part of urban design and architecture. Both youngest programmes are in Prague, one at the agricultural, the other at the technical university. Based on a shared history and development of all four current landscape architecture schools in Slovakia and Czechia, there is a strong potential for more intensive cooperation in education and research. An interesting, shared initiative of ČVUT Prague, ČZU Prague, MZLU Lednice (Czechia), and SPU Nitra (Slovakia) is the student design competition LAURUS, which started in 2021 with the aim to provide a shared platform for students and educators of all four schools. We hope that this initiative grows into a nice tradition and will be followed by other cooperation in landscape architecture education, such as an international master's, a joint degree, more intensive exchange of educators and students and others, which could benefit from shared history and traditions on one hand, and diversity on the other hand.

References

Benčať, F. (1982) *Atlas rozšírenia cudzokrajných drevín na Slovensku a rajonizácia ich pestovania*. Bratislava: VEDA, p. 359.

Benčať, F., Supuka, J., Vreštiak, P. and Hrubík P. (1979) *Analýza a návrh koncepcie tvorby a ochrany sídelnej zelene na Slovensku do roku 1990–2000*. Bratislava: SAV, p. 248.

Gál, P. (2015) 'Výučba krajinnej architektúry na FA STU', in Kristiánová, K. and Stankoci, I. (eds.) *Krajinná architektúra a krajinné plánovanie v perspektíve*. Bratislava: STU, p. 246.

Gál, P. and Kodoň, M. (1981) *Tvorba krajiny*. Bratislava: Slovenská vysoká škola technická.

Hruška, E. (1945) *Příroda a osídlení: Biologické základy krajinného plánování*. Praha: Ed. Grégr a syn.

Hruška, E. (1946) *Krajina a její soudobá urbanizace*. Praha: B. Pyšvejc.

Jureková, Z., Gregorová, H., Jakábová, A., Paganová, V., Pintér, E., Raček, M. and Valšíková, M. (2005) *Slovenské záhradníctvo, trendy vedy, praxe a vysokoškolského vzdelávania*. Nitra: Slovak University of Agriculture, p. 216.

Kavka, B., Ambrož, V., Čeřovský, J., Galuszka, E., Hruška, E., Kuča, O., Machovec, J., Říha, J. K., Scholz, J., Uličný, F. and Wagner, B. (1970) *Krajinářské sadovnictví*. 1st ed. Praha: SZN – Státní zemědělské nakladatelství, p. 580.

Kodoň, M. (1965) *Stavba miest. Krajina a technické dielo v nej*. Bratislava: Slovenská vysoká škola technická.

Kodoň, M. and Gál, P. (1989) *Parkové a sadové úpravy: Tvorba krajiny a parkové úpravy*. Bratislava: Slovenská vysoká škola technická.

Kristiánová, K. (2015a) 'Kontexty vzdelávania v krajinnej architektúre', *ALFA*, 20(3), pp. 4–13.

Kristiánová, K. (2015b) 'Tendencie výučby krajinnej architektúry – špecifiká bakalárskeho študijného programu krajinnej a záhradnej architektúry na FA STU', in *Krajinná architektúra a krajinné plánovanie v perspektíve*. Bratislava: Slovak University of Technology, pp. 204–220.

Kristiánová, K. (2016) 'Pioneers of landscape architecture education in Slovakia', in SGEM (ed.) *3rd International Multidisciplinary Scientific Conference on Social Sciences & Arts*. Sofia: STEF92 Technology, pp. 477–484.

Kristiánová, K., Putrová, E. and Gécová, K. (2017) 'Landscape Architecture for Architects – Teaching Landscape Architecture in the Architecture and Urbanism Study Programmes', *Global Journal of Engineering Education*, 19(1), pp. 60–65.

Kristiánová, K. and Stankoci, I. (2015) *25 rokov výučby krajinnej architektúry na FA STU*. Bratislava: Slovak University of Technology.

Mareček, J. (2017) '90 let zahradnického výzkumu v Průhonicích', *Zahrada – Park – Krajina*, 27(4), pp. 6–10.

Mareček, J. (2018) 'Zrození a rozvoj soudobého pojetí zahradní a krajinářské architektury', *Zahrada – Park – Krajina*, 28(4), pp. 8–11.

Mareček, J. (2019) 'Rámcová charakteristika rozvoje oboru zahradní a krajinářská architektura ve 2. polovině 20. století', in Zámečník, R. (ed.) *Záhradně-architektonická tvorba 20. století v Československu*. Lednice: Mendel University in Brno, pp. 25–31.

Otruba, I. (2002) *Zahradní architektura: Tvorba zahrad a parků*. Brno: ERA Group, p. 357.

Ottomanská, S. and Steinová, Š. (2017) *Stopy českých záhradních architektů na Slovensku*. Praha: Národní zemědělské muzeum, p. 200.

Steinhübel, G. (1990) *Slovenské záhrady a parky*. Martin: Osveta, p. 114.

Steinová, Š. (2018) 'Zahradní umění v letech 1918–1948', *Zahrada – Park – Krajina*, 28(4), pp. 18–23.

Supuka, J. (2019) 'Vývoj záhradnej a krajinnej architektúry na Slovensku v období 1939–1989 s dôrazom na vzdelávanie, výskum a prax', in Zámečník, R. (ed.) *Záhradně-architektonická tvorba 20. století v Československu*. Lednice: Mendel University in Brno, pp. 32–47.

Supuka, J. and Feriancová, Ľ. (eds.) (2008) *Vegetačné štruktúry v sídlach: Parky a záhrady*. Nitra: Slovak University of Agriculture, p. 499.

Supuka, J. and Tóth, A. (2019) 'The Historical Development of Landscape Architecture Education in Slovakia', in *Lessons from the Past, Visions for the Future: Celebrating One Hundred Years of Landscape Architecture Education in Europe*. Ås: Norwegian University of Life Sciences, pp. 143–144.

Supuka, J. and Vreštiak, P. (1984) *Základy tvorby parkových lesov a iných rekreačne využívaných lesov*. Bratislava: Veda, p. 226.

Supuka, J., Benčať, F., Bublinec, E., Gáper, J., Hrubík, P., Juhásová, G., Maglocký, Š. and Vreštiak, P. (1991) *Ekologické princípy tvorby a ochrany zelene*. Bratislava: Veda, p. 306.

Tlustý, J. (2017) 'Zahradníci 19. století III. Vzdělávání', *Zahrada – Park – Krajina*, 27(4), pp. 11–13.

Tomaško, I. (1967) 'Vedecké základy systému mestskej zelene rozpracované na príklade Bratislavy', in *Problémy dendrológie a sadovníctva*. Bratislava: VEDA, pp. 323–454.

Tomaško, I. (2004) *Historické parky a okrasné záhrady na Slovensku*. Bratislava: VEDA, p. 160.

Wagner, B. (1989) *Sadovnická tvorba 1*. Prague: SZN – Státní zemědělské nakladatelství, p. 336.

Wagner, B. (1990) *Sadovnická tvorba 2*. Prague: SZN – Státní zemědělské nakladatelství, p. 323.

Zámečník, R. (ed.) (2019) *Záhradně-architektonická tvorba 20. století v Československu I*. Brno: Mendel University, p. 288.

24
A LONG, YET SUCCESSFUL, JOURNEY

One hundred and fourteen years to implement a landscape architecture programme in Austria, 1877–1991

Ulrike Krippner and Lilli Lička

Rewriting the history of tertiary education in landscape architecture in Austria

The history of landscape architecture education in Austria is closely linked to the history of the University of Natural Resources and Life Sciences BOKU in Vienna, which has thus far been the only institution in the country to offer a tertiary education of this kind. The seeds were sown in 1877, five years after the foundation of BOKU, when the Viennese architect and garden architect Lothar Abel started offering a course on garden design within the agricultural programme. It took 114 years for a fully-fledged academic degree programme to be finally implemented at BOKU in 1991 (Table 24.1). So far, the history of landscape architecture education in Austria has only been described in brief, with a focus on particular parts of the story (Krippner 2015; Mang 2012; Schönthaler 1985; Welan 1997; Woess 1973). Scholarly works have discussed the oeuvres of landscape architects who taught at BOKU, such as Lothar Abel (Bacher 2006; Fuchs 2005), Albert Esch (Ludyga and Kosicek 1989; Strohmayr 1990), and Friedrich Woess (Maier 1990). However, there has as yet been no comprehensive analysis of the underlying discourse.

This chapter sets out to explore the complex history of landscape architecture education in Austria, and at BOKU in particular. We investigate crucial arguments and issues surrounding the implementation of the academic degree programme established by BOKU as a university of agriculture, forestry, and horticulture. As this lengthy process was essentially driven by a handful of protagonists and faced numerous setbacks along the way, the main focus of the chapter is on identifying the key individuals who promoted and directed this undertaking and on examining their specific professional backgrounds, design approaches, and understanding of landscape architecture in the context of societal and professional developments. As institutions and designers themselves become – along with design projects, practices, objects, and tools – "central components in a complex epistemic structure" (Mareis 2010, pp. 11–12), findings on the history of landscape architecture education will contribute to the overall history of the profession in Austria. If the various setbacks are part of the reason why both the early academic roots and the long and multifaceted discourse on the profession's nature are forgotten today, revealing the

Table 24.1 History of landscape architecture education at BOKU and professionalisation in Austria.

	2020	ÖGLA is renamed Österreichische Gesellschaft für Landschaftsarchitektur/Austrian Association of Landscape Architects
Title of programme is changed to "Landschaftsplanung und Landschaftsarchitektur/Landscape Planning and Landscape Architecture" Introduction of Bachelor and Master programmes	2004	ForumL is dissolved ÖGLA takes over the editorship of zoll+
Pilot programme is converted into a full academic programme "Landschaftsplanung und Landschaftspflege/Landscape Planning and Landscape Management". Implementation of four core professorships: Landscape Planning, Open Space Design, Landscape Management, and Landscape Construction	1995	Programme is accredited by the Federal Chamber of Architects and Engineer Consultants and the Austrian Chamber of Trade, Commerce, and Industry Change of title to Österreichische Gesellschaft für Landschaftsplanung und Landschaftsarchitektur/Austrian Association of Landscape Planning and Landscape Architecture ÖGLA
	1991	
	1990	Foundation of ForumL, an Austrian alumni association of landscape planning and architecture at BOKU
Studium irregulare is converted into pilot programme of the same name	1981	zoll+journal of Austrian Landscape Architecture is first edited by ForumL
	1978	ÖGA is renamed Verband Österreichischer Garten- und Landschaftsarchitekten/Association of Austrian Garden and Landscape Architects ÖGLA
Studium irregulare "Landschaftsökologie und Landschaftsgestaltung/Landscape Ecology and Landscape Design" at BOKU Vienna is accepted	1975	
Efforts to set up a joint programme "Landscape Ecology and Landscape Design" at BOKU and Technical University Vienna fail	1974	Congress of IFLA and IFPRA takes place in Vienna, titled "Naturally Designed Environment"
	1970	KOLG and ÖGA merge into Verband Österreichischer Garten- und Landschaftsarchitekten/Association of Austrian Garden and Landscape Architects ÖGA
Postgraduate course is converted to a specialism in the agriculture master's programme	1969	
Foundation of the Institute of Green Space Design and Gardening. Head: Friedrich Woess	1967	
Four-semester postgraduate course "Grünraumgestaltung/Green Space Design" is established	1964	
Efforts to set up a joint programme at BOKU and the Academy of Fine Arts fail	1963	
	1957	*Job title is changed to "garden and landscape planners" as "garden/landscape architect" is considered illegal*
Programme is disbanded. Friedrich Woess continues teaching garden design within the agricultural programme	1954	IFLA congress, titled "International Landscapes", takes place in Vienna
	1952	ÖGA becomes a member of IFLA. Foundation of a competing national association, titled Konsulentenverband für Landschafts- und Gartengestaltung/Consultant Association for Landscape and Garden Design KOLG

Table 24.1 (Continued)

One-year postgraduate course in horticulture and garden design is implemented. Albert Esch teaches garden design	1946	ÖGA is founded as Sektion Gartenarchitekten ÖGA der Berufsvereinigung Bildender Künstler Österreichs/Section of Garden Architects within the Austrian Association of Visual Artists
Attempts to implement an academic programme fail	1940	
	1939	Liquidation of VÖGA. Garden architects are obliged to enter the Reichskammer der Bildenden Künste. Jews are not accepted
Women are accepted as regular students	1919	
	1912	Vereinigung Österreichischer Gartenarchitekten/Association of Austrian Garden Architects VÖGA is founded
Classes in agricultural and garden architecture end	1896	
Lothar Abel teaches agricultural and garden architecture	1877	
Three year programme of forestry is implemented	1875	
Foundation of BOKU with a three-year programme of agriculture	1872	

Source: The authors.

long tradition might help combat the persistent misconception that landscape architecture is a young discipline. Taking a comprehensive view is essential if we are to contextualise, evaluate, and improve our understanding of the profession today and the education upon which it is built.

Research strategy and methods

In investigating the history of landscape architecture education in Austria, this chapter follows an inductive research approach and relies on descriptive strategies. The historiographic research is based on a critical examination of sources relating to the academic education at BOKU. Data is collected from historical publications and archival material as well as secondary sources, including 1) articles by key figures discussing the education, qualification, and tasks of landscape architects – these articles were identified in prior systematic research of professional journals published between 1910 and 1945; 2) course catalogues from BOKU dating from 1887 to 1991 when a full academic programme was implemented; 3) archival material on Albert Esch and Friedrich Woess whose collections are kept in the Archive of Austrian Landscape Architecture LArchiv at BOKU; 4) BOKU commemorative publications describing the history of the university; and 5) scholarly work on key figures who have taught at BOKU. This chapter does not explore the history of graduates, as the registers of the early 20th century have not yet been recorded and are not accessible for systematic research.

The collected data is classified and analysed with regard to 1) organisational parameters and the structuring of the education (initiatives to implement an academic degree programme, including potential partners; the shaping of the curriculum) and 2) professional parameters (the design approach, an understanding of landscape architecture as a profession, the tasks it involves, and the essential skills it requires). The design approach is then contrasted with design projects, which are selected from the collections kept at LArchiv. These analyses contribute to a comprehensive narrative, which sheds light on the long history of landscape architecture at BOKU and the struggles involved in implementing a full academic degree programme.

Lothar Abel: questions of style, taste, and social function in the late 19th century

Lothar Abel (1841–1896) started his career as an educator in 1868, at the age of twenty-seven. Still a student in the architectural programme at the Academy of Fine Arts in Vienna, he accepted a teaching position at the newly opened horticultural school of the Imperial Royal Horticultural Society (*k.k. Gartenbau-Gesellschaft*), where he instructed gardeners in garden design, planting schemes, landscape construction, and the building of greenhouses. Abel was no newcomer to "Garden Architecture", as the profession was called in those days. He could rely on the experience he had gained in his father's nursery (Bacher 2006, p. 13). The foundation of BOKU in 1872, which started with a three-year programme in agriculture and added a forestry programme in 1875, provided Abel with the opportunity to teach garden design at a tertiary level. From 1877 until 1896, Abel instructed prospective agronomists and farmers in "agricultural and garden architecture" as well as the "architecture, design of paths, and plantings on country estates" (Hochschule für Bodenkultur n.d.). Students were predominantly men, as women could only attend courses as non-degree students. This imbalance in higher education was only remedied in 1919, when women were finally accepted as regular students at Austrian universities. Abel felt that an academic programme in garden design should go beyond a solid practical and theoretical education in drawing, geometry, arithmetic, and natural sciences. It should foster and sharpen the artistic and aesthetic perception and understanding by offering a study of the history of architecture as well as of garden architecture (Abel 1895, p. 54). In education as well as practice, Abel expanded the range of professional tasks to include not only private commissions but also public projects such as city squares and parks. In addition, he advocated a modern approach to style and taste: While he did not reject the idea of designing large estate gardens in naturalistic landscape style, for gardens and parks in the urban context he called for an architectural design and repudiated an idealisation of nature (see Figure 24.1). This preference can certainly be traced back to Abel's education in architecture at the Academy of Fine Arts and his appreciation of Renaissance garden art. With an architectonic concept for the Maximilianplatz (1878), Abel opposed colleagues like Rudolf Siebeck, head of the Vienna garden department, who favoured a landscape style for the Vienna Stadtpark (1861) and Rathauspark (1873). The three projects were part of the Ringstrasse boulevard, which was realised after the dismantling of the city walls in the mid-19th century.

Abel did not only address questions of style and taste but in keeping with the *Volksgarten* idea also acknowledged the social and sanitary aspects of open spaces, both private and public, at a time of industrialisation and urbanisation. He demanded that garden art become a central issue in political programmes and called on municipalities to invest in public parks in order to improve the living conditions of the working class (Abel 1882, 1895, p. 14). When he applied for a position at the architectural programme at the Academy of Fine Arts in 1882, it was evident that he wanted to expand his teaching agenda from the agricultural context to the social and sanitary aspects of landscape design. His proposed curriculum extended beyond garden art and design and included the typology of urban open spaces as well as landscape planning (Bacher 2006, pp. 104–105). This broad understanding of landscape architecture corresponded

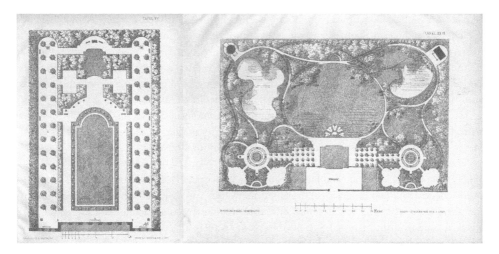

Figure 24.1 According to Abel, gardens in the urban context should show a symmetrical and architectonic design (figure to the left), whereas the layout of suburban villa gardens combines regularity with nature-like forms (figure to the right).

Source: Lothar Abel (1878) *Die Gartenkunst in ihren Formen planimetrisch entwickelt*, ill. XV and XXIII), http://rightsstatements.org/vocab/NoC-NC/1.0/.

with Abel's commissions, which ran the gamut from the design of private and public open spaces to land-use plans, for example, for the community of Reichenau in Lower Austria in 1882 (Fuchs 2005). Although Abel's application to the Academy of Fine Arts was rejected in 1885, he nevertheless influenced the modern understanding of public open spaces. In the Jugendstil period at the turn of the century, the architects Josef Hoffmann, Josef Maria Olbrich, and Robert Oerley followed his idea of an architectural garden, as did the garden architects Franz Lebisch and Titus Wotzy.

Challenges and setbacks in the early 20th century

In 1912, Austrian landscape architects founded the first professional association, *Vereinigung Österreichischer Gartenarchitekten*, VÖGA, which formulated groundbreaking standards for the profession in publishing competition rules and fee regulations (Krippner 2015, p. 52). However, there is no evidence that the organisation committed itself to education. It rather seems that the subject had all but disappeared from the academic stage after Abel's premature death in 1896 at the age of just fifty-five. As a consequence, the education at horticultural colleges gained in importance. The best-known institution was the *Höhere Obst- und Gartenbauschule* located in Eisgrub, today Lednice, Czech Republic, 50 kilometres north of Vienna (see Toth et al. in this volume). It attracted prospective garden architects from the entire Austro-Hungarian monarchy, including Titus Wotzy and Albert Esch (Recht 1976). Again, women were disadvantaged. Although Eisgrub accepted women from 1911 onwards, enrolment on its programme required a prior apprenticeship in gardening, which was difficult to obtain for women, who were often rejected as trainees. Yella Hertzka, née Fuchs, responded to this predicament by founding an advanced horticultural school for women in Vienna in 1913, the *Höhere Gartenbauschule für Frauen*. She successfully encouraged several women to enter the profession (Krippner and Meder 2015). In the interwar period, when garden architecture became a prosperous field, the demand for higher education in Vienna grew. Taking his cue from Germany, Armand Weiser, a Viennese architect, called for an academic education in "all branches of horticulture" at an Austrian agricultural university (Weiser 1927, p. 176). Albert Esch, who was teaching garden architecture at Hertzka's school in the late 1920s, envisioned a post-secondary institution for garden architecture in Vienna in the early 1930s. Esch regarded garden design as a multifaceted profession, which required architectural, spatial, horticultural, and aesthetic skills (Esch 1933, p. 38). Rather like Lothar Abel, he also supported a modern architectural approach and was critical of horticultural schools still relying on the 19th-century methods according to Gustav Meyer and his *Lehrbuch der schönen Gartenkunst*, first edition 1859/60.

The takeover by the National Socialist Party in 1938 changed the professional landscape completely. The new leaders expelled Jewish garden architects and closed Hertzka's school (Krippner and Meder 2016). The professional organisation VÖGA was dissolved, and teaching in Eisgrub was abandoned in 1942. At this point, efforts to implement an academic education in Vienna received support from National Socialist Germany. Anticipating an increasing number of commissions in landscape planning and design, the landscape architects attending the 1940 conference of the German Association of Garden Art (*Deutsche Gesellschaft für Gartenkunst, DGfG*) in Berlin called for the creation of an institute of higher education offering landscape architecture degree programmes that would supplement the work of the Berlin institute of landscape and garden design, headed at the time by Heinrich Wiepking (see Gröning in this volume). Although the attendees emphasised the predominately architectural character of planning and design, they preferred an affiliation to an agricultural university, which would provide in-depth education in natural sciences. As a "German cultural centre in the Eastern part of the German Reich", Vienna seemed the ideal location in which to offer landscape architectural education (Mappes 1941, pp. 23–33). At the conference, the Austrian delegate, Eduard Maria Ihm (1904–1971) welcomed the proposal, stressing that such a tertiary programme could be easily established at BOKU.

Efforts came to a standstill as the war dragged on, leaving behind a devastated country, brought to its knees both physically, by the destruction of its cities and landscapes, and morally, as a shattered society. Both of these aspects had major consequences for the profession: the expulsion of Jewish women landscape architects caused a professional rupture and a loss of role models for future generations. The gender ratio in garden and landscape architecture (as the profession was now called) in Austria only started to recover during the 1970s. In general, denazification was not an issue for the profession. Those who had sympathised with the National Socialists determined the landscape architecture profession and education in the post-war years. A new professional association was founded in 1946. That same year, BOKU started offering a one-year course in horticulture and garden design to graduates of agriculture and forestry. The curriculum included dendrology, plant sociology and cultivation, landscape construction, and garden design, the last of which was taught by Albert Esch (Maier 1990, p. 4). In 1954, when Esch died at the age of seventy, the programme was discontinued. Esch's course was taken over by Friedrich Woess, who was familiar with his predecessor's design principles, having completed the post-war course himself and practised at Esch's design studio.

A rocky, yet successful road to an academic programme

Woess (1915–1995) held a doctoral degree in botany, was a gifted watercolourist, and became an integral part of the fight for an independent degree programme at BOKU over the following four decades (see Figures 24.2 and 24.3). Much in the tradition of Albert Esch, Woess initially pursued the interwar understanding of landscape architecture with a focus on designing private gardens, where built structures in an "architectural" layout merged with "picturesque" plantings (Woess 1953a, 1953b, p. 95). In the 1960s, Woess broadened his programmatic understanding

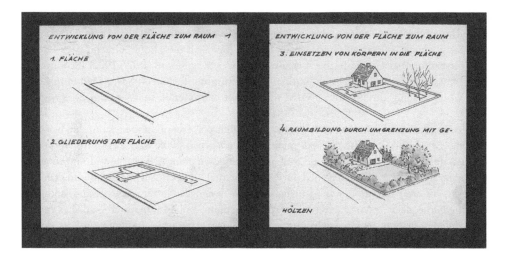

Figure 24.2 In the 1950s, Friedrich Woess taught primarily garden design, following a simple and traditional style, which referred to the Wohngarten of the interwar period.
Source: LArchiv BOKU Vienna, Woess collection.

Figure 24.3 These two undated teaching charts by Friedrich Woess explain the design principle "contrast": left, contrasting lines; right, contrasting light.
Source: LArchiv BOKU Vienna, Woess collection.

of the profession to include the idea of "green space design" covering 1) private gardens, 2) public green space in urban and rural areas, and 3) rural landscapes. Within this conception, he regarded green space design as an appropriate means of creating natural surroundings in a highly technological society in order to preserve a healthy human environment (Woess 1965, p. 3). In the following years, Woess supported two attempts to establish a joint landscape architecture programme, one at BOKU and the Academy of Fine Arts (early 1960s) and one at BOKU and the Technical University Vienna (early 1970s), which would cover the multiple tasks of landscape architects. When both projects failed, the next decades witnessed various attempts to find a compromise solution that would be solely anchored at BOKU: a four-semester-long postgraduate course in "Green Space Design" (1964–1969) and a "Green Space Design and Horticulture" specialism in the agriculture master's programme (1969–1992) (Schönthaler 1985; Woess 1973). The establishment of a specific Institute for Green Space Design in 1967, with Woess heading the institute, rooted this branch of academic study even more firmly at BOKU.

The environmental crises of the 1970s changed public perceptions of the relationship between man and nature, and of landscapes and public spaces (Dagenais 2008; Way 2015). In response to this dual shift, Woess added landscape ecology to the curriculum, which allowed him to apply the botanical experience he had gained during his doctoral studies in Vienna and Berlin Dahlem. However, his clear intention was to turn landscape design, landscape ecology, and landscape planning into equally important sections of the curriculum; and he anticipated graduates engaging in different fields of activity, such as urban and regional planning, community services, design studios, and organisations dealing with nature and landscape conservation (Anon 1971/72). Karl Paul Filipsky (1919–1976) – graduate of the BOKU post-war course and head of the professional association of garden and landscape architects ÖGA in the 1970s – envisioned landscape architects "as the sole advocates for nature-based and human-centred design at all scales of our unbuilt environment" (Filipsky 1973, p. 68, author's translation). The new socio-political understanding of nature and landscape resulted in the establishment of the independent study programme (or *studium irregulare*) in "Landscape Ecology and Landscape Design" in 1975. The programme was transferred into a pilot programme of the same name in 1981 and – finally – turned into a fully fledged academic degree programme in 1991. This new programme was called "Landscape Planning and Landscape Management", indicating a shift away from the ecology movement and suggesting that planning was again seen to have something to offer to society and students. Based on the idea of linking scientific, ecological, and technical skills, the curriculum was expanded to include social science and design subjects (Welan 1997, p. 224). Parallel to establishing the academic degree programme, the organisational structure at BOKU was extended to four institutes, each with a full professorship, covering landscape planning, landscape architecture, landscape management, and landscape construction. The resulting degree has been called "Landscape Planning and Landscape Architecture" since 2004 in order to emphasise the design aspect of the education and to accord with international standards. During its first decades of existence, the popularity of the academic degree programme increased enormously, from 112 students in 1981 to 1,218 students in 1991, a large proportion of them women. In 2021/22, 145 years after Lothar Abel's first class at BOKU and one hundred years after women were first admitted as regular students, women now outnumber men within the academic degree programme of landscape planning and landscape management (808 to 489) and at BOKU in general.

Conclusion

Challenges of implementing a landscape architecture programme

Today we can look back at a long and rich history of landscape architectural education at the University of Natural Resources and Life Sciences BOKU in Vienna, a history that is almost as old as the university itself. Over a period of 114 years the understanding of the profession and education has oscillated between arts, horticulture, ecology, design, and planning, the focus always responding to the societal and political context of the time. This oscillating understanding influenced the questions of whether an academic education should emphasise horticultural, architectural, or artistic skills and which university could best host the programme. Two initiatives in the 1960s and 1970s failed to go beyond the limits of a single university and implement a joint programme. However, the efforts to realise a degree programme were clearly pushed forward by the environmental crisis of the 1970s, when people were terrified by dire prognoses of limited resources and reports of impending nuclear disaster. The process, which had started at BOKU in 1877, had a successful finale there in 1991. While the natural sciences were already strongly represented, planning and design subjects initially needed to be expanded. However, joint courses and transdisciplinary education were and are of great relevance in coping with the diverse tasks of landscape architecture.

Key players and their backgrounds

The lengthy process of establishing a landscape architecture degree programme at BOKU was essentially driven by a handful of protagonists. In this chapter we have identified the three key players: Lothar Abel (1841–1896), who was the first to offer classes on garden design at BOKU in 1877; Albert Esch (1883–1954); and Friedrich Woess (1915–1995). Each of these educational pioneers had either an architectural or a horticultural background, but all acknowledged the multifaceted character of the profession of landscape architecture, even in pre-landscape architecture times. In the cases of Abel and Esch, their efforts to implement an academic degree programme were suspended after their deaths; so much depends on individual activity. Ultimately, Woess played the most significant role, as he taught garden and landscape design at BOKU for more than three decades, from 1954 to 1985, and committed his entire teaching career to the introduction of a degree programme at BOKU. It is essential to continue exploring his and other pioneer's influence on the education of landscape architects by investigating the rich holding of educational material within the Woess collection at LArchiv. Studies might investigate, for example, how the shift from garden architecture to green space design (1960s), landscape ecology (1970s), landscape planning (1990s), and landscape architecture (2000s) affected the character and scope of the educational agenda and programme. In addition, studies might analyse the transfer and impact of ideas and ideologies from one generation of instructors to the next.

Effects on the discipline of the delay in establishing education

In the process of establishing landscape architecture education in Austria, periods of innovation and success have alternated with setbacks. These ups and downs may contribute to the fact that we do not now perceive the long teaching tradition as a continuous process. Hence, much more research on the history and dissemination of the profession's narratives is needed to advance and promote landscape architecture in society as well as at the University of Natural Resources and Life Sciences BOKU in Vienna. Research would include, in particular, an extensive examination of the recent past following the establishment of the *studium irregulare* on "Landscape Ecology and Landscape Design" in 1975. Here, we can take advantage of the fact that some of the protagonists are still alive and can provide first-hand information about the latest development of tertiary education in landscape architecture.

References

Abel, L. (1878) *Die Gartenkunst in ihren Formen planimetrisch entwickelt*. Vienna: K.K. Gartenbau-Gesellschaft/Lehmann & Wentzel.
Abel, L. (1882) *Die Baumpflanzungen in der Stadt und auf dem Lande*. Vienna: Georg-Paul Faesy.
Abel, L. (1895) *Der gute Geschmack*. Vienna: A. Hartleben.

Anon (1971/72) Das aktuelle Interview – Prof. Woess. *boku blätter*, 11(2), p. 7.
Bacher, B. (2006) *Lothar Abel: Das gartenarchitektonische Werk*. PhD thesis, University of Natural Resources and Life Sciences (BOKU), Vienna.
Dagenais, D. (2008) Designing with Nature in Landscape Architecture. *WIT Transactions on Ecology and the Environment*, pp. 114, 213–222.
Esch, A. (1933) 'Wer hat die Berechtigung, den Titel Gartenarchitekt zu führen?', *Gartenzeitung*, 9(4), pp. 37–39, and 9(5), pp. 52–53.
Filipsky, K. P. (1973) 'Stellung des freischaffenden Garten- und Landschaftsarchitekten', in Studienrichtung Landwirtschaft an der Hochschule für Bodenkultur in Wien (ed.) *100 Jahre Hochschule für Bodenkultur in Wien, vol. 3, Vorträge der Studienrichtung Landwirtschaft, Studienzweig Grünraumgestaltung und Gartenbau*. Vienna: Hochschule für Bodenkultur, pp. 49–68.
Fuchs, B. (2005) *Die Stadt kommt aufs Land: Die gründerzeitliche Parzellierungsplanung von Lothar Abel in Reichenau an der Rax, NÖ und ihre Auswirkung auf die aktuellen landschafts- und freiraumplanerischen Qualitäten des Ortes*. PhD thesis, University of Natural Resources and Life Sciences (BOKU), Vienna.
Hochschule für Bodenkultur, later Universität für Bodenkultur (n.d.) *Vorlesungsverzeichnisse 1877–1991* [Online]. Available at: https://epub.boku.ac.at/obvbokvv (Accessed 31 January 2021).
Krippner, U. (2015) 'Landschaftsarchitektur auf dem Weg der Konsolidierung: Ein historischer Abriss', in Lička, L. and Grimm, K. (eds.) *Nextland – Zeitgenössische Landschaftsarchitektur in Österreich*. Basel: Birkhäuser, pp. 72–89.
Krippner, U. and Meder, I. (2015) 'Anna Plischke and Helene Wolf: Designing Gardens in the Early 20th Century', in Dümpelmann, S. and Beardsley, J. (eds.) *Women, Modernity, and Landscape Architecture*. Abingdon: Routledge, pp. 81–102.
Krippner, U. and Meder, I. (2016) 'Moderne Gärten für moderne Menschen: Jüdische Wiener Gartenarchitektinnen der 1920er und 1930er Jahre', *L'Homme: Europäische Zeitschrift für Feministische Geschichtswissenschaft*, 27(2), pp. 53–71.
Ludyga, A. and Kosicek, G. (1989) *Albert Esch und seine Gärten*. Unpublished seminar paper. TU Wien.
Maier, R. (1990) *Friedrich Woess: Gespräche über Leben und Werk*. Unpublished seminar paper, University of Natural Resources and Life Sciences (BOKU), Vienna.
Mang, B. (ed.) (2012) *Von der Gartenarchitektur zur Landschaftsarchitektur: Die Profession in Österreich 1912–2012*. Vienna: ÖGLA.
Mappes, M. (1941) 'Hochschulfragen – Fachschulfragen – Nachwuchsplanung: Bericht über die Tagung der Beiräte und Landesgruppenleiter der DGfG mit den Vertretern der Hochschule und der Fachschulen am 15. und 16. November im "Haus des deutschen Gartenbaues"', *Die Gartenkunst*, 54(2), pp. 23–33.
Mareis, C. (2010) 'Entwerfen – Wissen – Produzieren: Designforschung im Anwendungskontext', in Mareis, C., Joost, G. and Kimpel, K. (eds.) *Entwerfen – Wissen – Produzieren: Designforschung im Anwendungskontext*. Bielefeld: Transcript, pp. 9–32.
Recht, H. (1976) *Die höhere Obst- und Gartenbauschule und das Mendeleum in Eisgrub*. Vienna: Verlag des wissenschaftlichen Antiquariats H. Geyer.
Schönthaler, E. (1985) 'Entwicklung des Fachgebietes Landschaftsgestaltung an der Universität für Bodenkultur', in Asperger, B. and Woess, F. (eds.) *Beiträge zur Landschaftsgestaltung in Österreich: Festschrift für O. Univ. Prof. Dr. Friedrich Woess zur Vollendung seines 70. Lebensjahres*. Vienna: Inst. für Landschaftsgestaltung u. Gartenbau, pp. 13–14.
Strohmayr, L. (1990) *Albert Esch: Private Gartenanlagen; Auseinandersetzung mit Eschs Werken und seiner Schaffensweise, Bedeutung für die damalige Gartengestaltung, Interpretationsversuch einiger ausgewählter Beispiele*. Master's thesis, University of Natural Resources and Life Sciences (BOKU), Vienna.
Way, T. (2015) *The Landscape Architecture of Richard Haag: From Modern Space to Urban Ecological Design*. Seattle: University of Washington Press.
Weiser, A. (1927) 'Ausbildung von Gartenarchitekten', *Österreichische Bau- und Werkkunst*, 4(1), p. 176.
Welan, M. (ed.) (1997) *Die Universität für Bodenkultur Wien: Von der Gründung in die Zukunft 1872–1997*. Vienna: Böhlau.
Woess, F. (1953a) 'Gartenarchitektur – Gartengestaltung – Gartenkunst', *Illustrierte Flora*, 76(8), pp. 94–95.
Woess, F. (1953b) 'Die Ausbildungsgrundlagen der Gartengestaltung', *Illustrierte Flora*, 76(10), pp. 118–120.
Woess, F. (1965) 'Grünraumgestaltung ist angewandte Kunst', *Das moderne Heim*, 14(26), pp. 3–5.
Woess, F. (1973) 'Die Entwicklung des Fachgebietes Grünraumgestaltung an der Hochschule für Bodenkultur', in Studienrichtung Landwirtschaft an der Hochschule für Bodenkultur in Wien (ed.) *100 Jahre Hochschule für Bodenkultur in Wien, vol. 3, Vorträge der Studienrichtung Landwirtschaft: Studienzweig Grünraumgestaltung und Gartenbau*. Vienna: Hochschule für Bodenkultur, pp. 1–10.

25
LANDSCAPE ARCHITECTURE EDUCATION HISTORY IN THE GERMAN-SPEAKING PART OF SWITZERLAND

Sophie von Schwerin

It was in the early 1970s when two polytechnic institutes of higher education started to offer landscape architecture programmes in Switzerland. The first was in the French-speaking Canton Genève the traditional Châtelaine Horticultural School that established a Horticultural Technical College in 1971. Shortly after it moved to Lullier and merged in 2008 with the *Haute École du Paysage, d'Ingénierie et d'Architecture*, HEPIA, in Geneva, where the degree programme is still offered today. The second landscape architecture degree programme started in 1972, when an intercantonal technical college for technically oriented degree programmes opened in the city of Rapperswil, in the German-speaking part of Switzerland.

This chapter explores the evolution of vocational training and professional education in the German-speaking part of Switzerland. It begins with the time of vocational training before the programme in Rapperswil was established and continues by explaining how that programme developed during its 50-year history. To generate relevant information, primary sources pertaining to the different stages in the degree programme at Rapperswil were collected, samples were analysed and evaluated and several documents unveiled for the first time after their filing. Interviews on the bases of oral history add information to that gained through archival research. The chapter fills the gap between published knowledge about education and the current system of teaching landscape architecture in Rapperswil.

Years of vocational education and training

As Garden Art flourished in Switzerland during the period of Industrialisation, several garden design firms established. In the German-speaking cantons, these include the renowned family-owned businesses of Froebel and Mertens. Theodor Froebel (1810–1893), originally from Thuringia, had co-founded a commercial nursery in Zurich in the 1830s, which he ran alone from 1841. His son Otto Froebel (1844–1906) trained as a garden designer in France and Belgium, and subsequently and successfully developed numerous public and private gardens. Dutch-born Evariste Mertens (1846–1907) trained in Belgium, France and England, and like Froebel initially set up a horticultural business in Schaffhausen together with a partner whom he had met while studying at *Institut Horticole* in Gent, Belgium. He moved on to Zurich, worked together with the Froebel company for three years, and started his own firm in 1889, which his sons Walter (1885–1943) and Oskar (1887–1976) took over in 1907 (Bucher 1996, p. 45). Walter and Oskar Mertens chose different pathways of attaining higher education. Walter Mertens attended the Châtelaine Horticultural School and subsequently gained practical experience in England. Oskar Mertens studied at the Swiss Federal Institute of Technology in Zurich, completed internships in France, the Netherlands and England and continued studying at the School of Arts and Crafts in Düsseldorf in Germany. One of his instructors was the architect and designer Peter Behrens (1868–1940) (Bucher 1996, p. 60).

As illustrated by these examples, during the nineteenth century both, the first company founders and their successors largely conducted vocational training to obtain a formal degree, and they did so in European countries other than Switzerland. This still applies to designers of the first half of the twentieth century, who chose a variety of further education and training options. Examples of designers who had major influence on local garden culture are Gustav Ammann (1886–1955), the successor in Froebel's firm, alongside with Walter Leder (1892–1985) and Johannes Schweizer (1901–1983). After completing horticultural training in Zurich (Bucher 1996, p. 60), Ammann had gained practical experience in Germany with Ludwig Lesser (1869–1957), Leberecht Migge (1881–1935) and other garden designers who belonged to the avant-garde in the profession. Walter Leder embarked on a similar path; he initially was trained at the Mertens firm and, for further qualification, attended the Horticultural and Gardening School in Köstritz, Thuringia. He went on to work with Ludwig Lesser and subsequently Leberecht Migge until he returned to Zurich where he founded his own firm and became a very successful practitioner (Schwerin 2017, pp. 6–9). Johannes Schweizer came from a horticultural business in Glarus. He gained practical horticultural experience prior to studying in Berlin-Dahlem and practicing with garden architect Wilhelm Röhnick in Dresden, before he continued the family business in Switzerland (Bucher 1996, p. 61). As illustrated by these examples, one result of their international exposure was that the profession's progressive developments and reform movement in Germany influenced Switzerland's generation of the early twentieth-century garden designers.

When the Oeschberg Horticultural School established in 1920, a new opportunity for higher education became available. The school offered one-year courses and a curriculum of mainly gardening techniques but also subjects such as Geometry and Plan, Perspective and Nature Drawing. Figure 25.1 shows an example of the teaching method from the beginning time of the school. Design principles were also included, but played a subordinate role (Kantonale Gartenbauschule Oeschberg 2005, pp. 11–13). Numerous budding garden designers attended these one-year courses and later excelled as design practitioners. One example is Ernst Cramer (1898–1980), who trained with Gustav Ammann first, then attended the one-year course at Oeschberg in 1922–23, and subsequently gained international acclaim with his own office (Weilacher 2001, pp. 16–19).

However, as the profession grew and the complexity of tasks increased, still no academic education became available in Switzerland until the early 1970s. People who took up landscape architecture after World War II still needed to do part or all of their studies outside of the country. In Germany, they most frequently chose the programmes offered at the Technical University of Berlin, and the Universities in Kassel, Hannover and Weihenstephan (near Munich). It seems that, quite frequently, Swiss students would form small clusters within these degree programmes. Amongst each other, they exchanged experiences and recommendations. Some lived together and most returned home after graduation (Institut für Landschaft und Freiraum, unpublished interview series 2014–2018). During the 1960s, 70s and 80s, they shaped the Swiss landscape architecture community with their own offices or in offices where they were employed. Change, they felt, was needed. The first attempt made to establish a landscape architecture degree programme in Switzerland was in 1957 with the establishment of the Committee

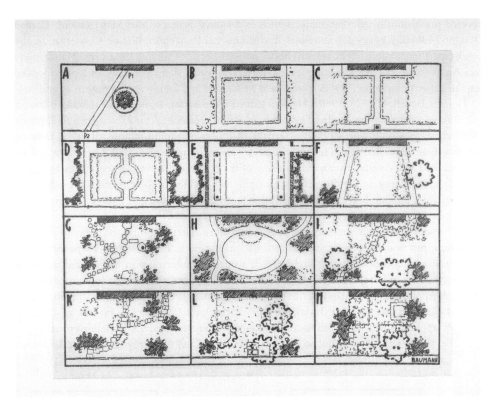

Figure 25.1 Illustrations of garden designs which Albert Baumann used during lectures at the Oeschberg Horticultural School in the 1920s.

Source: Albert Baumann, Drawing, 1924, Archiv für Schweizer Landschaftsarchitektur, Rapperswil.

for Higher Horticultural Education. There would be a number of further attempts that came to no avail. Finally, in 1972, the Intercantonal Technical College in Rapperswil successfully introduced a degree programme (Schubert 2020, p. 17).

The landscape architecture degree programme in Rapperswil, from the beginnings to the present day

The first programme at Rapperswil was called Green Planning, and Landscape and Garden Architecture. It was an undergraduate programme where students could earn a professional degree. It commenced in the winter semester of 1972. The objective was to offer education in ways that students became expert "in shaping the natural environment". After graduation, they were to find employment as executives or employees in cantonal or federal administrations or in planning groups, or to become independent entrepreneurs (Interkantonales Technikum 1972, p. 6). The six-semester programme was broad and diverse. It offered not only discipline-specific subjects but also various subjects of general and basic education such as languages, sports, civic and legal education, and financial and economic studies made up a considerable part of the programme. The mathematical and natural science subjects included Algebra, Geometry, the Use of Computers, the basics of Physics and Chemistry. More specific to the profession were the subjects of Soil Science, Botany, Garden Plants, and Landscape Ecology. The higher semesters covered professionally relevant subjects such as Material Science, Site Management, Garden Design (14 weekly hours per semester), Environmental Protection, Nature Conservation and Landscape Preservation (eight weekly hours per semester), as well as Landscape Management and Architecture (14 weekly hours per semester). Initially, there was only one professorship for the entire programme; it was in the field of Garden and Landscape Architecture and held by Helmut Bournot (1925–1980). He had gained a reputation mainly with making plans for large housing developments called *Siedlungen*, in Berlin. Teaching formats were mainly lectures. Lecturers were people who worked externally in offices or administrations and who took on teaching as an additional role. This had the advantage that, on the one hand, teaching was practice-oriented and, on the other hand, as many of them had studied in Germany, they were able to base their teaching on the theoretical principles they had learnt there (Institut für Landschaft und Freiraum, unpublished interview Christian Stern 2014). Figure 25.2 shows the structure of pre-conditions for the study program.

To be admitted to the programme, applicants had to meet a number of prerequisites such as previous practical experience and educational qualifications. Both were interdependent: Was the educational qualification of a higher level, a one-year internship in garden and landscape construction was required. In case school education was shorter, a three-year apprenticeship was required. Applicants had to pass an entrance examination, in which German, Arithmetic, Algebra and Geometry were tested (Interkantonales Technikum 1972, p. 16). For the first class of 1972, 30 applicants had registered for the entrance examination and 13 of them were accepted. This class was described as enthusiastic, eager to learn and full of energy (Archiv für Schweizer

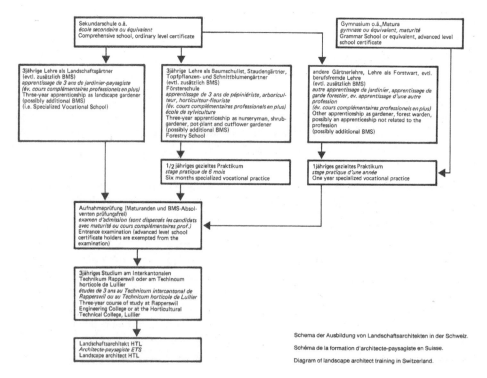

Figure 25.2 Admission requirements for the degree programme in the 1970s and early 80s.

Source: Schubert, B. (1980) *Die Ausbildung des Landschaftsarchitekten*. Anthos, 19 (3), p. 6.

Figure 25.3 Studying in Rapperswil in the early 1990s. The figure shows the working methods and the atmosphere by then.
Source: Photo: Christian Glaus, Archiv für Kommunikation, Ostschweizer Fachhochschule, Rapperswil.

Landschaftsarchitektur, unpublished files, letter 1972). Until 1980, an average of nine students per year successfully completed their studies (Schubert 1980, p. 3).

The curriculum gradually developed. In 1976 a professorship for Botany and Ecology was in addition established. The first substantial reform of the degree programme occurred in the early 1980s, mainly to better coordinate educational schedules in terms of time and content. For example, times where reduced for many basic subjects that students found were of little help to them while studying core areas of landscape architecture. Regarding educational formats, there was still little opportunity for interdisciplinarity and for study projects to cross disciplines (Archiv für Schweizer Landschaftsarchitektur, unpublished files, minutes of meetings 1980).

With the study programme established in 1982, interdisciplinary block teaching was introduced for students of the advanced semesters. The curriculum that used to have more than 40 different subjects was reduced to just over 30. Only during the early semesters did the programme now offer subjects of general and basic education, they specifically related to vocational subjects. The major subjects now already started with the first semester. With 40 weekly hours per semester, a clear emphasis of the degree programme was on Garden Architecture. The 40 hours spread across the entire course of study. Landscape Planning was introduced; it covered 18 weekly hours per semester, and Landscape Design and Ecology each covered ten weekly hours per semester. Botany was reduced from 12 to eight weekly hours per semester and Cultural History had seven weekly hours per semester (Interkantonales Technikum 1983, p. 28).

Figure 25.4 Professor Christian Stern during a presentation in the design studio.
Source: Photo: Christian Glaus, 1996, Archiv für Kommunikation, Ostschweizer Fachhochschule, Rapperswil.

The general requirements for admission to the programme remained the same as before. Before being admitted, applicants had to pass an entrance exam that included Algebra, Geometry and German. Depending on educational qualifications, however, exceptions to this rule were possible. Additional requirements were proof of a discipline-specific apprenticeship or an accelerated apprenticeship and/or a corresponding internship.

In 1991 there was another slight reform of the curriculum and during the following decade additional professorships were created. In 1991, Katharina König Urmi was the first woman appointed as professor. In 1994, seven professorship positions were filled. Most of the new professors were recruited from the existing pool of lecturers. While initially design subjects prevailed, gradually the number of staff of the Landscape Planning department also increased. Accordingly, the study programme was continuously adapted. The subjects Landscape Planning (19 weekly hours per semester) and Landscape Design (14 weekly hours per semester in 1993) grew to 20 weekly hours per semester in 1997. The degree programme was renamed to Landscape Architecture in 1993. From 1995 on, a seventh semester was added to the curriculum; it was exclusively designed for students to work on their final thesis (Interkantonales Technikum 1983–1999).

The Bologna Process that created the European Higher Education Area led to defining moments and additional reforms for the landscape architecture degree programmes all over Europe. In Rapperswil, the transition took effect in 2005, when the curriculum was narrowed down and teaching subjects focused on subjects that are considered core areas of disciplinary education. The sixth semester now includes the final thesis. The degree title changed from Diplom (FH) to Bachelor. The European Credit Transfer and Accumulation System (ECTS) was already adopted a bit earlier (Schubert 2020, p. 19). Generally considered an undergraduate degree, the Rapperswil BA in Landscape Architecture is recognised as a professional degree.

At present, the first year of study currently consists of exercises in Spatial Design and Elementary Design tasks. Drawing, Layout, Urban Development, Open Space Planning and the History of Gardening are as much a part of the portfolio as is Landscape Design, Landscape Ecology and Plant Knowledge. In the second year of study, the program is more condensed: In addition to Design, there are Open Space Planning, Landscape Design and Development, Landscape Ecology, but also Horticulture and Landscaping, as well as Constructive Design. The third year of study is divided into Design and a range of landscape planning subjects. The sixth semester is dedicated to the Bachelor thesis. During their studies, students can either deepen their knowledge in the fields of Landscape Development and Design, Design of Urban Open Spaces, or Landscape Construction and Management, or they can give equal weight to all subjects. The teaching staff includes 11 professors and a great number of lecturers, as well as research assistants with teaching licences. The team of professors comprises three women and eight men. In the curriculum, Design Subjects and Landscape Planning subjects are balanced equally. In recent years, there has been an increase in subjects that pertain to knowledge on Ecology and Bio-Diversity. The subject Use of Plants has also gained a stronger role. With the previous emphasis being entirely placed on plant knowledge, the focus is now primarily on the application or combination of plants in the design of various types of open space (Hochschule für Technik Rapperswil 2019).

Criteria for admission to the degree programme have been revised. There is no longer an entrance examination; instead, candidates need to have either completed a one-year internship or an apprenticeship in a related professional field, depending on their educational qualifications. On an average, some 50 students successfully complete their studies each year and, having earned their professional degrees, enter the labour market.

At present because of the good situation on the job market, only a few graduates continue studying landscape architecture beyond earning their BA degree, taking advantage of the opportunity to obtain the MA degree in Spatial Development and Landscape Architecture that the school in Rapperswil offers since 2016. In addition, the considerable scope for international exchange the Bologna Reform implemented throughout Europe was intended to allow for a general permeability regarding the choice of study location, graduate and postgraduate studies. To date, some of the Rapperswil students are from countries nearby Switzerland but it is an aim to extend that.

Conclusions and reflections

Over the period of half a century, the education in Rapperswil has evolved considerably. What started as small degree programme became a full-fledged landscape architecture education. What used to be small institutions became the University of Applied Sciences that it is today. The education that started as a blend of general education and vocational training has turned into an internationally calibrated specialist degree programme with only few general-education subjects remaining. Given the short duration of the programme, study schedules are largely predetermined. During early years, the programme had functioned partly as a field of experimentation for both students and lecturers, which left more room for individual interests, creativity and diversity to unfold. Initially, the focus was on Garden Design. When the second professorship was installed in the field of Botany and Ecology, the needs equalised. From the early 1980s two influential professionals in Garden Design held the positions, one was Dieter Kienast (1945–1998) and the other Christian Stern (*1935). For Open Space Planning Bernd Schubert (*1939) was engaged. The set up of the curriculum resulted not from political discussions, as has been the case with some degree programmes in other countries, but from preferences of the teaching staff and the professionals themselves. The growing demand for teaching Landscape Planning that became a new focal point was met by adding professorial positions. The number of professors grew from initially one to 11 today. In the beginning, most of the teaching staff were still engaged in their own offices and their teaching style was predominantly practice-oriented. For a large number of the professors today, the teaching appointment is full time. While the professors cover theoretical and methodological basics, lecturers contribute expertise from their professional practice. The strength of the degree programme lies in this combination of sound theory, method and practice oriented knowledge.

The number of students has increased markedly from an average of about 10 in the beginning to a number of wide more than 50 per year today. While, initially, students were mainly male, today just over half of the students are female. Admission requirements have been reduced significantly. They initially comprised an entrance examination, as well as an apprenticeship or at least one year of practical training. Today, either vocational training in a related discipline or a relevant one-year internship is sufficient. As in the degree programme, the admission requirements reflect a development away from accumulating broadly diversified knowledge towards gaining professionally focused expertise.

The landscape architecture degree programme in Rapperswil has strengthened the profession in Switzerland. Since the early beginning, a close cooperation between vocational training and practice has prevailed. Adding diversity, there continues a number of specialists, both in teaching and in the profession, who come to Switzerland from neighbouring countries. Largely, however, the profession in Switzerland is now self-generating and sustainable. At the same time, maintaining the BA as a final degree of a professional programme must be re-assessed. It would be enriching that more staff and students take advantage of opportunities the pan-European modular study system offers. At the graduate programme in landscape architecture that the ETH Zurich introduced in 2020, students can earn a Master's degree (see Richter in this volume). This programme is a valuable and needed addition to the Swiss educational landscape. It will also bring about change in the profession. For structural reasons, permeability between the two institutions is not quite yet possible. Ultimately, the objective of all education and training providers must be to fundamentally prepare graduates for their professional practice. Its tasks are diverse and require a variety of skilled professionals, which is precisely where education and training providers should get involved, coordinate their activities, and respond to the demands of the profession.

References

Archiv für Schweizer Landschaftsarchitektur, unpublished files, Bund Schweizer Gartengestalter.
Bucher, A. (1996) *Vom Landschaftsgarten zur Gartenlandschaft*. Zürich, vdf Hochschulverlag AG an der ETH Zürich.
Hochschule für Technik Rapperswil. (2000–2019) *Studienführer*. Rapperswil.
Institut für Landschaft und Freiraum. (2014–18) *Unpublished Interview Series*. Rapperswil.
Interkantonales Technikum. (1972–1999) *Studienführer*. Rapperswil.
Interkantonales Technikum. (1983) *Studienführer*. Rapperswil.
Kantonale Gartenbauschule Oeschberg. (2005) *Bericht zum Parkpflegewerk*. Koppigen.
Schubert, B. (1980) 'Die Ausbildung des Landschaftsarchitekten', *Anthos*, 19(3), pp. 1–7.
Schubert, B. (2020) 'OST. Das Zentrum der Landschaftsarchitektur-Ausbildung in der deutschen Schweiz', in *Rapperswil_made*. 2nd ed. Rapperswil, pp. 17–20.
Schwerin, S. V. (2017) 'Reformgärten für die Schweiz?', *Nike*, 32(5), pp. 4–9.
Weilacher, U. (2001) *Visionäre Gärten. Die modernen Landschaften von Ernst Cramer*. Basel, Birkhäuser.

26
THE HISTORY OF HIGHER LANDSCAPE ARCHITECTURE EDUCATION AT ETH ZURICH, SWITZERLAND

Dunja Richter

Since autumn 2020, the Department of Architecture at ETH Zurich (Swiss Federal Institute of Technology) has offered a two-year degree programme leading to a Master of Science in Landscape Architecture. After decades of effort, landscape architecture was now established as an independent master's programme at a Swiss university for the first time. This chapter first traces the historical development of landscape architecture education at ETH. It describes how the subject has developed into an integral part of the study of architecture since the foundation of the first professorship in landscape architecture in 1997 and the Institute of Landscape Architecture (ILA) in 2005. The chapter examines the ambitions to establish a university programme in landscape architecture. It focuses on the profile and objectives, the didactic structure and the main areas of the new study relating to a future-oriented landscape architecture which responds to the challenges of the 21st century. The chapter draws on primary and secondary sources, experiential knowledge and expert interviews.

Early initiatives to establish a study in landscape architecture at ETH Zurich

A few years after the founding of the Swiss Federation, the Swiss Federal Polytechnic School in Zurich (since 1911 *Eidgenössische Technische Hochschule Zürich, ETH Zürich*) opened as a national university in 1855. Unlike other universities, ETH Zurich has since been directly administered by federal government and not a canton.[1] For the emerging liberal industrialised country, the polytechnic school, with its technical and scientific orientation, became a decisive successful factor because it linked basic research and practical implementation and provided industry with up-to-date technical knowledge and well-trained specialists. The *Bauschule* – predecessor of today's Department of Architecture – was an important focus of the university and had a pioneering influence on the modern architectural profession. Led by German architect Gottfried Semper (1803–1879),[2] a pedagogy was developed still based on artistic design with form theory and architectural drawing, building construction and technology (Oechslin 2005, p. 56). The university today has around 22,000 students, including doctoral students from 120 countries at its campuses *Zentrum* and *Hönggerberg*.

While a modern, academic course had been available in architecture at ETH in Zurich since the middle of the 19th century, such a course in landscape architecture was established much later. The late 1950s saw the first initiatives to introduce landscape architecture at a higher academic level. The driving force was Emil Steiner (1922–2018). As editor of the journal *Schweizerisches Gartenbau-Blatt* (Swiss newspaper of horticulture), he maintained close ties with neighbouring countries (Schwerin 2014, p. 73), which already had incorporated similar degree programmes. In Switzerland he led a number of groups intending to introduce higher education in landscape architecture and horticulture. The landscape architects Richard Arioli (1905–1994), Willy Liechti (1918–1980), Pierre Zbinden (1913–1982) and Paul Zülli (1912–2001) were also members of these groups. This early involvement came mainly from heads of municipal garden and parks departments rather than from the private sector. The professional association *Bund Schweizerischer Garten- und Landschaftsarchitekten* (Swiss Association of Garden- and Landscape Architects – BSG, today BSLA),[3] founded in 1925, also campaigned for academic education from 1958 onwards (Schubert 2016, p. 19).

In the same year, the *Schweizerische Vereinigung für Landesplanung* (Swiss Association for National Planning – VLP), in collaboration with Pierre Zbinden, proposed a detailed educational programme to the ETH Board, intending to integrate a degree course in spatial planning with an emphasis on landscape design into the new *Institut für Orts-, Regional- und Landesplanung* (Institute for Local, Regional and National Planning – ORL) at ETH. The proposed disciplines Local and Regional Planning and Transport Planning were realised, but Landscape Design failed due to the attitude of other ETH departments, which considered such a course unnecessary (Schubert 1980, p. 4).

All attempts to introduce a degree course in landscape architecture in Switzerland had been unsuccessful because of the disagreement of those involved, scepticism within the profession, and resistance from ETH Zurich and the universities of applied sciences (Schubert 2016, p. 19). Such degree programmes were finally established at the higher technical colleges in Lullier in 1970 and in Rapperswil in 1972 (see von Schwerin in this volume).

From the first courses in architectural studies to the first professorship for landscape architecture in 1997

At the ORL Institute, founded in 1961, *Landschaftsgestaltung* (Landscape Design) became initially a diploma elective for the architectural programme. Christian Stern, who founded the office *Atelier Stern und Partner, Landschaftsarchitekten* in Zurich in 1974, lectured for ORL from 1970 to 1983 (Schwerin 2015, p. 72). The course covered topics relevant to urban and non-urban landscape design including landscape planning, reclamation planning or soil bioengineering aspects, as well as garden design.[4] Following Stern's appointment as professor at the Intercantonal Technical College in Rapperswil (ITR), Dieter Kienast (1945–1998) took over as lecturer at ETH in 1985. From 1980 Kienast, a landscape architect also from Zurich, managed the office *Stöckli & Kienast Landschaftsarchitekten* in Wettingen and Zurich with Peter Paul Stöckli. In addition to his practical and publishing activities, Kienast was a professor of Gartenarchitektur (Garden Architecture) at ITR from 1981 (Richter and Freytag 2014, p. 217).[5] At ETH, Kienast modified the diploma elective to better align with the architectural studies. He put the focus on open spaces in urban areas. In accordance with his own interests, the passing on of knowledge of history and theory of garden art and landscape architecture, its central ideas, the people, and concepts involved, was equal in importance to the imparting of creative and design knowledge. The elective proved more and more popular with students, and the number of diploma elective theses rose from 33 in the first academic year (1985/86) to 80 in 1990/91, prompting Kienast to co-supervise design studios and diploma theses of other professors as well.[6]

Dieter Kienast had earned his doctorate degree at the University of Kassel, Germany, in the field of Plant Sociology and under the supervision of Karl-Heinrich Hülbusch. Having gained insight in natural sciences, he countered Swiss design concepts that were under the influence of the natural garden movement, and he tried to bring form, symbolism and history into harmony with ecological processes. In doing so, he not only influenced an entire generation of students and practicing landscape architects in Switzerland and Germany (Freytag 2020, pp. 11–13), but also laid foundations for the study of landscape architecture at ETH, which persists to the present.[7]

A milestone was reached in 1997 with the establishment of the professorship for Landscape Architecture at the Department of Architecture at ETH, the first university chair for this discipline in Switzerland. After years of engagement, Dieter Kienast was appointed full professor on 1 April 1997 at ORL. The institute included the following disciplines: Landscape Architecture, led by Dieter Kienast; Architecture and Urban Design by Franz Oswald; Landscape and Environmental Planning by Willy A. Schmid; Spatial Planning by Hans Flückiger; and Regional and Environmental Economics by Angelo Rossi (ETH 1999, p. 12f.).

Over time, dealing with nature and the integration of landscape into the urban development process increasingly became the focus of the neighbouring disciplines of Architecture and Urban Design. The professorship for Landscape Architecture sought to educate architecture students in the conscious handling of urban and designed nature as well as in the basics of landscape architectural design in urban and rural areas. In their later professional assignments, graduates were expected to react competently to issues related to the design of open spaces in cooperation with landscape architects and planners (ETH 1999, p. 6f.).

In addition to the diploma elective, a compulsory lecture on the history of landscape architecture was introduced, together with semester theses as consolidation of the urban or architectural design project, seminar weeks, exercises and cooperative diploma supervision as a minor subject (ETH 1999, pp. 13, 26–31). Although an independent course of study had not yet been realised, the university considered the development of the Chair of Landscape Architecture to be the first step towards introducing a university level course in landscape architecture. In 1989, the professional association BSLA had submitted proposals to ETH Zurich, EPFL and the universities of Basel and Bern for the implementation of a university-level degree in landscape architecture. It criticised the generally conservative and defensive attitude towards landscape in Switzerland, which was strongly based on the aspects of protection and ecology. With a university anchoring, the discipline could be aimed towards future-oriented action and the development of visions, and achieve a needed substantial contribution to research in terms of theoretical foundations, cultural history, large-scale visions and development concepts. In 1998, the ETH Board saw the creation of a postgraduate course (NDS, now Master of Advanced Studies MAS) as a short-term solution, and this was introduced in 2003. As part of the restructuring of university education with the Bologna Reform, it proposed the establishment of a master's degree programme in Landscape Architecture as a long-term goal.[8]

The Institute of Landscape Architecture (ILA) as a centre of excellence for teaching and research since 2005

Dieter Kienast was only able to lead the professorship for a short time. He died in December 1998. In October 1999, Christophe Girot (born 1957 in Paris) took over as lecturer and from February 2001 as full professor of Landscape Architecture. Girot, from 1990 to 2000 professor at the École nationale supérieure de paysage de Versailles (ENSP) (see Blanchon et al. in this volume), expanded educational and research programmes at ETH. Influenced by the French School of Landscape Architecture, especially Bernard Lassus, Gilles Clément and Henri Bava, and having earned degrees from the University of California at Berkeley and Davis, Girot provided an international impetus and promoted the exchange with people from English-speaking countries.

Far-reaching changes took place at ETH, not only in terms of philosophy and human resources but also in terms of organisation. In 2002, the disciplines of Urban Design, Landscape Architecture and Spatial Planning were restructured. At the Department of Architecture (D-ARCH), the focus shifted to urban planning and design, in response to rapidly advancing urbanisation and the growth of agglomerations. The ORL Institute was dissolved and the research association Network City and Landscape (NSL) created, consisting of five institutes. The *Institut für Raum- und Landschaftsentwicklung* (Institute for Spatial and Landscape Development – IRL), which emerged from ORL, was affiliated to the Department of Civil, Environmental and Geomatic Engineering (D-BAUG) (Lendi 2002, pp. 225, 227). The field of landscape architecture remained within the D-ARCH and was strengthened in 2005 by the appointment of Günther Vogt (*1957) as professor of Landscape Architecture. Vogt originates from Liechtenstein and owns *Vogt Landschaftsarchitekten Zürich* with branch offices in Berlin, London and Paris.

In 2005, the two professorships for Landscape Architecture, led by Girot and Vogt, became a joint *Institut für Landschaftsarchitektur* (Institute of Landscape Architecture – ILA) within the Department of Architecture. ILA developed into an internationally recognised centre of excellence for learning and research in landscape architecture. Through research projects with international partners it is building scientific foundations in history and theory as well as in contemporary design in order to contribute to an aesthetically sophisticated and sustainable design of the environment, in relation to both teaching and applied research (see Freytag 2020; Kretz and Kueng 2016; Girot et al. 2016). Research is largely supported by doctoral students; many being graduates in architecture and art history from Swiss universities, as well as graduates in landscape architecture and other disciplines from abroad. The two professors initiate public debates on current and future-oriented aspects of landscape architecture via conferences, exhibitions, and publications (Kissling 2021; Girot and Imhof 2016; Girot et al. 2013; Girot and Wolf 2010; Institute for Landscape Architecture ETH Zurich 2005). These activities contributed to the profiling of landscape architecture within the department and to building up a name for the field in the world.

Design education in landscape architecture became established in its own right in the department. In the summer semester of 1990, landscape architecture was introduced as a minor subject to the 10-week diploma thesis at the D-ARCH and has been in high demand among students ever since.[9] The tasks were designed in consultation with the architecture professors in such a way that landscape architectural aspects played a significant design-determining role in the architectural or urban development projects. In the academic year 1998–99, around 150 students chose this (ETH 1999, p. 31). In 2005, the professorships for Landscape Architecture finally succeeded in launching an independent design studio, which, like that of their architecture colleagues, runs for one semester and comprises 16 hours a week.

In the design studios, compulsory lectures and elective courses, the professorship of Girot focuses on educating students on the history and theory as well as on specific methods, techniques and tools for analysis, design, visualisation and modelling in landscape architecture. Video, sound and photography as well as 3D point cloud technology were established early on (see ETH Zurich 2018; Girot 2016, 2013). Professor Vogt's programme understands landscape as the work of mankind, as a cultural construct that is often coded and perceived collectively. Design interventions are to be woven into the landscape in the sense of continuing the millennia-old history. Viewed thusly, the landscape itself becomes the driving force behind its change (Vogt and Kissling 2020a, 2020b, 2021; Vogt 2015). For years, the design work was carried out at the interface of the three disciplines of landscape architecture, architecture and engineering,

ETH Zurich, Switzerland

Figure 26.1 Final presentation: students present their results of the design studio "Designing a Dynamic Alpine Landscape, Bondo/Switzerland" in collaboration with Gramazio Kohler Research.
Source: ETH Zurich, Professorship of Christophe Girot, Fujan Fahmi 2018.

by establishing a collaboration with the professorship for Architecture and Design of Marcel Meili as well as the engineer Jürg Conzett.

From 2003 to 2016, the Master of Advanced Studies in Landscape Architecture under the professorship of Girot offered a two-semester postgraduate diploma programme for domestic and international graduates in landscape architecture, architecture, engineering and related fields. However, the demand for a university degree in landscape architecture continued to grow on the part of students as well as staff in planning offices, authorities, politics and science. The situation was exacerbated after the Bologna Reform, as no independent master's degree in landscape architecture is offered at Swiss universities of applied sciences.

The master's degree programme in landscape architecture at ETH Zurich since 2020

With the discipline gaining importance, the professorships led by Girot and Vogt received the support from ETH Zurich and the D-ARCH needed to launch a Landscape Architecture Master's degree programme. The start in autumn 2020 was preceded by an almost four-year coordination and planning process, which was closely supported as a degree programme initiative by the Department of Educational Development and Technology (LET) in terms of content and university didactics. As the programme manager, the landscape architect Dunja Richter led the interdepartmental process from the thematic development to compliance with the programme objectives. In spring 2019, eight workshops were held with international experts from science and professional practice to discuss learning goals, focal points, didactic concepts and learning formats for the curriculum.

Since January 2020, Teresa Galí-Izard joined the team as a full professor of landscape architecture. Galí-Izard (*1968 in Barcelona) had gained teaching experience in several countries including an associate professorship at University of Virginia from 2012 to 2018 and at Harvard Graduate School of Design from 2018 to 2019. With Teresa Galí-Izard, learning focuses on the interface between landscape architecture and agronomy and on the regenerative interactions between humans, animals and plants.

Institutional involvement and cooperation

The Master of Science ETH in Landscape Architecture is offered by the *Institut für Landschaft und Urbane Studien* (Institute of Landscape and Urban Studies – LUS). Since 2019, LUS, into which ILA was merged, has united the fields of landscape architecture, urban design and territorial planning at the Department of Architecture.

Practised multi-perspectivity, which shows different professionally and culturally influenced perspectives, is a central component in the master's degree. The students get to know, understand and take into account different theoretical positions, design approaches and working methods. This enables them to critically reflect on their own thoughts and actions. The international team of employees is also recruited from several areas such as landscape architecture, architecture, art history, music, geography as well as urban and spatial planning.

The students also benefit from the established cooperation between the various institutions of ETH. Staff of the departments of Architecture; Civil, Environmental and Geomatic Engineering (D-BAUG); Environmental Systems Science (D-USYS); Humanities, Social and Political Sciences (D-GESS) as well as Management, Technology, and Economics (D-MTEC) take part in the programme. External lecturers from the Zurich University of the Arts, the Art History Institute in Florence and the Institut de Recherche et Coordination Acoustique/Musique (IRCAM) in Paris are also involved.

Profile of the degree programme

The Master of Science in Landscape Architecture is a programme based on design and studio culture. The two-year master's degree programme is a non-consecutive course that offers an in-depth qualification in landscape architecture following a bachelor's degree in architecture for students of ETH, EPFL and USI Mendrisio as well as foreign universities. It also opens up a perspective for those graduates who have already completed their master's degree. Admission to the specialised master's programme in landscape architecture is regulated by an application procedure. The course is currently offered to a maximum of 20 students per year.

Its aim is to equip a new generation of designers, planners and decision-makers with the theories, methods and tools necessary for responding to increasingly complex societal and environmental scenarios. Rooted in the tradition of the field at the D-ARCH, it is part of the specific profile of the course that it offers highly experimental and prospective approaches that combine scientific knowledge with cultural awareness, and the living environment with technology. Through careful analytical observation, on-site exploration, design experimentation and technical synthesis, the programme proposes a methodological framework designed to address pivotal questions, both cultural and ethical, in the built environment.

ETH Zurich, Switzerland

The link with architecture studies is a defining feature of the Landscape Architecture course at ETH. Consideration of the relationship between city and landscape, architecture and open space, culture and nature is inherent in the pedagogic method. The programme aims to train graduates for both private and public institutions who are entering new fields of activity and research worldwide and are able to contribute their expertise in complex projects with neighbouring fields such as architecture, urban design, engineering and environmental sciences. An important task area can be seen in the progressing urbanisation, where new strategic approaches are needed to consider landscape as an integral system in the context of urban development in a visionary way. Dealing with the consequences of climate change requires a high level of technical and constructive competence, for example, the destruction of places as a result of natural disasters like landslides in the Alps. In addition to such regional transformation processes, challenges in design, which are important focal points in the education, also include climate adaptation and health, the built environment and historical context, social phenomena such as demographic change as well as landscape regeneration and the integration of new infrastructures into the landscape.

Main focus and structure of the studies

The master's degree programme covers design-planning, technical-constructive, natural science and humanities knowledge and skills. One hundred and twenty ECTS credits are acquired in four semesters and six months of internship. The study is bilingual (German and English) and

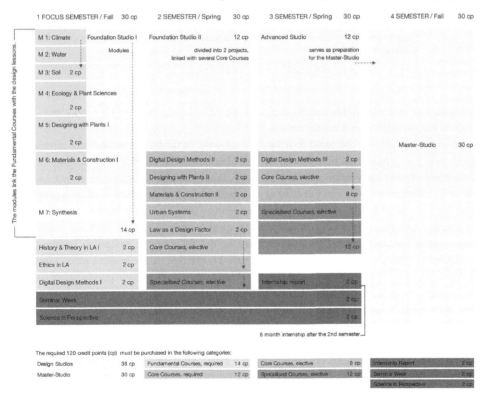

Figure 26.2 The master's degree programme in landscape architecture at ETH comprises four semesters and a vocational internship of six months between the second and third semester.

Source: ETH Zurich, LUS, Dunja Richter 2021.

Figure 26.3 Capture the acoustic dimension of the landscape: workshop "3D Landscape Mapping, Sounds and Point Clouds" in collaboration with Kyoto Institute of Technology, Shosei-en Garden in Kyoto/Japan.

Source: ETH Zürich, Professorship of Christophe Girot, Ludwig Berger 2016.

full time, and includes fundamental, core and specialised courses, seminar weeks and design studios. It concludes with a master studio.

In order to promote the rapid development of skills in landscape architecture among students, new forms of pedagogic methods and learning have been developed which interlink the acquisition of basic knowledge and in-depth application knowledge. A central, didactic principle of the education is the content-related and methodological connection of subject-specific learning content with one another and their integration into the design studios.

The focus semester responds with its compact module structure to the previous professional education of the students and creates the basis for the following semesters. It combines the fundamental courses, which convey a broad basic knowledge of landscape architecture, with design education. The first four modules deliver elementary skills in the natural science subjects climatology, hydrology, soil science as well as botany and ecology, in five weeks. Building on this, the next two modules educate on the basics of designing with plants as well as materials and construction in landscape architecture over a period of four weeks. The theoretical contents of each module are used in the respective design task, which is conceived as a short design exercise. The first semester concludes with a four-week synthesis module, which integrates the knowledge acquired thus far into a design task. Students quickly gain design skills in landscape architecture through this intensive training.

With this structure, the curriculum stands out from the classic bachelor's and master's degree programmes at ETH, which offer courses on a weekly basis throughout the semester. Although this requires a higher organisational effort, both students and lecturers benefit. The

Figure 26.4 Collaborative learning in the design studio: students are engaged in the analysis and study of different design variants in the project "Marseille – Maritime and Alpine Landscape".

Source: ETH Zurich, Professorship of Günther Vogt, Thomas Kissling 2019.

module structure allows greater flexibility in the design of the lessons, enabling the integration of short field trips or excursions lasting several days into the courses, thereby complementing theoretical lectures and studio work, practical observation and on-site experience. Restructuring the regular semester facilitates interdisciplinary collaboration with lecturers from other departments, federal research institutions, public institutions or private landscape architecture offices.

The second and third semesters follow a classic curriculum structure. The core courses build on the fundamental courses and provide broad knowledge in the main areas of landscape architecture in relation to design education. The selectable specialized courses offer students the opportunity to acquire in-depth knowledge in certain areas of landscape architecture, for example in planting design, historic garden preservation, river revitalisation or sustainable construction. In general, however, the education is broadly based. The integration of professionally relevant content is given preference over too narrow specialisation.

In addition to analogue, GIS- and rule-based approaches to planning and design, the use of advanced digital design methods such as 3D landscape visualisation and modelling, digital fabrication, robotics and landscape acoustics provides students with the operative tools to enable, conceive, model and represent a project. Students benefit from internationally recognised institutions such as the Landscape Visualization and Modeling Lab (LVML), founded in 2009 by the professorships of Christophe Girot at D-ARCH and Adrienne Grêt-Regamey at D-BAUG, as well as the AudioVisual Lab.

Studio culture – designing and planning as core competencies in landscape architecture

The experience of the professors from their professional practice and the results of theoretical and applied research flow into the design education. The studio work in the master's degree programme in landscape architecture comprises more than half of the 120 ECTS credits to be achieved. From the first to the third semester, a total of 38 ECTS credits are acquired and a further 30 ECTS credits in the subsequent master studio.

The tasks range from urban design and landscape analysis to the object level. The Foundation Studios I and II focus on the contemporary city, and include aspects of climate adaptation, urban hygiene and health. The Advanced Studio deals with the landscape as a dynamic system, working on complex design tasks in a large-scale landscape context, taking into account social, topological, hydrological and ecological aspects. The master studio, with a duration of fourteen weeks, in which the students demonstrate their independent design ability, is proof of the successful completion of their studies.

The design topics are set in the European landscape context. The idea is that students should first acquire a solid knowledge of landscape architectural design in a familiar cultural setting. This fundamental knowledge enables the later graduates to cope with work and project situations that are complex and unpredictable and require new solution models in a pluralistic and globalised society with rapidly changing framework conditions.[10]

Conclusion

After more than 50 years of commitment and perseverance by numerous individuals from science and practice, ETH Zurich is the first Swiss university that successfully introduced a master's degree programme in landscape architecture.[11] Based on the structures of the Swiss education system, the study provides a specialisation for the Bachelor in Architecture as a first step. In addition to the education offered at technical college level, the new programme contributes to strengthening the discipline as follows. Federal, cantonal and municipal departments are now able to fill important management positions with home grown university graduates. This makes it easier for employees in these offices to internally enforce the requirements of open space because they can argue and take action on a par with their colleagues from urban planning or construction. With a university degree, landscape architecture graduates from Switzerland finally have the right to obtain doctoral degrees with positive impulses for research. They also take positions in research and teaching at universities and universities of applied sciences as well as in private offices that, previously, depended on hiring graduates trained abroad when looking for highly qualified staff.

With the master's degree programme in landscape architecture at ETH, we are placing our hopes in a new generation of highly qualified individuals who actively play important roles in shaping and developing our living environment. These people will contribute to opening up fields of activity and research worldwide and to addressing societal challenges of the future.

Notes

1. In addition to the ETH Zurich, founded in 1855, the École polytechnique fédérale de Lausanne (EPFL) in French-speaking west Switzerland also became a federal institution in 1969.
2. Gottfried Semper made his impact on the ETH Zurich's *Bauschule* from 1855 to 1871. For further details cf. Tschanz 2015.
3. Today, *Bund Schweizer Landschaftsarchitekten und Landschaftsarchitektinnen* (Federation of Swiss Landscape Architects – BSLA).

4 ETH Zurich, gta Archive (NSL), NSL 16-S-2-3-2, letter from Dieter Kienast to Mario Campi/ Department of Architecture, ETH Zurich, 17 April 1990.
5 Dieter Kienast taught as a professor at ITR from 1981 to 1991. From 1992 to 1997, in addition to teaching at ETH, he was also professor at the University of Karlsruhe (Richter and Freytag 2014, p. 217). Further information on the biography and work of Dieter Kienast, cf. Freytag 2020.
6 Cf. endnote 4.
7 Regarding Kienast's landscape architectural approach, cf. Professur für Landschaftsarchitektur ETH Zürich 2002.
8 ETH Zurich, gta Archive (NSL), NSL 16-S-2-3-2, BSLA, Peter Wullschleger, protocol. 'Gespräch über die Möglichkeiten einer Hochschulausbildung für Landschaftsarchitektur in der Schweiz', 5 March 1998, pp. 2–3; BSLA Board Communication: 'Universitäre Landschaftsarchitekturausbildung in der Schweiz', July 1998, pp. 1–2.
9 Cf. endnote 4.
10 For information on the objectives and content of the MSc Landscape Architecture programme, see the website www.mscla.arch.ethz.ch; study guidelines (1 Oct 2019), www.arch.ethz.ch/en/studium/studienangebote/master-landschaftarchitektur.html [as of 20 Feb 2021]; ETH course catalogue, www.vvz.ethz.ch.
11 I would like to thank Christophe Girot, Günther Vogt, Thomas Kissling and Annemarie Bucher for the inspiring discussions on the history of the landscape architecture profession at ETH Zurich.

References

ETH Zurich, Institute for Landscape Architecture, Chair of Christophe Girot (2018) *Melting Landscapes. Sight and Sound Observations of the Morteratsch Glacier*. Zurich: ETH Zurich.
ETH Zurich, Institute for Local, Regional and National Planning (ORL) (1999) *Professorship of Landscape Architecture. [Brochure on pedagogy and research]*. Zurich: ETH Zurich.
Freytag, A. (2020) *The Landscapes of Dieter Kienast*. Zurich: gta Publishers.
Girot, C. (2013) 'Immanent Landscape', *Harvard Design Magazine*, 36, pp. 6–16.
Girot, C. (2016) *The Course of Landscape Architecture. A History of Our Designs on the Natural World, from Prehistory to the Present*. London: Thames and Hudson.
Girot, C., Freytag, A., Kirchengast, A. and Richter, D. (eds.) (2013) *Landscript 3. Topology: Topical Thoughts on the Contemporary Landscape*, series by the Chair of Prof. Girot, Institute of Landscape Architecture ILA ETH Zurich. Berlin: Jovis.
Girot, C., Hurkxkens, I., Melsom, J. and Werner, P. (2016) 'The Impossible View of the Saint Gotthard: Surveying, Modeling and Visualizing Alpine Landscapes', in Burkhalter, M. and Sumi, C. (eds.) *Der Gotthard/Il Gottardo: Landscape – Myths – Technology*. Zurich: Scheidegger & Spiess, pp. 249–287.
Girot, C. and Imhof, D. (eds.) (2016) *Thinking the Contemporary Landscape*. New York: Princeton Architectural Press.
Girot, C. and Wolf, S. (eds.) (2010) *Landscapevideo. Landscape in Movement*. Zurich: gta Publishers.
Institute for Landscape Architecture, ETH Zurich (2005) *Landscape Architecture in Mutation*. Zurich: Netzwerk Stadt und Landschaft.
Kissling, T. (ed.) (2021) *Solid Fluid Biotic. Changing Alpine Landscapes*. Zurich: Lars Müller Publishers.
Kretz, S. and Kueng, L. (eds.) (2016) *Urbane Qualitäten: ein Handbuch am Beispiel der Metropolitanregion Zürich*. Zurich: Edition Hochparterre.
Lendi, M. (2002) 'Zur Neuausrichtung der Forschung in räumlicher Entwicklung an der ETH Zürich', *Geographica Helvetica*, 57(3), pp. 225–228.
Oechslin, W. (2005) 'Das Departement Architektur (D-ARCH)', in Schweizerische Akademie der Technischen Wissenschaften (ed.) *Lehre und Forschung an der ETH Zürich. Eine Festschrift zum 150-Jahr-Jubiläum*. Basel: Birkhäuser, pp. 55–60.
Professur für Landschaftsarchitektur ETH Zürich (ed.) (2002) *Dieter Kienast – Die Poetik des Gartens. Über Chaos und Ordnung in der Landschaftsarchitektur*. Basel, Berlin and Boston: Birkhäuser.
Richter, D. and Freytag, A. (2014) 'Dieter Kienast', in Beyer, A., Savoy, B. and Tegethoff, W. (eds.) *Allgemeines Künstlerlexikon*, 80. Berlin: DeGruyter, p. 217.
Schubert, B. (1980) 'Die Ausbildung des Landschaftsarchitekten', *Anthos. Zeitschrift für Landschaftsarchitektur*, 19(3), pp. 1–7.

Schubert, B. (2016) 'Die HSR – das Zentrum der Landschaftsarchitektur-Ausbildung in der deutschen Schweiz', in Durand, V. H. (ed.) *HSRmade*. Rapperswil: University of Applied Sciences Rapperswil (HSR), Institute for Landscape and Open Space (ILF), pp. 19–22.

Tschanz, M. (2015) *Die Bauschule am Eidgenössischen Polytechnikum in Zürich. Architekturlehre zur Zeit von Gottfried Semper (1855–1871)*. Zurich: gta Publishers.

Vogt, G. (2015) *Landscape as a Cabinet of Curiosities. In Search of a Position*, edited by Bornhauser, R. and Kissling, T. Zurich: Lars Müller Publishers.

Vogt, G. and Kissling, T. (eds.) (2020a) *Mutation and Morphosis. Landscape as Aggregate*. Zurich: Lars Müller Publishers.

Vogt, G. and Kissling, T. (2020b) 'Regio Insubrica – The Stone Houses of Val Malvaglia', in Pedrozzi, M. (ed.) *Perpetuating Architecture. Martino Pedrozzi's Interventions on the Rural Heritage in Valle di Blenio and in Val Malvaglia 1994–2017*. Zurich: Park Books, pp. 5–16.

Vogt, G. and Kissling, T. (2021) 'Nature of the City – Landscape as Aggregate', *Chinese Landscape Architecture*, 37(4), pp. 6–13.

von Schwerin, S. (2014) 'Emil Schneider', *Anthos. Zeitschrift für Landschaftsarchitektur*, 53(3), pp. 70–73.

von Schwerin, S. (2015) 'Christian Stern', *Anthos. Zeitschrift für Landschaftsarchitektur*, 54(1), pp. 70–73.

PART III

Broadening the common ground

27
BROADENING THE COMMON GROUND
Education for the design of human environments

Diedrich Bruns and Stefanie Hennecke

The contributions to Part III exemplify how widening the geographic scope helps expanding thoughts beyond narrow patterns of established thought. With their contributions about China, Japan, the Middle East and also France and Italy, the authors in Part III open up new opportunities for gaining knowledge, enabling a reflection on broadening of what is to be regarded as the "common ground" of the discipline and its educational agendas.

Bernadette Blanchon, Pierre Donadieu, Chiara Santini, and Yves Petit-Berghem offer insights into educational history in France, tracing the path it took from gardening via horticultural engineering to landscape architectural education. The institutions and driving forces they describe identify the field as one of national importance. The "Chair of Garden and Greenhouse Architecture" taking up residence in Versailles in 1876 marks the beginning of a process of considerable disciplinary independence, one that increased towards the opening of the National School of Landscape Architecture at the iconic *Potager du Roi*. From the 1990s on, schools in Angers, Bordeaux, Lille, and Blois also established state-certified professional programmes for landscape architects and landscape engineers. Responding to societal developments in the course of 20th and beginning of the 21st centuries, these five schools are gradually broadening the scope of their curricula, emancipating themselves from but never ignoring horticultural, engineering, and garden design traditions.

Marking the other end of the spectrum of disciplinary independency, Francesca Mazzino and Bianca Maria Rinaldi discuss landscape architectural education in Italy, like France a country identified for her legacy of garden design.[1] Disregarding the growing success and world recognition of landscape architecture, education in Italy continues to lie in the hands of (five) well-established architectural programmes. (In addition, four schools of agriculture and forestry offer programmes dedicated to the design of urban green spaces.) Despite internationally accepted educational agendas and standards, the dominance of the *architetto integrale* continues. It is the tradition that prevails of an architect who, integrating being an "artist, technician and cultured person at the same time", is capable to experiment on all arenas and scales of spatial design (quoting the Italian engineer and architect Gustavo Giovannoni and his 1916 essay *Gli architetti e gli studi di Architettura in Italia*). Employing the metaphor of "fragmented patterns", Mazzino and Rinaldi illustrate how difficult it is for educators to get landscape architecture accepted as an independent area of knowledge and expertise, requiring specific education and training.

DOI: 10.4324/9781003212645-30

Jala Makhzoumi provides background and cases of schools and scholars who advance landscape architecture education in the Arab Middle East.[2] Most study programmes are affiliated with departments of architecture and planning (Baghdad, Bahrain, Lebanon, Palestinian Authority, and Saudi Arabia), one with Horticulture and Crop Sciences (Jordan), and the one in Beirut, the most independent department and the one where the author herself teaches, is called Landscape Design and Ecosystem Management. Reviewing programmes, Makhzoumi offers reflections on education in a region where the discipline is slowly recognised. This region is rich with historical references to gardens famously known as those of the Fertile Crescent. Landscape architectural education advanced after the region lost much of its indigenous landscape and ornamental garden heritage, and when "Western" concepts of "landscape" came to be introduced by colonial rule and the term "landscape" itself associated with colonial westernisation. More in some countries and less in others, the scope of the discipline continues to be predominantly viewed as of designing outdoor spaces and "beautifying" of cities.

Explaining how landscape architecture education developed in China, part of another region known for its wealth and long history of designed landscapes, Lei Gao and Guangsi Lin shed light on interfaces between China and the Western world. The career of 70 years of Professor Sun Xiaoxiang serves as an example to illustrate how educators strive to connect traditional and contemporary knowledge, to raise awareness for regional specificities and global developments with their students, and to define Chinese and Western landscape philosophy and aesthetics while conceptualizing landscape architecture in China. Resulting from these endeavours, and through the support of exchange programmes, Chinese students are interacting actively with landscape architecture education and practice worldwide. Currently, more than 280 universities offer undergraduate programmes, and about two dozen have graduate programmes in landscape architecture.

Chika Takatori provides insights into the history of garden and landscape design education in Japan where three educational systems co-exist side-by-side, with several links between them. The oldest one is the apprenticeship system that more recently merged into an *atelier* system. The second is the in-house system of government and city administrations. The third is the academic system of institutes of higher education. The example of Japan reminds us to think out of the institutional box, as it were. Good education happens not in academic institutions alone but also in design companies, private, government, and municipal offices where young people effectively learn the ropes of the trade from highly qualified professionals.

What are the driving forces to broaden the geographic scope? What motivation can universities have in regions with an independent, tried, and tested planning culture to consider adding a new design field to existing ones? Why, one might wonder, would institutes of higher education want to add landscape architecture to their portfolios if education and studies already exist that have strong regional ties? Italy, France, many countries of Arabia, Africa and Asia, such as China and Japan, are among the regions with the oldest tradition and history of landscape design, each under their own terminology and their own educational legacy. In the search for answers to the question of what advantages are associated with the establishment of landscape architecture courses, we come across some findings.

Several driving forces appear to be similar over the world (Birli 2022[3]). In Italy and France, universities build landscape education on existing strong traditions in garden art, as was the case in other European countries, albeit to a lesser degree. Universities embrace landscape architecture as an opportunity to widen the scope of that art. For students to acquire the ability to master challenges of growing populations, to meet the multitude of their needs, to provide for health benefits in town and country, to support tourism and industry, and to manage the landscape (Blanchon 2022). In Italy and other countries around the Mediterranean Sea, several schools of architecture and, in some cases of agriculture, came to introduce landscape design, first as courses and then as study programme. Both types of schools were intent on architecture claiming landscape architecture and adding to their own disciplinary realm. (In Italy, the discipline remains closely attached to architecture.) Universities and professional organizations collaborate in introducing landscape architectural education because they are aware of the discipline through the emergence of modern architecture and modern planning. In several cases, early beginnings of landscape architectural education can be traced to the introduction of Western-style city planning (by colonial authorities post-World War I and World War II) associated with the development of modern architecture, for example in the Middle East during the 1960s and 70s. This is the period when the first landscape design and architecture schools appear in Turkey and Iran. The intent was to emulate Western schools and to develop curricula to be in line with Western examples. During that time, educators and programme managers generally failed to look to the regional history and garden culture, be they ancient or vernacular (Makhzoumi 2022).

In Asian countries that have a long tradition in the artful shaping of human environments, such as Japan and China, designing gardens and landscapes is, for hundreds of years, carried out collectively by scholars who combine art and literature, and not by one single and independent discipline. The introduction of landscape as a new concept was part of "modernisation", part of looking to the "West" as the source of new things. For universities to embrace landscape architecture offers an opportunity to establish new study programmes and to include a modern discipline. In China, for example, that Ernest Henry Wilson labels in 1929 the *Mother of Gardens*, the introduction and development of landscape architecture is closely linked to the process that started in the late 19th century. It is the time when government targeted education for modernization, implemented through several waves of learning from "advanced" and "modern" countries, since the 1980s set in the context of the Open-Door policy. Especially after the 1990s, when China increasingly interacted with Western countries in terms of higher education and markets, landscape architecture appeared to be a new and demanding discipline that received more and more attention (Gao and Lin 2022).

Reading between the lines, as it were, helps reveal some of the values that governments and schools attach to landscape architecture, including the claim of superiority to Western design, a one-sided value and view that seem at odds with the contemporary emphasis on tolerance for and policy based on cultural equality and diversity. While "landscape" in mainstream landscape architecture of the 19th century remains dominated by "high culture" artistic appreciation, increasingly also by "advanced" scholarly attention, Western landscape concepts fail to include understandings of human environments outside of elitist worldviews. The chapters in Part III help broadening the common ground and help challenging dominant, latent, and arrogant ideologies.

Notes

1 French and Italian gardens are fixtures in the canon of references that students have been learning about in design history. For example, Geoffrey Jellicoe (1900–1996), the first president of the International Federation of Landscape Architects (IFLA), used a book on Italian Renaissance Gardens, which he and J.C. Shepherd co-authored and published in 1925.
2 There is no consensus as to what geographic area the term, often viewed as to be discriminatory and Eurocentric, constitutes. It commonly refers to the region spanning the Levant, Arabian Peninsula, Anatolia (including modern Turkey and Cyprus), Egypt, Iran, and Iraq. Following Jala Makhzoumi's suggestion, we use Middle East in this volume, to embrace a region with Iran to the east, lands of the eastern Mediterranean to the west and north, and the Arabian Peninsula in the south.
3 We thank chapter authors for providing information by personal correspondence. Authors are Barbara Birli, Bernadette Blanchon (11.02.2022), Jala Makhzoumi (14.02.2022), Lei Goa, and Guangsi Lin (10./11.02.2022)

28
LANDSCAPE ARCHITECTURE EDUCATION IN ITALY
Fragmented patterns

Francesca Mazzino and Bianca Maria Rinaldi[1]

Introduction

The year 2020 was a spirited time for landscape architecture education and profession in Italy. Two new graduate programmes were established, one in landscape studies at the University of Padua, the other in landscape architecture at the University of Palermo, both bringing evidence to the ever-growing interest and need for landscape-related education. At the same time, the National Chamber of Architects[2] put forward a controversial proposal to cancel the professions of planners, landscape architects, and conservators it currently registers separately. Somehow recalling ideas by the Italian engineer and architect Gustavo Giovannoni (1873–1947) of the *architetto integrale* (integral architect),[3] and Hans Hollein's 1966 famous manifesto *Alles ist Architektur* (Everything Is Architecture), the proposal seemed to advocate a return to old traditions: It suggested reuniting all professions related to the design of the built environment under the umbrella of architecture. With its proposal, the Chamber seemed to disregard the globally established role of landscape architects, the specificity of landscape architecture as a profession both in Italy and abroad, and the history of landscape architecture education in Italy and its growing success. These recent events, however, are emblematic of the difficulties landscape architecture has been encountering in Italy to establish itself and to be accepted both as a specific field of study and expertise, as well as an independent profession requiring specific competences and training. Taking the close relationship between the education and the profession of landscape architecture as its starting point, the chapter offers an overview of the history of landscape architecture education in Italy.

Approach and methods

Aiming to reveal specifics of the Italian system, the chapter analyzes discourses underlying the history of landscape architecture and education in the Italian cultural and social context and in relation to landscape architecture education in Europe and beyond. By discussing the key role of a number of educators, scholars, and practitioners, the chapter underlines their influence on educational approaches. By pointing out the role of the Chamber of Architects in leaving the differences in expertise between architects and landscape architects not clearly defined, the chapter argues how the ambiguity in acknowledging the specificity of the profession of landscape architecture led to a relatively recent tradition and still fragmented pattern of landscape architecture education.

Research methods include qualitative and quantitative data collection and analysis. Sources consist of 1) primary sources, including official orders, such as presidential decrees and decrees issued by the Ministry of Education; school curricula; official documents; lists of past and current programmes; textbooks; articles by scholars and practitioners; 2) secondary sources, such as publications issued by different schools; books and articles by scholars and educators. Qualitative textual analysis was performed, including content analysis and thematic analysis, to identify the key themes that accompany the evolution of landscape architecture education. An analysis of publications in various fields related to landscape studies and developed in 20th-century Italy provides an understanding of the scholarly context that formed the foundation for the Italian approach to landscape and landscape-related study methods. Interviews with scholars and experts in the field provide additional information.

Landscape architecture as the poor cousin

The proposal prepared in 2020 by the National Chamber of Architects was an unexpected result of the late and non-linear development the profession of landscape architecture had in Italy compared to other countries in Europe and abroad. It is rooted into the still blurred separation of technical competence of landscape architecture from architecture and into the perceived overlapping of expertise between landscape architects and other professional figures, such as agronomists and urban planners operating in the field of landscape planning.[4]

Traditionally, landscape architecture was perceived as subordinate to architecture and considered as one of the many diverse areas of expertise in that field. Only in 2001, as a result of the Italian Presidential Decree no. 328/2001, it was recognized as an independent profession. The Decree defined the generic competencies for landscape architects in park and garden design, landscape planning and natural parks planning, and historic park and garden restoration (D.P.R. 2001). It also introduced changes in the name and structure of the Chamber of Architects, giving it the inclusive denomination of *Ordine degli Architetti, Pianificatori, Paesaggisti, Conservatori* (Chamber of Architects, Planners, Landscape Architects and Conservators). The new Chamber was organized into two sections, one registering professionals with a bachelor degree in architecture and planning, the other registering professionals with a master's degree and was subdivided into four specific lists: architects, landscape architects, planners, and conservators.[5] Despite these efforts, the different levels of knowledge and expertise between architects, landscape architects, and landscape planners remained largely unspecified, thus opening the possibility for graduates in architecture to operate, without any specific qualification, in the professional field of landscape architecture (Mazzino 2002, 2019, p. 119; Corbari 2018, pp. 29–30). This incongruity resulted in the possibility for graduates in landscape architecture to register with a specific list at the *Ordine dei Dottori Agronomi e Forestali* (Chamber of Agronomists and Foresters), thus continuing a long-lasting approach that, instead of promoting professional specificity and collaboration among professions, increased confusion over fields of expertise. The accepted blurred boundaries between professions influenced the evolution of landscape architecture education over time.

History of landscape architecture education: the early stages

The close relationship between architecture and garden art has a long tradition in Europe. Eighteenth-century treatises of architecture, such as the *Essai sur l'Architecture* by Marc-Antoine Laugier, published in 1753, and *A Complete Body of Architecture* by Isaac Ware, published in 1756, include chapters focusing on gardens considering them an essential yet decorative feature of architecture, a complement to buildings.[6] While a new professional figure and specific field of study in garden and landscape architecture flourished in many European countries, in Italy the architectural tradition continued throughout the first decades of the 20th century, boosted by its noble origins. The role of architects in designing gardens in Italy is rooted in the idea of the "Italian Renaissance garden", a label that came to identify a garden typology whose spatial strategy was based on an intertwining of garden and buildings. As gardens and architecture were considered inseparable parts of a cohesive whole, architects were generally recognized as the professional figures most capable of shaping garden spaces. In addition to that, Italy had no tradition of gardeners who designed gardens or of estate owners capable themselves of designing their own gardens as it happened, for instance, in France, in Germany, or in England.

It is thus no surprise that when, in 1937, Pietro Porcinai (1910–1986) asserted the necessity of a new professional figure possessing the specific competencies he considered necessary for designing gardens, he coined the title *Architetti giardinieri*, "Architect-Gardeners" (Porcinai 1937a, p. 31). It was a new expression for a new professional figure he hoped to introduce. Nonetheless, its hybridity was deeply rooted in the architecture tradition.

Porcinai was himself a prolific designer of gardens and landscapes. He studied in an agricultural college (Latini 2016, p. 6) and had begun attending a school of architecture but did not complete the study programme because of the load of commissions he had and also as he considered the approaches and methods proposed not in line with his training needs. Nonetheless, he was often addressed as "architect" (Latini 2016, p. 5), a designation he disliked[7] but that bore evidence of the common opinion well established in the Italian consciousness that those who designed gardens had necessarily to be architects. It was only at the end of the 1940s that Porcinai begun to be addressed as an "Architect-Gardener" (Latini 2016, p. 9), the title he himself had coined.

Porcinai was one of the most fervent proponents of a system of education in landscape architecture in Italy. Giving tangible form to the systematic instruction he wished for, was a turbulent journey that extended for over fifty years. It started in the early 1920s, with a fragmented pattern of isolated courses offered within university curricula in architecture, thus reflecting the traditional, apparent close relationship between architecture and garden design.[8] Starting in 1921, renowned Italian architect, urban designer, and scholar Marcello Piacentini (1881–1960) introduced the course in *Edilizia cittadina e arte dei giardini* (City Building and Garden Art) at the newly founded *Regia Scuola Superiore di Architettura* (Royal School of Architecture) in Rome.[9] Taught together with architect Luigi Piccinato (1899–1983) from 1924 to 1930, the course was designed as a blending of architecture, urban design, and garden art, which Piacentini considered as a necessary body of knowledge to the formation of an architect capable of sensible urban transformations. In 1931 the course was renamed *Urbanistica* (Urban Planning) (Venturi 1924; Sangermano 2015, pp. 10–11), thus apparently diminishing the role of garden art as a participating discipline in the design of the city. A course with the same title, *Edilizia cittadina e arte dei giardini*, part of the curriculum of the newly founded *Scuola Superiore di Architettura* (School of Architecture) in Naples,[10] suffered the same fate: introduced in 1930 with Piccinato as instructor, in 1932 was transformed into a course in *Urbanistica* (Sangermano 2015, p. 18).

The simplistic and generic translation of landscape architecture into garden art characterized architectural curricula until the last decades of the 20th century. Courses in *Arte dei giardini* were offered as electives in several schools of architecture around the country. They mainly covered the history of Western garden art. However, in a few rare cases, they explicitly related garden history to design practice. A notable example was the course in Garden Art Francesco Fariello (1910–1973) outlined and offered by at the University of Rome from 1954 as a design history course for design students. It not only covered the traditional garden history, from Roman gardens to modern landscape architecture, but, showing a broader understanding of landscape architecture embedded in the European and North American context, Fariello introduced topics specifically related to design, which spanned from the spatial arrangement of urban public open spaces to the landscaping of infrastructures. Aware that teaching history to design students required investigating composition, spatial strategies and design elements, in 1956, Fariello expanded his teaching syllabus and material into a volume published under the titled *Arte dei giardini* (Fariello 1956).[11] Richly illustrated with photos, plans, and drawings which allowed a compositional reading of the gardens discussed, the book enjoyed editorial success and underwent several editions, as it was the first comprehensive Italian study on the history of gardens. While it mainly focused on Western garden art, it included a section on 19th-century North American urban open spaces and an entire chapter on gardens in Japan, as a reflection of the curiosity for Japanese culture favoured in Italy by the connections among the partners of the Axis alliance that preceded World War II. However, students considered the topics explored within the Garden Art course by Fariello bourgeois, particularly so in the years around 1968, and only a small number of them enrolled in the course.[12]

The necessity of landscape education

In the 1930s, Porcinai started questioning the generally accepted authority of architects as designers of gardens. According to him, architects did not possess the specific skills necessary to design green open spaces as, he wrote, they "act like a painter who wishes to paint without knowing the value of colours, or like an engineer unconscious of the materials to be used in constructions" (Porcinai 1937a, p. 31, translation here and in following citations by authors). In 1937, Porcinai argued for a new highly skilled professional figure, the "Architect-Gardeners", trained to be comparable to professionals operating in Germany, England, and the United States. Their knowledge and expertise had to extend across disciplines, bridging the spheres of design and natural sciences: The "Architect-Gardeners" Porcinai envisioned had to be "specialized for this purpose [of designing gardens] . . . [they] will have to master Architecture and Art History, but above all Nature and all the resources it is capable to give and donate" (Porcinai 1937a, p. 31).

Porcinai was a leading figure in the establishment of the landscape architecture profession in Europe as he was one of the founding members of the International Federation of Landscape Architects (IFLA) in 1948. His contacts with landscape architects outside Italy enabled him to develop a deep understanding of the state of the landscape architecture profession and education in Europe and the United States and to perceive the need to develop education systems at home.

Porcinai insisted on the necessity of specific schools preparing students for specific professional competences (Corbari 2018, p. 24) and proposed the German education system as a model. Germany was Porcinai's country of reference for garden culture. He developed a deep appreciation for German garden culture first during his professional learning periods in that country; later, his contacts with key figures such as Karl Foerster, Alwin Seifert, and Gustav Lüttge played an important role in his understanding of the German attitude towards landscape and garden design (Latini 2016, pp. 7–8; Revedin 1998, pp. 43–54). In an essay written in 1937, he argued:

> It is necessary that those who want to design or have gardens designed, consider the art of the garden not in a simplistic way, as it happens, but with the intention of having to engage art and technique. [. . .] If abroad, in Germany for example, gardens are the exclusive work of specialists who have been elevated and trained in special schools,

why shouldn't this not be the case with us [in Italy]? But the educational programmes of our Schools of Architecture neglect the teaching of gardening (However, it is clear from the size of these programmes that it is impossible to include such course). So for the rebirth of this art [of the gardens] in Italy the establishment of a special school for garden architects would be opportune and desirable.

(Porcinai 1937b, p. 38)

Perception of the backwardness of Italy compared to northern Europe both in the design of green open spaces and in establishing specific academic training programmes heightened in the 1950s and 1960s when it emerged in the architecture debate, both at national and international level. The discussion about the necessity of acquiring specific competences related to criticism for the urban planning and design strategies adopted during the Reconstruction and their indifference to open space design. It was also paired with concerns for a glaring lack of adequate policies and measures for the introduction of green open spaces within cities and for the preservation of natural landscapes. In 1952, on the pages of a special issue of the journal *L'architecture d'aujourd'hui* focusing on "Italy", Porcinai bitterly noted that while Italian urban designers and architects claimed that "cities and houses should be surrounded by greenery", "nothing was envisioned, nothing was realised" to construct more pleasant urban environments interspersed with parks and gardens as "public authorities do not provide the support and the guidelines to make it possible" (Porcinai 1952, p. 76). He denounced the inferior design and spatial quality of the urban green open spaces built in Italy in comparison with those designed in "Sweden, Denmark, the German-speaking Switzerland, Germany and, in general in all the Anglo-Saxon countries" (Porcinai 1952, p. 76). He considered it the result of constant interference in design practices of different professional figures and authorities lacking specific skills.[13] The inadequacy of urban green open spaces prompted architect Francesco Tentori (1931–2009) to provocatively define Italian cities as "metaphysical", thus evoking the desolated urban spaces depicted by Giorgio De Chirico and their absence of designed nature. In the 1964 monographic issue of *Casabella-continuità* on the "Necessity of green in Italy", Tentori attributed that particular quality of Italian cities to the "total lack of ideas about green [open spaces] that has characterized, in this century, their great expansion"[14] (Tentori 1964, p. 2). The absence of ideas about green open spaces was directly related to the absence of specific professional figures that could design them. Pondering the situation, in the same issue of *Casabella-continuità*, architect Pier Fausto Bagatti Valsecchi (b. 1929) lamented the different approaches to the design of new urban districts in Italy compared to

various European nations where the landscape architect is permanently placed side by side with the coordinating urban planner: this presupposes, and this is another shortcoming of our [Italian] situation, the existence of a professional qualification at the level of a specialized university program.

(Bagatti Valsecchi 1964, p. 3)

Possibly stimulated by that lively debate, in 1968, Porcinai developed a concept for an International Institute of Study in Garden and Landscape Architecture, an *Istituto internazionale di studi universitari per l'architettura del giardino e del paesaggio* at the university level, to be founded as the first educational institution designed for instruction of landscape architects (Corbari 2018, pp. 24–25). The catalogue of the institute curriculum included the following courses: Conservation of Natural Heritage; Landscape Ecology; History of Garden, Landscape, and Agriculture; Public Green Spaces in Urban Design, Green Open Space Design (Gardens, Parks, Cemeteries,

Playgrounds); Design of Green Areas within the Landscape (woods, highways, etc.); Landscape Planning; Landscape Economy; and Law (Porcinai 1968, 1986; Grossini and Giuntoli 2017, pp. 13–14). To reinforce his argument for the necessity of the professional figure of the landscape architect, Porcinai proposed landscape architecture as a form of moral elevation. In his (unpublished) document presenting the Institute, titled *Per l'insegnamento del "verde", del paesaggio e del giardino in Italia* (For the teaching of "green", landscape and garden in Italy), he suggested an intimate relationship between the quality of a landscape and the moral values of the society inhabiting it, thus manifesting the influential role German landscape culture played in his theories and in the education system he envisioned. He also affirmed the correlation between landscape and local identity and stated the crucial role of landscape architecture in preventing the destruction of natural and cultural landscapes (Porcinai 1968, 1986). Landscape architects were thus proposed as the rescuers of the Italian unique character, associated to its varied landscape.

Porcinai did not succeed in giving tangible form to his vision of a first Italian school for the education of landscape architects. Nonetheless, the interest for landscape and landscape studies was growing in the country. While Porcinai was developing his proposal for a landscape architecture academy, scholars from other disciplines, such as geography, ecology, and philosophy devoted themselves to landscape studies (Mazzino 2019, p. 119). Surpassing a picturesque approach to landscape, together with Porcinai they contributed to the understanding of landscape in Italy. The different perspectives and analytical tools they proposed shaped the cultural context that led to the development of the first landscape architecture study programmes and the holistic approach to landscape they envisioned.

Rosario Assunto (1915–1994), a Professor of Theoretical Philosophy at the University of Rome, focused on landscape aesthetic and on the role of experience in the perception of landscape and its subjective dimension (Assunto 1973). Influenced by the German romantic aesthetics, its conception of nature, and the relationship of man and nature it proposed, Assunto advocated sensible landscape transformations thus anticipating modern environmentalism and the ecological approach to landscape.

Eugenio Turri (1927–2005), a Professor of Landscape Geography at Politecnico di Milano, emphasized the anthropic dimension of landscape. In his celebrated studies on the transformation of the Italian landscape in the aftermath of World War II and the economic boom, Turri proposed a critical reading of landscape guided by semiotics, aimed at decoding the territory as a complex stratification of natural and artificial signs and symbols (Turri 1974, 1979).

Emilio Sereni (1907–1977), agronomist, proposed a concept of landscape as a result of social, economic, and ecological dynamics rooted in cultural Marxism. In his thorough investigations on the evolution of Italian agricultural landscape, he developed the idea of the historical "agricultural landscape", discussed agriculture as a form of adaptation to the natural landscape, and asserted the cultural value of rural landscape (Sereni 1961; Sereni and Litchfield 2014).

Botanist and ecologist Valerio Giacomini (1914–1981), who taught botany as a professor at the Sapienza University in Rome, conceived landscape as a "coherent constellation of ecosystems" (Giacomini 1971, p. 218), thus promoting a systemic approach to the study of landscape.

The studies by Sereni and Giacomini stimulated a growing interest for landscape and landscape studies within Schools of Agriculture. While Schools of Architecture continued to offer courses in Garden and Landscape History, and Landscape Design as elective courses, landscape architecture-related subjects were introduced within Schools of Agriculture. In 1968, agronomist Alessandro Chiusoli (1934–2022) established the course in *Floricultura e giardinaggio* (Floriculture and Gardening) at the University of Bologna and in 1983 he introduced the course in *Paesaggistica, parchi e giardini* (Landscaping, Parks and Gardens) (Chiusoli 1985). Courses with the same title were introduced at the Schools of Agriculture at the Universities of Ancona and

Bari. As the textbook by Chiusoli shows, the course in *Paesaggistica, parchi e giardini* discussed a variety of miscellaneous topics including landscape analysis, the idea of landscape, the history of garden art, design materials and techniques, and focused mainly on the use of ornamental plants (Chiusoli 1985).

The first attempts to give a unified form to this fragmented pattern of landscape studies were made in the early 1980s, with the establishment of the first intensive, structured courses of study in landscape architecture based on European and North American models. They were addressed to students with a graduate degree (awarded after the completion of a five-year study programme). In 1980, the School of Architecture at the University of Genoa launched a two-year (later extended to three-year) programme in Landscape Architecture for graduates, called *Scuola di Specializzazione in Architettura del Paesaggio*, later renamed as *Scuola di Specializzazione in Architettura dei Giardini, Progettazione e Assetto del Paesaggio*144[15] (Figure 28.1). It was the first programme in Italy based on an innovative multidisciplinary approach, following the model of landscape architecture schools in the United States (Corbari 2018, p. 26). The subjects taught spanned from landscape architecture history to landscape analysis, from landscape ecology to landscape planning, to historical parks and gardens conservation. The immediate success the course had prompted other universities in Italy to open similar programmes, which followed the same didactic approach developed at the University of Genoa. In 1996, the School of Agriculture and Forestry at the University of Turin established a two-year postgraduate programme *Parchi e giardini* (Parks and Gardens); the certificates of *Agronomo paesaggista* (Landscape Agronomist) or *Forestale paesaggista* (Forest Landscapist) students completing the programme were awarded bore witness to the emergence of a new hybrid professional figure. Finally, in the 1990s, the School of Agriculture at the University of Milan established a postgraduate programme in *Progettazione del verde negli spazi urbani* (Green Design in Urban Spaces); sponsored by the Lombardy Region it remained in operation for two years only and did not award any professional title.

Landscape architecture education: current patterns

Landscape architecture education in Italy gained momentum in the early 2000s, because of the implementation the principles of the Bologna Process. Following three decrees issued in 1999 and 2000 by the Ministry of Education (D.M. 1999, 2000a, 2000b) directing, among others, the institution of graduate programmes, several study programmes in landscape architecture flourished (Mazzino 2001, 2017; Corbari 2018, p. 27).[16] Within a few years, two-year graduate programmes in landscape architecture were established at several universities where students could now earn a master's degree in landscape architecture (Table 28.1). In addition, a variety of postgraduate programmes were launched, including one at the School of Architecture of the University of Naples "Federico II", and the other one established by the School of Architecture of the University of Florence and based in Pistoia.[17]

While the Ministerial Decree No. 509/1999 ruled against the introduction of undergraduate degrees in landscape architecture, landscape-architecture related curricula were introduced within undergraduate programmes at Schools of Architecture and of Agricultural Science (D.M. 1999) (see Table 28.1).

This dynamic period for landscape architecture education was short-lived. In 2004, the Ministerial Decree no. 270/2004 issued by the Ministry of Education determined the reduction of the number of study programmes and aimed to redesign existing curricula in order to give them more coherence within their main branches of study (D.M. 2004). The restrictions the

Landscape architecture education in Italy

Università degli Studi di Genova

3
ARCHITETTURA DEL PAESAGGIO
La scuola di specializzazione di Genova

Quaderni di Architettura

Figure 28.1 Cover of the 1984 issue of the book series *Quaderni di Architettura*, edited by the School of Architecture of the University of Genoa and focusing on the two-year programme in Landscape Architecture for graduates, called *Scuola di Specializzazione in Architettura del Paesaggio*, launched in 1980.

Source: De Fiore, G. et al., *Architettura del paesaggio. La scuola di specializzazione di Genova*, Quaderni di Architettura, 3, Genoa: Università degli Studi di Genova, 1984.

Table 28.1 First graduate programmes in Landscape Architecture and undergraduate programmes with Landscape-Architecture related curricula established in the early 2000s in Italy following the three decrees issued by the Ministry of Education (D.M. 1999, 2000a, 2000b).

Name of graduate/ undergraduate programme	University	Schools involved	Degree earned
Landscape Architecture (*Architettura del paesaggio*)	University of Genoa	School of Architecture	Master's degree in Landscape Architecture
Landscape Architecture (*Architettura del paesaggio*)	Sapienza University, Rome	School of Architecture 'Ludovico Quaroni'	Master's degree in Landscape Architecture
Landscape Architecture (*Architettura del paesaggio*)	University of Florence	School of Architecture + School of Agriculture and Forestry	Master's degree in Landscape Architecture
Gardens, Parks and Landscape Design (*Progettazione di giardini, parchi e paesaggio*)	Politecnico di Torino/University of Turin	II School of Architecture at Politecnico di Torino + School of Agriculture and Forestry at the University of Turin	Master's degree in Landscape Architecture
Techniques for Landscape Architecture (*Tecniche per l'architettura del paesaggio*)	University of Genoa	School of Architecture	Bachelor's degree in Architecture
Garden Architecture and Landscaping (*Architettura dei giardini e paesaggistica*)	University 'Mediterranea', Reggio Calabria	School of Architecture	Bachelor's degree in Architecture
Garden Architecture and Landscaping (*Architettura dei giardini e paesaggistica*)	Sapienza University, Rome	School of Architecture 'Ludovico Quaroni'	Bachelor's degree in Architecture
Territorial reorganization and landscape conservation (*Riassetto del territorio e tutela del paesaggio*)	University of Padua	School of Agriculture and Forestry	Bachelor's degree in territorial, urban, landscape and environmental planning
Ornamental Green Spaces and Protection of the Landscape (*Verde ornamentale e tutela del paesaggio*)	University of Bologna	School of Agriculture and Forestry	Bachelor's degree in Agricultural science and technology
Urban Green and Landscape Management (*Gestione del verde urbano e del paesaggio*)	University of Pisa	School of Agriculture and Forestry	Bachelor's degree in Agricultural science and technology
Planning and Management of Green Areas, Parks and Gardens (*Progettazione gestione di aree verdi, parchi e giardini*)	University of Catania	School of Agriculture and Forestry	Bachelor's degree in territorial, urban, landscape and environmental planning

Source: Mazzino and Rinaldi.

decree introduced resulted in the removal of landscape architecture courses within the curricula of undergraduate programmes in architecture and in the progressive weakening of the structure of landscape architecture education in general (Mazzino 2017). The University of Genoa lost the recognition by the International Federation of Landscape Architects it had previously gained for the courses in landscape architecture it offered. Some of the previously established study programmes in landscape architecture were stopped, while the Universities of Genoa, Milan, Turin and Politecnico di Torino joined forces to establish a new, interuniversity graduate programme in landscape architecture.

Currently, only five graduate programmes in landscape architecture exist. They are at the University of Florence, at the Sapienza University in Rome, at the University of Palermo, at Politecnico di Milano (offered in English as Landscape Architecture Land Landscape Heritage), while the Interuniversity Master Programme in *Progettazione delle Aree Verdi e del Paesaggio* (Green Open Space Design and Landscape Architecture) is a cooperation between the University of Genoa, University of Milan, University of Turin and Politecnico di Torino. All programmes share similar modes of instruction and learning, combining design studios and courses spanning the depth of landscape architecture and transcending the discipline's conventional boundaries. Among them, the Interuniversity Master Programme in *Progettazione delle Aree Verdi e del Paesaggio* has a peculiar format. The curriculum comprises a wide variety of elective courses, allowing students to design and define their own educational programmes according to their backgrounds, inclinations, and ambitions. A range of extracurricular activities, including lecture series, workshops, field trips, and different events, complement the educational model (Figure 28.2).

Figure 28.2 Second-year studio final presentations for students of the Interuniversity Master's Programme in *Progettazione delle Aree Verdi e del Paesaggio* (Green Open Space Design and Landscape Architecture) at Politecnico di Torino, AY 2019–2020.

Source: Photo by Bianca Maria Rinaldi.

An interuniversity undergraduate degree in *Pianificazione e Progettazione del Paesaggio e dell'Ambiente* (Landscape and Environmental Planning and Design) exists as a collaboration between the School of Agriculture at the University of Tuscia in Viterbo and the School of Architecture at the Sapienza University in Rome. However, students earn a bachelor's degree in landscape planning.

Conclusions

At present, a complete cycle of higher education in landscape architecture does not exist in Italy. Students can only enrol in graduate programmes awarding a master's degree in landscape architecture. In order to improve the structure of landscape architecture education in Italy, and to ensure comparability in the standards and quality of education between the different European countries, undergraduate programmes awarding bachelor degrees in landscape architecture must be introduced.[18] While the number of academics teaching landscape architecture is still low but slowly growing, the success of the graduate programmes in terms of the number of students enrolled, the growing field of the profession, with new young offices opening around the country, are reassuring signals for a promising future of both landscape architecture education and the profession in Italy.

Notes

1 The two authors contributed equally to the chapter and are listed alphabetically.
2 Its full Italian name is *Ordine degli Architetti, Pianificatori, Paesaggisti, Conservatori* (Chamber of Architects, Planners, Landscape Architects and Conservators).
3 "The 'integral architect' (*architetto integrale*) [was] a professional figure whose knowledge spanned artistic and technical knowledge and who worked for the responsible transformation of existing cities and territories". Cianfarani (2020).
4 As Karl Kullman observes, the overlapping of spatial design disciplines is a generalized phenomenon that occurred over the past two decades and that "has been interpreted as emblematic of landscape-based trans-disciplinary practice". Kullman (2016), p. 30.
5 The decree introduced new names for professional orders and defined the admission requirements for the state examinations as well as the contents of the state examinations. The two sections that organized the *Chamber of Architects, Planners, Landscape Architects and Conservators* from 2001 were: Section A, registering professionals with a master degree, subdivided into Sector A architects, Sector B planners, Sector C landscape architects, Sector D conservators; Section B, registering professionals with a bachelor degree, subdivided into Sector A architect 'junior' and Sector B planner 'junior'. As there was no undergraduate programme in landscape architecture in Italy, Section B of the new Chamber did not include landscape architects 'junior'. On the Italian Presidential Decree no. 328/2001 and the reform in academic teaching, see Mazzino (2005).
6 Chapter six of Laugier's *Essai sur l'Architecture* is entitled *De l'embellissement des Jardins*, see Laugier (1753), pp. 272–293. Book seven of Ware's treatise, entitled *Of Exterior Decorations*, included specific sections on gardens. See Ware (1756), pp. 636–656.
7 Franco Panzini, personal communication with Bianca Maria Rinaldi, 3 October 2021. For a comprehensive study on Porcinai, see Matteini (1991).
8 The section of the chapter that follows is loosely based on Mazzino (2019), but it considerably expands the topic with new research and results.
9 The *Regia Scuola Superiore di Architettura* was founded in 1919 as the first School of Architecture in Italy (Franchetti Pardo, 2001).
10 The *Scuola Superiore di Architettura* in Naples was founded in 1930.
11 The book is widely known with the title of its second edition, *Architettura dei giardini* (Rome: Edizioni dell'Ateneo, 1967). See Fariello (1967).
12 Franco Panzini, personal communication with Bianca Maria Rinaldi, 10 November 2020.

13 The close relationship between green open spaces and urban design strategies in Northern Europe was well known in Italy at the time, as projects for new urban districts were widely published in Italian journals of architecture and urban planning. It was the publication of those projects that contributed to raise awareness of the necessity for specific professional figures capable of designing urban landscapes and fostered the debate on the lack of landscape architecture education in Italy. In 1949, in the journal *Urbanistica*, urbanist Giovanni Astengo (1915–1990) published a long article on the General Extension Plan for Amsterdam (1928–1935) praising it as a model of reference for the generous presence of public green open spaces it envisioned for new residential areas; he also applauded the construction of a new large urban park, the Amsterdamse Bos. See Astengo (1949), pp. 27–42. The Amsterdamse Bos illustrated the cover of the journal of *Casabella* (issue 277, 1963) introducing an article by Antonio Cederna (1921–1996) who extensively commented on the role and variety of green open spaces in Amsterdam and opposed the quantity of green areas in the Dutch city to their scarcity in Rome. See Cederna (1963), pp. 34–49. In the early 1960s, the same journal published articles as well as monographic issues focusing on urban design strategies in Berlin, Vienna, Paris, and on regional planning in England and Denmark showing the wide presence of green open spaces, while *Urbanistica* gave space to recent landscape architectural projects in Sweden and Denmark, including green open spaces in residential areas, public parks, and playgrounds, see Aymonino (1960a), Aymonino (1960b), pp. 26–37, and Cederna (1965), pp. 69–88.

14 According to both Porcinai and Tentori, the only admirable project built during the Reconstruction was the system of expansive public green open spaces designed as part of the QT8 district, *Quartiere Triennale 8*, in Milan. Conceived during the first post-war Triennale held in 1947, the new urban development featured an artificial wooded hill called Monte Stella constructed with debris of the buildings bombed during World War II, which Tentori particularly praised. See Porcinai (1952), p. 76 and Tentori (1964), p. 2.

15 A fixed number of students could enroll to such programmes. The *Scuola di Specializzazione* in Genua, which remained in operation until 2000, accepted 25 students every year, as written in the school's statute book (Università degli Studi di Genova 1989). On the *Scuola di Specializzazione* in Genua, see De Fiore et al. (1984), Maniglio Calcagno (1989), Grassi and Meucci (1990), Mazzino (1993, 1997).

16 On the lively cultural debate on landscape education and, more generally, on landscape studies, in Italy in this period see Corbari (2018), pp. 27–30.

17 In this essay, we refer only to undergraduate and graduate programmes that provide state-recognized degrees to access the profession. We do not consider the variety of different courses in landscape architecture, called in Italy "Master", offered at different levels by both academic and non-academic institutions.

18 On the differences between Italy and Europe in the education and the profession of landscape architecture, see Mazzino (2018). For observations on current challenges for landscape architecture profession and education in Italy, see Corbari (2018), pp. 30–32.

References

Assunto, R. (1973) *Il paesaggio e l'estetica*. Napoli: Giannini.
Astengo, G. (1949) 'La lezione urbanistica di Amsterdam. Formazione storica e nuovi ampliamenti', *Urbanistica*, 2, pp. 27–42.
Aymonino, C. (1960a) 'Copenaghen. Sviluppo storico e P.R.', *Urbanistica*, 30, pp. 9–26.
Aymonino, C. (1960b) 'Copenaghen. Nuovi quartieri', *Urbanistica*, 30, pp. 26–37.
Bagatti Valsecchi, P. F. (1964) 'Il problema del verde', *Casabella-continuità*, 286, p. 3.
Cederna, A. (1963) 'Le attrezzature verdi di Amsterdam', *Casabella-continuità*, 277, pp. 34–49.
Cederna, A. (1965) 'Stoccolma. Il verde pubblico e parchi per il gioco per i ragazzi', *Urbanistica*, 44, pp. 69–88.
Chiusoli, A. (1985) *Elementi di Paesaggistica*. Bologna: Editrice CLUEB.
Cianfarani, F. (2020) 'The Fascist Legacy in the Built Environment', in Jones, K. B. and Pilat, S. (eds.) *The Routledge Companion to Italian Fascist Architecture: Reception and Legacy*. Abingdon: Routledge, pp. 10–53.
Corbari, V. (2018) *La formazione in architettura del paesaggio in Europa: riflessioni per un terreno comune, tra sintonie e divergenze. Casi di studio. Ricerca nell'ambito del programma Borsa di studio 2018 in Teorie e politiche per il paesaggio*. Treviso: Fondazione Benetton Studi Ricerche [Online]. Available at: www.fbsr.it/wp-content/uploads/2020/05/ViolaCorbari_CASI-STUDIO.pdf (Accessed 13 November 2020).

D. M., Decreto Ministeriale 3 novembre. (1999) n. 509, 'Regolamento recante norme concernenti l'autonomia didattica degli atenei', *Gazzetta Ufficiale*, 2, 4 gennaio 2000 [Online]. Available at: www.miur.it/0006Menu_C/0012Docume/0098Normat/2088Regola.htm (Accessed 10 November 2020).

D. M., Decreto Ministeriale 4 agosto. (2000a) 'Determinazione delle classi delle lauree universitarie', *Gazzetta Ufficiale*, 245, 19 ottobre 2000 [Online]. Available at: http://attiministeriali.miur.it/anno-2000/agosto/dm-04082000-%286%29.aspx (Accessed 10 November 2020).

D. M., Decreto Ministeriale 22 ottobre. (2004) non 270, 'Modifiche al regolamento recante norme concernenti l'autonomia didattica degli atenei, approvato con decreto del Ministro dell'università e della ricerca scientifica e tecnologica 3 novembre 1999, n. 509,' *Gazzetta Ufficiale*, 266, 12 novembre 2004 [Online]. Available at: www.miur.it/0006Menu_C/0012Docume/0098Normat/4640Modifi_cf2.htm (Accessed 10 November 2020).

D. M., Decreto Ministeriale 28 novembre. (2000b) 'Determinazione delle Lauree specialistiche', *Gazzetta Ufficiale*, 18, 23 gennaio 2001 [Online]. Available at: http://attiministeriali.miur.it/anno-2000/novembre/dm-28112000.aspx (Accessed 10 November 2020).

D. P. R., Decreto del Presidente della Repubblica 5 giugno. (2001) n. 328 'Modifiche ed integrazioni della disciplina dei requisiti per l'ammissione all'esame di Stato e delle relative prove per l'esercizio di talune professioni, nonché della disciplina dei relativi ordinamenti', *Gazzetta Ufficiale*, 190, 17 agosto 2001 [Online]. Available at: www.miur.it/0006Menu_C/0012Docume/0098Normat/1361Modifi.htm (Accessed 10 November 2020).

De Fiore, G., Maniglio Calcagno, A., Romani, V., Ingegnoli, V., Abrami, G., Mazzino F., Russo, R. E. and Sburlati, R. (1984) *Architettura del paesaggio. La scuola di specializzazione di Genova*, Quaderni di Architettura, 3. Genoa: Università degli Studi di Genova.

Fariello, F. (1956) *Arte dei giardini*. Rome: Edizioni dell'Ateneo.

Fariello, F. (1967) *Architettura dei giardini*. Rome: Edizioni dell'Ateneo.

Ferrara, G., Rizzo, G. and Zoppi, M. (eds.) (2007) *Paesaggio: didattica, ricerche e progetti*. Florence: Firenze University Press.

Franchetti Pardo, V. (ed.) (2001) *La Facoltà di Architettura dell'Università di Roma "La Sapienza" dalle origini al Duemila: Discipline, docenti, studenti*. Rome: Gangemi.

Giacomini, V. (1971) 'I Parchi Nazionali Italiani', in Contoli, L. and Palladino, S. (eds.) *Commissione di Studio per la Conservazione della Natura e delle sue Risorse (CNR), Libro bianco sulla natura in Italia*, Quaderni de la Ricerca scientifica, 74. Rome: CNR, pp. 281–302.

Grassi, M. T. and Meucci, D. (eds.) (1990) *Inaugurazione della Scuola di specializzazione in Architettura dei Giardini, Progettazione e Assetto del Paesaggio, 6–7 dicembre 1990*. Genoa: Università degli Studi di Genova.

Grossini, P. and Giuntoli, A. (2017) 'Una rilettura della sensibilità ecologica di Pietro Porcinai (Firenze, 1910–1986) nel progetto di giardino', *Bullettino, Periodico della Società Toscana di Orticultura*, 1, pp. 8–15.

Kullman, K. (2016) 'Disciplinary Convergence: Landscape Architecture and the Spatial Design Disciplines', *JoLA-Journal of Landscape Architecture*, 11/1, pp. 30–41.

Latini, L. (2016) 'A Life and Its Cultural Context', in Treib, M. and Latini, L. (eds.) *Pietro Porcinai and the Landscape of Modern Italy*. Furnham, and Burlington: Ashgate, pp. 2–41.

Laugier, M. A. (1753) *Essai sur l'Architecture*. Paris: Duchesne.

Maniglio Calcagno, A. (1989) 'La scuola di specializzazione in "architettura del paesaggio" di Genova', *Architettura del paesaggio: Notiziario AIAPP*, 13, special issue on *L'insegnamento dell'architettura del paesaggio in Italia*, pp. 43–44.

Matteini, M. (1991) *Pietro Porcinai architetto del giardino e del paesaggio*. Milan: Electa.

Mazzino, F. (1993) *Programmi. Scuola di Specializzazione in Architettura dei Giardini, Progettazione e Assetto del Paesaggio*. Genoa: Università degli Studi di Genova.

Mazzino, F. (1997) 'Scuola di specializzazione in architettura del paesaggio di Genova: formazione professionale e scambi europei', *Architettura del paesaggio: Notiziario AIAPP*, 22, pp. 71–80.

Mazzino, F. (2001) 'I primi corsi di laurea in architettura del paesaggio', *Architettura del paesaggio*, 7, p. 66.

Mazzino, F. (2002) 'Professione Architetto del paesaggio', in Longhi, G. and Mazzino, F. (eds.) *Professione Architetto del paesaggio: Problemi e prospettive della figura professionale in Europa e nel mondo alla luce della vigente legislazione. Atti del Convegno Internazionale Roma 18 maggio 2002*. Rome: AIAPP.

Mazzino, F. (2005) 'La riforma universitaria e il D.P.R. 328/2001', *Architettura del paesaggio*, 12, pp. 58–60.

Mazzino, F. (2017) 'La formazione degli architetti del paesaggio: una questione irrisolta', in Ministero dei Beni e delle Attività Culturali e del Turismo (ed.) *Rapporto sullo Stato delle Politiche per il paesaggio. Osservatorio Nazionale per la qualità del paesaggio*. Rome: CLAN Group, pp. 376–380.

Mazzino, F. (2018) 'Architettura del paesaggio in Italia. Disparità di diritto allo studio e al lavoro rispetto all'Europa', *Architettura del paesaggio*, 36(1), pp. 14–17.

Mazzino, F. (2019) 'History of Landscape Education in Italy. Many Antagonisms', in Gao, L. and Egoz, S. (eds.) *Lessons from the Past, Visions for the Future: Celebrating One Hundred Years of Landscape Architecture Education in Europe. Proceedings of the ECLAS and UNISCAPE Annual Conference 2019*. Ås: School of Landscape Architecture, Norwegian University of Life Sciences, pp. 117–119.

Porcinai, P. (1937a) 'Giardini privati', *Domus*, 115, pp. 30–37.

Porcinai, P. (1937b) 'L'Italia d'oggi e l'arte del verde', *Domus*, 110, pp. 33–40.

Porcinai, P. (1952) 'Jardins et espaces verts en Italie', *L' architecture d'aujourd'hui*, 41, p. 76.

Porcinai, P. (1968) *Per l'insegnamento del "verde", del paesaggio e del giardino in Italia*. Unpublished manuscript. Archivio Porcinai, Fiesole.

Porcinai, P. (1986) 'Per l'insegnamento del "verde", del paesaggio e del giardino in Italia', *Architettura del paesaggio: Notiziario AIAPP*, 10, special issue on *Pietro Porcinai. Architetto del giardino e del paesaggio*. pp. 50–56.

Revedin, J. (1998) 'Pietro Porcinai come progettista riformatore: la sua formazione nella Germania dell'inizio secolo', in Pozzana, M. (ed.) *I giardini del XX secolo: l'opera di Pietro Porcinai*. Firenze: Alinea, pp. 43–54.

Sangermano, S. (2015) *Luigi Piccinato, 1899–1983. L'impegno civile tra teoria e prassi: architettura, città, territorio*. PhD Dissertation, Università degli Studi di Napoli Federico II [Online]. Available at: www.fedoa.unina.it/10121/1/Tesi%20di%20Dottorato%20XXVII%20ciclo_Sandra%20Sangermano.pdf (Accessed 3 November 2020).

Sereni, E. (1961) *Storia del paesaggio agrario italiano*. Bari: Laterza.

Sereni, E. and Litchfield R. B. (2014) 'Preface', in Sereni, E. and Burr Litchfield, R. (eds.) *History of the Italian Agricultural Landscape*. Princeton: Princeton University Press.

Tentori, F. (1964) 'Questo numero', *Casabella-continuità*, 286, pp. 2–3.

Turri, E. (1974) *Antropologia del paesaggio*. Milano: Edizioni Comunità.

Turri, E. (1979) *Semiologia del paesaggio italiano*. Milano: Longanesi.

Università degli Studi di Genova. (1989) 'Statuto della Scuola di Specializzazione in Architettura dei Giardini, Progettazione e Assetto del Paesaggio', *Gazzetta Ufficiale*, 130 (266).

Venturi, G. (1924) 'La scuola superiore di architettura', *Architettura e Arti Decorative*, 2, pp. 107–125.

Ware, I. (1756) *A Complete Body of Architecture*. London: Osborne and Shipton.

29
THE TRAINING OF LANDSCAPE ARCHITECTS IN FRANCE

From the horticultural engineer to the landscape architect, 1876–2016

*Bernadette Blanchon, Pierre Donadieu, and Chiara Santini,
with Yves Petit-Berghem*

Introduction

How have landscape architects been trained in France for the last 140 years? How has the scientific and technical training of horticultural engineers been reconciled with that of landscape architects who design garden and landscape projects?

This chapter starts by evoking the beginnings. The training in "pre-landscape architecture times" begins with the creation, in 1876, of the *Chaire d'Architecture des Jardins et des Serres de l'École nationale d'Horticulture de Versailles* (Chair of Garden and Greenhouse Architecture at the National School of Horticulture of Versailles) and the *concours en loge* for landscape architects (closed preparatory workshops similar to those organised for the *Prix de Rome* competition) organised by the *Comité de l'art des jardins de la Société nationale d'horticulture de France* (Committee of Garden Arts of the National Horticultural Society of France). This was followed by the creation of the *Section du Paysage et de l'Art des Jardins* (SPAJ, Department of Landscape and Garden Art) of the National School of Horticulture in 1945, and with the opening in 1976 of the *École nationale supérieure du paysage de Versailles* (ENSP, National School of Landscape Architecture) on the historic heritage site *Potager du Roi*. From the 1990s, schools in Angers, Bordeaux, Lille, and Blois also established state-certified professional programmes for landscape architects and landscape engineers (*Paysagiste Concepteur* 2016).

Based on historical research conducted by the authors drawn from archives and interviews, the chapter describes the evolution of French expertise in garden and landscape project design (Santini 2021; Donadieu 2018–2020; Blanchon 1998, 2000, 2001, 2015, 2021). In resonance with urban and environmental issues of the second half of the 20th century, these five schools have a horticultural and garden tradition which they eventually broke away from.

1876–1945: training at the National School of Horticulture in Versailles

The history of education in "pre-landscape architecture" times dates from 1876 at the *École nationale d'Horticulture* (ENH, National School of Horticulture), and the creation of a course designed to provide "gardeners with all the theoretical and practical knowledge relating to the art of horticulture" (ENH 1874). Located at the *Potager du Roi*, established between 1678 and 1683 near the Château de Versailles, the new school showcased the development of French expertise and know-how in the cultivation of fruit trees.

Training in garden and greenhouse architecture

The term *Architecture des Jardins* (Garden Architecture), was commonly used at the time the first professional programme in the field began. The training provided by ENH had a three-year duration and included both theoretical and practical courses, including the course entitled *Architecture des Jardins et des Serres* (Garden and Greenhouse Architecture). This course, the first in France devoted to the principles of garden composition, was entrusted to the civil engineer Jean Darcel (1823–1907), one of Adolphe Alphand's main collaborators at the *Service des Promenades et Plantations* in Paris (SPP, the Department of Public Walks and Gardens). Jean Darcel had made a name for himself in this domain with a course open to the public at the *Société nationale d'Horticulture de France* (SNHF, National Horticultural Society of France) and the publication of the *Étude sur l'architecture des jardins* (Study in Garden Architecture) in 1875 (Darcel 1875). The course given by Darcel, who was replaced in 1877 by his student, the architect and engineer Auguste Choisy (1841–1909), consisted of approximately 20 lectures and a series of practical exercises. Centred on composition, it did not address the plant dimension of projects, which was entrusted to the other teachers at the ENH. For the final exam, students had to produce a garden project with an overall plan, a price estimate, and explanatory notes (Limido 2018).

Close ties with the Parisian administration

The creation of a course in garden architecture, and the appointment of Darcel, are evidence of the close ties between the new course and the SPP, on the one hand, and the SNHF on the other hand. From the Second Empire onwards, the development of the capital's public walks and gardens helped to transform Paris into a model city and to disseminate its planning principles. With Jean Darcel, two other teachers came from the SPP: the *architecte-paysagiste* (landscape architect) and botanist Édouard André (1840–1911), a former collaborator of Jean-Pierre Barillet-Deschamps who in 1892 took over the Chair of Garden and Greenhouse Architecture, and Auguste Pissot (1823–1883) the curator of the Bois de Boulogne who taught a course on forestry and ornamental tree growing. Thanks to their experience working with Alphand, they taught both the theoretical principles of the discipline and the technical procedures that had been developed in Paris (Santini 2021).

Towards professional recognition

Many Versailles horticulturists embarked on a career as *architectes-paysagistes* (landscape architects) although this title had not yet been given formal recognition. Some professionals worked freelance, others joined public administrations, and still others went to work abroad. Among the best known[1] were Henri Nivet (1863–1941) from the Édouard André agency and Henri Martinet, the designer of a project for the redevelopment of the imperial park in Tokyo (Durnerin 2001).

From 1901 onwards, the *architecte-paysagiste* (landscape architect) René-Édouard André (1867–1942) taught the class and doubled its hours. In 1934, when Ferdinand Duprat (1887–1976) took over from André, the chair provided 34 hours of classes and two days of practical exercises to which were added visits of gardens and classes in drawing, surveying, and levelling

Figure 29.1 ENH in Versailles. In the background, hanging on the wall, are the paintings used by Édouard André for teaching garden and greenhouse architecture.

Source: *Un examen*, photograph by Bargillon, éditions L. Garnier, Versailles, early 20th century (© ENSP fonds).

(Duprat 1945). However, according to Duprat, who was able to appreciate the importance of the garden arts in the Belgian, British, Swiss, German, and American schools of horticulture, the course was far too short to train *architectes-paysagistes* (landscape architects) properly. This was even more the case since, as the change in the title of the course in 1926 indicates, the teaching had become associated with the new discipline of *urbanisme* (urban planning).

Duprat was supported by the *Société française des architectes de jardins* (SFAJ, French Society of Garden Architects[2]) and the *Comité de l'Art des Jardins* (Garden Art Committee) of the SNHF,[3] which from 1902 set up the *"concours en loge"* for *architectes-paysagistes* (landscape architects), modelled on the workshops of the *Beaux-Arts*. He took the opportunity of the first *Congrès international des architectes de jardins* (International Congress of Garden Architects), which was held during the 1937 *Exposition Internationale des Arts et des Techniques appliqués à la vie moderne* (International Exposition of Art and Technology in Modern Life) in Paris, to embark on a collective discussion on the objectives and challenges in the training of *architecture des jardins* (garden architecture), a discipline which complemented that of architecture and yet was distinct from it. The Chair of Garden and Greenhouse Architecture, which later became the *Chaire d'Architecture des jardins et d'urbanisme* (Chair of Garden Architecture and Urban Planning), and the *SNHF* Committee of Garden Arts were therefore at the inception of the profession of landscape architecture, which, after the Second World War, underwent a spectacular development.

1945–1975: the department of landscape and garden arts of the National School of Horticulture in Versailles

In December 1945, General de Gaulle, the president of the provisional government of the Republic, signed a decree creating the SPAJ, *la Section du Paysage et de l'Art des Jardins* (Department of

Figure 29.2 ENH in Versailles. Students doing practical exercises in the *Potager du Roi*. Aquarelles painted for the *Exposition Universelle* in 1900.

Source: © AD78 – Archives départementales des Yvelines, École nationale d'horticulture fonds- 1W-dépôt).

Landscape and Garden Art). The SPAJ was created in response to repeated requests on the part of professionals who were advocating in favour of a specialised curriculum and diploma, notably since the creation of the French Professional Association of Architects in 1940. The Association in question, *L'Ordre des Architectes*, had forbidden the use of the title of *Architecte-Paysagiste* (landscape architect). This demand for artistic training in workshops, based on the same model as that of architects at the *École des Beaux-Arts*, had been supported since the beginning of the 20th century by the *Comité de l'art des jardins* (Committee of Garden Art), and especially by Jules Vacherot (1862–1925), Ferdinand Duprat, Achille Duchêne, and Robert Joffet.

This new one-year course, which was subsequently extended to two years, was open to horticultural engineers who had graduated from the ENH, or to students following a year of preparatory studies. Three periods can be identified.

The first period (1945–1960): workshops (ateliers) *on the model of the* Beaux-Arts

The first period gave eminence to figures such as André Riousse (1895–1952) and Théodore Leveau (1896–1971) who were *Architectes diplômés de l'École des Beaux-Arts et Urbanistes* (architects from the *Beaux-Arts*, and town planners) as well as others, some of whom had been trained in the *concours en loge* at the SNHF, as well as horticultural engineers such as Albert Audias who had trained with Duprat. With a small number of students who were mainly young horticultural engineers (four to six students per year), these teachers directed project workshops and transmitted the legacy of seasoned practitioners such as E. André and J.-C.N. Forestier (1861–1930). For this *Théorie de l'art des jardins et composition* class (Theory of Garden Art and Composition), students were required to provide one sketch per week and one project per month.

The title of state certified landscape architect, *Paysagiste Diplômé par le gouvernement*, DPLG, which was the official title from 1960 until 2015, made the distinction from the title of garden contractor and established the landscape architect's position in between the architect and the horticultural engineer. At the *Potager du Roi*, horticultural engineers and *paysagistes* (landscape architects) followed separate courses. The course at the *Section du Paysage et de l'art des Jardins* (department of landscape and garden art) consisted essentially of project workshops, so-called *ateliers* (including "Theory of Garden Art and Composition"), which it is difficult to dissociate from the lecture and applied technical teaching. The course was taught by temporary teachers some of whom were *paysagistes* (landscape architects) and foreshadowed the creation of the *École nationale supérieure du paysage* (ENSP).

Renewal (1963–1970): town planning and the environment

During the second period, young graduates in landscape architecture such as Jacques Sgard (born in 1929, SPAJ 1946) and Jean-Claude Saint-Maurice (1924–1989, SPAJ 1951), came to teach at the Department. They pointed to the emergence of public commissions and changed the focus of the workshops from the design of private gardens to those of urban public spaces. Trained in urban planning abroad they renewed the pedagogical approaches of the project workshop and with Bernard Lassus (born in 1929), a painter and artist, they integrated a visual and environmental approach to landscape design. At their request, the teaching of plant ecology was introduced by the botanist Jacques Montégut (1925–2007)[4] who taught at the ENSH.

At the end of the 1960s, in reaction to the limited financial resources available and the SPAJ's dependence on the ENSH, the teachers proposed a Landscape Institute project which did not come to fruition. Increasingly in demand and recognised, the few young graduates from the school took part in the building of new towns launched in 1965 and in major regional development projects particularly focused on tourism. This was why the Ministry of Agriculture considered reforming the SPAJ and the ENSH to meet the country's new needs in planning.

An educational change and a refocusing on design teaching in ateliers *(1970–1974)*

From 1971 to 1974, new teachers like Michel Corajoud, who were foreign to the world of horticulture and the garden arts, took over the SPAJ, which was in crisis and had become deserted following the departure of the former team. The failed project of the Landscape Institute led to the creation of a training centre which was as original as it was short-lived: the CNERP (*Centre national d'Étude et de Recherche du Paysage*, National Centre for Landscape Research) which was set up in Paris and later moved to Trappes near Versailles (1972–1979). With its own resources enabling it to conduct studies and research, inter-ministerial support (notably from the Ministry of the Environment created in 1971), and several teachers from the SPAJ, it was devoted to the planning and management of urban and rural landscapes, and was intended for engineers, landscape architects, and architects.

Within the SPAJ, the teachers Pierre Dauvergne (born in 1943) and Jacques Simon (1929–2015) asked landscape architect Michel Corajoud (1937–2014) to "save this training programme which was in decline" and share the knowledge he had acquired at the *AUA, Atelier d'Urbanisme et d'Architecture*, in Paris (Blanchon 2015). Corajoud recentred the training on the landscape project based on *ateliers* (design workshops) and educational trips; within an atypical setting at the time, these two elements took up most of the training time. With training focused on the drawing of plans and the designing of projects taking inspiration from the theoretical publications and methods of Kevin Lynch, Vittorio Gregotti, and Enrique Ciriani, the vision of landscape architects as mediators between mankind and nature thus shifted to that of landscape

Figure 29.3 SPAJ. The design studio (*atelier*) in 1965–1967. Professor Leandro Delgado standing – a former pupil of Roberto Burle-Marx. On the wall: program for the study trip in Spain.

Source: © Marguerite Mercier (student of the 1965–1967 class sitting in the middle of the view).

architects imposing their mark on the existing environment. Some of the students subsequently became major figures in landscape architecture and planning in France from the 1980s onwards.

The courses at the SPAJ and the ENSH were terminated by the Ministry of Agriculture in August 1972 with students completing their diplomas until 1974, and reframed in 1975.

1975–2016: landscape architecture schools

The ENSP: 1975–1995, a pioneering school

In 1976, the ENSP was created next to the ENSH at the *Potager du Roi* in Versailles. This new school, which increased the duration of the professional landscape architectural programme from two to four years, with a selected access after two years of bachelor studies, replaced the CNERP and the SPAJ.

Created without a full professor and dependent on the ENSH administration, the ENSP initially used the full professors of the ENSH for scientific and technical subjects. It set up departments for the other courses dedicated to the living environment, human and social sciences, techniques in project design, the plastic arts and rendering. However, like the SPAJ, the ENSP placed a great deal of emphasis on *ateliers*, project and design studios which took almost half the teaching schedule during the first three years and were run by Michel Corajoud and Bernard Lassus – respectively in charge of the Le Nôtre and the Dufresny studios (*Atelier Le Nôtre et Atelier Dufresny*). The former focused on landscape materiality and the use of design to address landscape transformation, while the latter gave precedence to a "creative analysis" based on a cultural approach to the site.

Figure 29.4 ENSP, Versailles, 2019. Students doing ecological field exercises at the *Potager du Roi* (the King's Kitchen Garden).

Source: © ENSP.

In 1983, a crisis between the two schools led to the departure of the teachers of the ENSH and completed the break with horticulture. This event triggered off a new project to set up a French landscape institute, which also failed in 1985. However, this last failed attempt did allow for the recruitment of the first tenured professors, including the landscape architect Michel Corajoud.

Under the direction of M. Corajoud,[5] with landscape architect and horticultural engineer A. Provost and the horticultural engineer M. Rumelhart, an important educational reform redefined the pedagogical relations between workshops and teaching departments, which had become too compartmentalised. At the end of the 1980s, the *Potager du Roi* was used for the teaching of plant ecology and the plastic arts while the ENSH prepared to move to Angers. This evolution was an indication of the intention to define the landscape architect's competence technically (soil and plants) as well as artistically.

From 1995: a strengthening and transformation of the training of landscape architects in France

The 1990s were marked by a political and regulatory context that was particularly favourable to the recognition of the professional value of landscape architects. The role of the landscape architect in the implementation of public policies for the development of land and public spaces, town planning, and heritage conservation was emphasised by the State. In response to the training needs of landscape professionals and bolstered by the signing of the European Landscape

Convention in Florence in 2000, in just under 15 years, three new schools awarding a diploma in landscape architecture were established.

Two training programmes in landscape architecture were set up at the schools of architecture in Bordeaux in 1991 and in Lille in 2005. Another programme had been set up at the *École nationale supérieure de la nature et du paysage* (National Nature and Landscape School) in Blois in 1993 to which was added the landscape engineering programme at the *Institut d'horticulture et de paysage* (Horticultural and Landscape Institute) in Angers in 1997.

During this period, attendance increased, so did the proportion of female students (Dubost 1983; Blanchon 2021). At the same time, new tools, and a new regulatory framework for urban planning practices (landscape plans, atlases, and charters) were experimented with and helped to stimulate business among landscape agencies.

The ENSP in Versailles took over the management of the *Potager du Roi* after the departure of the ENSH in 1995, for training purposes (gardening and art classes) and heritage conservation (collections of species and forms of fruit trees). The vegetable and ornamental garden was opened to the public in 1991, and in 1993 a research laboratory for landscape projects (LAREP) and a doctoral programme were set up at the *Potager du Roi*, with a major in "Landscape sciences and architecture" delivered by the *AgroParisTech* doctoral school. Similar changes were made at the other schools of landscape architecture which thus shifted more towards academic research, resulting in the creation of new degree titles, such as a Master degree in landscape architecture which provided access to doctoral schools (Davodeau 2016). It was during these years that the schools initiated international exchanges and the ENSP set up a Mediterranean branch in Marseille. It was also during this period that the historical research this chapter is based on began.

Diversification of pedagogical methods

From 1986 to 2016, during the fourth year of training, ENSP replaced internships in agencies, which had proved to be unsatisfactory, with 10 *ateliers pédagogiques régionaux* (regional educational workshops) financed by public commissions.

In all of the schools, this dual system of training through practice in the field and in workshops and academic research (Doctorate) had made it possible to gradually train the professionals needed to respond to public commissions as well as provide the teacher-researchers for the schools. It continued with success in Versailles until the state certified diploma in landscape architecture was reformed in 2016 to be replaced with a new diploma (*Paysagiste DE, Diplômé d'État*) and the creation of the professional title of "*paysagiste concepteur*" (landscape architect) awarded after three years of study, ending with a master's degree. This title was recognised by the law 2016–1087 of 8 August 2016 for the restoration of biodiversity, nature, and landscapes, which included a new vision of the inclusion of the natural dimension in landscape project design (Petit-Berghem 2017). This law also reflected a change of trend in public commissions and design programmes that started in the years 2000. Techniques in landscape architecture were combined with techniques in ecological engineering to be deployed at several geographical levels.

At the school in Versailles, the teaching of ecology encompassed the ecological paradigms of the design of natural spaces such as ecological networks, ecological continuities and functions, and new environmental requirements imposed on contracting authorities (Dacheux-Auzière and Petit-Berghem 2017). This change involved landscape architects and ecologists, and practitioners and researchers coming together to improve operational collaboration (Morin et al. 2016) while remaining at the *Potager du Roi*, which also enabled training to evolve at a practical level. In the design of landscape projects, ecological approaches gave prominence to the multifunctional use of space and emphasised the importance of biodiversity as a factor of economic and social development. In order to integrate the new environmental challenges of public orders, the training of landscape architects always enabled them to adapt to the expectations of the contracting authorities and the development departments of local authorities. This pedagogical evolution thus contributed to the renewal of a training offer that decompartmentalised knowledge and combined different intuitive, aesthetic, and scientific approaches throughout the project process.

Conclusion

During three periods, the ENSH, followed by the ENSP, together with their supervisory authority, the Ministry of Agriculture and Forestry, adapted the training courses in Versailles to the changing demands placed on garden contractors and landscape architects in France and abroad.

During the first period, courses gave priority to the sciences and techniques of horticulture at the expense of time spent on design. This lack of a balanced approach proved to be detrimental to the development of the creative skills which had, until the economic crisis of 1929, become almost exclusively focused on private gardens.

During the second period, teachers at the *Beaux-Arts* and the French school of decorative arts gradually acknowledged the importance of *ateliers*, project and studio workshops, in addressing professional requirements. The teachers at the SPAJ then shifted away from horticultural training to broaden the educational scope to embrace the territorial and urban dimensions of landscape planning.

They completed this rather radical change during the third period, with the move of the ENSH to Angers in 1995. With the appointment of full professors, the ENSP replaced the SPAJ and the CNERP and acquired international recognition in the field of landscape architecture and urban planning. The model thus experimented in Versailles was adapted to other schools under the direction of the Ministry of Culture or of the Ministry of Higher Education, including the Landscape and Horticultural National Institute in Angers (today the Landscape Department of *Agrocampus Ouest Angers-Rennes* – Ministry of Agriculture).

Thus, from 1993, the public landscape policy of the Ministry of the Environment (landscape plans, charters, atlases, and observatories), which included the *Grands Prix du Paysage et de l'Urbanisme* (Grand Prizes in Landscape and in Urban Planning), was able to source the acknowledged multi-level and multidisciplinary skills needed for its implementation.

The break with the scientific training provided by the School of Horticulture fostered the development of a teaching approach that encompassed the *atelier*, project workshop (50%), and the visual arts. At the same time, it made it possible to adapt to changes in public commissioning, initially in the case of public parks and subsequently in the defining of public policies governing local authority planning. It also made it possible to find original solutions to the teaching of biotechnical (gardening by students), historical and socio-geographical disciplines, and finally to reappropriate the cultural heritage of the Potager du Roi.

The successive contributions of Jacques Sgard on large scale landscape and planning, of Michel Corajoud on the design of the urban project and the revelation of the site, the conceptual research of Bernard Lassus, and the early ecological teachings of Gilles Clément among others, represent different facets of a teaching method which encompasses them all in spite of their differences and which are applied according to the specific methods of each establishment. The research developed by the school's teacher-researchers made use of the landscape project (project-based research) or methods borrowed from the social and human sciences such as ethnosciences, social geography, and the history of architecture, gardens, and landscape architecture. They addressed contemporary issues relating to the skills of landscape architects in the domains of ecology, energy transition, and climate change, in disciplines such as agroforestry and landscape planning, and responded to societal changes relating to the role of women, heritage conservation, and agriculture.

Current and future research focuses on two major questions: what are the fundamental skills and practices for professionals planning and designing landscapes and what are the forms and roles of public action and regional management in the production of landscapes? The challenge, in both cases, is to reconstruct a broader history of the profession and of educational system and approach, including the role of women, social inclusion, the collections in archives, the emerging role of the contracting authority and an observatory of landscape architecture and the capitalisation of practices. Researchers and practitioners are part of a common dynamic stretching from the past to the future and which is nurtured by an iterative process involving research and the teaching and practice of landscape architecture.

Notes

1 In her chapter on Portugal, Ana Duarte Rodrigues (see Duarte Rodrigues in this volume) writes how the French designer Henri Lusseau won the public competition for the Park of Liberty in Lisbon and he himself as landscape architect in his correspondence with the Lisbon City Council. Lusseau was certainly a well-known landscape architect, as between 1880 and 1890 he obtained prizes and recognition alongside Édouard André, Eugène Dény, and Henri Duchêne. However, unlike these three others, his name appears much more rarely in the specialised press.
2 The *Société française des architectes de jardins* (French Society of Garden Architects) was founded in Paris in 1933 by a group of landscape architects (Ferdinand Duprat, Maurice Thionnaire, Louis Deny, Jacques Gréber, Louis Decorges, Prosper Péan, Albert Séret, Anatole-Guy Otin, and André Riousse). Since its creation, it had been chaired by Achille Duchêne (1866–1947).
3 This committee was created in 1888 within the SNHF.
4 Montégut also trained Marc Rumelhart (IH 1971), later head of the ecology department in ENSP and Gilles Clément (IH 1965, SPAJ 1967), who became world famous for his early innovative ecological approach.
5 Corajoud has always been driving project teaching (design studio) in Versailles. First in the last years of the section du Paysage (1970–74 SPAJ), then in the École du Paysage ENSP. In the first period of the school (1976–1986) it was in parallel with Bernard Lassus (students choose between the two different studios). Then (1986–1996) he was the main leading figure of the studio teaching and also the whole school (voted for and) followed his proposal for a pedagogical program. In the same year (1986), Bernard Lassus left Versailles to create a doctoral program in landscape architecture at the school of Architecture of Paris-La Villette ("Jardins, Paysages, Territoires") – which later (2005) became a research post master and doctoral degree "Architecture, Milieu, Paysage"). Lassus then taught abroad (Canada, USA, Germany, Italy).

References

Blanchon, B. (2000) 'Les paysagistes français de 1945 à 1975, l'ouverture des espaces urbains', *Les annales de la recherche urbaine*, 'Paysages en villes', n° 85, pp. 20–29.
Blanchon, B. (2001) 'Paysagiste, naissance d'une profession', in Texier, S. (ed.) *Les parcs et jardins dans l'urbanisme parisien*, exhibition catalogue, Collection B. de Andia. Paris: Délégation à l'Action Artistique de la Ville de Paris, pp. 258–266.
Blanchon, B. (2015) 'Jacques Simon et Michel Corajoud à l'AUA, ou la fondation du paysagisme urbain', in Cohen, J. L. with Grossman, V. (eds.) *Une architecture de l'engagement: l'AUA 1960–1985*. Paris: Editions Carré, Cité de l'Architecture et du Patrimoine, pp. 214–225.
Blanchon B. (2021) 'The Activism of Women Landscape Architects in France, Since 1945: Reference Points and Profiles', in Bouysse-Mesnage, S., Dadour, S., Grudet, I., Labroille, A. et Macaire, É. (eds.)

Proceedings of the Webinar Conference, Gender Dynamics and Practices in Architecture, Urbanism and Landscape Architecture (February 4th–5th 2021), Paris: ENSAPLV. Available online at https://let.archi.fr/spip.php?article11583

Blanchon, B. with Audouy, M. and Thibault, E. (1998) *Pratiques paysagères en France de 1945 à 1975 dans les grands ensembles d'habitations*, Research report, Versailles-Paris, ENSP/Plan Construction et Architecture, Ministère de l'Équipement, des Transports et du Logement. Programme Architecture des espaces publics modernes. Distributed by Ville-Recherche- Diffusion (1998), Versailles, ENSA V [Online]. Available at: https://topia.fr/travaux-de-chercheurs/publications-en-ligne/#BB1 (Accessed 15 November 2020).

Dacheux-Auzière, B. and Petit-Berghem, Y. (2017) 'Quelle écologisation de la pratique des paysagistes concepteurs?', *Projets de paysage*, 16 [Online since 8 July 2017]. Available at: http://journals.openedition.org/paysage/5536; https://doi.org/10.4000/paysage.5536 (Accessed 15 November 2020).

Darcel, J. (1875) *Étude sur l'architecture des jardins*, extrait des *Annales des ponts et chaussées* (IIe semestre, tome X). Paris: Dunod.

Davodeau, H. (2016) 'Le réseau des écoles de paysage françaises: enjeux pédagogiques, scientifiques et professionnels', *Sud-Ouest européen*, 38 [Online since 18 March 2016]. Available at: http://journals.openedition.org/soe/1599 (Accessed 15 November 2020).

Donadieu, P. (2018–2020) *Histoire de l'ENSP* [Online]. Available at: https://topia.fr/2018/03/27/histoire-de-lensp-2/ (Accessed 15 November 2020).

Dubost, F. (1983) 'Les paysagistes et l'invention du paysage', *Sociologie du travail*, n°4, Les professions artistiques, pp. 432–445.

Duprat, F. (1945) 'L'enseignement de l'art des jardins', *Revue Horticole*, p. 292.

Durnerin, A. (2001) 'Quelques figures d'anciens élèves de l'École nationale d'horticulture de Versailles au temps d'Édouard André', in André, F. and de Courtois, S. (eds.) *Édouard André (1840–1911). Un paysagiste botaniste sur les chemins du monde*. Paris: Éditions de l'Imprimeur, pp. 302–308.

ENH Final Brochure. (1874) *Journal de la Société centrale d'horticulture de France* (tome VIII), pp. 321–325.

Limido L. (2018) 'The *Architecture des Jardins et des Serres* at the École Nationale d'Horticulture de Versailles (1876–1892): The Pedagogy of the Project According to Jean Darcel and Auguste Choisy', *Studies in the History of Gardens & Designed Landscapes*, pp. 1–17.

Morin, S., Bonthoux, S. and Clergeau, P. (2016) 'Le paysagiste et l'écologue: comment obtenir une meilleure collaboration opérationnelle?', *VertigO – la revue électronique en sciences de l'environnement* [Online since 10 June 2016]. Available at: http://journals.openedition.org/vertigo/17356; https://doi.org/10.4000/vertigo.17356 (Accessed 15 November 2020).

Petit-Berghem Y. (2017) *Écologie et paysage. Réinterroger le vivant*. Toulouse: UPPR Éditions.

Santini C. (2021) *Adolphe Alphand et la construction du paysage de Paris*. Paris: Hermann.

30
REFLECTIONS ON LANDSCAPE AND LANDSCAPE ARCHITECTURE EDUCATION IN THE ARAB MIDDLE EAST

Jala Makhzoumi

Introduction

Reflecting on my own learning trajectory, my first encounter with 'landscape' was in a design studio while studying architecture. There I learnt that landscape implies outdoor spaces that must be designed, rendered useful, and 'beautiful'. There wasn't a single landscape architect in Iraq at the time in the 1970s, and it took years of research and practice for me to realise that landscape was not only the domain of designers and that it was so much more than outdoor space and beautification.

This is the background and context for my review of progress made in landscape education in the region called the Middle East. The term is problematic because it was coined in reference to European colonial powers, and there is no consensus as to what geographic area it constitutes. In the absence of a readily recognised reference to the region, I use Middle East in this chapter to embrace countries roughly corresponding to what was historically known as the 'fertile crescent', in reference to the first agrarian, irrigated landscapes in Mesopotamia, Iran to the east, lands of the eastern Mediterranean to the west and north, and the Arabian peninsula in the south. The region gave birth to all three Abrahamic religions and the concept of *paradise*, a productive garden and place of shade and water – the antithesis of the barren lands that characterise much of the region. The region is rich with historical references to gardens, be it agrarian landscapes, pleasure gardens of the Assyrians, the hanging gardens in Babylon, the enclosed, geometric gardens of Persia, or Ottoman palace gardens.

The agrarian landscape and ornamental garden heritage was for the most part lost when the modern concept of 'landscape' came to be introduced by colonial rule in the Middle East following World War I. Thereafter, the meaning of landscape came to be narrowly associated with modernity and colonial westernisation of Middle Eastern cities. And although landscape architecture is slowly being recognised in the region as a profession independent from architecture, more in some countries than others, the professional scope continues to be predominantly of designed outdoor spaces and 'beautification', mirroring my own encounter so long ago.

Landscape architecture practice and education in the Middle East is intertwined with what I call the 'colonial legacy'. One challenge facing landscape architecture in the region is that not only was the cultural and ecological continuity with the past disrupted, but western language, culture, and values came to replace local ones. Another challenge is the absence of a word and concept of landscape that is rooted in local cultures and ecologies. Yet another challenge is that landscape architecture is a profession in the making. These are the ideas I hope to explore in this chapter to contextualise landscape architecture education in the region. After a brief overview of landscape architecture scholarship and practice, I will turn to countries of the Arab Middle East to review programmes offered in the Kingdom of Saudi Arabia, Lebanon, Bahrain, and Palestine. As an example, I present the landscape architecture education at the American University of Beirut, Lebanon, and its contribution to a contextualised discourse on landscape. Far from a comprehensive survey, my aim is to provide the background and showcase the contribution of scholars to advance landscape architecture education in the Arab Middle East.

The narrative I present draws on primary sources, correspondence with academics and professionals in the region, secondary sources, including books and journal articles, and my own experience of teaching and practicing in Iraq and Lebanon for over three decades. I am especially grateful to academics and professionals who head landscape architecture programmes and/or departments in the countries discussed in this chapter for responding to my questionnaire and providing related information. Listed in alphabetical order, these include Rawaa Abbawi, Associate Professor, Department of Architecture, University of Technology, Baghdad, Iraq; Yaser Abunaser, Associate Professor, Department of Landscape Design and Ecosystem Management, Faculty of Agricultural and Food Sciences, American University of Beirut; Wafa Almadani, Assistant Professor and Program Coordinator, the Department of Architecture and Interior Design, the University of Bahrain, Bahrain; Samar Al Nazer, Assistant Professor, Department of Architectural Engineering, Faculty of Engineering, Birzeit University, Palestinian Authority; Meryem Atik, Professor and chairperson (2017–2020), Department of Landscape Architecture, Faculty of Agriculture, University of Akdeniz, Antalya, Turkey; Malik Al-Ajlouni, Associate Professor, Landscape and Floriculture Program Coordinator, Department of Horticulture and Crop Sciences, University of Jordan, Amman, Jordan; Hassan Bitar, Associate Professor, founding chair, Landscape Architecture Program, Urban Planning Institute of ALBA, Académie Libanaise des Beaux-Arts, University of Balamand; Armin P. Rad, Executive Director, Iranian Society of Landscape Professionals (ISLAP); Salma Samaha, Associate Professor and chair (2013–17, 209–21), Department of Landscape and Territory Planning, Agriculture Faculty, Lebanese University; and Mamdouh M.A. Sobaihi, Associate Professor and chairperson (2011–2019), Landscape Department, Faculty of Architecture and Planning, King Abdulaziz University, Jeddah, Kingdom of Saudi Arabia. I am also indebted to Shelley Egoz, professor emerita of landscape architecture, who was the first to alert me of the landscape education programme in Bir Zeit University, and Ramzi Hassan, Associate Professor, Urban Planning and Landscape Architecture, Norwegian University of Life Sciences.

Scholarship, practice and education in landscape architecture

Landscape is an idea that is culturally informed and place specific. The English word evolved over centuries, gathering multiple meanings from the early Anglo-Saxon meaning which implied 'a patch of cultivated ground', small-scale and lived-in, to a definition "that [corresponds] to the larger political spaces of those with power", for example, 'territory' and 'country'. The meaning changed once more to acquire its current meaning tied to a particular 'way of seeing' that prioritises scenery (Bender 1993, p. 1). Of all the meanings, it is the latter, landscape as scenery, that has come to dominate the meaning in the English-speaking world (Lorzing 2001).

The word resurfaces in the late nineteenth century in the title of a new profession, 'landscape architecture'. Since the 1990s, scholarship across disciplines has enhanced the discourse of landscape architecture professional practice. And while the sciences, ecology, and landscape ecology, informed of the inner workings of landscape, as dynamic and evolving, connecting local landscapes with regional and global ones, social scientists, specifically cultural geographers, focused on the social, cultural, and political dimensions of landscape. These influences expanded the scope of landscape architecture and the role of landscape architects. An initiative by the International Federation of Landscape Architects (IFLA), to update the professional definition as listed by the International Labour Organization (ILO) is clear indication of the changing professional scope. While the focus of the older definition was on 'product delivery', namely the design, construction, and maintenance of outdoor spaces (ILO 2012, p. 122), the proposed definition, yet to be approved by the ILO, aims to demonstrate that landscape architects address issues of "ecological sustainability, quality and health of landscapes, collective memory, heritage and culture, and territorial justice".[1]

Transformations in the professional definition and scope invariably influenced landscape architectural education as it prepares for growing complexity and unpredictability of the world we inhabit. Inspired by the call for urgent changes in teaching and learning proposed by the Organisation for Economic Co-operation and Development, Fetzer (2019, p. 15), president of the European Council of Landscape Architecture Schools (ECLAS), argues that "landscape and landscape architecture education provide an ideal context" to teach 'transformative' and 'anticipatory' competences, and one might add, coordinate interdisciplinary collaboration. The future-looking, "action-oriented approach to planning, management and protection of landscapes" introduced by the European Landscape Convention (ELC), has been another significant influence on landscape architecture education and practice (Herlin and Stiles 2016, p. 175). Since 2000, the ELC has enriched the discourse on landscape at a continental scale as an "essential factor for individual and communal well-being and an important part of people's quality of life" (ibid.). The ELC's definition of landscape recognises that people and their perception and valuation of landscape are the foundation for landscape policy, planning and management, and that these vary from one place to another and are far from being universal (see Fetzer in this volume).

The ELC contributes to the discourse on landscape architecture and inspires landscape architecture education and practice in Europe and elsewhere in the world. In education especially, landscape architecture programmes increasingly focus on participatory design and Community Based Learning (CBL). In parallel, the overlap of 'nation', 'country', and 'landscape' by the ELC has served as a dynamic platform for the practice of citizenships, rights, and the construction of heritage and identity, highlighting the political dimension of landscape. These influences have found their way to landscape architecture education and are slowly being recognised in the Middle East where the challenges are different.

'Landscape' in the Middle East: the colonial legacy

The word 'landscape' was introduced into the Middle East as part and parcel of Western colonisation. French and British authorities transformed the medieval fabric of Middle Eastern cities in an attempt to westernise and modernise the urban landscape. Orthogonal streets and wide tree-lined boulevards, traffic roundabouts, and municipal parks formed what came to be known as a 'public realm'. This "explains why the two terms [urban] public realm and landscape continue to be used interchangeably and why the meaning of 'landscape' today is closely associated with urban open spaces and municipal parks" (Makhzoumi 2021, p. 187).

Just as big a challenge is to overcome repercussions of the colonial mindset that permeates thinking and values in the Middle East. Whether intentionally or inadvertently, the colonial legacy continues to impact the discourse on landscape in the region. National borders, or 'lines in the sand', imposed by Western colonisers disrupted cultural and ecological continuities and undermined the shared community pride in local values and heritage. Failure to recognise vernacular horticulture practices, rural garden culture, and traditional management of scarce water resources, were not recognised as 'heritage'. Rather, colonial authorities focused on the region's biblical heritage and ancient archaeologies (Maffi and Daher 2014). As a result, rural landscapes that would have served as inspiration for ecologically responsive design and water-efficient management came to be undervalued which, in parallel, denuded the contemporary understanding of 'landscape' of its cultural and ecological significance.

Also problematic was the denigration of vernacular landscape culture and the century-old traditional practices abound in the region. Many traditional practices were not documented and are threatened, for example: soil conservation through stone terraces; intercropping date palms in arid Iraq, olive orchards in the eastern Mediterranean; rainwater harvesting ponds in Lebanon; water wheels, Syria and Iraq; Qanat irrigation, Iran and northern Iraq; and *aflaj* irrigation, United Arab Emirates. Villages in southern Lebanon converted village rainwater collection ponds into formal municipal gardens because the idea of a 'park' is perceived as landscape 'modernisation' (Makhzoumi and Shibli 2017). Similarly, the concept of village house gardens, a productive landscape combining use and pleasure rooted in the Middle East and eastern Mediterranean culture (Stordalen 2000), is discarded in favour of western style, ornamental gardens that are energy and water intensive and devoid of cultural significance (Makhzoumi 2015).

Such denigration of vernacular practices is not unique to the region. Colonial authorities were similarly blind to aboriginal people, cultures, and landscapes elsewhere, for example in Australia. Jones (2021) calls for a recasting of the 'Terra Nullius' by uncovering close affinities between landscape architecture and an indigenous concept of 'Country'(see Bryant et al. in this volume). The same challenge the profession faces in Africa (see Young in this volume). Doherty (2021, n.p.) argues that professionals need to consciously overcome the colonial mindset, not only to avoid "being agents, however unwittingly, for the extension of colonialism", but in order to "contribute to the discipline of landscape architecture and enrich it, rather than by myopically apply[ing] western-centric landscape architecture without a critical translation".

Another challenge is the narrow, outdated translation of the English word 'landscape'. While Europe and western countries are moving away from 'landscape' as scenic setting as discussed earlier, the meaning continues to dominate the local discourse in many Middle Eastern countries. The response to the impossibility of translating the accumulation meanings of the English word has varied (Lorzing 2001). In Turkey, for example the French word *peysage* is used, while in Iran the Farsi word *manzar*, is used – the word's Arabic origin implying 'scenery'. The translation in most English-Arabic dictionaries is similarly *manzar*, and more recently *mashhad*, respectively 'scenery' and 'scene' (Makhzoumi 2002). Borrowing the English or French word makes it difficult to strip them of their Western association. Similarly, borrowed words and ones with archaic meaning, hinders development of a contextualised understanding that is anchored in local histories, traditions, and values.

As a profession, landscape architecture came to prominence in the last decades of the twentieth century. Architects filled in the professional absence, doubling up as landscape architects, handling design and planning concepts and the hardscape, while agricultural engineers took the responsibility of planting and irrigation, the softscape – both professions, architecture and agricultural engineering, well established by the 1950s. As landscape architecture education picked up in Turkey and Iran, the professional body grew consistently in these two countries, but not in Arab countries in the region. IFLA affiliated national associations is one indication of the professional status and the challenge of landscape education in the Arab Middle East (see Figure 30.1).

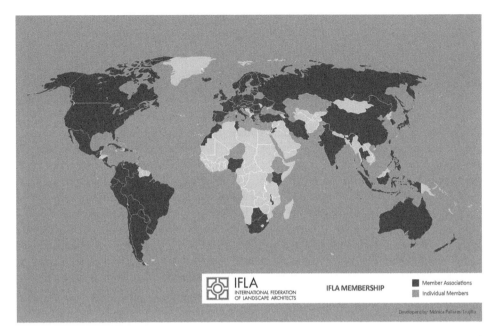

Figure 30.1 Landscape architecture is conspicuously absent in countries of the Middle East and North Africa in a map of national associations of the International Federation of Landscape Architects.

Source: Credit: Mónica Pallares, IFLA.

Landscape architecture education in the Middle East

Building on the foundations of the long-established garden culture in Turkey and Iran, the earliest academic programmes, ones that preceded landscape architecture, were introduced in the 1960s, with a focus on 'ornamental plants', gardens, and horticulture. There was no ambiguity in the word and idea of garden in academic and professional spheres and by the public at large. It took several decades for the title landscape architecture to appear, at first in faculties of agriculture and forestry, where courses were taught by "a mix of horticulturalist, agronomist and foresters" (Ortacesme et al. 2019, p. 161). In Iran, of the 42 programmes offering undergraduate qualification in landscape architecture, 13 are in departments of agriculture or forestry, the remaining are offered by departments of architecture, fine arts and design. Turkey has made great strides in the subject, with 72 landscape architecture departments, all ratified by the Higher Education Council. Yet again the absence of a word for landscape in Turkish and Farsi is problematic. In Turkey, the French word is used in the professional title, *Peyzag Mirmarlik*, while in Farsi, the title is *Mimar Manzar* – *mimar* and *manzar*, respectively, 'architect' and 'scenery'. The title 'landscape engineering' is translated into Farsi as 'green space engineering'.

The fluid nature of landscape architecture education, a field that breaches sciences and agriculture faculties, the arts and architecture, makes it possible for programmes to be located in either. Moreover, both professions, agricultural engineering, and architecture, are publicly recognised throughout the Middle East. Agricultural schools ensure a good foundation in ecology and earth sciences, horticulture, and irrigation. Architectural schools excel in design studio teaching, even if they fail to tackle "the intricate relationship between natural resources, the built environment, and human beings" to incorporate the "unpredictability of how ecological processes, social systems, and political dynamics will affect landscapes" (Egoz 2019, p. 84).

As a rule, professional titles offered by science faculties are 'agricultural engineer' or 'landscape engineer'. Programmes in schools of engineering and architecture generally confer the title 'landscape architect'. In Iran where landscape architecture education is well-advanced, 15 of the 19 universities offer a Bachelor of Science in Landscape Engineering, the remaining four granting a Bachelor of Science in Green Space Engineering. 'Landscape architecture' is reserved for graduate degrees, with 16 universities offering master's degrees in landscape architecture. Lebanese universities follow a similar pattern. The Department of Landscape and Territory Planning, established in 1998 by the Faculty of Agriculture at the Lebanese University, offers the degree of Agriculture Engineer with two specialisation tracks: Master in Landscape Heritage and Territory Engineering, and Master in Landscape Engineering. The Institute of Fine Arts, also at the Lebanese University, focuses on architecture and urbanism programmes and offers a Master in Landscape Architecture and Environment. Variations in the title of academic degrees conferred influenced the naming of IFLA National Associations in Iran and Lebanon, respectively the Iran Society for Landscape Professionals (ISLAP), and the Lebanese Landscape Association (LELA) – the intentional omission of 'architecture' allow membership to those qualifying in 'landscape engineer'.[2]

Landscape architecture education in the Arab Middle East

Landscape education in the Arab Middle East is still lagging in comparison to Turkey and Iran. Apart from landscape courses offered as part of agricultural engineering degrees in Jordan and landscape architecture departments in Iraq, there are only four programmes in the Arab Middle East that offer a degree in 'landscape architecture'. They are King Abdulaziz University, Kingdom of Saudi Arabia (KSA); Balamand University and the American University of Beirut, Lebanon; and very recently, the University of Bahrain. The absence is understandable in Iraq, Syria, and Yemen, countries that have experienced prolonged wars and continue to suffer the repercussion of civil strife. In the oil-exporting Gulf States the reasons are harder to fathom, due most probably to a lack of awareness of the new profession and its scope.

The Department of Landscape Architecture, Faculty of Engineering, King Abdulaziz University, established in 1976, was the first in the Arab Middle East to offer a Bachelor of Landscape Architecture. The four-year programme has 155 credit/hours, with design studio as a core subject in the last three years. Understanding of the programme by the public, academics and co-professional is far from secure. Questions continue to be raised regarding the status of an independent department with attempts to re-locate landscape under architecture, a profession well recognised in the KSA. Another challenge is securing accreditation from the National Centre for Academic Accreditation and Evaluation in the Kingdom. As in other Arab countries, the process is laborious and complicated further because the profession is new – the Bachelor of Landscape Architecture was formally accredited in 2021, a little under 50 years after the programme was launched. Holders of the degree are admitted to the Saudi Council of Engineers, which provides an umbrella for professional practice. Vision 2030, a strategic framework to reduce the nation's financial dependency on oil, emphasising greening and quality of life in Riyadh,[3] offers an opportunity to shed light on the profession and provides employment for young landscape architecture professionals.

The Urban Planning Institute (IUA), of the Académie Libanaise des Beaux-Arts (ALBA), University of Balamand, offers a four-year Bachelor of Landscape Architecture and a Master of Landscape Architecture, established respectively in 2006 and 2012. Landscape education benefits from IUA's membership with ECLAS and from its position as a founding member of the Association pour la Promotion de l'Enseignement et de la Recherche en Aménagement et Urbanisme (APERAU). The programme aims for an education that enables students to become specialists in emerging areas of sustainability, cultural diversity, and global environmental practices across three full-time and 20 part-time educators who teach undergraduate and graduate programmes, with both MLA and BLA programmes being recognised by the Lebanese Ministry of Higher Education. It is regrettable, however, holders of the BLA don't qualify for graduate membership of the Lebanese Order of Engineers and Architects. The programme falls in the cracks, neither architecture nor agricultural engineering, presenting another challenge for landscape education in the region.

Bahrain has led the way in the Gulf States. The Bachelor of Landscape Architecture was launched by the Department of Architecture and Interior Design, at the University of Bahrain, in 2017, with nine students enrolling in the academic year 2018–19. The programme continues to face challenges: the title of the discipline in Arabic creates confusion because of the range of possible translations of the degree title; it is difficult to recruit teaching staff with qualifications in landscape architecture; and students and their parents are concerned about future job opportunities.[4] Just as problematic is the relentless need to justify the programme, how it differs from architecture and urban design – established professions in Bahrain – and why architects and urban designers can't do the work of landscape architects.

Progress made in landscape education in the Palestinian State warrants attention here, considering the extreme political living conditions in this small corner of the Middle East. Attempts since the late 1980s by Samar Al Nazer include a two-year programme leading to a diploma at the Palestine Polytechnic University, Hebron, and another programme proposed and launched in 1996 at the Palestine Technical Community College in Ramallah. Both programmes were terminated because there were no applicants. The Bachelor degree in Spatial Planning Engineering launched by the Department of Architectural Engineering, Bir Zeit University (BZU), succeeded where the other two programmes failed. This was in large part due to an agreement for academic cooperation between BZU and the Department of Landscape Architecture at the Norwegian University of Life Sciences (NMBU).[5] The Norwegian Council for Higher Education, Centre for International Cooperation, endorsed and funded two agreements that were signed in 1998 and 2003, the latter securing scholarships in landscape architecture for two doctoral and four master's degree candidates. Supporting qualified landscape architecture academics was the surest way to address the main obstacle to landscape architecture education in the region. In parallel, cooperation with NMBU supported the preparation of educational programmes offering MS degrees in Landscape Architecture in the Department of Architecture, BZU, the programme officially launching in 2007. The graduate programme was open to applicants with undergraduate degrees in architecture or civil engineering, tuition covered by the cooperation agreement with NMBU. Yet again there was controversy regarding the degree title. The curriculum and courses were planned for a Bachelor of Landscape Architecture, however, during approval, the Ministry of Higher education requested that the title be changed to Bachelor in Planning and Urban Design Engineering. Although this is the official title, the one used in the programme is Urban Planning and Landscape Architecture.

Landscape education at the American University of Beirut

Encouraged by the demand for the landscape services provided by agricultural engineers in the KSA and Gulf States, the Bachelor of Science in Landscape Architecture and Ecosystem Management (BS LDEM) was launched in 2000 by the Department of Plant Sciences, Faculty of

Agricultural and Food Sciences (FAFS), American University of Beirut (AUB). The four-year programme offered a dual degree, a BS LDEM, and the *Diploma of Ingénieur Agricole*, the French name of the degree recognised by the Ministry of Education, entitling holders of the title membership in the Order of Engineers and Architects in Beirut.[6] The location in a science department/faculty offered several advantages: a well-established focus on extension, now referred to as community service; pioneering research in community-led biodiversity conservation, led by professor Salma Talhouk; and applied projects in ecosystem management and sustainable food systems, pioneered by professor Rami Zurayk. Spring term residency in the third year LDEM in FAFS Agricultural Research Education Centre in the Beqaa Valley was another advantage, exposing predominantly urban students to the diverse culture in rural Lebanon. There were also drawbacks: agricultural science courses outweighed design studio credit/hours; the notable absence of an established design studio culture alongside a low teacher/student ratio, a prerequisite in design studio teaching; and the difficulties of recruiting teaching staff. The curriculum was gradually amended to increase the credits for landscape design studio and tailor science courses to landscape architecture education rather than agriculture. To compensate for the absence of educators with academic qualification and experience in teaching landscape architecture, interdisciplinary design studio teaching was adopted, whereby scientists and architects shared their perspective and knowledge with students and, just as importantly, with each other.

Curricular changes and programme logistics aside, the biggest challenge was the prevailing perception in Lebanon of the profession, limiting the scope of landscape architecture to gardening. One way of showcasing the programme and demonstrating the professional scope was through Service Based Learning, whereby student skills were applied to address the needs of AUB community and the outstanding landscape of the AUB campus (see Figure 30.2). In successive years, SBL was applied in studio teaching to address social, economic, and environmental needs throughout Lebanon. The cap-stone project was another opportunity to demonstrate the scope of the profession to include community-driven nature conservation, urban infrastructural landscapes, landscape

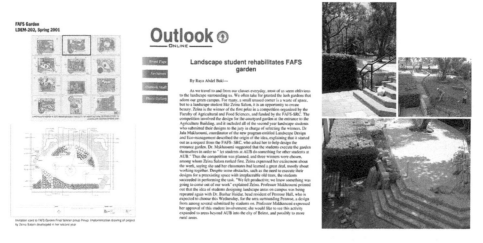

Figure 30.2 Winning landscape design by first-year student Zeina Salam for the faculty garden. Implementation drawings completed in her second year. The project is profiled in the university bulletin, an opportunity to showcase landscape architecture education.

Source: Images by Jala Makhzoumi.

heritage protection, and sustainable urban greening. Successive graduating classes helped promote the profession in Lebanon and the Gulf states, where many sought job opportunities.

Growing recognition of the programme prompted FAFS in 2007 to establish the Department of Landscape Design and Ecosystem Management. This facilitated planning and managing the programme and aligning it with advances in landscape architecture elsewhere. In 2012, the name of the degree was changed to Bachelor of Landscape Architecture, as a first step in seeking accreditation from the Landscape Architecture Accreditation Board (LAAB) in the USA.[7] LAAB accreditation was granted in the October 2021, which undoubtedly adds to the credibility of landscape architecture education at AUB and helps showcase the profession.

The curriculum of the BLA as it stands today is divided as follows: 27 credits in electives and language courses under General Education; nine from electives within the Faculty; and 108 required from BLA courses within the department. The combined total of 144 credits over four years and three summers provides a fundamental education that integrates the mission, goals, and objectives at all levels of the curriculum. Composition of the teaching staff reflects the overlap between architecture and landscape architecture in the region. Of the seven full-time teaching staff, four have undergraduate degrees in architecture and academic qualification in landscape architecture (PhD, MS/MA), and two have undergraduate degrees in agricultural engineering, specialising in landscape horticulture and nature conservation, and ecosystem management, soil, and irrigation. A roster of 14 part-time educators from different disciplines contribute to teaching. As the programme evolves, the operational challenges lie in the continued need to explain the profession to the university administration, to keep the programme running, balancing enrolment with budgetary restrictions amplified by Lebanon's economic crisis in 2020–21.

In the LDEM programme's 20 years, students and teachers have attempted to contribute towards a contextualised discourse on landscape architecture. Addressing the shortcomings of the colonial legacy is implicit to the discourse, realised through responsiveness to the political, social, cultural, and ecological context in Lebanon. The institutional set up at AUB, the exceptional campus landscape and physical proximity of departments and academic faculties has made possible collaboration across academic faculties and AUB research centres. A landscape approach was applied in projects on community-led biodiversity conservation in collaboration with the AUB Nature Conservation Centre, with attempts to link biodiversity and heritage and local identities in Lebanese villages.[8] Recovery following the war between Israel and Lebanon in 2006 was another opportunity to apply a landscape approach to repair disrupted landscapes and affected communities, an approach that came to be institutionalised in the AUB Center for Civic Engagement and Community Service.[9] More recently, landscape has led the path in tackling human rights and post-conflict displacement, namely the challenge of Syrian Refugees (Trovato 2018).

While anchored in a science faculty, landscape education at the AUB, has from its inception reached out to the Department of Architecture and Design (DArD), Faculty of Engineering and Architecture. Over two decades, a mutually beneficial collaboration has been established that includes joint design studio teaching and shared theory courses. The collaboration came to expose architects and landscape architects to each other's design approaches, specifically showcasing the potentials of an expansive landscape framing to architecture students and teachers. The most recent example is the Post-Disaster Community Management intensive programme that addresses the recovery in the aftermath of the Beirut blast of August 4, 2020. The international workshop advances the role of higher education in promoting social change and builds interdisciplinary collaborations between landscape students, LDEM, and students from the Business School, AUB, with architecture, landscape architecture, and urban design students from different universities to work on one of the most urgent, devastating, and emotionally charged challenges affecting the future and the livelihood of Lebanese people generally and the local community more specifically.[10] The workshop used intercultural partnerships to advance and broaden the discourse on landscape architecture and put the field at the forefront of urban and community resilience and recovery in times of uncertainty and change.

At the graduate level, cooperation between the two departments has focused on thesis supervision and graduate course teaching. This has been an opportunity to expose graduate students enrolled in the Master of Urban Design (DArD), architects, and landscape architects to a holistic landscape framing and the methodology of ecological landscape design (Makhzoumi 2019). The collaboration will flourish with plans approved by AUB administration to consolidate all design programmes under the School of Design, that will operate with an independent administration but continue to benefit from the resources of both faculties.

Conclusion

Like my own learning trajectory, landscape education is slowly transforming the prevailing conception of landscape beyond 'scenery', as landscape architecture students and young professional test new frontiers. Equipped with critical thinking and creative design sensibilities they respond to the political injustice, social adversity, and cultural specificity of the Middle East. The discourse on landscape architecture today has evolved to reflect the tensions between two contrasting meanings and conceptions of landscape. On one hand is the totalising landscape of state authorities and corporate property developers, where landscape continues to imply 'scenic' setting, a product. On the other hand is the meaning of landscape as lived in place, polity, and community, which emerges from landscape scholarship and practice and equally from activism (Makhzoumi 2021). These early beginnings are likely to overcome misconceptions resulting from poor translations of 'landscape' and form the beginning in the search for contextualised conceptions of landscape.

Regarding the rift in landscape education between science faculties and architecture, and the titles they confer, respectively 'landscape engineer' and 'landscape architect', it is important not only to accept but celebrate the fluidity of landscape education, that the discipline bridges the sciences, social sciences, art, and architecture. Reflecting on the ECLAS 2019 conference celebrating 100 years of landscape architecture education, Andreucci (2019, p. 96) points out "that the unique position of our discipline lies in that the very sphere it operates in allows for the enriching influence and contamination of various others, including but not limited to ecology, engineering and social sciences". Interdisciplinary cooperation across faculties at AUB was, as we argue, mutually beneficial. If we accept that the underlying aim of landscape education is to enable an expansive, dynamic, and scalar framing of the world we inhabit, to recognise global landscape interconnectedness while conceding to cultural specificities, then the 'conversion' of architects to the 'landscape' approach should be encouraged. After all, architecture is an established profession, with large numbers in practice and education, that can benefit from the landscape approach to contribute to a better future at the local, regional, and global levels.

I conclude by emphasising the role of academic institutions in the global north. Their support of landscape architecture education throughout countries of the global south is necessary to meet the many challenges argued in this chapter. Support can come in many forms. Academic support from Norway is one example that enabled Palestinian academics and students to successfully establish the foundation for landscape architecture education in this small part of the Middle East. Another form is to support programmes secure accreditation, as with the American University of Beirut. Accreditation provides credibility to this 'new' profession, at once raising the profile of landscape architecture education programmes and allaying concerns of students on future employment opportunities.[11]

Notes

1. The author was a member of the IFLA Task Force (2019–2020) which was tasked by IFLA to develop a new definition for landscape architecture for the ILO. The new definition was approved by IFLA Council is now posted on the IFLA website, available at www.iflaworld.com/the-profession (08.08.2021).
2. 'Architecture' has been adopted in naming the newly established Jordanian Association of Landscape Architecture (JALA), though Jordanian universities have yet to establish landscape architecture programmes.
3. See Green Riyadh City www.rcrc.gov.sa/en/projects/green-riyadh-project (accessed 27.07.2021).
4. Correspondence with Wafa Almadani.
5. The late professor Karsten Jørgensen of NMBU was directly responsible for initiating the academic collaboration with BZU in Palestine and for successful follow up for close to two decades (see Jørgensen and Hassan 2014). Shady Ghadban, professor of landscape architecture, BZU, who has since retired, was responsible for the cooperation and programme development.
6. Membership of the Order of Engineers and Architects offers health insurance coverage for the member and their family, which is necessary considering the absence of a national health system.
7. The American University of Beirut, previously known as the Syrian Protestant College, was registered in the New York State Education Department in 1866. The BS LDEM was registered in 2003.
8. See Baldati Bi'ati – Biodiversity Village Award, https://aub.edu.lb/natureconservation/Pages/Baldati.aspx (accessed 16.09.2021). The collaboration is discussed in Makhzoumi 2016.
9. Rabih Shibli, presently director of the Center for Civic Engagement and Community Service (CCECS), taught landscape design at the Department of Landscape Design and Ecosystem Management, AUB, (2006–2015). He spearheaded the early initiatives that defined the CCECS and continues to create opportunities for teachers and students at AUB to service disadvantaged communities in Lebanon. www.aub.edu.lb/ccecs/Pages/default.aspx (16.09.2021).
10. The project was initiated by Nayla Al-Akl, Assistant Professor, Department of Landscape Design and Ecosystem Management, AUB. The project is part of the Middle East Social Innovation Lab (MESIL) programme funded by DAAD in partnership with Nürtingen-Geislingen University, the German Jordanian University, Birzeit University and the Royal Society for the Conservation of Nature.
11. The efforts of IFLA Education and Academic Affairs is considerable in this respect, and will provide welcome support once approved to landscape education in regions where the profession is still new. See http://iflaworld.com/education-and-academic-affairs (12/08/2021)

References

Andreucci, M. (2019) 'ECLAS Conference 2019. Lessons from the Past, Visions for the Future: Celebrating 100 Years of Landscape Architecture Education in Europe', *Journal of Landscape Architecture*, 14(3), p. 96.

Bender, B. (1993) *Landscape Politics and Perspectives*. Oxford: Berg.

Doherty, G. (2021) 'Culture, Design and Two Yoruba Landscapes', *African Journal of Landscape Architecture*, 1, n.p. [Online]. Available at: www.ajlajournal.org/articles/culture-design-and-two-yoruba-landscapes (Accessed 2 August 2021).

Egoz, S. (2019) 'Landscape Is More Than the Sum of Its Parts: Teaching an Understanding of Landscape Complexity', in Jørgensen, K., Karadeniz, N., Mertens, E. and Stiles, R. (eds.) *The Routledge Handbook of Teaching Landscape*. London: Routledge, pp. 84–95.

Fetzer, E. (2019) 'Landscape Education: Our Path Towards Responsible Citizenship', in Gao, L. and Egoz, S. (eds.) *Lessons from the Past, Visions for the Future: Celebrating One Hundred Years of Landscape Architecture Education in Europe. Proceeding of the ECLAS Conference*, 16–17 September. As: School of Landscape Architecture, Norwegian University of Life Sciences, p. 15.

Herlin, I. and Stiles, R. (2016) 'The European Landscape Convention in Landscape Architecture Education', in Jørgensen, K., Clemetsen, M., Thoren, K. and Richardson, T. (eds.) *Mainstreaming Landscape through the European Landscape Convention*. London: Routledge, pp. 175–186.

ILO. (2012) *International Standard Classification of Occupations. Volume 1 Structure, Group Definitions and Correspondence Tables*. Geneva: International Labour Office.

Jones, D. (ed.) (2021) *Learning Country in Landscape Architecture. Indigenous Knowledge Systems, Respect and Appreciation*. Singapore: Palgrave Macmillan.

Jørgensen, K. and Hassan, R. (2014) 'Capacity Building in Landscape Architecture in Palestine', in Wolschke-Bulmahn, J., Fischer, H. and Ozacky-Lazar, S. (eds.) *Environmental Policy and Landscape Architecture*. Munich: Akademische Verlagsgemeinschaft, pp. 223–236.

Lorzing, H. (2001) *The Nature of Landscape: A Personal Quest*. Rotterdam: 010 Publishers.

Maffi, I. and Daher, R. (eds.) (2014) *The Politics and Practices of Cultural Heritage in the Middle East. Positioning the Material Past in Contemporary Societies*. London: I.B. Tauris.

Makhzoumi, J. (2002) 'Landscape in the Middle East: An Inquiry', *Landscape Research*, 27(3), pp. 213–228.

Makhzoumi, J. (2015) 'Borrowed or Rooted? The Discourse of 'Landscape' in the Arab Middle East', in Bruns, D., Kuhne, O., Schonwald, A. and Theile, S. (eds.) *Landscape Culture-Culturing Landscapes: The Differentiated Construction of Landscapes*. Wiesbaden: Springer Verlarg, pp. 111–126.

Makhzoumi, J. (2016) 'From Urban Beautification to a Holistic Approach: The Discourse of 'Landscape' in the Arab Middle East', *Landscape Research,* 2(10), pp. 1–10.

Makhzoumi, J. (2019) 'In-depth, Dynamic Understanding of Context: Ecological Landscape Design Method Applications in Graduate Urban Design Research', in Gao, L. and Egoz, S. (eds.) *Lessons from the Past, Visions for the Future: Celebrating One Hundred Years of Landscape Architecture Education in Europe. Proceeding of the ECLAS Conference*, 16–17 September. As: School of Landscape Architecture, Norwegian University of Life Sciences, pp. 90–91.

Makhzoumi, J. (2021) 'Beirut's Public Realm and the Discourse of Landscape Citizenships', in Waterman, T., Wolf, J. and Wall, E. (eds.) *Landscape Citizenships*. London: Routledge, pp. 182–204.

Makhzoumi, J. and Shibli, R. (2017) 'Beyond Reconstruction', *Topos*, 99, pp. 66–73.

Ortacesme, V., Uzun, O., Atik, M., Karacor, E., Yldrime, E. and Senik, B. (2019) 'Evolution of Landscape Architecture Education – Celebrating Its 50th Anniversary in Turkey', in Gao, L. and Egoz, S. (eds.) *Lessons from the Past, Visions for the Future: Celebrating One Hundred Years of Landscape Architecture Education in Europe. Proceeding of the ECLAS Conference*, 16–17 September. As: School of Landscape Architecture, Norwegian University of Life Sciences, pp. 161–162.

Stordalen, T. (2000) *Echoes of Eden. Genesis 2–3 and symbolism of the Eden Garden in Biblical Hebrew Literature*. Leuven: Peeters Publishing.

Trovato, M. G. (2018) 'A Landscape Perspective on the Impact of Syrian Refugees in Lebanon', in Asgary, A. (ed.) *Resettlement Challenges for Displaced Populations and Refugees*. Cham: Springer.

31
LANDSCAPE ARCHITECTURE EDUCATION IN CHINA
The pioneering work of Sun Xiaoxiang

Leï Gao and Guangsi Lin

Introduction

Chinese garden design inspired English landscape gardens in the 18th century, which influenced the works of Frederick Law Olmsted, founder of the modern profession of landscape architecture. Despite its early reputation, landscape architecture in modern China and the education of its professionals remains largely unknown to the world. How did modern landscape architecture education develop in China? Which connections exist between traditional and modern China, and between China and the Western world? This chapter is addressing both questions through a case study involving the example of Professor Sun Xiaoxiang (1921–2018),[1] an important education pioneer. Through thematic analysis of both primary sources, such as books and articles written by Sun Xiaoxiang and interviews with Sun conducted by the authors, and secondary sources, such as studies and recollections of Sun by his relatives, friends, colleagues and students, the authors aim to reveal the contribution of Sun in the development of landscape architecture education in contemporary China.

Sun Xiaoxiang was one of the founders of the first landscape architecture programme in the People's Republic of China (PRC). A talented landscape architect who integrated traditional and modern knowledge, and who had a good vision and global awareness, Sun played an important role in East–West communication since the 1980s. He helped introduce Chinese landscape philosophy and aesthetics to the Western world, and he was innovative in conceptualising landscape architecture in China. He encouraged students to study abroad, and many of them have become key figures in landscape architecture practice and education in China today. His career of 70 years spans and reflects the development of landscape architecture in China. He was one of the two Chinese landscape architects who received the IFLA Sir Geoffrey Jellicoe Award (SGJA) (IFLA 2014). The other awardee, Yu Kongjian, was a student of Sun during the 1980s at Beijing Forestry University.

The emergence of landscape architecture education in China

In Chinese culture and language, the terms 'landscape' and 'landscape architecture' are borrowed words. The Chinese word 山水 (*shan shui*, literally translated as 'mountain and water') is a commonly accepted translation for 'landscape', seeing in various landscape-related terms such as landscape painting (山水画, *shan shui hua*, literally translated as 'mountain-and-water painting'), landscape poetry (山水诗, *shan shui shi*, literally translated as 'mountain-and-water poetry') and so on.[2] However, the translation of 'landscape architecture' required more efforts, as it defines a profession that did not exist in ancient China. After its introduction into China in the early 20th century, several versions of the translation coexisted, and debates about the different translations continued until 2011, when 风景园林 (*feng jing yuan lin*, literally translated as 'scenery/landscape and garden') became the official translation for landscape architecture (Ministry of Education of the PRC 2011, p. 6).

China has a written history of landscape design of more than twenty centuries.[3] Traditionally, landscape and garden making relied on masters of planning and designing, and on craftsmen who were responsible for implementation and sometimes detailed designing. Masters included all who have control over the garden, such as property owners and their artist friends, governmental officials and even emperors themselves, often also the educated elite. Craftsmen included carpenters, tile and brick artisans, gardeners, artificial mountain artisans and so on (Ji and Hardie 1988; Yu and Padua 2006, p. 12). They were educated in practice through family heritage or apprenticeship.

Chinese garden traditions continued until the end of the 19th century, when Chinese students started to pursuit design education overseas, and when modern education systems gradually began to emerge. In 1917, the Canton Christian College, founded by American Christians, opened the course 'Landscaping' (renamed as 'Landscape Gardening' in 1920), taught in English by George Weidman Groff, a teacher who graduated from Pennsylvania State University and worked in the Canton Christian College since 1907 (Lin and Huang 2020, pp. 129–131). In 1920s, a few universities opened a landscape gardening course under their horticulture departments, such as Southeast University, Zhejiang University and a few others (Lin 2005, pp. 1–2). Students and teachers returning from Japan, the USA, France and other countries introduced Western landscape architecture concepts to China (Zhao 2008, p. 5). In 1935, Chen Zhi (1899–1989), a graduate from Tokyo Imperial University in Japan published the book *An Introduction to Landscape Architecture* (造园学概论), the first of such kind in China, in which Chen investigated the meanings, history and types of landscape architecture worldwide, and argued to establish an independent landscape architecture programme in China (Chen 1935).[4] Due to wars and changing regimes in the following decades, his vision came true only in 1951, two years after the founding of the PRC.

In 1951, Beijing Agricultural University launched the first landscape architecture programme (Lin 2012, p. 51). The programme transferred to Beijing Forestry College (renamed Beijing Forestry University in 1985) in 1956, and it remained one of a handful of landscape architecture programmes in China during the following decades. Along with rapid urbanisation since 1990s, and especially with university expansions since the late 1990s, landscape architecture education has gained increasing societal attention and numbers of programmes grew fast. Between 1998 and 2009, the number of universities that offered undergraduate degrees in landscape architecture had been growing 14% annually and 20% for those that offered graduate degree programmes (Lin 2014, p. 43). In 2006, for example, about 140 universities had undergraduate landscape architecture programmes, and twenty-four had graduate programmes in landscape architecture in Mainland China (Lin 2007, pp. 9 and 12). In 2014, the numbers were over 280 and over 120 respectively in Mainland China (Lin 2014, p. 43). Today, Chinese professionals and students are interacting actively with landscape architecture education and practice worldwide.

Sun Xiaoxiang and the early period of landscape architecture education in the PRC (1950s–1970s)

Sun Xiaoxiang was born in 1921 into a farming family in the Zhejiang province. He received primary education through home schooling before attending a modern school for secondary education. Such a combined way of learning equipped him with both traditional Chinese culture and modern scientific knowledge. Sun had shown talents in poetry, painting, calligraphy and theatrical play ever since his teenage years (Zhu 2019, pp. 11–12). In 1942, Sun enrolled at Zhejiang University, initially studying agriculture, moved to agricultural chemistry in the second year, and graduated from the Department of Horticulture in 1946, when he became a teaching assistant at Zhejiang University and taught floriculture and landscape architecture (Wang et al. 2007, p. 27).[5]

In 1949, when the PRC was established, landscape professionals became urgently needed to provide expertise for city reconstruction and expansion projects. In 1951, during a committee meeting of the Bureau of Construction at Beijing Municipal Government, the landscape architect Wang Juyuan (1913–1996) from Beijing Agriculture University and the architect Wu Liangyong (born 1922) from Tsinghua University proposed that the two universities should initiate a joint landscape architecture programme. The plan was submitted to their universities first and then to the Ministry of Education. It was quickly approved. In autumn 1951, the landscape architecture programme was launched on a pilot basis. Ten second-year students from the Department of Horticulture at Beijing Agriculture University enrolled. In addition to their two years of horticulture, they took two years in the Department of Architecture at Tsinghua University. Eight of them graduated in 1953 and became the first landscape architecture professionals trained in the PRC (see Table 31.1) (Wu 2006, p. 1).

In 1953, the landscape architecture programme returned to Beijing Agriculture University under the reorganisation of the educational system in China which aimed to turn comprehensive universities into specialised universities (Wu 2006, p. 2). With weakened teaching support from Tsinghua University, Wang Juyuan had to 'borrow' teachers from other universities. Sun Xiaoxiang was one of them. Seconded for the programme from Zhejiang University, he taught floriculture and garden art. In 1954, Sun was sent to Nanjing Institute of Technology to study architecture. One year after, he was formally employed as a lecturer at Beijing Agriculture University, teaching garden art and landscape design (Meng 2019, p. 15). In 1956, after further adjustments made according to higher educational models from the Soviet Union, the landscape architecture programme, together with all its teachers and students, were relocated to Beijing

Table 31.1 The curriculum for the 1951–1953 landscape architecture programme, jointly established by Beijing Agriculture University and Tsinghua University.

Urban planning (including urban greening)	Plant classification
Garden design	Forestry
Architecture design	Surveying
History of Chinese gardens	Architecture
History of Western gardens	Ornamental trees and flowers
Introduction to architecture	Garden art
Cartography perspective	Garden engineering
Drawing (sketches, watercolours)	Landscape management
	Chinese architecture

Source: Yang and Zheng 2019, p. 37.

Forestry College (renamed Beijing Forestry University in 1985) and renamed Department of 'Greening of cities and inhabited places' in 1957 (Chen 2002, p. 5; Lin 2012, p. 52). The name changed again to Department of Urban Landscape Architecture in 1964. In 1965, the Department, together with the landscape architecture programme, was abolished and not restored until 1974. Since 1979, it was renamed as Department of Landscape Architecture (Lin 2012, p. 52).

To improve living environments for socialist workers, including both recreational and educational opportunities, urban parks, botanical gardens and the like were especially needed in the newly founded PRC. By the time Sun came to Beijing, he had led two landscape projects in Hangzhou, Zhejiang province. Both were very successful and had great influence on landscape architectural design in China. One project is the park Viewing Fish at Flower Harbour, designed in 1951–1954 and completed in 1956, by the historic scenic site of West Lake. The other is Hangzhou Botanical Gardens, designed in 1952–1953 and the first part opened in 1957. Within the next decade, Sun designed five more botanical gardens. With his growing practical experience, Sun quickly became a popular teacher, inspiring students with vivid examples about using traditional garden art for contemporary landscape design (Meng 2019, p. 15). Sun became the head of the landscape design teaching office in the landscape architecture programme at Beijing Forestry College in 1957 (Wang et al. 2007, p. 28).

At that time, no textbook was available for teaching. Sun wrote two manuscripts, *Garden Art* and *Landscape Design*, and published them in the form of mimeographed lecture notes in 1958. In the following decades, he combined the two manuscripts into one volume that he updated several times and eventually published in 1986. Entitled *Garden Art and Landscape Design*, the book since then has remained a classic textbook. The latest edition appeared in 2011 (see Table 31.2). This book reflects Sun's perception of landscape architecture in the early period of his career, where the focus was mainly on landscape design of public parks and green spaces, embedded with traditional Chinese aesthetics.

What made Sun quickly stand out from his contemporaries were his profound skills in Chinese art (especially painting, poetry and calligraphy), his professional competence solidly rooted in horticulture, his broad interests in different fields, as well as his passion and creativity (Hu 2019, pp. 27–30). Not formally trained as a landscape architect, he learned landscape design from different sources. For example, he got inspirations on landscape composition from Chinese paintings, and learned how to use contour lines by surveying public parks in Shanghai designed by English landscape architects (Meng 2019, p. 16). He synthesized such knowledge into his

Table 31.2 Table of contents of Sun Xiaoxiang's book *Garden Art and Landscape Design*.

Part	Chapter
Part 1. Theories of garden art	1.1 Features of garden art
	1.2 Principles of garden design
	1.3 Spatial design and progression arrangement
	1.4 Colour design
Part 2. Planting design	2.1 Lawns and ground cover plants
	2.2 Formal planting design
	2.3 Natural planting design
Part 3. Landscape design	3.1 Public parks
	3.2 Botanical gardens
	3.3 Zoological gardens

Source: Sun (2011) (table drawn and translated by the authors).

landscape projects, including the Viewing Fish at Flower Harbour park. Historically the site was a well-known scenic spot but turned derelict at the time when Sun successfully re-created the historical scene of *fish kissing flower petals floating on water*. At the same time, the new design satisfied contemporary needs to function as a public park, with lawns and various thematic gardens grouped in an area of 11 hectares for recreational and educational purposes. Sun's design took inspiration from both traditional Chinese gardens and from English and Japanese gardens. Soon after its completion, the work was regarded as the best public park in the country. It was included in exhibitions both in the Soviet Union and the first International Congress in Parks and Recreation in the United Kingdom in 1957 (Zhao 2008, p. 85).

Public park design in China in the 1950s adopted the Soviet principle of 'socialist content/values and national styles', which were largely interpreted as public parks based on traditional garden art and serving the needs of socialist workers. In *Garden Art and Landscape Design*, Sun summarised the guiding principles specifically for public parks and gardens, botanical gardens and zoological gardens. These gardens were to serve the working class; they are to be appropriate and useful, economical and beautiful; creating socialist gardens and parks by critically inheriting traditional Chinese garden culture and selectively studying foreign landscape architecture; and paying attention to the relationship between design, construction and maintenance (Sun 1986a, pp. 5–8). In an article published in 1962, Sun explained how traditional garden design was consistent with contemporary guiding principles. For example, as stated in a 17th-century Chinese garden book *Yuan Ye*, using existing features on site and borrowing views from the surroundings could save money and enrich the experience (Sun 1962, pp. 84–86).

During the Cultural Revolution period (1966–1976), Sun's career was halted, and so was the landscape architecture programme at Beijing Forestry College (Chen 2002, p. 5). It was not until the late 1970s that landscape architecture education resumed.

Sun Xiaoxiang and landscape architectural education after the 1980s

In 1979, the 'Reform and Openness' policy was launched, and a series of cultural and scientific exchanges were promoted to re-establish communication between China and Western countries. For Example, in 1981, a Chinese delegation of landscape architects visited National Parks and urban green infrastructure in North American cities. Sun was one of the delegates. In his first experience abroad, Sun saw great differences among landscape professions and their education in China and in the West. Upon his return to China, Sun published two articles in the Journal of the Beijing Forestry College, describing the national park system in the United States and reflecting on the management of scenic areas and historical sites in China (Sun 1982a, 1982b). This visit and visits during the following years expanded Sun's understanding of the core issues of contemporary landscape architecture, deepened his thinking on the significance of Chinese gardens for the contemporary urban environment, and initiated Sun's development of theories, including those of ecosystem engineering of urban green spaces that he called 'Earthscape Planning' (Wang et al. 2007, p. 31).

In autumn 1981, Darwina Neal, vice president of the American Society of Landscape Architects, met Sun at Beijing Forestry College and introduced the International Federation of Landscape Architects (IFLA). Sun joined the IFLA World Congress in 1982. He 'charmed those with whom he spoke and showed [. . .] beautiful watercolour paintings he had done of Chinese landscapes' (Darwina 2011, p. 23). In 1983, Sun became an individual member of IFLA and since then had been a featured speaker at several IFLA conferences. He continued to work on the Chinese Society of Landscape Architects joining IFLA, which became reality in 2005 (Darwina 2011, p. 23; Richard 2011, p. 19).

Between 1985 and 1991, Sun made several academic visits to Australia and the USA[6] (Lin n.d.; Sun 1994, p. 22). He gave lectures on Chinese gardens at universities and to the public, held solo exhibitions on Chinese paintings, absorbed new ideas in landscape architecture and introduced them to China, gradually building more bridges between China and the West in the greater world of landscape architecture.

Interpreting Chinese garden art and evaluating Chinese landscape philosophy

Starting around the 1980s, one of Sun's endeavours was to introduce Chinese gardens, their philosophy and art, to the Western world, and to uncover their contemporary meanings for his students, for the landscape architecture profession and for society in general. As an educator and practicing professional, one of his messages was the Chinese reverence for nature. This message came at a time of growing environmental concern in Western countries. Different from Western views, in traditional Chinese perception, not man but '(N)ature is the ruler and purpose of the world and man is only a member of the community of living creatures in nature and the world' (Sun 1986b, p. 484). 'Chinese love of, and sensitivity towards, the aesthetics of nature were widely expressed in the areas of philosophy, art, literature and painting for more than 3000 years' (Sun 1986b, p. 484, 1994, p. 22). Therefore, it is not surprising to see how 'China was the first country to pursue a natural approach to landscape design' (Sun 1986b, p. 484, 1994, p. 23).

In the 1980s, the challenges of rapid urbanisation were topics of several conferences Sun attended[7] (IFLA 2012, p. 44; Sun 1985, p. 252). He contributed by discussing to rethink human relationship with nature and to answer questions such as 'how to build modern cities without destroying the pleasure and ecosystem of nature'. Sun pointed out that 'the pleasures of wild nature and the heritage of art should be mingled with the planning of the modern city' (Sun 1994, p. 27), and 'the traditional Chinese approach of creating cities rich with the pleasure of wild nature may be of some use to the modern landscape planners and designers' (Sun 1994, p. 22).

To uncover Chinese garden design principles, based on his understanding of classical garden art and practice of modern landscape projects, Sun proposed 'three aesthetic levels': the level of natural beauty and beauty of life, the level of artistic beauty, and the level of spiritual beauty. The first level is to find prototypes from nature and life. Beautiful landscapes in China provide rich sources of inspiration. When designing a garden, one should always think about the life in it: reading, drawing, listening or playing music, drinking tea, playing chess, meeting friends and so on. After collecting prototypes from nature and life, the second level is to use the painter's eyes and musician's ears to make the garden artistically admirable. Sun introduced various ways to achieving this. For example, selecting and surveying the site is a key step to make good use of natural elements and surroundings. Chinese painting theories can be used in guiding water, rockery and plant arrangement; the succession of the seasons should also be considered to create all-seasonal enjoyment. The third level is, with the love of a poet, the wisdom of a philosopher and a childlike innocence, to create the spirit of the garden and make the garden a personal utopia and paradise. Sun believes that the third level reflects the state of a civilisation and makes a garden a real piece of art (Sun 1984, pp. 53–55, 2001, pp. 62–68, 2005b, pp. 36–38). Sun disseminated this theory through articles, lectures and conference speeches in China and abroad, and demonstrated his theory in various design projects with his students.

Redefining landscape architecture education

Taking part in the World Conference on Education for Landscape Planning at Harvard University in 1986 left a profound impact on Sun (Wang 2011, p. 17). He embraced the idea of Landscape Planning as 'an interdisciplinary field which focuses on issues related to land use and natural resources management, the development and change of rural regions, landscape ecology and the urban and metropolitan landscape' (Sun 2006, p. 11). These insights lead to reforms in landscape architecture education in China. Before this conference, Sun once summarised 'five legs' that support an outstanding landscape architect: painting, ecology, horticulture, architecture and poetry (Sun 1986b, p. 486). At the conference, Sun renewed his idea and said

> if I want to be an outstanding modern *landscape planner*, these five legs are still not enough. In that case, I would require a knowledge of economics and sophisticated technologies such as aerial survey techniques, remote sensing and computer-aided design methods.
>
> (Sun 1986b, p. 486)

Driven by increasing environmental concerns, landscape architects had started (re)exploring landscape at larger scales. In 1969, Ian McHarg published his seminal book *Design with Nature*, creating a rational, systematised design process. In his book *Landscape Planning: an Introduction to Theory and Practice*, Brian Hackett (1971) set up professional and technical foundations for landscape planning. The perception of landscape architecture was renewed, and its boundaries expanded (Seddon 1986, pp. 340–341).

Recognising gaps in landscape architectural education between China and the West in the late 1980s, Sun saw that, in educating landscape planners, China must learn new skills from other countries. He recommended to his talented students that they take graduate studies abroad. Among them were Hu Jie who studied at University of Illinois Urbana-Champaign 1990–1995, Wang Xiangrong who studied at University of Kassel in Germany 1991–1995 and Yu Kongjian who studied at Harvard University 1992–1995. All returned to China and became influential landscape architects and educators at Tsinghua University, Beijing Forestry University and Beijing University respectively.

At the turn of the 21st century, along with the speedy urban expansion in China, environmental issues became increasingly severe. Recognising the growing responsibilities of landscape architecture professionals, Sun appealed to widen the scope of the profession to embrace 'Earthscape Planning' (Sun 2002, 2006). Sun defines Earthscape Planning as a macro-scope strategy beyond territorial boundaries in the management of a broad range of spheres, reaching from inner crust to outer space, a layer 60 km in thickness. This means that integrated planning ought to consider all resources in both continents and oceans, deep under and high above the surface of the Earth (Sun 2005a, p. 7). The scale of landscape architecture, therefore, stretches from designing landscapes from one end to the other of Earthscape Planning. The central tasks for landscape architects, as described by Sun, are first to protect the wild nature (first nature) that has not been exploited by humans, and secondly to wisely plan the second nature (such as agriculture land, industrial land, cities and villages and so on) to ensure a sustainable development (Wang et al. 2007, p. 33).

In order to encourage new generations of landscape architecture professionals to obtain multidisciplinary knowledge and skills for such responsibility, Sun proposed to establish the Graduate School of Earthscape Planning at Beijing Forestry University (Sun 2006, pp. 12–13).

Table 31.3 A comparison between the existing and the envisioned structures of landscape architecture-related institutions at Beijing Forestry University.

Current structure		Envisioned structure by Sun Xiaoxiang		
School of Landscape Architecture • Landscape Architecture • Landscape Gardening • Urban Planning • Ornamental Horticulture • Tourism Management	Research Institute for Ecological Habitat Environment of Beautiful China	School of Landscape Architecture • Landscape Architecture • Urban Design Architecture • Landscape engineering	School of Environmental Horticulture • Environmental Horticulture • Management and Maintenance of Urban Park and Green Space System • Garden plants Protection • Generic Engineering of Garden plants • Green Space Tourism	Graduate School of Earthscape Planning

Source: *Table drawn by* the authors, with references to the official website (Beijing Forestry University 2020) and Sun's article (Sun 2006, p. 13).

Although such vision could hardly be realised or even be understood at the beginning, its value has been gradually revealed. In 2008, Yu Kongjian led a planning research project 'National Ecological Security Pattern Plan' commissioned by the Chinese Ministry of Environmental Protection (Saunders 2012, p. 192). In 2018, an interdisciplinary Research Institute for Ecological Habitat Environment of Beautiful China was established at Beijing Forestry University, led by Wang Xiangrong (see Table 31.3) (Beijing Forestry University 2018). In 2020, a group of scholars from multiple disciplines, led by the landscape architect Yang Rui at Tsinghua University, published 'Cost-effective priorities for the expansion of global terrestrial protected areas: Setting post-2020 global and national targets' in *Science Advances* (Yang et al. 2020). Large-scale strategic planning led by landscape architects, as Sun appealed since the late 1980s, are gradually taking shape in China.

Would Sun have been satisfied with these developments? If we look at the three aesthetic levels he proposed, we might find a place for 'Earthscape Planning' at the first aesthetic level, which aims at keeping nature alive, or say environmental sustainability. Nevertheless, there are higher levels that landscape architects should attain. As Sun said, to create a paradise on earth, the third level – the realm of spiritual beauty – is needed. 'A garden is a paradise in the human world. This paradise should be full of love. The highest level is selflessness'. To further explain, Sun quoted Confucius: 'In practicing the rules of propriety, it is harmony that is prize' and argued that the ultimate strategy for a sustainable future of humanity lies in this ancient philosophy (Wang 2011, p. 17).

Conclusion

The history of modern landscape architecture education in China has been shaped by generations of pioneers, such as Chen Zhi, who comprehensively introduced the discipline of landscape architecture and called for its establishment in Chinese higher education, and Wang Juyuan and Wu Liangyong, who promoted and created the first landscape architecture programme in China, to name but a few. Amongst the pioneers, Professor Sun Xiaoxiang is a unique figure.

His life-long experiences went in parallel with the development of the PRC and well reflected the development of landscape architecture education in contemporary China. His early studies of traditional Chinese art and literature laid the foundation for Sun's understanding of Chinese gardens and cultivated a heart that reveres nature. His subsequent exchanges with Western landscape architects and educators broadened his horizons and facilitated his reinterpretation of the contemporary meanings of Chinese gardens and views of nature. In the field of landscape architecture education, he is a bridge between the past and today, between China and the West.

Notes

1 Except for the two authors of this chapter, all Chinese names in this chapter appear as surname first, followed by first name(s).
2 The appreciation of natural landscape appeared in Chinese culture much earlier than that in Europe. The earliest existing landscape painting in China, entitled 'Spring tour' and painted by Zhan Ziqian (ca. 545–618), dates to the 6th century, for example.
3 For example, in the *Book of Poetry*, the Chinese classics dating from the 11th to 7th centuries BC, there is a description of the King of Zhou's park. The park was planned by the King and was constructed by his people in a short time. The park contained at least an earthen terrace, a fishpond, trees and various animals. The place was used for ritual ceremonies and gatherings (Chinese Text Project n.d.).
4 The English books that Chen Zhi referenced in *An Introduction to Landscape Architecture* include Samuel Parsons (1915) *The Art of Landscape Architecture*, O. C. Simonds (1920) *Landscape Gardening* and Frank F. Waugh (1927) *Formal Design in Landscape Architecture*, and others.
5 The Chinese term for the landscape architecture course that Sun taught is 造园学 (*Zao yuan xue*, literally translated as 'studies on garden making'), a term that Chen Zhi used for his book *An Introduction to Landscape Architecture* published in 1935.
6 Sun held the 1985 Haydn Williams Fellowship at the Curtin University of Technology, Western Australia, and made a lecture tour to fifteen universities of Australia's major cities, jointly sponsored by the Australian Institute of Landscape Architects (AILA) and the Garden History Society, and financed by a grant from the Design Arts Board of the Australia Council. He served as visiting scholar at the Graduate School of Design, Harvard University, from September 1989 to February 1990, and made a lecture tour to ten American universities under the sponsorship of Harvard University (Lin n.d.; Sun 1994, p. 22).
7 For example, the topic of IFLA's Third Eastern Regional Conference in Hong Kong in 1984 was 'Urban explosion in Asia'; the topic of Australian Institute of Landscape Architects (AILA)'s Annual Conference in 1985 was 'the Cityscape'. Sun Xiaoxiang attended both conferences and gave speeches.

References

Beijing Forestry University. (2018) *Beijing Forestry University Beautiful China Habitat Ecological and Environment Institute Construction Seminar Held* [Online]. Available at: http://sola.bjfu.edu.cn/english/events/281992.html (Accessed 22 November 2020).
Beijing Forestry University. (2020) *Introduction of Schools: School of Landscape Architecture* (学院简介：园林学院) [Online]. Available at: http://sola.bjfu.edu.cn/chinese/gaikuang/xyjj/ (Accessed 22 November 2020).
Chen, Y. (2002) 'For the 50th Anniversary of the Establishment of LA Division (Discipline) (纪念造园组 [园林专业]创建五十周年)', *Chinese Landscape Architecture*, 18(1), pp. 4–5.
Chen, Z. (1935) *An Interduction to Landscape Architecture* (造园学概论). Shanghai: Commercial Press.
Chinese Text Project. (n.d.) *Book of Poetry: Ling Tai* [Online]. Available at: https://ctext.org/book-of-poetry/ling-tai (Accessed 11 November 2020).
Darwina, L. N. (2011) 'Prof. Sun Xiao-Xiang: 60-Year Career', *Landscape Architecture*, (3), pp. 22–24.
Hackett, B. (1971) *Landscape Planning: An Introduction to Theory and Practice*. Newcastle upon Tyne: Oriel Press.
Hu, J. (2019) 'Missing Professor Sun Xiaoxiang, Our Mentor (怀念导师孙筱祥先生)', *Landscape Architecture*, (10), pp. 26–32.

IFLA. (2012) *IFLA Green Book: 50 Anniversary Book* [Online]. Available at: https://issuu.com/ifla_publications/docs/ifla_green_book (Accessed 23 November 2020).
IFLA. (2014) *SGJA Awardee 2014: Professor Sun Xiao Xiang* [Online]. Available at: www.iflaworld.com/sgja-2014-winner (Accessed 10 September 2020).
Ji, C. and Hardie, A. (trans.) (1988) *The Craft of Gardens*. New Haven: Yale University Press.
Lin, G. (2005) 'Review and Prospect – A Study of the Landscape Architecture Education in China (1) (回顾与展望 – 中国LA学科教育研讨[1])', *Chinese Landscape Architecture*, 21(9), pp. 1–8.
Lin, G. (2007) 'The Survey and Analysis of Landscape Architecture Disciplines and Specialities Setting in Mainland China 1951–2006 (1951–2006 年中国内地风景园林学科与专业设置情况普查与分析)', *Chinese Landscape Architecture*, 23(5), pp. 7–13.
Lin, G. (2012) 'Establishment and Development of LA Specialty in Beijing Forestry University (北林风景园林学科创办及发展)', *Landscape Architecture*, (4), pp. 51–54.
Lin, G. (2014) '30 Years of Landscape Architecture Education Development in China (中国风景园林教育发展30年)', *Garden*, 2014(10), pp. 42–44.
Lin, G. (n.d.) *Chronology of Sun Xiaoxiang* (unpublished manuscript).
Lin, G. and Huang, W. (2020) 'Early Development of Landscape Architecture Education in Lingnan University (岭南大学风景园林教育的早期发展)', *Chinese Landscape Architecture*, 36(2), pp. 129–133.
McHarg, I. (1969) *Design with Nature*. Garden City: The Natural History Press.
Meng, Z. (2019) 'Most Memorable Professor Sun Xiaoxiang (难忘的孙筱祥先生)', *Landscape Architecture*, 26(10), pp. 15–16.
Ministry of Education of the PRC. (2011) *Catalog of Degrees and Talent Development (2011)* (学位授予和人才培养学科目录[2011年]) [online]. Available at: www.moe.gov.cn/srcsite/A22/moe_833/201103/t20110308_116439.html (Accessed 30 October 2021).
Richard, L. P. T. (2011) 'A True Citizen of the World', *Landscape Architecture*, (3), p. 19.
Saunders, W. S. (2012) 'Reinvent the Good Earth: National Ecological Security Pattern Plan', in *Designed Ecologies: The Landscape Architecture of Kongjian Yu*. Basel: Birkhäuser, pp. 192–199.
Seddon, G. (1986) 'Landscape Planning: A Conceptual Perspective', *Landscape and Urban Planning*, 13, pp. 335–347.
Sun, X. (1962) 'A Discussion on Traditional Chinese Garden Design Methods (中国传统园林艺术创作方法的探讨)', *Journal of Horticulture*, 1(1), pp. 79–88.
Sun, X. (1982a) 'Scenic Areas and Sites of Historic Interest of China (中国风景名胜区)', *Journal of Beijing Forestry College*, (2), pp. 12–16.
Sun, X. (1982b) 'National Parks in the United States of American (美国的国家公园)', *Journal of Beijing Forestry College*, (2), pp. 43–49.
Sun, X. (1984) 'Let a Hundred Flowers Blossom!: Classical Gardens in China', in Leander, B. (eds.) *Cultures – Dialogue between the Peoples of the World 34/35*. Paris: UNESCO, pp. 51–58.
Sun, X. (1985) 'Professor Sun Xiaoxiang Shines in Australia', *Landscape Australia*, 7(3), p. 252.
Sun, X. (1986a) *Garden Art and Landscape Design* (园林艺术及园林设计). Beijing: Beijing Foresty University Press.
Sun, X. (1986b) 'The Aesthetics and Education of Landscape Planning in China', *Landscape and Urban Planning*, (13), pp. 481–486.
Sun, X. (1994) 'The City Should Be Rich in the Pleasures of Wild Nature', *Keisitics*, 61(364–365), pp. 22–28.
Sun, X. (2001) 'Art: An Aesthetic Theme in Chinese Scholars' Gardens', in Anagnostopoulos, G. L. (ed.) *Art and Landscape*. Athens: Panayotis and Effie Michelis Foundation, pp. 55–71.
Sun, X. (2002) 'Landscape Architecture. From Garden Craft, Garden Art, Landscape Gardening to Landscape Architecture, Earthscape Planning (风景园林：从造园术、造园艺术、风景造园到风景园林、地球表层规划)', *Chinese Landscape Architecture*, (4), pp. 7–12.
Sun, X. (2005a) 'The Ecological Engineering of Modern Urban Garden and Green Space System and the Sustainable Development of Cities (abridged) (现代城市园林绿地生态系统工程与城市可持续发展[选登])', *Landscape Architecture*, (1), pp. 3–8.
Sun, X. (2005b) 'Art: An Aesthetic Theme in Chines Scholars' Gardens (艺术是中国文人园林的美学主题)', *Landscape Architecture*, (2), pp. 32–39.
Sun, X. (2006) 'Sun Xiaoxiang's Column on Modern Landscape Architecture & Landscape Planning. Fourth Topic: A Suggestion on the Establishment of the 'Earthscape Planning' Discipline on Basis of the Improvement of the 'Landscape Architecture' Discipline (孙筱祥谈大地规划与风景园林　第四

谈：关于建立与国际接轨的"大地与风景园林规划设计学"学科，并从速发展而建立"地球表层规划"的新学科的教学新体制的建议)', *Landscape Architecture*, (2), pp. 10–13.
Sun, X. (2011) *Garden Art and Landscape Design* (园林艺术及园林设计). Beijing: China Architecture and Building Press.
Wang, S., Lin, G. and Liu, Z. (2007) 'Solitary Cultivation and Silent Dedication – Prof. Sun Xiaoxiang's Great Contribution to Landscape Architecture Subject and Its Profound Influence (孤寂耕耘 默默奉献 – 孙筱祥教授对"风景园林与大地规划设计学科"的巨大贡献及其深远影响)', *Chinese Landscape Architecture*, (12), pp. 27–40.
Wang, X. (2011) 'I Am an Old Student of Landscape Architecture – Interview with Professor Sun Xiao-Xiang, a Well-Renowned Chinese Landscape Architecture Educator and Landscape Architect ("我是风景园林学的一名老学生" – 访我国著名风景园林教育家和设计师孙筱祥教授)', *Landscape Architecture*, (3), pp. 16–17.
Wu, L. (2006) 'Recalling Mr. Wang Ju-Yuan and the Establishment of Earliest Landscape Architecture Subject in China (追记中国第一个园林专业的创办 – 缅怀汪菊渊先生)', *Chinese Landscape Architecture*, (3), pp. 1–3.
Yang, R., Cao, Y., Hou, S., et al. (2020) 'Cost-Effective Priorities for the Expansion of Global Terrestrial Protected Areas: Setting Post-2020 Global and National Targets', *Science Advances*, 6(37). Epub ahead of print 9 September 2020. https://doi.org/10.1126/sciadv.abc3436.
Yang, R. and Zheng, X. (eds) (2019) *66 Years of Excellence – Landscape Architecture at Tsinghua University*. Beijing: Applied Research and Design Publishing.
Yu, K. and Padua, M. (eds.) (2006) *The Art of Survival: Recovering Landscape Architecture*. Beijing: China Architecture and Building Press.
Zhao, J. (2008) *Thirty Years of Landscape Design in China (1949–1979): The Era of Mao Zedong*. PhD thesis, University of Sheffield, Sheffield.
Zhu, C. (2019) 'Let Life Be Beautiful Like Summer Flowers and Death Like Autumn Leaves: Speech at the Memorial Meeting on the Anniversary of the Loss of Professor Sun Xiaoxiang (生如夏花，逝如秋叶 – 孙筱祥先生逝世周年追思会感言)', *Landscape Architecture*, (10), pp. 11–13.

32
LANDSCAPE DESIGN EDUCATION IN JAPAN
The Meiji, Taisho, and Showa Periods

Chika Takatori

Introduction and overview

Aiming to provide insights into the history of garden and landscape design education in Japan, this chapter spans the time of the Meiji and Taisho (1868–1925) and the Showa periods (1926–1989). Recognizing three educational systems, the oldest one is the apprenticeship system that, after around 1912 merged into an atelier system. The second is the in-house system of government and city administrations. The third is the academic system of institutes of higher education. With apprenticeship, atelier and in-house systems learners are company and office staff members and learning occurs on the job; learners meet requirements that individual masters and agencies set. At schools, colleges, and universities, students meet requirements of formal study programmes and curricula. In Japan, several boundary-crossing links exist between the three systems, either through individuals who work in more than one system, or through ideas that flow from one system to another.

Locating and sampling primary sources, reviewing journal papers and books, and interviewing people involved in studying landscape education history, this chapter is based on connecting sources from different disciplines, including landscape design and history. It identifies how, through their life-long achievements, individuals connect the three systems of apprenticeship, in-house, and academic education, and links knowledge about the three systems, which have only been studied separately so far.

During the Meiji and Taisho periods (1868–1925), in-house garden designers worked inside administrative and city organizations such as the Imperial Household Agency (up until World War II the "Imperial Household Ministry"), the Ministry of Interior, Tokyo City, Osaka City, etc. They employed garden design techniques they had learned by conventional apprenticeship. To incorporate ancient garden design knowledge into the modernizing academic system, connecting garden design with horticultural studies played a major role, also the combination with the newly established field of forestry. The Imperial Household Agency "Takumi Dormitory" was instrumental in adding elements of westernization to traditional Japanese garden design and gardening techniques of which court landscape artists were the masters. The Imperial Household Agency thus contributed to the creation of horticultural studies at universities in Japan. In addition, the Meiji Jingu Building Bureau, Ministry of Interior created the basis for modern horticultural education, by combining advanced knowledge about forestry as field of study imported from Western countries. Many landscape architects who rose in the ranks of the Meiji Jingu Landscape Bureau turned to work in government offices and universities to teach and train younger generations of designers. In the reconstruction plan after the Great Kanto Earthquake, modern parks became widespread and the Tokyo Higher Landscape School was created, which became Japan's first full-scale institute of higher landscape design education. In the meantime, the traditional apprenticeship system that existed long before the Meiji era continued as a separate educational system during the Meiji and Taisho periods. Regarding public landscape design, such as the Meiji Jingu construction for example, scholars from the academic system and ministerial and public agency staff took on all planning and designing tasks. Private contractors did not undertake any public design, but artisans and tradesmen carried out private commissions.

During the Showa period (1926–1989), the number and scale of landscape design grew. The demand for public landscapes increased and highly educated designers were needed to create them, particularly after the war and during the period of reconstruction and economic growth. The staff working in public agencies, small in number, was no longer able to meet demands. Private firms began to fill the gap in both planning and designing, and in construction. It was the birth of the design atelier, the landscape design office. University graduates started to work in the new atelier, and many landscape architects received their professional education under the teacher-apprentice system inside the atelier (Awano 2016). In the atelier, there was a distinction made between a "disciple" who graduated from a university and a "craftsman" who was considered a professional technician. In this way, as the private atelier expanded design tasks from private gardens to public landscapes, and from basic planning to construction management, generated a new type of master-apprenticeship system. The director of the atelier acts as master, and the young office member as apprentice, as it were. From the teacher-apprentice relationship, many large-scale private and public landscapes emerged. The new designs were works of masters who added modern and contemporary interpretations to traditional Japanese garden techniques. On the other hand, since the 1960s, in the academic system, landscape planning departments developed in arts and engineering universities, and in schools of higher education all over the country, Since the 1970s, with the introduction of landscape planning, universities were aiming for the education of specialists who would be able to help secure green and healthy environments. The development of the academic system of landscape design and planning was a response to global and urban environmental problems associated with strong economic growth and urban expansion. Landscape architecture became an academic field with tendencies of increasing differentiation.

In this way, during the Meiji and Taisho (1868–1925) and the Showa periods (1926–1989), the apprenticeship system of garden design and the emerging academic system for public landscape design developed side-by-side for a while, with some of their areas of professional activities overlapping. What follows is an account of the history of landscape education (see Figure 32.1).

Meiji and Taisho periods (1868–1925)

Historically, Japanese landscape design education was a form of apprenticeship system. Most gardens of temples and shrines, aristocrat and daimyo gardens are created on private property. The gardener-designer is the master who surveys the ground and decides what kind of garden to make, in consultation with the owner. All knowledge of garden making is passed on from master to apprentice, including deciding on the character of materials to use, such as stones

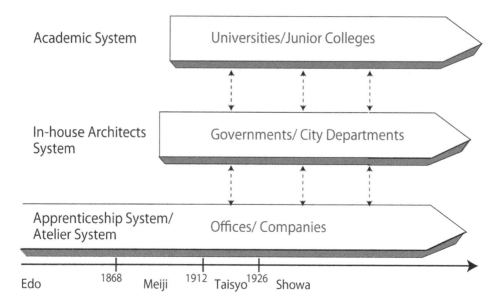

Figure 32.1 Three design education systems developed side-by-side in Japan with links between them, the apprenticeship, the in-house, and the academic system.

Source: Visualized by Chika Takatori.

and plants, how to arrange them, how to construct the garden, and how to maintain it over time.

As reference, the idea of Japanese garden design is described in *Sakuteiki*, published circa 1075 and said to be the oldest landscape book in the world (Shirahata 2014). The author, Tachibana no Toshitsuna (1028–1094), is believed to have been a garden master maker himself. The most important point he makes in the book is the attitude of learning naturally, which is a major feature of the Japanese garden that has not changed ever since (Shinji 2005). Learning includes the act of feeling the movement of nature, and assembling its energy is the act garden making and maintaining.

While the gardens created until the Edo period have Chinese and Korean influence, the gardens created after the Meiji period are characterized by Western influence (Akasaka 2006). For instance, Josiah Conder (1852–1920), an architect who came to Japan from England in 1877 served as a professor of architecture at the Imperial College of Engineering, a forerunner of Tokyo University, to train Japanese architects. (He also worked as a consultant to the Ministry of Engineering.) In 1893, he published the book *Landscape Gardening in Japan*, which comprehensively explains Japanese gardens with reference to several older books on landscape design (Conder 1893). While he introduced Western landscape studies and techniques in Japan, he modernized the traditional gardening techniques developed in the Edo period such as describe in the book *Tsukiyama Niwatsukuriden* (Ryoshimaken 1918), and reconstructed them by using modern gardening techniques (Suzuki 2021). Conder showed a deep understanding of the Japanese culture, such as flower arrangement, painting, and garden design, while, in practice, creating a mainly Western-style garden using rose gardens, flower beds, fountains, orthogonal ponds, and submerged gardens. He practiced this style from the late Meiji to the Taisho period. Gardens that his students will have known are the Iwasaki Yanosuke Garden and Furukawa Toranosuke Residence Garden in Tokyo.

Apprentice system and the introduction of naturalistic design styles

The wave of Europeanization in the Meiji era brought ideas of "naturalism" to Japan, which, in landscape design, became a trend that is different from conventional Japanese garden styles. Aritomo Yamagata is a master who, in the apprentice system, gave expression to these new ideas. To his apprentices, he imparted ideas of "full-scale nature" such as those reminiscent of mountain streams and mountain villages that could be recreated in the garden. Katsugoro Iwamoto, of the fourth Negishi gardener generation, was in charge of the construction of the Chinzan-so Garden, the masterpiece of Aritomo Yamagata (Suzuki 2005, Awano 2021). It was the seventh Jihei Ogawagardener generation whose popular name is Ueji who took Yamagata's ideas to maturity. (In Japan, names of ancestors, parents, masters, and other predecessors, are customarily bequeathed to successors, particularly in occupations that require classical performing arts and traditional special techniques, and Katsugoro Iwamoto and Jihei Ogawa also inherited their names that way.) As a garden maker and master, the seventh generation of Jihei Ogawa is considered a pioneer of modern Japanese gardens. He was only 35 years old when he started with the *Murin-an Garden* at Nanzenji Temple, which was constructed under the guidance of Yamagata. (Other works include the Kyoto Imperial Palace, Katsura Imperial Villa, Nijo Castle, and the Iwasaki House in Tokyo.) (Suzuki 2013).

Links between court landscape architects and formal education

In-house architects of the Imperial Household Agency's "Takumi Dormitory" contributed to the development of the in-house system and of horticulture in the higher education of Landscape Studies. During the early Meiji era, it became a status symbol for the privileged class (Imperial family, royal family, etc.) as well as the leaders in the growing industrial world to have Western-style buildings and gardens (Awano 2005). Designers who worked on these new and prestigious gardens continued to teach apprentices, and they created links between the Imperial Household Agency's "Takumi Dormitory" and Institutes of higher education. One example is Yoshichika Kodaira who, a child of the garden of the Edo Shogunate, entered the ministry in 1870 and became the first court landscape architect of the "Naishaku Dormitory" (Awano 2008). Other designers who merged Japanese and Western styles are Hayato Fukuba, who served as the head of the inner garden in the latter half of the Meiji era, and Yukio Ichikawa, who designed pure Western-style gardens such as the one at the Shrine Museum Garden. To educate future landscape designers in the now fashionable Western landscape style, Hayato Fukuba gave courses at the Tokyo Agricultural and Forestry School.

Landscape studies in institutes of higher education

The idea of "landscape" is foreign to the Japanese culture. It was first adopted at institutes of higher education where scholars had brought it from Europe and North America. In the Meiji era, the idea of public space incorporating Western concepts of landscape lead to a broadening of landscape architectural tasks, expanding from garden to public space design. Subjects at schools of higher education included the Japanese park system (that started in 1873) and parks created by government offices (Sato 1977). These subjects continued to be taught in the in-house system.[1]

With the increased demand for public landscapes, urban authorities have come to seek the guidance of authoritative scholars. Concepts of landscape entered educational programmes of agricultural, forestry, and horticultural fields during the Meiji to Taisho periods. For the first time in Japan, garden design education was then officially included, under the title of gardening education, in the curriculum of this school. In the Meiji period, the Tokyo School of Agriculture and Forestry (predecessor of the Faculty of Agriculture of the University of Tokyo) was established in 1886. In April 1908, Tokyo Metropolitan School of Horticulture was established. This school was the first institution for horticultural education in Japan. Chiba Prefectural Horticulture College (currently the Faculty of Horticulture, Chiba University) was established in 1909.

Initially, the "landscape studies" of higher education had a strong influence by the horticultural field in which Fukuba Hayato specialized. Ever since Seiroku Honda, who was a prominent landscape scholar of that period, designed Hibiya Park in 1903, landscape education began to be conducted in the fields of horticulture and forestry (Shirahata 1982). He entered the Tokyo Sanrin School (later the Faculty of Agriculture, University of Tokyo) in 1884. In 1890, he enrolled at the *Forstliche Hochschule Tharandt* (Royal Forestry School) in Saxony (now part of the School of Environmental Sciences of Dresden University of Technology) and earned his doctoral degree at the University of Munich, Bavaria. In 1892, he returned to Japan to become assistant professor and professor at the University of Tokyo. In his teaching, Honda referred to Western concepts of landscape that he had learned in Europe. Ever since Seiroku Honda had become professor at Tokyo Imperial Agricultural University, landscape education began to be conducted in schools of gardening and forestry (Honda 1940).

Landscape architectural education in schools of forestry and horticulture

During the Showa period to the prewar period, the Meiji Jingu construction and the earthquake-reconstruction park project after the Great Kanto Earthquake played major roles in the development of landscape architecture in Japan (Parks & Open Space Association of Japan 1978). Administrative staff and engineers who had earned their degrees at a university were in charge of planning and design.

For the construction of the inner and outer gardens of Meiji Jingu, landscape experts gathered from all over the country to practice urban green space design at hitherto unprecedented scale and vision. For example, the large Meiji Jingu forest in Inner Garden, developed in 1920, is famous for having been created with the aim to reach completion after 100 years and with thoughts of forest ecology as design foundation (Mizuuchi 2019). In the construction of Meiji-Jingu Outer Garden, there was a conflict between forestry and agriculture regarding which of them should be responsible for landscape. There was, at Tokyo Imperial University Agricultural University, a fierce debate going on between Hiroshi Hara, the leader of agricultural science who gave horticultural classes, and Honda, the leader of landscape who gave forestry classes (Nishikawa 2018). Despite such conflict, large numbers of landscape artists graduated at the Meiji Jingu Landscape Bureau, which was established in 1918, from forestry (with Seiroku Honda, Takanori Hongo, Tsuyoshi Tamura, and Keiji Uehara as teachers) and from agriculture (with Hiroshi Hara and Yoshinobu Orishimo as teachers).

For example, after graduating from the Department of Forestry, Tokyo Imperial University Agricultural University, Takanori Hongo (1877–1949) had earned his doctoral degree in Munich, Germany, in 1910. He became an assistant in the forestry class where his teacher was the innovative Seiroku Honda. He participated in the Hibiya Park construction project and majored in landscape design. He became a lecturer in the Department of Forestry of the Agricultural University. Since 1913, Hongo has given the first lecture on garden studies in the regular course of higher education at the invitation of the principal of Chiba Prefectural Higher

Horticulture School (predecessor of the Faculty of Horticulture, Chiba University). Among the graduates of the Chiba School are Kannosuke Mori who later worked on landscape education at that school, and Yoichi Aikawa who supported Kiyoshi Inoshita as an engineer in the Tokyo City Park Division. In 1915, Tsuyoshi Tamura, a graduate of the Department of Forestry, Faculty of Agriculture, Tokyo Imperial University, took over as a successor to participate in the construction project of Meiji Jingu. In 1920, Hongo concurrently served as an engineer for the Tokyo Regional Committee for Urban Planning, Ministry of Home Affairs, and participated in the establishment of the Garden Association in the same year. In addition, from 1918 to the end of the war, he worked as a commissioned member of the Shrine Bureau of the Ministry of Home Affairs to develop plans for the forest gardens of shrines and shrines in Japan and overseas. Tsuyoshi Tamura published the book *Introduction to Landscape Design* in 1918 (Tamura 1918), and earned a doctoral degree with his dissertation *On the Development of Japanese Gardens* in 1920. Tamura established the Garden Association renamed Japan Garden Association in 1925 (or "Japanese Society of Landscape Architecture" in English). He set up a landscape study class (later the Forest Landscape Planning Laboratory) in the Forestry No. 2 Laboratory of the Forestry Department, and as an educator, he lectured on landscape studies in the forestry department as well as in the architecture class and the horticultural science course of the Faculty of Agriculture.

Keiji Uehara graduated from the Department of Forestry, Faculty of Agriculture, Tokyo Imperial University in 1914 and went on to graduate school. Uehara later received a doctorate in forestry and went to Europe and the United States to study landscape architecture (Uehara 1974).

Education needs after the Great Kanto Earthquake

After the Great Kanto Earthquake of 1923, a large number of parks were built at once as a disaster recovery project. As modern parks and their design concepts became popular with the public, Keiji Uehara (1889–1981) who was involved in the reconstruction plan for Tokyo, felt the need for more and better education in the field of public landscape design. Uehara, considered the "father of landscape architecture in Japan", established the Tokyo College of Landscape Architecture, the forerunner of the Department of Landscape Architecture Science, Tokyo Agriculture University, established 1924. Uehara became the first principal of the school and involved in the research plan commissioned by Tokyo City and Yokohama City (Ishikawa 2001).

Uehara published the book *General theory of landscaping* in 1924, which analyzes and systematizes landscape in the scholarly way practiced at the Tokyo School (Uehara 1924). This became Japan's first book used for full-scale landscape education. In 1925, researchers, educators, and practitioners involved in landscape got together and established the Japanese Institute of Landscape Architecture as an incorporated association. The magazines of the institute are "Landscape Magazine" and "Landscape Studies", both continue until today.

During the Showa Period (1926–1989) the systematic promotion of the understanding of landscape science and technology accelerated. The publications of the Japanese Garden Association mark important themes of the time, particularly the 24 volumes of the Landscape Series" (1928–29). Themes and authors include "Urban Beauty" (Vol. 2, Tomonobu Kuroda), "Natural Park" (Vol. 3, Seiroku Honda), "Park Design" (Vol. 4, Kiyoshi Inoshita), "Landscape and Civil Engineering" (Vol. 11, Tetsusaburo Tanimura), "Exercise Play Equipment" (Vol. 13, Yoichi Aikawa), and "Namiki (Roadside Trees)" (Vol. 19, Ihachiro Miura). Landscape begins to feature in book publications. Activities to form an international professional association for landscape began in the late 1930s. The Japan Landscape Association was established in 1938 as a professional group of landscape designers. It merged with the Japan Garden Association in 1977 (Suzuki 2021).

Atelier system

The Tokyo School graduated many landscape design experts who joined private companies and landscape design offices. These experts played a major role in connecting the design of private gardens and public projects, both of which had been separate until then.

With numbers of public commissions increasing, landscape design experts established private offices, or ateliers. Among the new landscape design offices are the Uehara Landscape Research Institute that Keiji Uehara founded in 1918, and the Tono office that Takuma Tono had established, the first in Japan to obtain a master's degree in landscape architecture overseas. The scope of projects handled by the atelier and landscape design offices soon went far beyond the boundaries of private gardens and addressed designs for a variety of public landscapes (Awano 2012).

In addition to universities establishing study programmes, the old apprenticeship system continued, the range of education expanding to include public landscape design. Graduates from the academic system merged with those of the apprentice system, and a new permeability between the systems developed.

Examples for links between different systems are the personal connections between Juki Iida and Kenzo Ogata. Juki Iida trained as apprentice under the second Ikujiro Matsumoto and the fourth Katsugoro Iwamoto generations, who were in charge of the construction of the Chinzan-so Garden of Aritomo Yamagata. He enjoyed Japanese culture such as ikebana, basket making, tea scoop bonsai, and calligraphy. He established the *Iida Landscape Design Office* in Mejiro, Tokyo, in 1918, and promoted the idea of *natural gardens* and the concept of natural-style tree planting. Kenzo Ogata (1912–1988) graduated from Chiba Horticultural School in 1945 and worked in the Park Division of the Tokyo City Health Bureau from 1934. He became very interested in natural landscape gardens and, as he wanted to learn more about the natural style, became a member of Juki Iida's landscape office from 1932. Juki Iida introduced natural garden design styles to him. Ogata expanded the application of this style to include the design of public space, and he used it not only in Japan but overseas. While inheriting Iida's style, Ogata developed the theory and interpretation of naturalistic design further, as a modern garden style that became a reference for the future. With every step of forming new trends in design, each designer was instrumental in the training of their successors. For example, Iida established the company Tokyo Gardener in 1946 and, in 1950, handed it over to a designer who had trained with him. In 1957, Iida founded the new company Tokyo Garden. In 1958, Ogata became president, and Iida himself continued as advisor.

Other individuals that learned the art of garden and landscape design through links between different systems include Michiaki Akimoto, Takeo Inagaki, and Takuma Tono from Tokyo Garden, Kinya Taira from Tokyo Agriculture University, Hoichi Kurisu from Takuma Tono, and Hachiro Sakakibara from Tokyo Agriculture University (Awano 2012).

In this boundary-crossing way, garden and landscape designers of the Showa period expanded their fields of activity considerably, soon to include urban development projects, public space open to the people, and so forth. In the design offices, the "personal connection continuity" provided the basis for "design schools" to form. These "schools" play a major role in Japan to the present.

Landscape architecture programmes

Chronicling the development of the academic system, education of horticulture had started in thirty agriculture schools by 1942. Twenty-six of these schools included subjects such as garden, gardening and, eventually, landscape architecture in their curricula. Six of these schools began to offer formal courses either in garden design or in landscape architecture. In several other types of institutes of higher education, landscape architectural education also became one of the items covered in horticultural and agricultural courses (Kobayashi 2009). The first department in the field that later became landscape architecture established in 1941 at the Tokyo Metropolitan School of Gardening that offered two programmes, horticulture and landscape architecture. At the time, this is the only agricultural school with an independent landscape architecture programme in Japan.

After World War II, several institutes of higher education started to offer landscape design courses, including universities, vocational schools, and vocational training schools. These include more than 80 universities and faculties related to horticulture, arts and engineering.

The postwar reconstruction and economic growth period were challenging times for academic landscape architecture in Japan. Due to the rapid increase in housing demand, urban green spaces such as parks and green belts were targeted for development one after another. On the other hand, the awareness for environmental issues grew since the end of the 1960s, and the environmental conservation movements with it. Addressing the quality of the urban environment, securing and enriching green open space, Japanese universities, research institutes, and government cooperated in developing a system for landscape evaluation and planning. Ecological planning developed, technically supported by GIS, and remote sensing technologies. Scholars who had studied in Germany, introduced the concept of *Landschaft*, and landscape education was extended to include landscape planning. In this way, since the 1960s, landscape design and planning projects are carried out, commissions including scales from site to region.

Conclusion

Re-examining the complicated and fragmented history of landscape education in Japan, this chapter focuses on the Meiji and Taisho periods (1868–1925), when the modernization of Japan began, and the Showa era (1926–1989), which spans the times of war and subsequent reconstruction, and economic development. Three major educational streams, or systems, developed. One is the apprenticeship system, another the atelier or office system, and the third the academic system of institutes of higher education. As these systems shaped the history of landscape design education, they relate to one another, links forming between them through the activity of individual designer-teachers. As the tasks of professional work expanded to include private and public landscapes, university graduates were employed in offices and ateliers where design knowledge was traded on through a succession of master designers. The apprenticeship system continued to offer conventional garden design, but in a new form of teacher-apprentice relationship and with new styles such as naturalistic planting design that also became applied to public open space. In the face of global and urban environmental issues, the role of the landscape design field that established in academic systems is changing and expanding significantly. Current challenges include institutes of higher education to install systems of accreditation and quality assurance, for example under the umbrella of the IFLA Asia-Pacific education programme, and for landscape scholarship to contribute to the Continuing Professional Education system (that started in 2005) to include Education for Sustainable Development.

Note

1 For example, in 1903, Yasuhei Nagaoka was involved in the landscape of Itsukushima-park, and he was the earliest in Japan to be involved in park design. He was active mainly in the Meiji era. Subsequently, Seiroku Honda (1866–1952) designed Hibiya Park as the first park created by the government, and after the project, he has been involved in park projects all over the country (Yamashita and Miyagi 1995). These projects were basically carried out directly by government and city administrations where they served as examples for in-house learning.

References

Akasaka, M. (ed.) (2006) 造園がわかる本 (A Book That Understands Landscape). Tokyo: 彰国社 (Shokokusya).
Awano, T. (2005) 'Style and Space of Semi-western Gardens in Modernized Residential Spaces in the Meiji Tokyo', *Journal of the Japanese Institute of Landscape Architecture*, 68(5), pp. 381–384.
Awano, T. (2008) 'Yoshichika KODAIRA as the Modern Garden Designer, and His Works', *Journal of Japanese Garden Society*, 19, pp. 65–70.
Awano, T. (2016) 'The Trend of Modern Gardens and the Genealogy of Garden Artists in the Showa Era (40th Anniversary Special Feature on the Future of Gardens)', in *NIWA (magazine), No.225 "40th Anniversary of the First Issue a Group of Gardeners and Landscape Architects Coloring the Showa Era"*, pp. 21–24.
Awano, T. (2021) 'Katsugoro Iwamoto, the 4th Generation, Is the Bearer of the Naturalistic Garden Trusted by Aritomo Yamagata, a Pioneer of Modern Gardens', *NIWA (Magazine)*, No. 245, pp. 43–45.
Awano, T. and Suzuki, M. (2012) Modern Landscape Designer and Formation of Their School in Postwar Japan', *Journal of the Japanese Institute of Landscape Architecture*, 76(2), pp. 127–130.
Conder, J. (1893) *Landscape Gardening in Japan*. Tokyo: Kelly and Walsh Ltd.
Honda, S. (1940) 林学の発達とその思出 —造林学を中心として (Development of Forestry and Its Thoughts-Focusing on Forestry). Tokyo: 全国農業学校長協会編「日本農業教育史」("History of Japanese Agricultural Education" edited by the National Association of Agricultural School Directors), pp. 719–726.
Ishikawa, M. (2001) 都市と緑地—新しい都市環境の創造に向けて (City and Open Space –Towards the Creation of a New Urban Environment). Tokyo: 岩波書店 (Iwanami Shoten, Publishers).
Kobayashi, F. and Shinji, I. (2009) 'A Study on Landscape Architecture Education at Japanese Agricultural Schools', *Tokyo Agriculture University Agricultural Bulletin*, 54(1), pp. 59–70.
Mizuuchi, Y. (2019) 'Landscape Elements of Meiji Jingu in Thoughts and Its Transition of Planning Concept', *Online Journal of the Japanese Institute of Landscape Architecture*, 12, pp. 50–61.
Nishikawa, R. (2018) 'Study About the Cognition of "City Planning" by Silviculturist Before WW2 – Focus on Works and Words by Seiroku Honda, Tsuyoshi Tamura and Keiji Uehara', *Journal of the City Planning Institute of Japan*, 53(3).
Parks & Open Space Association of Japan. (1978) 日本公園百年史-総論・各論 *(100-Year History of Japanese Park-General and Each Theory)*. Tokyo: 日本公園百年史刊行会 (Japan Park Centennial History Publishing Association).
Ryoshimaken, A. (1918) 築山庭造伝 後編上 *(Tsukiyama Niwatsukuriden, Part 2)*. Tokyo: 建築書院 (Architectural Shoin Publishing).
Sato, A. (1977) 日本公園緑地発達史 *(History of Open Space Development in Japan Park, Tokyo)*. Tokyo: 都市計画研究所 (City Planning Institute).
Shinji, I. (2005) 日本の庭園—造景の技とこころ *(Japanese Garden-Scenery and Heart)*. Tokyo: 中公新書 (Chuko Shinsyo, Publishers).
Shirahata, Y. (1982) '近代化のなかの「公園」-日比谷公園の誕生と海外情報 ("Park" in Modernization-Birth of Hibiya Park and Overseas Information)', 京都大学人文科学研究所『人文学報』 *(Journal of Humanities)*', 53, pp. 213–245.
Shirahata, Y. (2014) 『作庭記』と日本の庭園 *("Sakutei-ki" and Japanese Garden)*. Kyoto: 思文閣出版 (Shibunkaku Publishing).
Suzuki, H. (2013) 庭師 小川治兵衛とその時代 *(Gardener Ogawa Jihei and His Time)*. Tokyo: 東京大学出版会 (University of Tokyo Press).
Suzuki, M. (2021) 'Science/Technology and Profession of Landscape Architecture', *Journal of Agriculture Science, Tokyo University of Agriculture*, 65(4), pp. 87–94.
Suzuki, M., Awano, T. and Inokawa, W. (2005) 'Aritomo YAMAGATA's Image and View of Gardens and Chinzan-so', *Journal of the Japanese Institute of Landscape Architecture*, 68(4), pp. 339–350.
Tamura, T. (1918) 造園概論 *(Introduction to Landscape)*. Tokyo: 成美堂書店 (Narumido Shoten).
Uehara, K. (1924) 造園学汎論 *(General Theory of Landscape)*. Tokyo: 林泉社 (Rinsensya).
Uehara, K. (1974) 造園大系第7巻 風景・森林 *(Landscape System vol.7, Landscape/Forest)*. Tokyo: 加島書店 (Kashima Shoten Publishing).
Yamashita, H. and Miyagi, S. (1995) 'Spatial Characteristic of Hibiya Koen Park Found in Numerous Design Proposals and Reserved Through a Series of Renovations in the Later Years', *Journal of the Japanese Institute of Landscape Architecture*, 58(5), pp. 13–16.

PART IV

Aiming for justice, reconciliation, and decolonization

33
INNOVATING EDUCATION POLICY

Justice, reconciliation, and decolonization

Diedrich Bruns and Stefanie Hennecke

Part IV brings together findings from educational research in regions that have experienced substantial forms of social and political transition in recent history, such as change in government, legal system, and also natural and man-made disasters, all of which are relevant to landscape architecture education. Recognizing three different types of transformation and change, one includes countries that experienced socio-political system changes, such as the building of republics from imperial and communist past, as it happened in Europe during the 20th century and is currently happening in several parts of the world. Another type are countries whose development is marked by processes of colonization and, more recently, decolonization, as it is happening in several parts of Africa, Latin America, Australia, and Oceania. The third type includes effects of natural and man-made disaster, such as earthquakes, floods, storms, fire, terrorism, and war. It is important, while incorporating dimensions of transition into education policy, to pay respect to identities and self-determination rights of ancestral cultures that colonial and war powers partly or largely ignore. Chapter authors are providing suggestions for critical scholarship perspectives. There is much potential in education for cultural and power awareness to grow, for justice, diverse cultures, inclusive competences to develop. We must continue developing these perspectives and potentials, not least by linking historic awareness and responsibility, international networking, and sustainable development.

Zydi Teqja, Katarzyna Rędzińska, and Agnieszka Cieśla illustrate developments in two countries, Albania and Poland, which serve as examples of landscape architectural education in former communist countries after the fall of the Iron Curtain and joining the European Union. (Similar transformations occurred in the Czech Republic, Romania, Croatia, Hungary, and Slovakia presented elsewhere in this volume). Teqja analyzes educational challenges and potentials in Albania, where politicians became aware of the role and importance of planning and design experts as public pressure on environmental and landscape issues increased. Based on findings from analyzing discourses among practitioners and educators, Rędzińska and Cieśla chronicle the history of landscape architecture education in Poland. They discuss the co-evolution of education and professional practice and the challenges both are facing regarding changing legal and political contexts.

Graham A. Young offers an overview and examples of landscape architecture programmes in different African countries, such as South Africa, Kenya, Nigeria, Morocco, Tunisia, and Ethiopia and discusses the influences that came to bare on their establishment. Looking to the future, he introduces a discussion "seeking a post-colonial approach to education". As examples for Latin America, Gloria Aponte-García, Cristina Felsenhardt, Lucas Períes, and Karla María Hinojosa De la Garza present developments in Chile, Argentina, and Mexico. In both regions, Africa and Latin America, landscape education is characterized by heterogeneous development consequent with their cultural, political, and physiographic diversity. Educational approaches, subjects, and course contents vary considerably. Alice Lewis, Sue Anne Ware, Martin Bryant, Jen Lynch, Penny Allan, and Katrina Simon present the Australian case. Taking two programmes as examples, the authors explain how each supports experimentation in curricula, provides distinctive pedagogical approaches, and fosters the development of design techniques to address imminent issues of reconciliation and climate change. For Aotearoa-New Zealand, Simon Swaffield, Jacky Bowring, and Gill Lawson describe and interpret the first 50 years of landscape architecture education at Lincoln University. They are focused upon the context, events, and actions that have shaped the diploma and degree programmes described here as Landscape@Lincoln, shaped during major social and economic transformations with neoliberal public policies, diverse ethnicities, increasingly bi-cultural governance, and the experiences of major earthquakes, floods, wildfire, and terrorist attack.

The chapters in Part IV provide accounts suited to comparison with wider international developments in landscape architecture education during times of societal transition. Not least, they point at the relevance of social and institutional developments for landscape architecture education. Crosscutting themes include innovating educational policy to meet societal challenges of justice, reconciliation, and decolonization. One topic is the introduction and establishment of courses of study in times when old power relations and systems are disappearing and new ones are emerging. It is about models that help educators to learn to assert themselves between different currents and interest groups. For example, for us to understand relationships between education, landscape, and democracy, Zydi Teqja suggests applying democracy indices to rate how democracy values are becoming integrated in landscape architecture education (Teqja compares 338 study programmes worldwide for levels of democracy).

Another issue associated with system changes are the far-reaching transformations in development and planning policy. Rędzińska and Cieśla discuss the need for landscape architecture students to learn dealing with changes in legal and administrative framing of planning, understanding processes of people moving from rural to urban areas, how such processes lead to a boom in urban development and construction, which in turn affect urban, suburban, and rural landscapes. As landscape changes are linked with development practices, students must appreciate the different forms of power that lead to changes and to apply principles of environmental justice in landscape design projects, for example in the context of neoliberal development practice.

Another theme is to place importance on Critical Scholarship Perspectives in the development of university profile and study content, to build up potentials for growing cultural awareness, for justice, diverse cultures, and inclusiveness. In particular, where educators defer to North American or European influences when establishing new programmes, such as in Africa, Latin America, and so on, they now strive to deal with post-colonial education challenges. Asking the question how to develop *post-colonial* forms of education it is important to discuss how such forms refer less to North American and European but more to "indigenous" understandings of knowledge, particularly as it applies to the spatial and landscape theories which underpin the discipline. Higher education experiences shift in approaches to design which recognize and consider the identity and self-determination of ancestral peoples and the particular ways global challenges affect them. For example, rather than insisting on Western understandings of landscape, educators in Australia specifically aim to understand concepts and practice of "Country";

in this case the term "Country" is an all-encompassing ideal that describes "Indigenous Australian's" highly complex and layered relationship to land. In the same vein, imperatives for future development in Aotearoa-New Zealand include strengthening partnerships with the indigenous people. Lessons learned from these examples are transferrable to other regions where "landscape" is not part of the vernacular vocabulary and equivalent indigenous ideas and concepts exist.

Together, authors contributing to Part IV present landscape architectural education that is regionally grounded and outward looking. They argue that educational development must express adaptions to wider social, cultural, and economic transformations and crises, and that institutional, regional, national, and global context are key factors. Can the rest of the world learn from the current experiences of African, Latin American, Australian, and Oceanic academics and teachers? This question we raise not because we have good answers but to point at the urgency with which we must find some. We need to find them within the framework of ongoing curriculum development and landscape architectural research and teaching across large parts of the world. At the same time, the deepening and widening of design and planning responses to challenges of war, terror, justice, climate change, human health, urbanization, and so on, remain, all within a context of sustainable development. For landscape architecture, linking these issues is essential to ensure enduring capacity in practice, education, and research to meet the needs of future generations.

34
LANDSCAPE ARCHITECTURE EDUCATION IN ALBANIA AFTER THE FALL OF THE IRON CURTAIN

Zydi Teqja

Introduction

In 1989, Albania still had the most authoritarian regime in Eastern Europe. After the fall of the Iron Curtain the country experienced dramatic changes and reforms that completely transformed the Albanian economy and society. At the beginning of 21st century, the country became fully oriented towards integration with the European Union. In 2020, the European Council adopted the decision of the General Affairs Council of the EU to open negotiations with Albania.

Starting in the 1990s and lasting for over a decade, Albania saw a boom in construction and rise of informal settlements in urban, suburban and rural areas. The urban pressure coming from people moving from rural to urban areas, the issuing of building permits without updating zoning plans first, and the rise of private car transport, all led to an environment of poorly planned spaces and to what has been perceived by many as degradation of the landscape (Pojani and Maci 2015). After 2010 the public awareness on environmental issues increased. People became aware of the role and importance of green areas and the benefits that gardens and urban parks have for their health and well-being (Teqja and Kopali 2012; Lekaj and Teqja 2019). A gradual improvement of the urban landscape can be seen in all municipalities, a development that is reflected in new local plans.

Addressing the peculiarities, challenges and potentials of landscape architecture education that occurred during the dramatic changes after the fall of the Iron Curtain, this chapter explores, for the first time, the evolution of educational programmes in Albania and other former communist countries. It also addresses relationships between the development of landscape architecture education and democracy. It points at needs for social and institutional developments that provide the framework necessary for landscape architecture education. Objectives are to:

(a) chronicle the landscape architecture educational history in Albania, starting around the time of the fall of the Iron Curtain; (b) present and analyse educational challenges and potentials that are specific to the Albanian context; (c) explain some of the reasons why landscape architecture education started late in Albania and other former communist countries.

To present an account of landscape architecture education history in Albania, documents were identified and analyzed that are available in the library of the Agricultural University of Tirana (AUT), in archives of the Polis University of Tirana and the Department of Horticulture and Landscape Architecture of AUT. The documents were studied to collect data on the professional profile and education of the people involved in the design and management of green areas. From the archives of the Department of Horticulture and Landscape Architecture of AUT and Polis University data were collected regarding important dates for landscape architectural education and the debates that occurred about it. To study the relationship between education history and democracy in Albania and other former communist countries, a list of landscape architecture schools worldwide was compiled. The list was drawn from membership archives of the European Council of Landscape Architecture Schools (ECLAS),[1] of the European Network of Universities dedicated to landscape studies (UNISCAPE)[2] and of the Council of Educators in Landscape Architecture (CELA).[3] The list was extended by searching additional resources. A database of 338 landscape architecture programmes worldwide was compiled, 23 of them in former communist countries. To measure the level of democracy of different countries, two resources were used: One is the *2019 Democracy Index*, compiled by the Economist Intelligence Unit,[4] and the other is the *Freedom House Report from 2020*.[5] To measure the level of landscape architecture education development the number of university programmes per 10 million people was calculated.

Chronology of landscape architecture education in Albania

The tradition of living in green places has a long history in Albania. It has influenced the way villages and cities are built and organized. Similar to other European countries, the need for urban planning related to landscape became evident at the beginning of 20th century, especially during the 1920s and 1930s, when urban parks were designed in the centre of major cities (Menghini et al. 2012; Mauro 2012; Pojani 2014). Italian architects in the 1930s and early 1940s envisioned the capital city Tirana as a so-called Garden City (Menghini et al. 2012). In the 1930s, Pietro Porcinai is perhaps the first professionally qualified landscape architect working in Albania and in Tirana in particular (Giusti 2012; Teqja and Dervishaj 2020). During Communism, landscapes experienced significant changes that are rooted in a totalitarian ideology prevailing in most countries east of the Iron Curtain. Some of the major public squares and parks for cities were created on the basis of general urban plans prepared by state-owned agencies while there was no private activity in the field of planning. The majority of specialists dealing with green spaces and landscape during communist time had graduated from the Agricultural University as agronomist and forest engineers (Teqja and Dervishaj 2020).

Starting around the middle of 20th century, courses in gardening with simple elements of garden design were offered in the Agricultural University of Tirana. Trends of broadening areas of professional competence and to include environmental subjects were part of the restructuring processes that occurred at the beginning of 21st century. By that time, education began to target the preparation of specialists and scientists who are able to contribute to environmental protection and to sustainable development of the country. Part of this development was the introduction in Agricultural Master of Science studies of new disciplines like Landscape Ecology and Architecture of Environment. The last one was, in 2010, replaced by Fundamentals of Landscape Architecture.

Fundamentals of Landscape Architecture have been taught as part of the architectural curricula at the Polytechnic University of Tirana since beginning of 1970s, and at private universities established after the fall of the Iron Curtain. In 2009, Vasil Jani published his textbook *Landscape*

Architecture: The Art of Greening in Urban Environments and in Natural Landscapes (translation by author). Jani is the designer of some of the most well-known urban parks established during communist time. Graduated in 1960 from the University of Sofia, Bulgaria, he considers himself an engineer, and a specialist in the architecture of landscapes (Jani 2009).

In 2011, Polis University (a new private university) started a two-semester professional graduate programme in landscape and urban design that leads to a master's degree. This programme provides knowledge on concepts in landscape and urban design of urban/natural territories and public space. The courses fit into two categories: landscape architecture, green and gardening (smaller scale) and landscape planning (larger scale), including the design of public space and park design.[6]

As student interest and market demand increased, AUT invited the Department of Architecture of Polytechnic University to join efforts for a professional graduate programme in landscape architecture, thus trying to optimize in-country professional recourses. To support this initiative, an agreement was signed between the two universities. The department of architecture took the responsibility to cover landscape construction and graphic design disciplines while AUT covered plant and environmental-related disciplines. The programme was designed for students coming mainly from undergraduate horticultural studies or related fields, so the programme was horticulture oriented and at the beginning it had no studio classes. Later on, a basic design studio was added. The objective was to offer an educational programme that provided students with the necessary tools to develop the profession of landscape architecture, integrating knowledge and competences in the field of landscape design and management with the ability to collaborate with other professions in the field of architecture, engineering, natural and agronomic sciences. The professional graduate programme leading to a master's degree started in 2013. Initially it was a three-semester programme. In 2017 it was extended to four semesters. Special attention is paid to aims and visions of sustainable development (Teqja and Dennis 2016a).

One year after starting this programme, the Department of Horticulture was renamed Department of Horticulture and Landscape Architecture, thus creating the first landscape architecture department in the history of Albania. Students graduating from this programme contribute significantly to promoting the profession of landscape architecture in the country. In September 2016, the Albanian parliament approved the law adopting and implementing the *European Landscape Convention, ELC*. In doing so, Albania is committed, among others, to the education of specialists in landscape appraisal and operations, and to providing educational resources such as university courses that address the values attached to landscapes.

Based on these developments, the success of the professional graduate programme and responding to the increasing need for professionals in landscape architecture, AUT established a three-year undergraduate programme leading to a bachelor's degree in landscape architecture. The programme was offered for the first time in 2018. A number of documents gave guidance in this process, in particular documents provided by the International Federation of Landscape Architects (IFLA World and IFLA Europe). Their educational documents outline the key elements to be included in landscape architectural education (IFLA Europe 2019). Participation in ECLAS activities and studying the *ECLAS Guidance on Landscape Architecture Education* (ECLAS 2010) proved to be useful and effective in the process of reaching consensus and building the new landscape architecture education in Albania. The first students of this programme graduated in 2021.

Educational challenges and potentials

Since 2010, there has been an ongoing discussion among members of the Albanian academic community about which institution should offer landscape architecture education. Options include the Faculty of Architecture and Urban Planning of the Polytechnic University and the Faculty of Agriculture and Environment of the Agricultural University. In addition, decisions needed to be made what department landscape architecture should be affiliated with. To address these questions, experience gained in Europe and the USA was analyzed. A nine-month Fulbright Scholarship (2015–2016) at the Department of Landscape Architecture of the University of Wisconsin, USA, helped to conduct the analysis. Taking a closer look at who designed and managed landscapes in Albania, most of the people engaged in green area and landscape design and management during the last decades graduated from the Agricultural University of Tirana (Teqja and Dervishaj 2020). This finding served as an argument to support the initiative for starting an undergraduate landscape architecture programme at AUT.

Two challenges to be addressed in the process are how to deal with effects of the former educational system, and how to integrate a design programme into the environment of a life sciences university. Remnants from the time before the fall of the Iron Curtain are still noticeable. That system was ideologically charged and mainly focused on transmitting knowledge, rather than allowing students to be creative (Alhasani 2015; Teqja and Dennis 2016b).

The second challenge for AUT is to develop a design programme that fits in a school with an emphasis in life and environment sciences and a tradition of having common courses during first year of undergraduate studies. The argument for maintaining both elements is that there are good employment opportunities for graduates in developing countries like Albania when they have a wide set of knowledge.

Through many debates after reforming the professional graduate programme and a successful application for a graduate program where students are able to earn a Master of Science degree, in 2021 the landscape architecture curriculum is designed to offer courses that include studio based learning at a level of about 40% of student's workload aiming to reach the ECLAS standards in the near future. AUT continues working to provide resources needed for studio teaching, including sufficient consultation and tutoring capacity, adequate studio rooms, and resources to train IT-related skills and competences. Courses based on standard lectures and seminars take the major part of the first and second years, while the third year of undergraduate and graduate programmes include more elective and design courses at an advanced level. The intention of design courses is to apply the newly acquired knowledge of theoretical courses in the previous semesters and increase students' critical thinking and design skills in the process.

The basic skills and theory courses are combined in studio classes that require design solutions commonly addressed in landscape architecture practice. Studio based learning is supported by technical classes related to plants, construction materials, irrigation and other mechanical systems. A 5 hectares didactic farm with landscape plants and some small gardens around the university campus support teaching and learning. To acquire professional competences, internships mainly supported by the Tirana Municipality are included into the landscape architecture programmes. Experience gained during periods of practical training outside of the university create the opportunity for students to have a practical work experience in the public or private sector and to develop research skills that should be reflected upon by writing a term paper, or by preparing the final thesis.

Table 34.1 SWOT analysis related to landscape architecture education in Albania and specifically at Agricultural University of Tirana.

Strengths	Weaknesses
– Unique program, not only at AUT, but in the entire higher education system in the country. • Qualified academic staff, mainly professors. • Academic staff with many years of experience in teaching and research. • Academic staff with good publishing activity. • Academic staff involved in many national and international research projects. • Good teaching infrastructure, with design studios and computer labs as well as demonstration gardens. • Cooperation agreements with domestic and foreign universities. • Support of the European Association of Landscape Architects and ECLAS.	• The program opens for the first time in Albania and carries unexpected things in its implementation. • The library is still poor with literature in the field of landscape architecture. • Not very high quality of students entering the AUT. • Not very long experience of the staff for some specific disciplines of the new program. • Impacts of the old educational system.
Opportunities	**Threats**
• Incentives from a market in demand for landscape architects and studies in this field. • Participation for studies for third parties. • Trainings for continuing education. • Opportunity to offer new study programs, in accordance with the requirements of the time and the labor market. • Expanding the activity of units dealing with greenery in all municipalities of the country. • Significant increase of residential areas which need landscape architects. • Employment possibilities not just in Albania but also in region and EU countries.	• Difficulties of employment of graduated students from AUT. • Lack of funds for research. • Population aging trend. • A tendency of young people to leave Albania without graduation.

Source: Zydi Teqja.

A SWOT analysis is used to identify educational challenges and potentials that are specific to the Albanian context. A summary of strengths, weaknesses, opportunities and threats related to landscape architecture education in Albania and specifically at Agricultural University of Tirana is presented in Table 34.1.

Landscape architecture education and democracy

The following presents findings from an analysis of experiences from former communist countries in Eastern Europe. Findings indicate how the progress in establishing democratic institutions and civil rights create the environment necessary for the development of the landscape architecture profession. Private property rights, increased public awareness for the value of public space, and the public's willingness to participate in the democratic processes are seen to be supporting the development of higher education for this profession (Teqja and Karaj 2022).

To verify the hypothesis that landscape architecture development in Albania and other former communist countries is related to developments in democracy, civil liberties, pluralism and freedom, the current distribution of landscape architecture study programmes worldwide is juxtaposed with indexes of democracy and freedom. One is the Democracy Index compiled by the *Economist Intelligence Unit*. The index categorizes countries into one of four regime types: full democracies, flawed democracies, hybrid regimes and authoritarian regimes.[7] The other is Freedom in the World 2020. It evaluates the state of freedom during calendar year 2019. The indicators are grouped into the categories of political rights (0–40) and civil liberties (0–60), whose totals are weighted equally to determine whether the country has an overall status of Free, Partly Free, or Not Free.[8]

The analysis of landscape architecture programmes according to the state of democracy and freedom indicates how from 338 landscape architecture programmes worldwide about 87% are offered in democratic countries, 9% in hybrid regimes and 4% in authoritarian regimes; 82% are offered in free countries, 13% in partly free countries and 5% in not free countries. A conclusion drawn from the analysis is that progress in developing democratic institutions, freedom, pluralism and civil rights create the necessary environment for the development of landscape architecture education. The evolution that happened during the second half of 19th century from garden design to landscape design and landscape architecture seems to be not just a change in scale but a significant change in content and process.

In the case of Albania, the country has experienced, since the collapse of communism in the beginning of the 1990s, enormous changes and a transition towards a democratic society and a market economy began to take hold (Teqja et al. 2000). Analyzing the way Albania has moved from a closed and authoritarian regime to an open and democratic society, it can be seen that after 2001 the democracy index ranks Albania among hybrid regime countries but still quite close to flawed democracy where most Western countries are also placed. Regarding the economy, Albania moved from a mixed to capitalist status. An analysis of democracy indices for former communist countries from East Europe indicates that, from 28 countries, 23 have landscape architecture programmes, 14 at university level (Table 34.2).

Twenty-one are in democratic countries, one in a country with hybrid regime (Albania), and one in a country with authoritarian regime, Russia. Twenty programmes are in countries that are EU members and three in candidate countries. About half of the universities had established landscape architecture programmes before the time when these countries fell under communist regimes. In some of them the beginnings of landscape architecture go back to the beginning of the 20th century (Perekovic et al. 2019; Supuka and Tóth 2019). Long tradition helped them to prevail during communist time and get revived after the fall of the Iron Curtain. Several universities are established after the fall of communist regimes. The establishment of the European Foundation for Landscape Architecture (EFLA) and the first meeting of The European Council of Landscape Architecture Schools (ECLAS) in 1989 helped to promote landscape architecture education all over Europe, including the former communist countries.

Correlations between degrees of democracy, Freedom house score and numbers of landscape architecture programmes appears to be strong. It seems that landscape architectural education often starts just after a society has reached a specific level of democracy as expressed by the index and by levels of the freedom score (democracy index higher than 4 and freedom score higher than 50; for Albania these indicators are respectively 5.9 and 68) (Figure 34.1).

After the fall of the Iron Curtain there were many changes in curricula, names of the programmes but also in the names of the departments, and consequently to the title given to those who graduate. This diversity grew during the 1990s when new schools of horticulture,

Table 34.2 Democracy indexes and number of landscape architecture programmes for former communist countries.

No.	Country and political regimes	Overall score	Civil liberties	Political participation	Functioning of government	Electoral process and pluralism	No. of LA Prog	LA Prog/ 10ml
1	Turkmenistan	**1.72**	0.00	0.79	5.00	0.59	0.00	0.00
2	Tajikistan	**1.93**	0.08	0.79	6.25	0.88	0.00	0.00
3	Uzbekistan	**2.01**	0.08	1.86	5.00	0.88	0.00	0.00
4	Belarus	**2.48**	0.92	2.00	4.38	2.35	0.00	0.00
5	Azerbaijan	**2.75**	0.50	3.21	3.75	3.53	0.00	0.00
6	Kazakhstan	**2.94**	0.50	2.14	4.38	3.24	0.00	0.00
7	Russia	**3.11**	2.17	1.79	2.50	4.12	1.00	0.07
	Authoritarian regimes						1.00	**0.07**
8	Bosnia and Herzegovina	**4.86**	6.17	2.93	3.75	5.88	0.00	0.00
9	Kyrgyzstan	**4.89**	6.08	2.93	3.75	5.00	0.00	0.00
10	Georgia	**5.42**	7.83	3.21	4.38	5.59	0.00	0.00
11	Armenia	**5.54**	7.50	5.36	3.13	5.59	0.00	0.00
12	Montenegro	**5.65**	5.67	5.36	4.38	6.76	0.00	0.00
13	Moldova	**5.75**	6.58	4.64	4.38	7.06	0.00	0.00
14	Albania	**5.89**	7.00	5.36	5.00	7.65	1.00	3.47
15	Ukraine	**5.90**	7.42	2.71	6.25	6.47	0.00	0.00
16	North Macedonia	**5.97**	7.00	5.36	3.75	7.06	0.00	0.00
	Hybrid regimes						1.00	**3.47**
17	Serbia	**6.41**	8.25	5.36	5.00	7.35	2.00	2.28
18	Romania	**6.49**	9.17	5.71	4.38	7.65	2.00	1.03
19	Croatia	**6.57**	9.17	6.07	5.00	7.06	1.00	2.42
20	Poland	**6.62**	9.17	6.07	4.38	7.35	3.00	0.79
21	Hungary	**6.63**	8.75	6.07	6.25	7.06	2.00	2.07
22	Bulgaria	**7.03**	9.17	6.43	4.38	7.94	1.00	1.43
23	Slovakia	**7.17**	9.58	7.14	5.63	7.94	2.00	3.67
24	Latvia	**7.49**	9.58	6.07	6.88	8.82	1.00	5.24
25	Lithuania	**7.50**	9.58	6.43	6.25	9.12	2.00	7.25
26	Slovenia	**7.50**	9.58	6.79	6.25	8.24	1.00	4.81
27	Czech Republic	**7.69**	9.58	6.79	6.88	8.53	3.00	2.81
28	Estonia	**7.90**	9.58	7.86	6.88	8.53	1.00	7.54
	Flawed democracies						21.00	**41.33**

Source: (Teqja and Karaj 2022).

urbanism or ecology started to establish landscape architecture degree programmes. In some cases, students who graduated from these programmes were practically horticulturists, urban planners or environmental engineers with some knowledge on landscape architecture (Tudora 2011). Recently most of the departments tend to include in their names landscape architecture

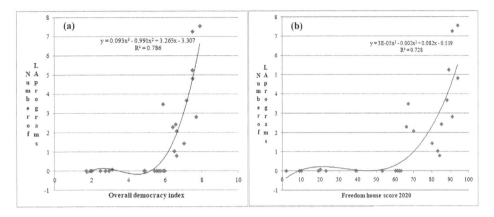

Figure 34.1 The relationship between overall democracy index (a), Freedom house score (b) and number of landscape architecture programmes for 10 million people in former communist countries.

Source: (Teqja and Karaj 2022).

and to adjust their programmes as much as possible to ECLAS guidelines. But still it can be noticed that in former communist countries in distinction to contemporary landscape architecture curriculums, there is a higher portion of natural sciences and of horticultural subjects. Though most of the academic community is convinced that it is outdated, there are arguments that there is a larger space for employment beneficial for these countries if students get a wider set of knowledge.

Since 2010 some ECLAS projects are contributing to address the problem of wide diversity and trying to encourage understanding and cooperation between different landscape architecture schools (see Stiles in this volume). Among them are: LE:NOTRE and EU-teach (ECLAS, 2010, 2011; Teqja and Dennis 2016a). Such programmes need to continue to help landscape architecture programs in former communist countries upgrade to the highest contemporary standards.

Conclusions

In Albania, higher landscape architectural education began to develop after the fall of the Iron Curtain by organizing professional graduate programmes in horticulture or architecture and planning. The collaboration between the Agricultural University of Tirana (AUT) and the Polytechnic University of Tirana resulted in a good and effective promotion of the profession in the country. It also helped in meeting growing demands for specialists in the field of planning, designing, constructing and maintaining residential green areas and public parks.

Findings indicate how the progress in establishing democratic institutions and civil rights create the environment necessary for the development of the landscape architecture profession. Private property rights, increased public awareness for the value of public space and the public willingness to participate in the democratic processes seem to be supporting the development of higher education for this profession.

With the progress made in the process of EU integration and Albania's accession to the European Landscape Convention and the Albanian Government commitment to implement the recommendations of the Convention opened the way for AUT to create the potential of starting a full and independent set of landscape architecture education. Limitations in human resources are overcome by employing qualified staff and through good collaboration with other universities in Albania and abroad.

Notes

1 https://www.eclas.org/
2 https://www.uniscape.eu/
3 https://thecela.org/
4 www.eiu.com/topic/democracy-index
5 https://freedomhouse.org
6 www.universitetipolis.edu.al
7 www.eiu.com
8 https://freedomhouse.org

References

Alhasani D. M. (2015) Educational Turning Point in Albania: No More Mechanic Parrots but Critical Thinkers. *Journal of Educational Issues*, 1(2), pp. 117–128.
European Council of Landscape Architecture Schools (ECLAS). (2010) *ECLAS Guidance on Landscape Architecture Education*. Available at: www.eclas.org/eclas-education-guide (Accessed 16 March 2022).
Giusti, M. A. (2012) '"Villa Reale" di Tirana: architetture, giardini, arredi, opere d"arte, dai progetti del ventennio al progetto di restauro', in *Architetti e Ingegneri Italiani in Albania. A cura di Ezio Godoli e Ulisse Tramonti*. Florence: Edifir Edizioni Firenze s.r.l., pp. 137–143.
IFLA Europe. (2019) *IFLA Europe Charter on European Landscape Architects* [Online]. Available at: www.iflaeurope.eu/assets/docs/Charter_on_European_Landscape_Architect_2019.pdf (Accessed 15 October 2020).
Jani, V. (2009) *Landscape Architecture: The Art of Greening in Urban Environments and in Natural Landscapes (In Albanian)*. Tirana: UFO Press, p. 213.
Lekaj, E. and Teqja, Z. (2019) 'Users' Evaluation for Public Parks: Influences of Location, Season, Gender and Age', *European Academic Research*, VII(1), pp. 603–624.
Mauro, E. (2012) 'L'architettura dei giardini del Ventennio in Albania', in *Architetti e Ingegneri Italiani in Albania. A cura di Ezio Godoli e Ulisse Tramonti*. Florence: Edifir Edizioni Firenze s.r.l., pp. 129–135.
Menghini, A. B., Pashako, F. and Stigliano, M. (2012) *Architettura moderna italiana per le città d'Albania. Modelli e iterpretazioni*. Bari: Politecnico di Bari, Facolta di Architettura, Dipartimento ICAR, p. 258.
Perekovic, P., Kamenecki, M., Reljic, D., Hrdalo, I. and Zmire, I. (2019) 'Landscape Architecture in Croatia 1900–1990', in *ECLAS and UNISCAPE International Conference: Lessons from the Past, Visions for the Future: Celebrating One Hundred Years of Landscape Architecture Education in Europe*. Oslo: Norwegian University of Life Sciences, pp. 55–56.
Pojani, D. (2014) 'Urban Design, Ideology, and Power: Use of the Central Square in Tirana During One Century of Political Transformations', *Planning Perspectives*, 30, pp. 67–94.
Pojani, D. and Maci, G. (2015) 'The Detriments and Benefits of the Fall of Planning: The Evolution of Public Space in a Balkan Post-socialist Capital', *Journal of Urban Design* 20(2), pp. 251–272.
Supuka, J. and Tóth, A. (2019) 'The Historical Development of Landscape Architecture Education in Slovakia', in *ECLAS and UNISCAPE International Conference: Lessons from the Past, Visions for the Future: Celebrating One Hundred Years of Landscape Architecture Education in Europe*. Oslo: Norwegian University of Life Sciences, pp. 143–144.
Teqja, Z, Beka, I. and Shkreli, E. (2000) 'Albanian Agriculture – Dramatic Changes from a Very Centralized Economy to Free Market. A Strategy for Future Development', *Medit*, 1, pp. 21–29.
Teqja, Z. and Dennis, S. (2016a) 'A Landscape Architecture Program for Albania', *Educational Alternatives*, 14, pp. 530–542.
Teqja, Z. and Dennis, S. (2016b) 'Creative Thinking, Critical Thinking and Systemic Thinking – Key Instruments to Deeply Transform the Higher Education System in Albania: The Case of Landscape Architecture', *Educational Alternatives*, 14, pp. 543–555.

Teqja, Z. and Dervishaj, A. (2020) 'Landscape Architecture Education in Albania – The Challenge of Having a Studio and Research-Based Programme', *European Academic Research*, VII(10), pp. 4866–4881.

Teqja, Z. and Karaj, A. (2022) 'Landscape Architecture Education and Democracy', *Projets de paysage* [Online], Special issue/ 2022. Online since 5 July 2022. Available at: http://journals.openedition.org/paysage/27792; DOI: https://doi.org/10.4000/paysage.27792

Teqja, Z. and Kopali, A. (2012) 'A Study on the Influences of Trade Centers on Urban Lifestyle of Tirana Population', in *Proceedings Book: International Conference on "Towards Future Sustainable Development"*. Shkodra: Luigj Gurakuqi University.

Tudora, I. (2011) 'Teaching Landscape Architecture: Tuning Programs in Europe for a Common Policy. The Romanian Case', *First International Conference "Horticulture and Landscape Architecture in Transylvania", Agriculture and Environment Supplement*, pp. 283–297.

35
LANDSCAPE ARCHITECTURE EDUCATION IN POLAND

Katarzyna Rędzińska and Agnieszka Cieśla[1]

Introduction

This chapter offers insights into landscape architectural education in relation to professional practice in Poland. It aims to explore the role that education has not only in passing on knowledge for its application but also in improving opportunities for graduates in practice. Landscape architecture education and practice in Poland serve as an example for countries that, during recent decades, experienced three political system changes, each associated with far-reaching transformations of development and planning policy. The three revolutionary socio-political system changes, which took place in the 20th century, were regaining independence, socialism, and neoliberalism after 1989. They highly influenced the development of the landscape architecture education and profession in Poland by transforming the context of landscape architects' activities.

Due to the lack of continuity and shifts of paradigms, as well as the silo approach, there is a lack of cohesion between existing regulations related to landscape shaping, protection, and management in Poland (Giedych 2004). A similar situation is found in the Czech Republic and Slovakia (Kozová and Finka 2014), where legal instruments addressed to landscape planning are missing.

This chapter presents findings from the discourse analysis of the Polish Landscape Architecture Forum on the background of the discipline evolution in terms of education and professional practice and their current determinants. The Forum is an event where educators, researchers, and practitioners regularly meet and discuss current issues relating to landscape. Forum discussions offer opportunities for researchers to study past and current discourses pertaining to landscape. They also serve as platforms for landscape architects to meet with decision-makers, representatives of public administration, planners, and designers whose activities have landscape impact. To get a comprehensive understanding of the main discourses and narratives emerging from the analysis, this study encompasses a literature review and an analysis of legal regulations, university statistics, and curricula.

History of landscape architecture education in Poland

The history of Polish landscape architecture education spans nearly 100 years (Table 35.1). It grew out of the centuries-old tradition of planning gardens and parks. In the development of

Table 35.1 The stages of the development of landscape architecture education in Poland according to Wolski (2007), modified by authors based on Wolski (2015), Böhm and Sykta (2013), Niedźwiecka-Filipiak (2018), Sobota and Drabiński (2017).

1. **1923–1930: Pioneer teaching at the ornamental studies, at the Warsaw University of Life Sciences (SGGW) Faculty of Gardening**
   1930–1954: Founding the Departments teaching landscape architecture at the leading state universities

Warsaw University of Life Sciences (SGGW)	Faculty of Gardening – Department of Landscape Architecture and Parks Science (1930); after World War II, transformed into the Chair of Landscape Shaping and Decorating headed by Prof. Franciszek Krzywda-Polkowski then by Prof. Alfons Zielonko. In 1952, the specialisation course in the development of green areas was introduced.
Warsaw University of Technology	Faculty of Architecture – the Chair of Interior and Landscape Designing (1932). After World War II, it transformed into the Department of Landscape Architecture, managed first by Prof. Franciszek Krzywda-Polkowski, and next by Prof. Witold Plapis in the years 1955–1968.
Cracow University of Technology	Faculty of Architecture, the Department of Spatial Planning, the Institute of Planning of Green Areas (1952). In 1963 transformed into the independent Chair of Landscape Planning and Green Areas, founded and headed first by Dr. Zygmunt Novák, and after 1961 by Prof. Gerard Ciołek, as well as the Department of Landscape Planning and Preservation, headed first by Dr. Zygmunt Novák, and, after 1967, by prof. Janusz Bogdanowski.

2. **1954–1995: Foundation of Standards of landscape architecture teaching and publication of seminal books, lack of separate field of studies**

Warsaw University of Life Sciences (SGGW)	1954 – founding of the Section of Greenery Shaping at the Faculty of Gardening. During 1954–1959, the studies were two-stage, and after 1959 single-stage. Two chairs existed: the Chair of Green Area Designing and the Chair of Green Area Arranging. After 1970, the curricula, were adjusted to landscape architecture syllabuses of the Western universities. It also provided a basis for determining the minimum requirements for landscape architecture master studies, established in 1999, following draft standards of landscape architecture teaching. 1988 – Establishing (for the first time in Poland) the landscape architecture field of study. It was taught only in the form of specialisation of gardening; transformation of the Section of Green Area Shaping into the Division of Landscape Architecture.
Cracow University of Technology	After several changes in the organisational structure of the Faculty of Architecture, the Institute of Landscape Architecture was established in 1992, headed by Prof. Janusz Bogdanowski. Landscape architecture was taught to architecture and urban planning students, and it was not a separate field of studies.

(Continued)

Table 35.1 (Continued)

3. **1995–2018: Official, legal establishment of the profession, the foundation of the separate field of studies**

1995	Legal establishment of the profession
1999	Landscape architecture as a separate field of study at Warsaw University of Life Sciences and Cracow University of Technology; after 2000 at numerous state universities and some private colleges; first Landscape Architecture Forum discussing existing forms and possibilities of landscape architects education in Poland;
Stages of the development of the curricula:	
1999–2006	Uniform five-year master's program
2007–2011	The three-cycle system of studies as part of the Bologna Process. The teaching standards determined the curricula in terms of appropriate proportions of subjects and their substantive content.
2012–2018	Curricula adjusted to the National Qualification Framework (NQF) in 2012, modified to the 8-level Polish Qualifications Framework (PQF) in 2016, which were adequate to the European Qualifications Framework (EQF) (2017) as well as to the qualifications levels in individual European countries. 2012 – Foundation of the "Union of Universities for the Development of Landscape Architecture Studies" (the agreement between 11 universities providing education in landscape architecture signed by the rectors: the Catholic University of Lublin, Białystok University of Technology, Cracow University of Technology, Warsaw University of Life Sciences – SGGW, University of Life Sciences in Lublin, Poznań University of Life Sciences, the Wrocław University of Environmental and Life Sciences, the University of Agriculture in Cracow, University of Rzeszów, University of Warmia and Mazury, West Pomeranian University of Technology)
Since 2019	Curricula of the field of studies need to be adjusted to the scientific disciplines of the teaching Universities and Departments according to The Higher Education Reform Act form 2018

landscape architecture education, three stages can be identified: 1) after 1918, emerging with the revival of the Polish state; 2) after 1945, embedded in the socialist system; and 3) after 1989, formed within the neoliberal system.

The beginning of landscape architecture education is associated with the revival of the Polish state (Łukaszkiewicz et al. 2019). Universities became the centres of intellectual life, mapping new scientific perspectives and approaches. In the inter-war period, four orientations were identified (Michałowski 2000):

- Utilitarian, with practical experience as a testing ground, represented by Franciszek Krzywda-Polkowski (1881–1949) from Warsaw.
- Natural sciences, represented by Władyslaw Schafer (1886–1970) and Walery Goetel (1889–1972) from Krakow, and Adam Wodiczko (1887–1948) from Poznań.
- Architectural-landscape, represented by Tadeusz Tołwiński and Witold Plapis (1905–1968) from Warsaw, Zygmunt Novák (1897–1972) from Krakow, and Władysław Czarnecki (1895–1983) from Poznań.
- Conservation, represented by Gerard Ciołek (1909–1966) from Warsaw and Krakow.

World War II interrupted development of the discipline and, for a long time, eliminated landscape architects from participating in developing Polish landscapes through planning and design (Zachariasz 2018). Under the socialist regime, Poland underwent rapid urbanisation and industrialisation. From 1948 to 1988, the urban population in Poland grew by 15.5 million people, and 7.5 million flats were built, mostly in prefabricated systems. Spatial planning was entirely reorganised based on Soviet principles. Because of prioritising industry, concerns about the environment and natural resources were largely ignored. There was no long-term orientation in development policy and spatial development, and the immediate effect of planning had priority, without regard to social, architectural, and landscape consequences. The universities had a limited impact on real-world development. They focused on the development of theoretical principles for shaping landscapes and their protection. This approach is unfortunately valid also for today, when the research achievements are rarely implemented in reality.

Out of the four previously mentioned orientations, during the socialist period, the Warsaw (Ursynów) School of Landscape Architecture strengthened its position, and the Krakow School of Landscape Architecture soon followed (Michałowski 2000). Both schools had different specialties. While the Krakow School focused on protecting the cultural landscape heritage and on theory-building, the Warsaw (Ursynów) School was more practice-oriented. Its creators were renowned designers with extensive professional achievements: Franciszek Krzywda-Polkowski (1881–1949), Alfons Zielonko (1907–1999), Alina Zofia Scholtz-Richert (1908–1996), Ludwik Lawin (1908–1984), Longin Majdecki (1925–1997), Władysław Niemirski (1914–2001), and Edward Bartman (1929–2018). Their projects enrich urban and rural landscapes to this day, and their practical knowledge and experience were shared with several generations of graduates (Łukaszkiewicz et al. 2019). Contrarily, Krakow representatives wrote landscape architecture handbooks: the first one was written by Novák (1950), and two others by Bogdanowski (Bogdanowski et al. 1973; Bogdanowski 1976). The latter remain the basic Polish reference books used by landscape architecture students (Böhm and Kosiński 2007).

After 1995, due to the legal establishment of the profession and the forming of landscape architecture degree programmes, the popularity of the profession grew. Standardised, rigid curricula were replaced by the National Qualifications Framework (NQF) in 2012, which, after 2016, was reformed as the Polish Qualifications Framework, (PQF) to fit the European Qualification Framework (EQF) (European Parliament & Council of Europe 2017). Also in 2012, the universities offering landscape architecture programmes founded The Union of Universities for the Development of Studies in Landscape Architecture to strengthen the position of the landscape architecture profession and improve its educational quality and practical use. The founding agreement was signed by 11 state universities with the aim to maintain commonality in the teaching of landscape architecture. According to the agreement learning outcomes of the bachelor studies are to be composed of: 45% of the technical sciences; 45% of agricultural, forest, and veterinary sciences; and 10% of the arts (Zachariasz 2020).

Current status of landscape architecture as a field of study and profession in Poland

Current educational framework

In 2020, 17 institutes of higher education were offering degree programmes in landscape architecture (Perspektywy 2020). Ten state universities provided undergraduate and graduate programmes, four state universities were offering undergraduate programmes, and three private universities were offering undergraduate and graduate programmes. With the growing number of units offering the landscape architecture field of study located at different university types (agricultural, technical, and artistic), differentiation has been increasing (Zachariasz 2018). This process was strengthened by the higher education reform introduced in 2018. Since then, every faculty member, curriculum, and researcher needs to be assigned to at least one scientific discipline. As a result, the diversity of educational profiles grew considerably, and nowadays, they cannot be associated clearly with a given university specialty. The curricula analysis of 13 among 17 universities offering landscape architecture education in Poland in 2020 showed that these units usually have one leading and, normally, two to three additional disciplines, while two universities declared more disciplines (Table 35.2).

Six universities (three of life sciences, one of technology, and three general universities) chose technical sciences as a leading discipline (environmental engineering, mining, and energy, as well as architecture and town planning). Five universities (two universities, two of life sciences, and one of technology) chose natural sciences as a leading discipline (discipline in agriculture and horticulture). One university chose biological sciences. Among complementary disciplines, two dominated: architecture and town planning (10 universities – both stages, all types) and visual arts and art preservation (10 universities – mostly only at the bachelor level). Seven universities chose a discipline in agriculture and horticulture (both stages), and 15 other disciplines appeared in five schools' curricula. The diversity of the disciplines proves that the specificity of the landscape architecture education is no longer associated with the type of university as it was until the higher education reform. Such a situation leads to a growing differentiation of the landscape education in Poland and will be a challenge in maintaining good education quality.

Additionally, a decreasing number of students due to demographic change in the future will force faculties to compete and to modernise their curricula. Growing global challenges like climate change, pandemics, and digitalisation are likely to determine the direction for modernisation.

Current professional framework

The emergence of a landscape architect profession in Poland dates back to the beginning of the 20th century. However, it was officially introduced in 1995 by the Minister of Labour and Social Policy in the ordinance called Classification of Occupations and Qualifications and Structure

Table 35.2 Curricula and scientific disciplines mapping in 2020.

lp	University	Degree (bachelor I, Master II)	Full-time studies	Part-time studies	Year of accreditation	Leading disciplines — Natural sciences: Agriculture and horticulture	Leading disciplines — Technical sciences: Environmental engineering, mining and energy	Leading disciplines — Technical sciences: Architecture and town planning	Leading disciplines — Biological sciences	Complementary disciplines — Natural sciences: Agriculture and horticulture	Complementary disciplines — Technical sciences: Architecture and town planning	Complementary disciplines — Technical sciences: Environmental engineering, mining and energy	Complementary disciplines — Technical sciences: Civil engineering and transport	Complementary disciplines — Arts: Visual arts and art. preservation	Others
1	University Of Warmia And Mazury In Olsztyn	I	1	1	PKA 2019	55%					32%			13%	
		II	1	1	x	52%					48%				
2	Cracow University of Technology	I	1	1	PKA 2020			v		v		v	v	v	
		II	1	1	IFLA			v		v		v	v	v	
3	University of Agriculture in Krakow	I	1		PKA 2020		52%			21%	27%				
		II	1	1	x		56%			16%	28%				
4	University of Ecology and Management in Warsaw	I	1	1	PKA 2020			v		v					
		II	1	1	x			v		v					
5	SGGW – Warsaw University of Life Sciences	I	1	1	PKA 2019		51%			14%	25%			10%	
		II	1	1	x		54%				46%				
6	Poznań University of Life Sciences	I	1	1	PKA 2016	53%					23%	10%		14%	
		II	1	1	x	55%					37%	8%			
7	University Of The Arts Poznan	I	1			–	–	–	–	–	–	–	–	–	–
		II	1			–	–	–	–	–	–	–	–	–	–

(Continued)

Table 35.2 (Continued)

lp	University	Degree (bachelor I, Master II)	Full-time studies	Part-time studies	Year of accreditation	Leading disciplines — Natural sciences — Agriculture and horticulture	Leading disciplines — Technical sciences — Environmental engineering, mining and energy	Leading disciplines — Technical sciences — Architecture and town planning	Leading disciplines — Biological sciences	Complementary disciplines — Natural sciences — Agriculture and horticulture	Complementary disciplines — Technical sciences — Architecture and town planning	Complementary disciplines — Technical sciences — Environmental engineering, mining and energy	Complementary disciplines — Technical sciences — Civil engineering and transport	Complementary disciplines — Arts — Visual arts and art. preservation	Others
8	West Pomeranian University of Technology in Szczecin	I	1	1	PKA 2019	55%					40%			5%	5%
		II	1	1	x	55%					40%			5%	5%
9	Gdańska Szkoła Wyższa	I	1	1											–
10	The John Paul II Catholic University of Lublin	I	1		PKA 2020			55%		39.50%				2%	5%
		II	1		IFLA			56%		35%				1.50%	9.50%
11	University of Life Sciences in Lublin	I	1	1	PKA 2018	60%					30%			10%	
		II	1	1	x	60%					30%			10%	
12	Białystok University of Technology	I	1	1			v				v				–
		II	1	1			v				v				–
13	Sopocka Szkoła Wyższa	I	1	1	PKA 2013	–					–			–	–
14	University of Opole/Universitas Opoliensis	I	1						53%	1%	17%			8%	21%
		II	1						52%		25%			1%	23%
15	Wrocław University of Environmental and Life Sciences (UPWr)	I	1				v			v	v			v	
		II	1				v				v				
16	University of Rzeszów	I	1	1	PKA 2018	64%					27%			9%	
		II	1	1		63%					31%			6%	
17	University of Science and Technology in Bydgoszcz	I	1			–					–			–	–

for Organising Information on Labour (Forczek-Brataniec 2018; Sobota and Drabiński 2017; Wolski 2015, 2017).

In the current Classification of Occupations, the profession of landscape architect is in the same group as architects, planners, surveyors, and designers (Wortal 2020). However, the frameworks of performing the landscape architect profession are weaker than of the other professions in this group. Formally, landscape architects in Poland can perform a wide range of activities: programmes and projects for the protection and shaping of landscapes; projects for the construction, renewal, and restoration of landscape architecture facilities such as parks, squares, gardens, and other green areas under the author's supervision; carrying out studies on landscape; and assessing the expected impact on the landscape by land-use plans and planned investments (Forczek-Brataniec 2018; Wortal 2020). There are, however, two main obstacles that limit their professional practice.

Firstly, the profession is neither regulated nor protected in Poland (Forczek-Brataniec 2018; Sobota and Drabiński 2017; Wolski 2015, 2017). The building code includes landscape architecture projects for which only chartered architects or civil engineers can sign application documents. Landscape architects are included in neither the architects' nor the civil engineers' chambers (Wolski 2015, 2017).

Until 2014, landscape architecture graduates who completed postgraduate studies in architectural, urban, or spatial management programmes could apply for admission to the chamber of town planners. However, this chamber was abolished by the Deregulation of the Professions Act in 2014 (Forczek-Brataniec 2018). It can be stated that landscape architects are equal participants in the spatial planning process; however, they cannot fully use their skills, as the landscape is weakly incorporated in the spatial planning instruments.

The second obstacle limiting career prospects of landscape architects is the fact that the Polish Building Code of 1994 does not define the landscape architect's role and creates no basis for the landscape architect's participation in any part of the investment process (Wolski 2017).

Surprisingly, these barriers in the professional practise do not negatively affect the popularity of the landscape architecture field of study. According to the alumni statistics, the number of landscape architecture graduates has been growing consistently over the years. Almost 700 graduates finished their studies in 2018 (KSI 2020). Prof Barbara Szulczewska, the former Dean of Student Affairs at the Warsaw University of Life Sciences, stated that the popularity of the landscape architecture field of study was triggered by a TV series that started in 2000, where a leading character was a landscape architect. Additionally, growing awareness among young people of climate change challenges and the importance of green infrastructure as an ecosystem services provider contributes to attracting students to the landscape architecture profession.

However, partly due to the obstacles mentioned earlier, specialists from other fields perform many tasks linked with landscape architecture: architecture (over 1,600 graduates each year), urban planning (over 1,300 graduates), geography, and geodesy (KSI 2020). As a result, the demand for landscape architects is limited, and graduates face problems with finding a job, particularly a well-paid one. An experienced, qualified landscape architect earns half of the average wage in the region where he works (PGTS 2020).

Main forum discourses

The universities offering landscape architecture education have been organising the Landscape Architecture Forum since 1998 annually until 2019. Due to the pandemic, the Forum was cancelled in 2020. A year later, it took place in an online mode, and a subsequent Forum is already planned. For the purposes of this analysis, documents from Forums up to 2019 were considered.

The first Forum events concentrated on didactics, as a consequence of the legal foundation of the profession in 1995 and the beginning of landscape architectural education as a separate degree programme in 1999. Following further development, the Forum started to deal with other scientific, professional, and organisational themes (Table 35.3), and it became a platform for professional meetings and discussions. The Forum formats included lectures, discussions, field trips, and occasional workshops. The number of participants ranged from 60 to more than 200.

Method of analysis

The approach to synthesising data is a triangulation of findings using critical discourse analysis (Jørgensen and Phillips 2002). The conference proceedings, books of abstracts, programmes, and web pages were analysed. Conference presentations' titles served as the higher order for classifying data. In case of uncertainties about their meanings, keywords, abstracts, and publications were considered. The analysis includes data from 1,113 presentations, primarily given in the form of lectures. The majority concentrated on scientific issues; however, 155 related to landscape architecture in terms of education and practice. They form the basis for further analysis and are classified into topics and clusters of topics. For each cluster, the word frequency was analysed by NVivo software based on lecture titles. The analysis produced three clusters: education, professional practice, and the position of landscape in spatial planning systems. The article focuses on the exploration of the education and professional practice clusters in detail, including the position of landscape in planning system as part of the professional practice cluster.

Results

While topics and clusters related to education and the profession were mainly featured in the 1998, 1999, 2000, 2007, 2012, and 2017 Forums, contributions concerning the position of landscape in planning have been increasing since the adoption in 2006 and implementation in 2015 of the European Landscape Convention (ELC) by Poland. Narratives and discourses of Landscape Forums are presented in Figure 35.1.

Landscape architecture forum discourses on landscape architecture education

The cluster related to the landscape architecture education encompassed 77 lectures, focused on general considerations (19 lectures), didactic methods (14), courses (12), foreign context (12), curricula (11), history of landscape architecture education (5), and other issues (4). The first three Forums of 1998, 1999, and 2000 concentrated on teaching approaches, including presentations of particular curricula of Polish universities in 1998, 1999, and 2000, and international ones in 1999. The aim was to standardise the education programmes in the field of landscape architecture due to the emergence of the profession (1995) and the field of education (1998). The landscape architecture education issues were discussed again in 2007, 2011 was dedicated to didactic approaches and teaching methods, and 2017 concentrated on courses.

The discussions were also important to define the tradition and the Polish educator community's identity. The identification and promotion of Polish landscape architecture education and practice legacy was one of the most significant topics. Such topics were presented during Forums in 2007, 2009, and 2019. They also highly contributed to the formation of landscape architecture as a specific area of knowledge.

Table 35.3 Themes of the Landscape Architecture Forum, 1998 to 2019.

Forum (year of event)	Title of conference	Hosting university/institution	Lectures number
I (1998)	"The First Didactic Forum for Landscape Architecture"	Krakow University of Technology	16
II (1999)	"II Didactic Forum of Landscape Architecture" (international conference)	Krakow University of Technology	12
III (2000)	"New Ideas and development in Landscape Architecture in Poland"	Warsaw University of Life Sciences (WULS-SGGW)	44
IV (2001)	"Landscape as regional identity image (threats, protection and shaping"	Silesian University of Technology Upper Silesia Space Foundation	26
V (2002)	"Landscape transformations in Suburban Zones. Sustainable Development Directions"	Agricultural University of Wroclaw Wroclaw University of Science and Technology	50
VI (2003)	"Countryside landscape and garden"	The Botanical Garden of the University of Wroclaw The John Paul II Catholic University of Lublin, University of Life Sciences in Lublin, Lublin Village Museum	66
VII (2004)	"Landscape without borders"	Upper Silesia Space Foundation City Hall in Katowice, Bielsko-Biała Town and Commune Office	28
VIII (2005)	"Landscape shaped by agriculture"	University of Warmia and Mazury in Olsztyn, Cultural Community Association "Borussia" Starostwo Powiatowe in Olsztyn	84
IX (2006)	"Water in the landscape"	Agricultural University in Szczecin	0
X (2007)	"International congress of polish landscape architects: The Art of Environmental Protection and Management. Profession – Theory – Teaching"	Cracow University of Technology Capitula of Landscape Architecture	100
XI (2008)	"International Landscape Architecture Forum: Recreation development versus landscape protection and shaping"	Poznań University of and Life Sciences	53
XII (2009)	"Landscapes of Europe – planned economy or generating chaos?" "Landscaping and protection of river valleys"	Wroclaw University of Environmental and Life Sciences	14

(Continued)

Table 35.3 (Continued)

Forum (year of event)	Title of conference	Hosting university/institution	Lectures number
XIII (2010)	"Nationwide scientific conference. Horizons of Landscape Architecture. Subject. Method. Language"	Warsaw University of Life Sciences (WULS – SGGW)	67
XIV (2011)	"The landscape a-new"	Wrocław University of Environmental and Life Sciences	96
XV (2012)	"Landscape planning"	University of Life Sciences in Lublin	38
XVI (2013)	"Land-scape of the future"	Uniwersytet Warmińsko–Mazurski w Olsztynie	59
XVII (2014)	International Scientific Conference: "Landscape valorisation – when does landscape become attractive?"	West Pomeranian University of Technology in Szczecin	28
XVIII (2015)	"Landscape identity"	Białystok University of Technology (BUT)	47
XIX (2016)	"Polish landscape. Praise others. Protection and shaping of the native landscape"	Rzeszów University	54
XX (2017)	"Professional practice – academic research – didactics"	Cracow University of Technology	104
XXI (2018)	"Landscape architecture challenges"	Warsaw University of Life Sciences	71
XXII (2019)	"Difficult landscapes"	Wrocław University of Environmental and Life Sciences	56
Total			1113

Source: Author's own.

Following are examples of titles of lectures qualified for this cluster:

- The following lectures were classified as general considerations: "A few thoughts on the development of the European dimension of education in the field of landscape architecture"; "Protection and creation of beauty, how to teach it?"; "Teaching a profession: educating professionals"; "Teaching: conditions and difficulties".
- The teaching methods were discussed in the following lectures: "The role of the theory of Open Form in landscape architecture: prototyping of space and contemporary tools to support design"; "Popularization of the principles of landscape composition using the tool of architectural and landscape drawing: social education".
- Curricula were the subject of the subsequent lectures: "Teaching the subject of landscape architecture at the faculty of . . ."; "Exemplary set of learning outcomes for the field of landscape architecture".
- Several lectures dealt with the course presentation: "Delimitation of historic parks and gardens: a new course in the field of Landscape Architecture at the University of Agriculture in Krakow".
- Foreign context was also a subject of some lectures: "Teaching landscape architecture in the United States: issues and directions".
- The history of landscape architecture education was also a subject of lectures: "Teaching landscape architects in Poland".
- Many other issues were also discussed in this cluster like presentation of the activities of student research groups or didactic gardens.

The analysis showed that the general review of curricula, taking into account the foreign context, took place 20 years ago, and the teaching methods were discussed 10 years ago.

Landscape architecture forum discourses on professional practice

The professional practice cluster of topics included 78 lectures, which we divided into several topics: the history of the profession (6 lectures), the legal situation and position of the profession in the planning system (22 lectures), the legal position of the landscape in the planning system (15), available tools for landscape protection (12) and design (8) or both (5), and general considerations about spatial planning systems in reference to landscape planning (10).

The lack of the possibility for landscape architects to belong to a professional chamber is an essential element in Forum discussions. In order to strengthen the position of landscape architects in Poland, professional associations founded the Federation of Landscape Architecture Associations (Forczek-Brataniec 2018; Sobota and Drabiński 2017). Their members are also educators and researchers. Professional associations offer vocational trainings and fund an award for outstanding diplomas in the field of landscape architecture. These issues were vital for adjustment of education offers to the job market and forming foundations for the career prospects of graduates. Vivid discussions were held around the topic on how to perform the landscape architecture profession in the system that lacks specifically addressed tools and instruments for landscape planning and design.

Based on the Forum discussions, the development of the landscape architecture profession was the second most important driver for community building next to the history of education.

- Examples of papers qualified for the cluster concerning the position of the profession are as follows: "The beginnings of the landscape architect profession"; "The role of a landscape

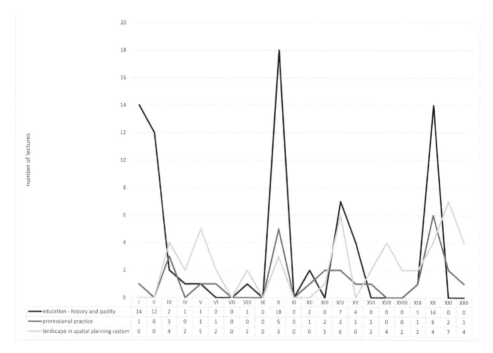

Figure 35.1 The frequency of the clusters of topics related to landscape architecture education, professional practice and landscape in the spatial planning system during the Landscape Forum, 1998 to 2019.

Source: Visualised by the authors.

architect"; "Legal and organizational conditions of Polish landscape architecture in 1998–2017"; "The situation of the landscape architect profession in Europe based on the work of a team working within IFLA EU, which deals with supporting the regulation of the profession in European countries (Professional Regulation Assistance – PRA – IFLA EU)".

- The adoption of the ELC highly fuelled the discussions on the position of landscape in the planning system. The following titles of lectures are examples of papers qualified for this cluster: "Landscape act"; "New legal and organizational conditions related to the shaping and protection of landscape"; "The impact of the binding building code on landscaping"; "Local land use plan as a tool to protect the identity of the urban landscape"; "The usage of urban indicators in documents on spatial development in communes"; "Identification and assessment of landscape as the basis for shaping space on a local scale"; "Contribution to the discussion on the landscape audit"; "Planning tools for shaping and protecting the urban landscape"; "Spatial planning and landscape architecture"; "Landscape and planning"; "Possibilities of adaptation of French legal and planning solutions in the field of landscape protection to Polish conditions".

Conclusions

Landscape architecture education and professional practice have a century-old tradition in Poland. Overcoming setbacks from several socio-political changes, practitioners and educators are working hard to embed their field into evolving legal and planning systems and to raise societal awareness of the profession. Analysing Landscape Architecture Forum discourses, we identified three groups of themes related to education, professional practice, and science. For education, the analysis revealed that substantial reviews of national and international curricula date back about 20 years and that the last full discussion of didactic methods occurred about 10 years ago. Needing to cope with current societal challenges, new discussions on teaching methods and approaches are called for, with educators and practitioners contributing.

The digitalisation challenges and opportunities in terms of education and professional practice were also weakly considered and need more attention. The technological progress provides new tools and working methods. It requires specific knowledge enabling cooperation with specialists of different fields, openness to new methods and didactic tools, and constant adjustment of the educational effects.

In professional practice, the legal regulation of recognition and licensing is important. Findings from our analysis suggest how the lack of regulations limits professional development, progress in education, and graduates' future prospects. For example, Polish landscape architects do not officially have the authorship rights to typical landscape architecture projects like parks. However, they are entitled work in planning and provide services such as land-use plans.

For both education and professional practice, the identification and promotion of legacy were crucial. This concerned the appreciation for the founders of landscape architecture in Poland, seminal books, outstanding projects, and their completion. Based on work done by the Universities Union and professional organisations, the quality of education and landscape architectural practice will continue to improve. High-quality education will have a significant impact on improving the situation for graduates.

Complex and changing societal and political transformation are forces that have shaped the profession from the start. In countries like Poland that have experienced several system changes in recent history, landscape architecture has been among the most adaptive and interdisciplinary of professions. The field is proving to be flexible in addressing emerging societal challenges, such as demographic and climate change, and in adjusting to changing policy and legal conditions. These challenges and transformations provide the context for educators, researchers, and the practitioner community to develop didactic approaches to introduce flexible curricula and innovative educational methods, making use of new digital technologies and tools.

More than before, the field is becoming interdisciplinary because of the broadness of 21st-century planning tasks. The need for practice to be interdisciplinary and flexible highly influences the development of landscape architecture education as it becomes involved in addressing real-world challenges.

Note

1 Acknowledgements: We would like to acknowledge Barbara Szulczewska, Przemysław Wolski, Jacek Rybarkiewicz, and Irena Niedźwiecka-Filipiak for sharing archives and for consultation.

References

Bogdanowski, J. (1976) *Kompozycja i planowanie w architekturze krajobrazu*. Kraków: Wydawnictwo Polskiej Akademii Nauk.

Bogdanowski, J., Łuczyńska-Bruzda, M. and Novák, Z. (1973) *Architektura krajobrazu*. Kraków: Państwowe Wydawnictwo Naukowe PWN.

Böhm, A. and Kosiński, W. (2007) *Founders of Polish Landscape Architecture* [Online]. Available: https://riad.pk.edu.pl/~a-8/kongres/rejestra.htm (Accessed 19 August 2020).

Böhm, A. and Sykta, I. (eds.) (2013) *1992–2012: The Twentieth Anniversary of the Institute of Landscape Architecture Cracow University of Technology*. Kraków: Politechnika Krakowska.

European Parliament & Council of Europe. (2017) 'Recommendation of the Council, of 22 May 2017, on the European Qualifications Framework for Lifelong Learning and Repealing the Recommendation of the European Parliament and of the Council of 23 April 2008', *Official Journal of the European Union*, C 189, pp. 15–28.

Forczek-Brataniec, U. (2018) 'Situation of Landscape Architect Profession in Europe Based on the Teamwork Supporting the Regulation of the Profession Professional Recognition Assistance (PRA) IFLA Europe', *Architektura Krajobrazu*, 2, pp. 4–23.

Giedych, R. (2004) 'Legal Basis of Landscape Protection, Planning and Management in Poland in the Light of European Landscape Convention', *The Problems of Landscape Ecology*, 13, pp. 29–34.

Jørgensen, M. W. and Phillips, L. J. (2002) *Discourse Analysis as Theory and Method*. London: SAGE.

Kozová, M. and Finka, M. (2014) 'Landscape Development Planning and Management Systems in Selected European Countries', *The Problems of Landscape Ecology*, XXVIII, pp. 101–110.

KSI. (2020) *Kierunki Studiów.info* [Online]. Grupa studentnews. Available at: www.kierunki-studiow.info/studia/2st (Accessed 18 April 2020).

Łukaszkiewicz, J., Fortuna-Antoszkiewicz, B. and Rosłon-Szeryńska, E. (2019) 'Ursynowska szkoła architektury krajobrazu – mistrzowie i ich dzieło (cz. 1, cz. 2)', *Acta Scientiarum Polonorum. Architectura*, 18, pp. 133–146, 125–138.

Michałowski, A. (2000) 'Landscape Architecture at the End of the 20th Century. Significance and Tasks', in Wolski, P. (ed.) *III Forum of Landscape Architecture*. Warsaw: The Centre for the Preservation of Historic Landscape. A National Institution for Culture, pp. 27–39.

Niedźwiecka-Filipiak, I. (2018) 'Education and Promotion of the Profession of a Landscape Architect in Poland', *Architektura Krajobrazu*, 2, pp. 24–41.

Novák, Z. J. (1950) *Przyrodnicze elementy planowania regionalnego i udział w nim architekta*. Kraków: Politechnika Krakowska.

Perspektywy. (2020) *Top-wyszukiwarka* [Online]. Available at: http://perspektywy.pl/portal/index.php?option=com_content&view=article&id=2&Itemid=113 (Accessed 8 April 2020).

PGTS. (2020) Polish Graduate Tracking System [Online]. *Ministerstwo Edukacji i Nauki*. Available: https://ela.nauka.gov.pl/pl (Accessed 8 June 2020).

Sobota, M. and Drabiński, A. (2017) 'Legal and Organizational Aspects of Polish Landscape Architecture in the Years 1998–2017', *Architektura Krajobrazu*, 4, pp. 32–53.

Wolski, P. (2007) 'Nauczanie architektów krajobrazu w Polsce', *Czasopismo Techniczne. Architektura*, 104, pp. 26–30.

Wolski, P. (2015) 'Us – Outside the Law', *Architektura krajobrazu. Studia i prezentacje*, 2015, pp. 72–76.

Wolski, P. (2017) 'Rola architekta krajobrazu – w świetle prawa', *Topiarius. Studia krajobrazowe*, 2, pp. 65–76.

Wortal. (2020) Wortal Publicznych Służb Zatrudnienia [Online]. *Ministerstwo Rozwoju, Pracy i Technologii*. Available at: https://psz.praca.gov.pl/ (Accessed 19 August 2020).

Zachariasz, A. (2018) 'Beginnings of Landscape Architecture in Poland', *Landscape Architecture and Art*, 13, pp. 115–127.

Zachariasz, A. (2020) 'Professional Licensing of Landscape Architects in Poland', *World Transactions on Engineering and Technology Education*, 18(4), pp. 444–449.

36
LANDSCAPE @LINCOLN

Place and context in the development of an antipodean landscape architecture programme

Simon Swaffield, Jacky Bowring, and Gill Lawson

Introduction

What shapes a School of Landscape Architecture in a small country a long way from anywhere? What challenges has it faced, what strategies have been adopted and what distinctive features have emerged? This chapter describes and interprets the first 50 years of landscape architecture education at Lincoln University in Aotearoa-New Zealand (Aotearoa-NZ). It is focused upon the context, events, and actions that have shaped the diploma and degree programmes described here as Landscape@Lincoln and aims to provide an account suited to comparison with wider international developments in landscape architecture education. It draws upon the Lincoln University 'Living Heritage' archive, articles published in *The Landscape* magazine, the website of Tuia Pito Ora – The New Zealand Institute of Landscape Architects, media reports, and unpublished reports and personal papers.

The co-authors are current or past heads of school. We trace a history that is locally grounded and outward looking, with teaching at BLA, MLA, and PhD levels, underpinned by research and innovation. We argue that the development of Landscape@Lincoln expresses a series of strategic adaptions to wider social, cultural, and economic transformations and crises, and that institutional, regional, national, and global contexts have been key factors in the story.

Imperatives for future development include strengthening partnerships with Māori, the indigenous people of Aotearoa-NZ, and the deepening and widening of design and planning responses to climate change, within a context of sustainable development. These are essential to ensure enduring capacity in landscape architecture teaching and research to meet the needs of future generations.

Context

The Lincoln landscape architecture programme was established in 1969 at Lincoln University College of Agriculture, in a rural setting in Te Wai Pounamu – the New Zealand South Island. It was the first landscape architecture programme in Aotearoa-NZ, which at the time had no formally recognised landscape architecture profession. There was only one professionally accredited School of Architecture (Auckland University), and there are still none in the South Island. In 1990 Lincoln College became Lincoln University, focused upon land-based

DOI: 10.4324/9781003212645-40

disciplines. These origins required and enabled a high degree of disciplinary autonomy in Landscape Architecture.

Aotearoa-NZ is an isolated country in the Southwest Pacific, with a small population – only 2.8 million when landscape architecture education was started, and 5.2 million in 2020. Originally governed by independent indigenous tribes, who with arrival of Europeans defined themselves collectively as Māori, its modern constitution is grounded in the Treaty of Waitangi. Signed in 1840 by Māori chiefs and representatives of the British Crown, the Treaty includes agreements over sovereignty, citizenship, governance, land ownership, and resources (Orange 2015).

Landscape@Lincoln was established at a time of political, economic, social, and environmental tensions in Aotearoa-NZ. Māori were pressing the NZ government to honour the terms of the Treaty; a global 'oil shock' and UK entry into the European Common Market in 1973 destabilised the Aotearoa-NZ economy; while protests about sports relationships with apartheid South Africa revealed deep social divisions. Growing environmental awareness such as the 'Save Manapouri' campaign (1969–1972) helped the establishment of landscape architecture, but the emerging socio-economic tensions created challenges for both education and practice.

A change of government in 1984 (just 15 years after landscape architecture teaching began) was the catalyst for major social and economic transformations described by Kelsey (1995) as 'The New Zealand Experiment'. This reorientated the country from an economically highly regulated former colony to become a global trading nation, with neo-liberal public policies, diverse ethnicities, increasingly bi-cultural governance framed by Treaty obligations, and a reformed environmental planning system. These transformations changed the nature of professional practice in landscape architecture and imposed new requirements on universities.

Over the past decade, the regional context has also created challenges and opportunities for Landscape@Lincoln. In 2010–11 the nearby city of Christchurch experienced major earthquakes, requiring large-scale urban reconstruction. Floods and wildfire followed in 2017, and a major terrorist attack on two city mosques in 2019. Most recently, the global Covid-19 pandemic has affected student recruitment and teaching, highlighting both advantages and consequences of geographical isolation. Separated from neighbours and international partners by major oceans and time zones, Landscape@Lincoln has always needed to be well connected yet self-reliant.

Content

We present our history of Landscape@Lincoln in four periods. First, the Landscape 'Apostles' (1969–1982), a period establishing and promoting the inaugural landscape architecture programme and a new profession. Describing new graduates as 'Apostles' for landscape architecture is credited to programme founder 'Charlie' Challenger (Scoop News 2007). Second, 'No 8 Wire' (1983–1992), adapting to 'The New Zealand Experiment' and introducing a multi-level degree framework. 'No 8 Wire' is a Aotearoa-NZ colloquial expression referring to improvisation with materials at hand. Third, 'Overseas Experience' (1993–2006), focused on building international relationships to enhance research capacity and student numbers. Overseas experience (OE) refers to young New Zealanders travelling internationally. Fourth, 'Future Shock' (2007–2020) involves local adaption to successive regional and global crises. The title is from Toffler (1980). Table 36.1 summarises the development periods and key events.

Landscape @Lincoln

Table 36.1 Summary of key events and developments, 1960s to 2020.

Regional, national & global context	Development period	Landscape@Lincoln imperatives and strategies	Landscape@Lincoln events
*NZ environmental movement and concerns about landscape impacts of development. *UK entry to EEC impacts NZ economy. *Māori pressure on government to honour the obligations of the Treaty of Waitangi. *Growing social divisions and debate over national identity and future directions.	1960s–1982 The Landscape 'Apostles'	*Educating, creating and promoting a NZ profession of landscape architecture* *Appoint and train a programme leader. *Introduce a landscape architecture qualification. *Establish a new professional institute. *Create the Landscape Consulting Service. *Publish a professional magazine.	1969: Certificate in Landscape Design 1969: Diploma in Landscape Architecture 1972–73: New Zealand Institute of Landscape Architects founded 1976: "The Landscape" commences publication 1982: Programme founder Charlie Challenger retires
*New government elected in 1984 and begins 'The New Zealand Experiment' in public policy, with a focus on economy, and reform of public service and environmental law. *Lincoln College becomes Lincoln University in 1990.	1983–1992 'No. 8 Wire'	*Developing the framework for a NZ discipline and adapting to 'The NZ Experiment'* *Introduce degrees in Landscape Architecture. *Collaborate with other degrees and disciplines. *Achieve greater organisational autonomy for Landscape Architecture.	1985: MLA introduced 1988: BLA introduced to replace Dip LA 1989: Move to expanded facilities 1992: Department of Landscape Architecture created
*Government reforms extend to education, research, and social policy. *Māori and Government negotiate over Treaty claims, and in 1998 Ngai Tahu settle their claim over much of Te Wai Pounamu – The South Island.	1993–2006 'Overseas Experience'	*Looking outward to build research capacity & student numbers* *Enhance engagement with Māori. *Develop staff research skills and undertake funded research collaborations. *Extend teaching to PhD level. *Engage with overseas landscape educators. *Offer study abroad and exchange opportunities. *Establish an Australasian peer reviewed journal. *Increase international student enrolments. *Expand BLA delivery offshore.	1993 onwards: Participate in AELA, CELA, ECLAS 1995: Landscape Review begins publication 1995: LOLA conference @ Lincoln 1998: First PhD in Landscape Architecture awarded 2000: Singapore BLA teaching starts 2004: CELA annual conference hosted @ Lincoln 2007: LU closes entry to the Singapore BLA 2003; 2006: Success in Performance Based Research Fund

(Continued)

Table 36.1 (Continued)

Regional, national & global context	Development period	Landscape@Lincoln imperatives and strategies	Landscape@Lincoln events
Global Financial Crisis 2007–08 Canterbury Earthquakes 2010–11 Christchurch Floods & Port Hills Fires 2017 Christchurch Mosque attacks 2019 Covid-19 (2020 onwards)	2007–2020 'Future Shock'	Creative local adaption to successive crises *Contribute to regional imperatives. *Reorientate studio towards 'design inquiry'. *Build technological & remote teaching capability. *Combine School with related discipline. *Revise MLA programme.	2009: School of Landscape Architecture (SOLA) & new landscape architecture building opens 2014–19: DesignLab 2012; 2018: Success in Performance Based Research Fund 2019: SOLA 50 year celebrations 2020: Applied Computing joins SOLA

Source: The authors.

1960s–1982, the landscape 'Apostles': creating, educating and promoting an Aotearoa-NZ profession of landscape architecture

The imperative to establish landscape architecture education in Aotearoa-NZ was the perceived threat to the Aotearoa-NZ environment from development (Salmon 1960), and a consequent need for landscape architectural expertise, particularly in agencies responsible for policy, design, and implementation of public infrastructure and management of public land. In 1966 Lincoln College (as it then was) sent English horticulture lecturer 'Charlie' Challenger back to the UK (Newcastle University) to train as a landscape architect. In 1969 the College then introduced a postgraduate Diploma in Landscape Architecture (DipLA) under his leadership, supported by New Zealander Frank Boffa, a recent BLA graduate from the University of Georgia, USA.

The Lincoln DipLA curriculum drew significantly on the UK model of postgraduate landscape architecture education at that time (Densem 1990). Two years of intense studio teaching took students through increasingly complex projects at different scales, supported by lectures in landscape sciences, history, design, planting, and construction. Small cohorts of mature and committed students (5–10 entries per year) enabled group work and one to one teaching with enthusiastic staff (see Figure 36.1), concluding with a dissertation and individual design project. Teaching focused on contextually appropriate design and planning decisions based upon landscape and design principles.

Based in the Department of Horticulture, the DipLA was supported by other land-based disciplines at Lincoln, including soil science, ecology, and natural resources engineering. Landscape students also spent time at the University of Canterbury School of Fine Art. The content and character of the programme therefore expressed landscape architecture as a discipline bringing together arts and sciences. Students included employees sent to Lincoln by government agencies (e.g.: Ministry of Works and Development, Department of Lands and Survey, New Zealand Forest Service), and others enrolled independently following diverse undergraduate degrees. Alongside the professionally focused postgraduate DipLA the Landscape

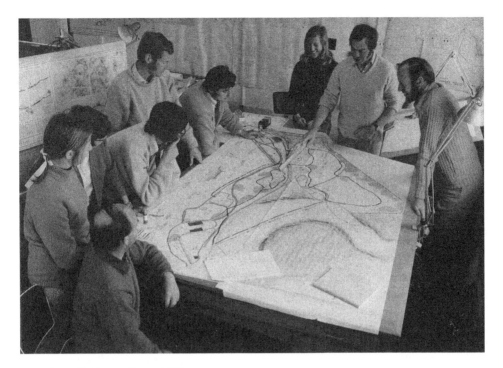

Figure 36.1 Design studios ca. 1970.

Source: Credit: Lincoln University Living Heritage. Item no. 0003451. '1970 Landscape Architecture student group'. Accessed from https://livingheritage.lincoln.ac.nz/nodes/view/3927. Licensed under Creative Commons Attribution 3.0 New Zealand License.

staff also taught into the undergraduate Diploma in Horticulture and one-year Certificate in Landscape Design.

Development of a landscape architecture profession in Aotearoa-NZ was vital during this establishment period. Colonial New Zealand had notable landscape gardeners (Tipples 1989), and by the 1960s a few European and US trained landscape architects were employed in government and private practice (Challenger 1990; Adam and Bradbury 2004; Falconer et al. 2020). Some affiliated with the Australian Institute of Landscape Architects (founded in 1966), and Charlie Challenger became an individual member of the International Federation of Landscape Architects (IFLA) (Aitken 1982). Keen to promote their own new professional voice and identity, Lincoln staff and graduates led the establishment of the New Zealand Institute of Landscape Architects (NZILA) in 1972–73, with Lincoln DipLA graduate Tony Jackman as founding president. The NZILA adopted a code of practice and membership protocols, promoted application of landscape principles to design and planning, advocated for employment of qualified landscape architects, and became active within IFLA.

Another initiative was the Lincoln College Landscape Consultancy Service, through which Lincoln staff undertook projects on behalf of the College for external clients, demonstrating the values and scope of landscape architecture. Frank Boffa and Associates subsequently established a new private practice based on landscape architecture, which has continued in different forms to the present day.

Lincoln also launched a newsletter, which evolved into a professional magazine, *The Landscape*, documenting and showcasing professional projects and techniques. By the early 1980s

Landscape@Lincoln was well established, and the new profession was broad in scope, active in garden, park, and housing design, major infrastructure, tourism and conservation, landscape planning, and assessment. However, the context and conditions for education and practice were about to change dramatically.

1983–1992, the spirit of 'No. 8 Wire': adapting to 'The New Zealand Experiment' and developing a robust framework of qualifications

The national transformation described previously as 'The New Zealand Experiment' had immediate implications for landscape architecture education. The national economy shifted to a more competitive market orientation and by 1990 all government agencies that formerly sponsored students and employed landscape architecture graduates from Lincoln had been restructured or disestablished. Demand from school leavers for environmental and design education was growing, and between 1983 and 1992 there were several strategic initiatives and adaptions at Lincoln.

Most fundamentally, the academic structure of Landscape@Lincoln was reconfigured from postgraduate and undergraduate diplomas into a hierarchy of degree qualifications – Bachelor of Landscape Architecture (BLA), Master of Landscape Architecture (MLA), and Doctor of Philosophy (PhD). A four-year undergraduate BLA introduced in 1988 enabled school-leavers to progress directly to an accredited professional qualification, which could be awarded with honours for high achieving students. The DipLA was phased out and replaced by a two-year graduate entry BLA which provided an accelerated route to professional qualification for students with relevant undergraduate degrees. The MLA could be entered from both undergraduate and graduate BLA's and comprised advanced coursework and/or research. The BLA and MLA degrees therefore provided several different entry and qualification pathways. The PhD was based on the UK model of three years of independent research. It first enrolled students in the 1990s when suitably qualified supervisors became available.

Like the previous DipLA, the BLA had a core progression of studios culminating in an independent design project. Studios were linked to teaching in landscape architecture theory and practice, and supported by subjects such as Biogeography and Environment, Geomorphology, Ecology, and New Zealand Society, taught by related disciplines. These were shared with other Lincoln degrees. The BLA structure from 1988 has largely continued to the present, with refinement of subjects, credit weighting and sequences (Table 36.2). Degree minors were later developed to enable landscape architecture students to use elective subjects in other disciplines to develop a specialisation such as tourism, parks and recreation, landscape ecology, or environmental management.

The approach to theoretical teaching became more 'critical' in the BLA, as the earlier focus on 'landscape principles' drawn from landscape history changed to comparative examination of different forms of landscape knowledge and ideals (Swaffield 1990), exemplified in the NZILA conference 'The Critical Path' (NZILA 1991). An increasing professional emphasis upon urban landscape challenged Lincoln's traditional rural orientation, and urban landscape design became integrated into the new BLA degree as a studio course.

The growing discipline of landscape architecture also achieved greater organisational autonomy within Lincoln. In 1990, Lincoln College became Lincoln University. At the same time, growing student numbers on the BLA degree programme required larger teaching studios, and the Department of Landscape Architecture was established in 1992 in a converted library. Departmental status gave Landscape Architecture an independent voice at Professorial Board – the key decision-making body for the university.

Table 36.2 Evolving academic content of landscape architecture qualifications at Lincoln.

Year	PG Diploma in Landscape Architecture 1971	Bachelor of Landscape Architecture (graduate entry) 1991 (24 units)	Bachelor of Landscape Architecture (undergraduate) 1991 (36 units)	Bachelor of Landscape Architecture (undergraduate) 2020 (32 units)
One	Basic Design (studio)	ECOL 101 NZ Biogeography and Environment 2 units	ECOL 101 NZ Biogeography and Environment 2 units	DESN 101 Digital Tools for Design (studio)
	Landscape Science	LASC 101 Introduction to the Cultural Landscape	LASC 101 Introduction to the Cultural Landscape	DESN 102 Introduction to 3D Design (studio)
	Landscape Design Principles Ia (studio & practical)	LASC 102 Design Introduction (studio)	LASC 103 Introduction to Design 2 units (studio)	DESN 103 Visual Communication (studio)
	Landscape Design Principles Ib (studio)	LASC 105 Landscape Interpretation & Analysis 2 units (studio)	LASC 105 Landscape Interpretation & Analysis 2 units (studio)	DESN 104 History of Design & Culture
	Landscape Design Studio I	LASC 107 Introduction to Landscape Practice	LASC 107 Introduction to Landscape Practice	ENGN 106 Landscape Surfaces, Water & Structures
		BIOS 204 Biology LASC201 Principles of Landscape Design I	SOCI 101 Introduction to Human Sciences	PHSC 107 Introduction to Earth and Ecological Sciences
		LASC 206 Landscape Planting Practice LASC 202 Site Design A 2 units (studio)		*Plus two electives*
Two	Landscape Design Principles IIa (studio & practical)	LASC 201 Principles of Landscape Design II LASC 203 Site Design B (studio)	LASC 201 Principles of Landscape Design I LASC 202 Site Design A (studio)	LASC 215 Landscape Analysis, Planning & Design 2 units (studio)
	Landscape Design Principles IIb (studio & practical)	LASC 205 Landscape Construction LASC 307 Landscape Project Implementation	LASC 203 Site Design B 2 units (studio) LASC 205 Landscape Construction	LASC 216 Site Design (studio) LASC 217 Design Detail
	Landscape and Land Use	ECOL 301 Ecology SOCI 201 Aotearoa NZ Society	LASC 206 Landscape Planting Practice SOCI 201 Aotearoa NZ Society	LASC 206 Landscape Planting Practice LASC 218 Landscape & Culture
	Professional Practice	LASC 402 Professional Practice in Landscape Architecture	SOSC 203 Landform & Soils	LASC 211 Planting Design & Management
	Landscape Design Studio II	LASC 401 Theory & Planning LASC 405 Design & Planning 2 units (studio) LASC 407 Major Design 2 units (studio)	BIOS 204 Biology	*Plus 1 elective*
Three			LASC 201 Principles of Landscape Design II	DESN 301 Design Theory

(Continued)

Table 36.2 (Continued)

Year	PG Diploma in Landscape Architecture 1971	Bachelor of Landscape Architecture (graduate entry) 1991 (24 units)	Bachelor of Landscape Architecture (undergraduate) 1991 (36 units)	Bachelor of Landscape Architecture (undergraduate) 2020 (32 units)
Four			ECOL 301 Ecology	LASC 312 Landscape Ecology
			LASC 303 Rural Landscape Design 2 units (studio)	LASC 316 Innovative Design A (studio)
				LASC 321 Structure Plans (studio)
			LASC 305 Urban Landscape Design 2 units (studio)	LASC 322 Sustainable Design & Planning 2 units (studio)
			LASC 307 Landscape Project Implementation	LASC 393 Applied Landscape Practice
			Plus 2 electives	Plus 1 elective
			LASC 401 Theory & Planning	LASC 318 Landscape Assessment & Planning
			LASC 402 Professional Practice in Landscape Architecture	LASC 415 Landscape Architecture Professional Practice
			LASC 405 Design & Planning 2 units (studio)	LASC 406 Complex Design 2 units (studio)
			LASC 407 Major Design 2 units (studio)	LASC 409 Major Design 2 units (studio)
			Plus 3 electives	Plus 2 electives

Source: Lincoln College Calendar 1971, Lincoln University Calendars 1991 and 2020. Accessed from https://livingheritage.lincoln.ac.nz/nodes/view/56.

All taught subjects from 1991 are one unit of study unless otherwise noted. There are eight units in a standard undergraduate year at Lincoln.

Greater autonomy facilitated curriculum development, including increasing engagement with Māori communities. A vital part of the political transformation of Aotearoa-NZ in the 1980s and 90s was a process enabling *iwi* (Māori tribes) to make claims against the Crown for breaches of the Treaty of Waitangi. This also highlighted the obligation of public institutions to engage in partnerships with those who exercise *mana whenua* (Māori territorial rights and guardianship for land), and to provide educational opportunities for Māori. In 1987, a NZILA Conference 'E rua nga iwi, kotahi ano te whenua = Two cultures, one landscape' (NZILA 1987) examined the profession's obligations and opportunities. Building on this, in 1993 a new staff appointment in the Department of Landscape Architecture was made with specific responsibility for developing greater Māori engagement. A significant number of Māori students have subsequently completed landscape architectural qualifications at Lincoln and joined the NZILA (which is now known as Tuia Pito Ora NZILA) and collaborative projects have become vital to landscape pedagogy and practice.

The 'New Zealand Experiment' also included major reform of environmental management and planning. A significant change for landscape architecture was the new Resource

Management Act 1991 (RMA), replacing a complex mix of existing laws and regulations with a single framework based on principles of sustainable management of natural and physical resources. A new 'effects based' approach drawing on EIA procedures and protections for outstanding natural landscapes brought landscape issues into the core of much environmental decision making. This emphasised the importance of integrating landscape assessment, planning, and design within landscape architecture education and practice.

1993–2006, 'Overseas Experience': building international networks and partnerships

The 'New Zealand Experiment' continued during the 1990s and extended the market-based policy approach into public education. Universities became competing 'providers' of tertiary education, and in 2003 the new Performance Based Research Fund (PBRF) added requirements for academic staff to be independently graded for their individual research performance. This reinforced pressure upon landscape academics to become productive and recognised researchers.

Financial requirements imposed on Lincoln University also resulted in management pressure to expand student numbers in landscape architecture (see Figure 36.2), particularly international students. The more competitive approach between tertiary institutions led to new landscape architecture programmes being introduced in Auckland (Unitec) and later Wellington (VUW), both associated with architecture schools. These developments all demanded new strategies at Lincoln.

The first year BLA was reconfigured as an open entry introductory year. Years two to four were defined as 'professional' years, with entry requiring a defined grade point average in critical subjects. This enabled growth in enrolments while ensuring the competencies needed to progress through the design studios. Increasing student numbers also needed adjunct teachers,

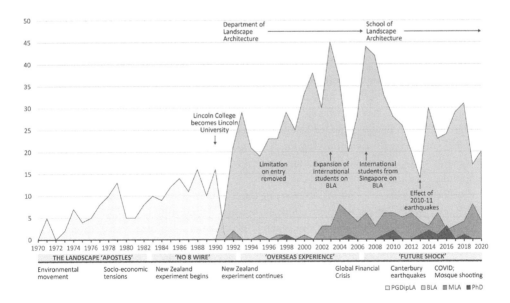

Figure 36.2 Landscape@Lincoln graduation numbers, 1971–2020.

Source: Content, the authors; data source, Lincoln College and Lincoln University Graduation Booklets 1970–2020. Accessed from https://livingheritage.lincoln.ac.nz/nodes/view/56; graphic production, Hanley Chen.

drawn from local practices, which maintained contact with the profession while full time academics became more focused on research. Enrolment of international students with English as a second language created teaching challenges, and for a period a Mandarin speaking tutor was engaged.

Other landscape initiatives developed staff research skills and activity. International networks were built with educators and researchers in Australia (Australasian Educators in Landscape Architecture (AELA)), North America (Council of Educators in Landscape Architecture (CELA)), and Europe (European Council of Landscape Architecture Schools (ECLAS), and LE:NOTRE (initially an EU funded project, now an Institute)). Landscape@Lincoln staff contributed to multidisciplinary research teams on topics such as tourism and forestry. An academic peer reviewed journal was launched (*Landscape Review*), projecting a distinctive Lincoln voice, and international conferences were hosted, including 'Languages of Landscape Architecture' (1995 and 1998), and 'The Global and the Local' (2004), the first CELA annual conference held outside North America.

International teaching partnerships were developed. An exchange and study abroad programme attracted eight to ten students annually, particularly from North America, but the most ambitious initiative was the Lincoln BLA in Singapore, commencing in 2000. Delivered in collaboration with a local business partner, it was taught part-time to graduates, and later undergraduates. The Singapore BLA aimed to meet a need in Singapore itself and to also improve access for Lincoln to student markets in India and South-East Asia. The vision was complementary teaching in two locations, Aotearoa-NZ offering a temperate rural and lower density urban perspective, and Singapore a tropical high-density urban focus. Student numbers grew steadily, but offshore teaching was particularly demanding on Lincoln staff. Following business management issues the Singapore-based teaching closed to new entrants in 2007.

2007–2020, 'Future Shock': creative local adaption to successive crises

In 2008 the School of Landscape Architecture (SOLA) was established in a new purpose-built facility. This provided opportunity to refine studio teaching, and a new SOLA-based *DesignLab* placed design inquiry more centrally within the research and pedagogy of the School. Design inquiry and landscape research have subsequently become incorporated into two new University Centres of Excellence (*DesignLab* was disestablished in 2019).

The new SOLA facility included IT laboratories and expanded digital technology capability, including audio-visuals, laser scanning for 3D modelling, augmented reality modelling, virtual reality immersion, and media for remote supervisory meetings. These were invaluable in enabling studio teaching to continue off-campus during the Covid-19 pandemic. New technology also enabled an open access digital resource at Lincoln University Library, including landscape architectural theses, research reports, and a landscape architecture archive. *Landscape Review* changed to digital publication, using an Open Journal Systems platform with a searchable PDF back catalogue.

International partnerships continued and expanded, bilaterally with partner universities in China, and multilaterally in Europe through the Euroleague for Life Sciences (ELLS) network. SOLA hosted and organised conferences and events designed to stimulate new relationships, including 2007 'Globalisation of Landscape Architecture' (St Petersburg), 2013 'Design for Conservation' (Lincoln), 2016 'Integrated Urban Grey and Green Infrastructure' (Lincoln), and 2018 'The Story of New Zealand's National Parks' (Beijing).

However, establishment of SOLA and its new facility also coincided with the 2007–08 Global Financial Crisis, followed by a succession of unexpected regional and international events which required continuing adaptations of landscape architecture teaching and research. Most dramatically, Canterbury suffered a series of earthquakes that literally shook the region to its foundations. In early September 2010 a 7.1 earthquake only 20 km from Lincoln caused serious structural damage to historic buildings on campus, but the region was relatively unscathed. Almost six months later, on 22nd February 2011, two major aftershocks occurred in quick succession close to the Christchurch Central Business District. Several large buildings collapsed and there was extensive structural damage across the city from tremors, landslides, and ground liquefaction. Thousands were injured and made homeless, and 185 people died.

There were profound consequences for SOLA. Students and staff were directly affected by the disaster and the campus closed. In subsequent months and years, SOLA contributed in multiple ways to recovery and rebuild efforts, memorials to those lost, and rebuilding of identity. The lessons learned all provided new teaching and research opportunities. However, student enrolments at Lincoln declined (Figure 36.2), based on perceived safety concerns. Subsequent financial pressures on the University placed SOLA's autonomy under threat, with proposals to amalgamate Landscape Architecture with other accredited professional programmes such as Planning and Property Valuation.

In 2017 two more major natural events – flooding in Christchurch, and a wild-fire on the city fringe – highlighted longer-term landscape implications of climate change for the Canterbury region, and Aotearoa-NZ more generally, and stimulated further new research and design projects. Two years on, another unforeseen event shook the region and Lincoln. The March 2019 Christchurch Mosque attacks on Muslim worshippers at Friday prayers shocked the nation. Fifty-one people were killed, and the attacks posed a major challenge to the region's sense of community and placed new security demands upon public institutions including universities. The subsequent need for commemoration and community rebuilding was ethnically sensitive and required carefully considered engagement and design responses. International student enrolments again suffered because of security concerns.

The shocks continued. In January 2020, rumors arrived with Chinese students returning to Lincoln about a mysterious Wuhan coronavirus. By the end of March 2020 all of Aotearoa-NZ was at the highest national alert level 4, workplaces and educational facilities closed, and everyone confined to their homes. The country's now advantageous geographical isolation enabled early government action to control the Covid-19 virus spread, but with significant socio-economic impacts and major implications for education. Closure of national borders and controls on face-to-face meetings required emergency remote teaching (ERT) and online pedagogy, and again cut international student enrolments. Budget cuts were severe across the higher education sector.

Several new strategic initiatives were undertaken over this period. The graduate entry BLA had not been attracting sufficient students academically able to progress to the MLA, and an accredited, direct entry MLA option was introduced, designed for domestic and international 'career changers' interested in studying at advanced level. It also better met the profession's need for mature graduates able to move into landscape planning and assessment positions.

In 2021 Lincoln's academic staff in Applied Computing joined SOLA, bringing complementary skills and knowledge, and the combined disciplines stimulated new research collaborations. A new Master of Applied Computing significantly increased postgraduate student enrolments, which, with the new MLA option, gave SOLA a teaching profile that better met University expectations and sustained its administrative autonomy.

The Covid-19 emergency remote teaching created major challenges for studio-based and experiential field learning. New IT capability enables online interactive student presentations, group discussions, and individual critique sessions to ensure continuity of studies, and conversion of delivery, assessment and feedback to digital systems provides flexibility to respond rapidly to future crises. However, remote learning and teaching involve long hours of screen work, and lack of face-to-face engagement. When combined with financial uncertainty this makes the mental and physical health of staff and students a significant issue. Nonetheless, the experience has prompted new research and student projects on therapeutic design, design challenges of containment, and implications of pandemics for urban design. The succession of crises over the decade 2011–21 has also stimulated broadening interest in design for community resilience, and for sustainable grey and green infrastructure to better manage environmental hazards and climate change.

Figure 36.2 summarises the 50-year history we have described, showing the numbers of students graduating in professional degrees in relation to the development periods and key events.

Conclusion: future pathways

We began by asking what contexts and events have shaped Landscape@Lincoln, what strategies have been adopted, and what features emerged? We now look ahead. Three linked imperatives will shape the programme's future.

Within Aotearoa-NZ, Treaty partnership with Māori in the reimagination and regeneration of landscape education, landscape practice, and the landscape itself represents a major opportunity and obligation for both the landscape profession and educators in Aotearoa-NZ (Swaffield 1990). The issues are examined in detail in Hill (2021).

The existential global imperative for landscape educators is how to effectively address climate change in teaching and research. Lincoln alumnus, Craig Pocock, used the term 'carbon landscape' to challenge landscape architects internationally to place carbon budgets at the heart of their design process (Pocock 2007). Climate action and the creation of adaptive and resilient communities have been an increasing focus of SOLA teaching and research in subsequent years and will inevitably become ever more pressing.

As 'a small country a long way from anywhere', international collaboration is vital to Aotearoa-NZ and to Landscape@Lincoln, and this will continue. Lincoln University is part of the Global Challenge University Alliance, a network of universities with a shared vision of sustainable, global development framed through the United Nations Sustainable Development Goals (UNSCGs) (Swedish University of Agricultural Sciences 2021). A recent new partnership for Lincoln University features landscape architecture as part of this initiative.

The 'local' Aotearoa-NZ imperatives of the Treaty of Waitangi and resilient communities, and 'global' imperatives regarding climate change and the UNSDGs together provide a powerful ethical and practical framework to interweave knowledge, skills, and responsibility into undergraduate and postgraduate landscape architecture education.

References

Adam, J. P. and Bradbury, M. (2004) 'Fred Tschopp, Landscape Architect: The American Practice 1938–1970', *Landscape Review*, 9(1), pp. 40–44.
Aitken, N. (1982) 'The First 10 Years', *The Landscape*, 14, p. 2.
Challenger, C. (1990) 'Viewpoint: A Retrospect', *The Landscape*, 45, p. 4.
Densem, G. (1990) '21 Years of Landscape at Lincoln', *The Landscape*, 45, pp. 5–9.

Falconer, G., Harvey, B. and Harvey, R. A. (2020) *Harry Turbott, New Zealand's First Landscape Architect*. Wanaka: Blue Acre Press.

Hill, C. (ed.) (2021) *Kia Whakanuia Te Whenua: People, Place, Landscape*. Auckland: Mary Egan Publishing.

Kelsey, J. (1995) *The New Zealand Experiment: A World Model for Structural Adjustment?* Auckland: Bridget Williams Books with Auckland University Press.

NZILA. (1987) *E rua nga iwi, kotahi ano te whenua = Two cultures, one landscape. Proceedings of the NZILA Annual Conference*. Wellington: NZILA.

NZILA and NZIA. (1991) *The Critical Path. Proceedings of a Joint Annual Conference*. Wellington: NZILA.

Orange, C. (2015) *The Treaty of Waitangi*. Auckland: Bridget Williams Books.

Pocock, C. (2007) 'The Carbon Landscape', Presented to the IFLA World Congress, Kuala Lumpur [Online]. Available at: www.landscapearchitecture.nz/landscape-architecture-aotearoa/2019/1/30/the-carbon-landscape-managing-the-impact-of-landscape-architecture (Accessed 8 March 2021).

Salmon, J. T. (1960) *Heritage Destroyed: The Crisis of Scenery Preservation in New Zealand*. Wellington: AH & AW Reed.

Scoop News. (2007) *Father of NZ Landscape Architecture Dies* [Online]. Available at: www.scoop.co.nz/stories/CU0709/S00272/father-of-nz-landscape-architecture-dies.htm (Accessed 13 March 2021).

Swaffield, S. (1990) 'Directions in Diversity', *The Landscape*, 45, pp. 10–13.

Swedish University of Agricultural Sciences. (2021) *About GCUA2030* [Online]. Available at: www.slu.se/en/collaboration/international/slu-global/global-challenges-university-alliance/about-gcua/?submenu=open (Accessed 7 June 2021).

Tipples, R. (1989) *Colonial Landscape Gardener, Alfred Buxton of Christchurch, New Zealand 1872–1950*. Lincoln College: Department of Horticulture and Landscape.

Toffler, A. (1980) *Future Shock*. New York: Collins.

37
LEARNING TO PRACTICE CREATIVELY
Emergent techniques in the climate emergency

Alice Lewis, Sue Anne Ware, Martin Bryant, Jen Lynch, Penny Allan, and Katrina Simon

Climate change, country, and landscape architecture

Australian landscape education is on a tipping point as Australia's natural systems begin to falter. Economic growth, fuelled by two centuries of unrelenting extraction of resources, has degenerated the ancient and biodiverse landscapes. Future threats are daunting: rising oceans, marine heatwaves, extreme heat patterns and longer fire seasons. In this moment, when the past's problematic legacies and the climate emergency converge, landscape architects are questioning the discipline's reach and the ability to effect change. As they do so, Australian landscape architecture educators are rethinking the discipline's practices and the techniques to enable them.

Overlapping this moment of reflection is the presence of Indigenous knowledge systems that inextricably connect land, culture and climate. The term 'Country' describes Indigenous people's highly complex and layered relationship to land. Palyku woman and writer, Ambelin Kwaymullina, explains:

> For Aboriginal peoples, Country is much more than a place. Rock, tree, river, hill, animal, human – all were formed of the same substance by the Ancestors who continue to live in land, water, sky. Country is filled with relations speaking language and following Law, no matter whether the shape of that relation is human, rock, crow, wattle. Country is loved, needed, and cared for, and Country loves, needs, and cares for her peoples in turn. Country is family, culture, identity. Country is self.
>
> (Kwaymullina 2021, p. 8)

Country is also, in this moment, experiencing a re-awakening. In a not too-distant past, European-led settlement dispossessed First Nations people of their land, and then reorganised and reimagined it. Now, non-Aboriginal Australians are beginning to recognise the value held in Indigenous knowledge of Country, which, amongst other things, has endured the uncertainty of different climates across tens of thousands of years. Australian landscape architecture programmes have redoubled their embrace of Indigenous knowledge. Together with First Nations people they are exploring ways to instil students with skills to regenerate the faltering landscape, and respond to the impending climate emergency.

One discourse to surface amongst educators is around design techniques that address uncertainty in the climate emergency. The foregrounding of design technique, rather than issues or theory, emphasises processes of designing, over the 'end' product. These investigations into design technique are, in themselves, a type of research: a creative practice research, which interprets design process as a source of new knowledge that has potential applications in the discipline. This chapter discusses the research into some of the experimental design techniques that have emerged from student studio explorations in Master of Landscape Architecture programmes in Australia. These design techniques, we argue, cultivate new ways of working, new roles and scopes, and new practices for living in Country, all of which has the potential to make an impact in the climate emergency that this moment is demanding.

The emergence of creative practice research in landscape architecture education

While Indigenous people always practiced forms of landscape management and care, the earliest landscape architectural projects in Australia emanated from architects in the early to mid-20th century (Saniga 2012). An environmental awakening followed in the late 1960s and 1970s, when the discipline was characterised by a 'Bush School' approach of restoring and enhancing endemic ecosystems with 'native' plants and local materials. Widely published projects from this era not only raised the profile of the discipline at that time, they also triggered a need for a differentiation from architecture in educational approaches.

The first four-year degree took enrolments in 1974 at the University of New South Wales' (UNSW), led by British practitioner Peter Spooner, with British landscape sensibilities that were attuned to Sydney's 'Bush School' practitioners. Nearly 50 years later, eight Australian universities offer landscape architecture programmes. They are spread across the nation's state capital cities: UNSW and University of Technology Sydney (UTS); Royal Melbourne Institute of Technology (RMIT) and Melbourne University; Brisbane's Queensland University of Technology; Perth's University of Western Australia; the University of Adelaide; and the University of Canberra. These programmes vary in scope and cover an array of landscape architectural educational approaches. All are comprehensive, and all distinct from but in proximity to architecture schools. Rather than surveying these eight institutions for their pedagogical leanings, this chapter focuses on new techniques that have recently emerged from experimentation in landscape architecture curricula at RMIT and UTS, where creative practice research is prominent.

In Melbourne, RMIT's landscape architecture programme became a full four-year undergraduate programme from 1980. Jim Sinatra, the programme's founder and a graduate student of Ian McHarg at the University of Pennsylvania, imported layered mapping techniques, and localised them by working directly on Country, initiating embedded experiences in the Australian bush with Indigenous communities in the landscape programme (Raxworthy and Bauer 2000). The RMIT programme then began to distinguish itself as firstly Modernist, then Post-Modernist, then Deconstructionist, then Landscape Urbanist, and now by working with adaptive and emergent phenomena. In each phase, RMIT openly embraced contemporary theory and paradigms, seeking to be at the forefront of landscape architectural discourse. This remains evident through the annually published *Kerb Journal of Landscape Architecture*, a student-led journal that has maintained a critical and creative commentary on landscape architecture since 1994.

The hallmark of the RMIT programme is its engagement with creative practice research. This is nurtured in the choice-based structure of the '3 plus 2' undergraduate and master's courses, where students elect vertical studios, rather than share curricula across cohort groups, and in doing so they curate their own unique and individual trajectories, exercising autonomy and self-direction as they develop their own creative practice. These course-based landscape architecture programmes at RMIT are also intertwined with the School of Architecture and Urban Design's creative practice PhD programme, which was one of the first in Australia to focus on the ways that practitioners practice. Structured around twice-yearly gatherings of practitioner/candidates on the RMIT campus in Melbourne, and then in various locations in Europe and Asia as the programmes expanded, the PhD Practice Research Symposia (PRS) foster a culture of critical debate around disciplinary experimentation in technique. This has enabled an openness to alternate ways of working, and the development of a shared discourse around the value of design practice as a means for generating innovation and new knowledge.[1]

In comparison, Sydney's UTS's history is short. It established its landscape architecture programme in 2014, and since 2017 its direction has been cultivated by educators whose creative practice training at RMIT has been translated into Sydney's context, which has begun to sprawl into drier and hotter hinterlands where the benefits of creative design research are compelling. Embedding contemporary local, urban and regional challenges within landscape discourses, UTS's four-year undergraduate degree draws upon climate studies, advanced technologies, and Indigenous knowledge as the springboard for its design studios. Its master's degree intensifies explorations into design methodologies, attracting post-professional and career-changing non-cognate enrolees interested in re-skilling with advanced technologies and new knowledge, in order to shift, or re-invent, the ways in which they practice. The pedagogy helps them to do so by reframing traditional landscape architectural ways of seeing into, being in and caring for the world in response to climate change and the knowledge systems of First Nations peoples.

In addressing climate and indigeneity, the UTS and RMIT master's studios share an engagement with landscapes as open works (Eco 1989) that are always in a process of becoming. They draw out landscape's multiple and competing histories, and contemplate futures where human inhabitation is inextricably interconnected to change in landscape. Following is a discussion of some techniques that have emerged in these studios. The techniques have all coupled conventional design processes with new ones: one speculates on a way of climate storytelling; another finds use for advanced technologies in mapping the new-normal of disasters; a third calibrates cartographic precision with ambiguity; and the concluding example is a technique for sharing knowledge, drawing on James Corner's rhizomatic mapping in unique gatherings of thought typified by 'communities of practice'.

Fish, floods and umbrellas: new storytelling

The UTS MLA studios critique traditional landscape architectural techniques which carve up space using the same tools of observation and measurement that paved the way for whaling, colonisation and extraction. To reframe old arguments, the programme is experimenting with new ways of drawing, using the tools at hand which mark, map and measure – the survey, the aerial photograph and the plan and section – in ways that are less hierarchical, less stridently outcome-focused, more curious, more intimate and more humble. They provoke questions rather than answers, reimagine the role of the landscape architect in an expanded disciplinary field, and offer the potential to carve out a new niche for landscape architectural practice.

One of the experimental techniques is storytelling. Used skilfully storytelling can be open-ended or ambiguous, subtle, hypnotic and inclusive. It agitates, inducts, animates or complexifies (Shann 2000). It empowers designers' "agency in acknowledging temporary interests while conveying the urgency of slow narrative of geological time" (Sennett 2014). Storytelling is useful in design studio because it holds complexity, but very loosely, so that new connections become apparent, and patterns begin to emerge.

The recent 'Hyperdensity' design studio, specifically interrogated the human/non-human relationship which is shaping paradigm shifts in climate debates. It did so in the context of dense urbanity. The studio began during the student protests in Hong Kong and was completed in Sydney's Chinatown, and speculated on the hypothetical need, in some not-too-distant future, for an expanded notion of 'public' space: one where humans, material processes, plants, animals and microorganisms would co-exist in intense proximity. Students employed drawing techniques and different genres of storytelling to suit their individually identified subject matter within the studio.

One student looked at the trade in rare and endangered fish (see Figure 37.1 top). Equal parts dystopian graphic novel, speculative fiction and cautionary tale, its drawing style slightly hallucinogenic, the fish story exposed the dirty reality of our obsession, by siting a market for the consumption of exotic and rare fish next to a prominent casino in Sydney's waterside gambling district.

Another student investigated the rituals of human relationships with water. In the manner of a creation story, the project described a time when, in an act of daring and optimism about the future, the inhabitants of the city flooded the bay, establishing a new set of rituals about the care of water and the multiplicity of species that would depend on it.

A third student looked in Hong Kong, at how the patterns of umbrellas during a protest signalled attitudes of strategy, defence and care. In Sydney this evolved into a science fiction story about its future citizens who, in order to exploit a 'strange but true' legislative loophole prohibiting acts of dissidence on all but cemetery grounds, constructed a sprawling underworld network of the dead woven into the fabric of the city as way of claiming the legal right to protest on public urban lands (see Figure 37.1 middle). In each case, students used stories to highlight the fatal flaws in human behaviour that have led, for example, to the extinction of one species for the benefit of another, the exploitation of resources, and the imposition of draconian laws that deny basic human rights.

In another studio, a student explored another type of fish narratives – alongside all the other stories of myriad creatures and vegetation – in their collaboration with the Arakwal First Nations people of the Bundjalung Nation. Like many Aboriginal communities, they are attached to landscapes that are contested. The site, Tallow Creek, is a sacred space for the Arakwal, but in 2019 the local government, pressured by residents whose backyards were flooding, artificially opened the mouth of the creek with very little warning. Thousands of mullet, bream, whiting, flathead, and eels died of asphyxiation. It was a traumatic event, driven by the authorities' ignorance of the complex ecosystems and spiritual significance of the creek. Over many months, the student acted as a kind of visual translator, exploring the interface between Indigenous ecological knowledge and climate impacts, drawing stories with Arakwal to highlight the dynamic nature of this sacred space (see Figure 37.1 lower).

Storytelling, or yarning, is an important practice in Indigenous communities for intergenerational transfer of values, and is known to have passed knowledge of Country in the changing climate of past Ice Ages (Massy 2020). These new design approaches to storytelling may be essential for the students' future practice, and the discipline's longevity, but most significantly, they provide examples of techniques for critically reframing the key questions of our time, and offer new ways to advocate and persuade.

Of moss and krill: new mapping with advanced technologies

UTS landscape students, with the aid of advanced technologies, are also approaching cartography in critical ways, subverting the tools they have inherited to reframe the way we relate to

Figure 37.1 Narratives of the future: The dystopic Fish Market of endangered fish (top), a drawing by Thomas Woodhead; an underground public space that supports a venue for protest (middle), a drawing by Yousef Ghazal; the fish life of a coastal lagoon, mapped through Indigenous stories (lower) drawings by Nathan Galluzzo in collaboration with Uncle Norman Graham, an Arakwal elder.

Learning to practice creatively

Figure 37.2 New cartographies to bring distance closer: the 2019 New Year's Eve bushfire (top), a drawing by Sam Clare; Deception Island in Antarctica (lower), a drawing by Faid Ahmad.

the world. They have, for instance, investigated the ongoing legacy of the 2019–2020 Australian bushfires, an extreme landscape that is moving closer and closer to the supposed 'normality' of Australian suburban domestic lives. They used a wide range of data-gathering techniques to document the fire event, to register the devastating loss of biodiversity and to suggest alternatives to the status quo (see Figure 37.2 top). They tested ways communities might live with fire rather than defend private property by cutting down trees, a strategy currently mandated by national legislation. They worked at territorial scales using topography, wind, firebreaks, fuel loads, forestry management regimes and riparian corridors to alter the intensity and path of a fire. And at a local scale the mapping inculcated the forest in the lives of the community, and facilitated a re-thinking away from private property protection to co-management of landscapes held in common.

The students have also worked in the extreme landscapes of Antarctica. Here, the microscopic worlds of krill challenge them to develop techniques that will allow them to practice care remotely and at a territorial scale, and thereby offer an alternative and intimate perspective on global climates (see Figure 37.2 lower). To recalibrate the space between physical distance and intimacy with technologies that map the world of microscopic moss in the Antarctic McMurdo Dry Plain, they created parallel worlds at home of growing, tending and encountering moss, rock, ice, air, moisture and wind. These techniques catalysed a collaboration with Antarctic scientists in Tasmania, in which the students translated the abstractions of scientific data in qualitative ways, and re-envisioned technical terms like 'glacial melt', 'biomass' and 'knotted pressure' through drawings, models and film.

The advanced technology used in the cartographic techniques for exploring fire-threatened landscapes and the Antarctica was crucial: it provided a cutting edge to have a dialogue with the interested communities about making distant landscape less remote and more relevant.

Shifting sands: exploring the agency of ambiguity

There is an ambiguity in regard to what 'landscape' is and does. At a time when traditional concepts of meaning, control and power in relation to landscapes are deeply contested, it can be tempting to try to reduce ambiguity[2] and clarify meaning. A design research seminar course developed at RMIT called 'The Agency of Ambiguity' adopted a different position. It focused on the ways that ambiguity can have productive agency in the ongoing formation of the lived world. In practice this entails repurposing the mechanisms of control as productive techniques for understanding and engaging with the dynamic nature of landscapes, and landscape architectural design.

'The Agency of Ambiguity' started with mapping, a foundational but fraught way of working in the discipline. Traditional mapping practices can reduce the uncertainty of landscape and make its constituent parts more explicit and precise, but at the same time, they can also ambiguate our understanding of landscape through different forms of projection and generalisation (Simon 2012). Out of ambiguity, the complex and interconnected relationships between physical and political territories surface; and the intervisibility of different forms of landscape inscriptions – images, picture, maps and the territory itself – offer mutual, reciprocal relationships (Minca 1995; Casey 2002). The seminar experimented with alternate mapping techniques to draw attention to inherently ambiguous aspects of maps and their landscape medium. One of the mapping techniques entailed the use of traditional tools for overlaying, slicing and smudging to re-organise maps depicting students' home environments in new ambiguous hierarchies and configurations. Another used the self-organising tendencies and independent subjective logic of material media such as ink and sand to expose the role of representational tools in apprehending and shaping territory.

One student deployed techniques from this seminar to come to terms with unpredictable human and non-human forces of global sand movement (see Figure 37.3). Having developed skills in material design, representation and critique through the vertical studios, the student went on to embrace and work with the self-organising tendencies of sand and oceans as agents of form production. The student explored a once-dynamic riverine inlet on Victoria's Surf Coast which is now anthropocentrically and expensively frozen in a stasis following ceased mining operations. Building on the divergent abundance that was cultivated by Indigenous people, the project's mapping challenged the need for control over landscape processes (and the tools we currently use to measure, monitor and work with them). The student developed techniques of design that relinquished control to open-ended processes that open the river inlet to non-human movements. The agency of ambiguity facilitated this outcome.

The seminar shows not just that ambiguity is ever-present in landscape, but that designers have the potential to perpetuate it. Ambiguity, rather than being perplexing, can open systems

Learning to practice creatively

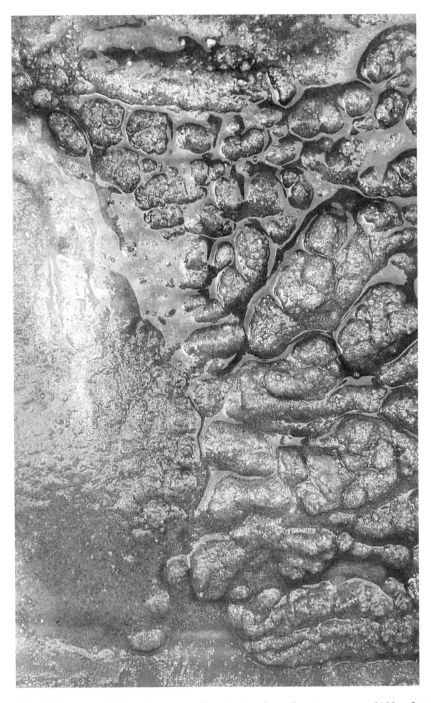

Figure 37.3 Explorations of the performance of sand, using dye and water to expose hidden flows and convergences, a drawing by Conrad J. Cooper MLA.

to new possibilities and avoid the temptation to pursue fixed 'solutions' to prescribed problems. When students become comfortable with ambiguity, they are ready to engage with rapidly changing landscape resulting from changing climates and a future of uncertainty.

Rhizomatic mapping: a technique for fostering communities of practice

Experimental practices are bolstered through the collective sharing of knowledge between those seeking to work in similar emergent areas. In professional practice, one way of enabling and progressing this sharing of practice techniques is through a 'community of practices'. By working together in a community in diverse and different ways, landscape architects can progress practice beyond traditional scopes and "venture into unknown territories" (Downton 2003, p. 6). The process of establishing a community of practice is a requirement within the RMIT PhD model as it provides a foundation from which a practitioner may explore and expand knowledge of practicing. But the point of a community of practice is not primarily to elevate knowledge: it is to situate oneself in practice. Developing a community of practice for students helps to situate practices so that students become interested and invested in who else is exploring and thinking in similar ways. They see their own techniques, compare them to peers and external practices, and in so doing visualise and critique contemporary landscape architecture discourses in relation to the complex conditions of landscape and their individual ethical concerns. This builds a support network for creative practice experimentation and thereby cultivates a position for novel practice trajectories that can contribute innovative ideas and approaches for landscape architecture.

In supporting students to seek out new ways of practicing in emergent landscape conditions, another RMIT seminar course called 'Ways of Working' used a critical approach to mapping to enable students to build their own community of practice. The agency of mapping as a technique lies in its capacity to reveal and make present "realities previously unseen or unimagined" (Corner 2014, p. 214). More specifically, the seminar course engages with the concept of 'rhizomatic mapping', which describes an open-ended and iterative process of mapping non-planimetric elements that are without a geographical base. The community of practice mappings appearing in PhD dissertations are examples. They bring ideas, methods, and other unmappable phenomena, such as cultural values and traditions, into relations, but they remain highly flexible, thus accommodating changes and alterations determined by the author as practices become more refined.

The seminar course worked with Corner's rhizomatic mapping operations. 'Field-making' established the preliminary scope of the map through collection-based research. 'Extracting' outlined the 'deterritorialisation' of information "to be studied, manipulated and networked with other figures in the field" (Corner 2014, p. 230). 'Plotting' draws out "new and latent relationships" (Corner 2014, p. 230). By pairing mapping operations with communities of practices, students were able to collect and interrogate techniques, and to populate distinct but sometimes overlapping groups that formed communities of related design practices. The visualisation of communities of practice through the rhizomatic map revealed the students' position vis-a-vis existing discourses and practices and, in making this explicit, provided a surface upon which to project emergent practices. The open, evolving nature of this framework supported new ways of working because it showed that it can respond to a multiplicity of agendas, experiences and cultures while also taking in and responding to new material. It therefore encouraged thinking around the way landscape architecture can be continuously re-interpreted to respond to the complex conditions of landscapes locally and globally.

Learning to practice creatively

Figure 37.4 Rhizomatic mapping (top) and detail (bottom) exploring the interconnected communities of a practice of 'collecting', drawing by Louella Exton MLA.

One student, in the process of 'field making' recognised 'collecting' as a key technique used in the student's own practice. Through iterative mapping (see Figure 37.4), the student identified a range of aligned communities that 'collect' in diverse ways – the 'list makers', the 'travelling project designers', 'the material samplers', and so on – towards different ends. By surfacing a new way of seeing one's own practice and the communities of related practices, the mapping technique supported the student's emerging practice trajectory, and speculated on the ethical and collective forms of practice. Thus, by using an iterative process of visual mapping, the community of practice supports the exploration and development of individual ways of working and new approaches to the complexities of landscape architectural design that both critique and are supported by existing practice.

The emergent techniques in the climate emergency

At both UTS and RMIT, the techniques discussed uncover and communicate design processes as a way of generating knowledge that is situated in ways of doing, ways of knowing and ways of understanding. The techniques grapple with the traditional conventions of landscape architecture pedagogy and practice, altering or re-purposing them in response to the changing landscape, striving not for single solutions, but for a multiplicity of iterative scenarios based on various contingencies and questions.

They have been developed to enable graduates with skill in complex design processes, to instil confidence in critically reflective practices, and to help them discover their agency. They ensure that future practitioners can ground and explicate their project work in relationship to contemporary and often urgent discourse. The students who use them can actively contribute to critically transforming landscape architectural practice in Australia. The evidence of this shift is not yet apparent in the built work of landscape architectural practice, and it may take a while before it is. Like any emergent situation, it needs nurturing and space to grow. As these techniques develop and mutate, we anticipate and expect shifts to follow.

But the shifts are beginning in the attitudes of professional practitioners. Evidence for this is occurring in the advocacy by RMIT alumni members of the Australia Institute of Landscape Architects (AILA) to initiate a focus on Connection to Country, driving the development of the AILA's first Reconciliation Action Plan (2018), the elevation of 'Connection to Country' as a Strategic Value of its National Strategy (2021), and the appointment of Aboriginal Cultural Ambassadors to advise on the profession's trajectory. The impact of this advocacy is being reflected back in projects by practices across the country. There are the movements away from the omniscient master plans where simplified ecologies are sometimes statically choreographed, towards tactics that create spaces of ambiguity, happenstance, and room for human and non-human complexity.

As the discipline starts to negotiate its future it recognises the need to equip students with the knowledge to develop modes of practice that will allow them to remain provocative, active and engaged. They need to change hearts and minds, and the way politicians operate. After all, there's no point in having good ideas or knowing the right thing to do if nobody is listening.

Radical hope

As we write, the Covid-19 pandemic and the climate emergency continue to disrupt. Online teaching, with its propensity to restrict encounters with landscape, to decontextualise and to diminish intertextuality, encumbers the potential to address climate change.

Perspectives such as Jonathan Lear's radical hope reminds us that humans inhabit a way of life inside a given culture and since a way of life is vulnerable, so is the human condition. Lear invites us to understand how we live with shared vulnerability and cope with civilisation's possible end – an idea that many find incomprehensible. Lear has confidence that cultures will find ways to sustain themselves, they will endure and adapt discovering something salvageable he calls radical hope from the wreckage (Lear 2008).

The emergent techniques discussed here have the potential to engage with the radical hope for human and non-human futures, which challenges our authority over the planet and our intentions for other living and non-living entities. The current transformation of our discipline, our professional practice, and our pedagogical approaches in education requires the equivalent resolution that Lear describes. It contends with losing our current way of life even when we cannot understand what this really necessitates.

Notes

1 The explicit discussion of what this knowledge is and how it has been created or discovered is available through the publicly accessible examination and dissertation archive, https://practice-research.com/.
2 Ambiguity is "having or expressing more than one possible meaning, sometimes intentionally" (Cambridge University Press, n.d.).

References

Casey, E. (2002) *Representing Place: Landscape Painting and Maps*. Minneapolis: University Minnesota Press.
Corner, J. (2014) 'The Agency of Mapping: Speculation, Critique and Invention', in *The Landscape Imagination: Collected Essays of James Corner 1990–2010*. Princeton: Princeton Architectural Press.

Downton, P. (2003) *Design Research*. Melbourne: RMIT University.

Eco, U. (1989) *The Open Work*. Cambridge: Harvard University Press.

Kwaymullina, A. (2021) *Meaning of Land to Aboriginal People – Creative Spirits* [Online]. Available at: www.creativespirits.info/aboriginalculture/land/meaning-of-land-to-aboriginal-people (Accessed 20 April 2022).

Lear, J. (2008) *Radical Hope: Ethics in the Face of Cultural Devastation*. Boston: Harvard University Press.

Massy, C. (2020) 'Second Nature', *Granta: Magazine of New Writing*, 153.

Minca, C. (1995) 'Humboldt's Compromise, or the Forgotten Geographies of Landscape,' *Progress in Human Geography*, 31(2), 179; with reference to Augustin Berque, *Les Raisons du Paysage* (Paris: Hazan).

Raxworthy, J. and Bauer, K. (2000) 'Paradise Is Just Where You Are Right Now: A Subjective History of Landscape Architecture Education at RMIT', *Landscape Australia*, 22. [Online]. Available at: www.jstor.org/stable/45146190 (Accessed 20 April 2022).

Saniga, A. (2012) *Making Landscape Architecture in Australia*. Sydney: University of New South Wales Press.

Sennett, R. (2014) 'Narrative and Agency', *Topos: The Narrative of Landscape*, 88.

Shann, S. (2000) *Mating with the World: On the Nature of Story-Telling in Psychotherapy*, PhD thesis, unpublished, University of Western Sydney.

Simon, K. (2012) *Image, Territory, Picture, Map: The Slipperiness of Landscape Inscriptions*. Unpublished PhD Thesis, Sydney College of the Arts, University of Sydney, Australia.

38
LANDSCAPE ARCHITECTURE EDUCATION IN AFRICA

Graham A. Young[1]

This chapter presents the history of the establishment of landscape architecture programmes in Africa and discusses the influences that came to bare on their establishment. It begins with the focus on the pioneering role that the University of Pretoria played and then moves to other programmes in South Africa and across the continent. Most of these programmes deferred to North American or European influences, but have been refined, to a degree, to deal with local challenges and issues. The primary source of the material for the chapter is derived from previous writings of mine as well as discussions and correspondence with current educators in the landscape architecture programmes across Africa.

The chapter ends with the question around post-colonial education, which arguable today, must give way to an 'indigenous' understanding of knowledge, particularly as it applies to the spatial theories which underpin the discipline. It also poses the question as to whether the rest of the world can learn from the current experiences of African academics and teachers. These questions have, however, not been answered and have been raised primarily to focus the need to urgently address them within the framework of ongoing curriculum development and landscape architectural research and teaching across Africa.

Landscape architecture programmes in South Africa

The pioneering role of the University of Pretoria[2]

Landscape architecture in South Africa barely existed as a profession, when in the early 1960s, Joane Pim (1939–1974)[3] brought the role of the landscape architect to the attention of a wider audience. She was hired by Anglo-American's property division to be part of the professional team to plan and design the mining town of Welkom, Orange Free State. In May 1962 Pim, along with Ann Sutton, Peter Leutscher, and Roelf Botha, founded the Institute of Landscape Architects of South Africa (ILASA) and set in motion the next phase of the development of the profession, namely the establishment of a university course. In 1970, their work came to fruition when the University of Pretoria received a substantial donation from Sentrakor, a property development company, to be used to establish a landscape architecture programme.

In 1971, Pim consolidated her thinking about the significance of the profession in her book *Beauty Is Necessary* (1971). She highlighted the importance of landscape architecture and forcefully called for immediate action to halt the environmental degradation taking place in South Africa by advocating the use of good environmental planning principles. Also in 1971, a Chair in Landscape Architecture at the University of Pretoria was initiated under Prof Roelf Botha. Botha was well known in the fields of architecture and landscape architecture. He had received his degree in architecture from the University of Pretoria and a degree in landscape architecture from the University of California, Berkeley.

Even in its early years, the landscape architecture programme played an integral role in many notable events which shaped environmental planning and management policy in South Africa. The most noteworthy event occurred in September 1973 when ILASA, along with university personnel, hosted an environmental symposium in Pretoria. The theme of the two-week long international symposium was 'Planning for Environmental Conservation'. Speakers from outside South Africa included Australian John Oldham, the vice-president of the International Federation of Landscape Architects (IFLA), his co-vice-president Hans Werkmeister from Germany, Hubert Owens of the University of Georgia and IFLA President (1974–76), Brazilian master painter and landscape architect Roberto Burle-Marx, and Ian McHarg, (Wiesmeyer 2007, p. 68) then head of the Department of Landscape Architecture and Regional Planning at the University of Pennsylvania and author of the celebrated book *Design with Nature*. It was noticeable that after the conference the Department of Environment Affairs started making use of the services of landscape architects and began to systematise its approach to environmental impact assessment. An important outcome being the Integrated Environmental Management procedure (IEM), which directed the discussion toward the environmental legislation and the establishment of the Environment Conservation Act (Act 73 of 1989) (Stoffberg et al. 2012b, p. 14).

Initially, the landscape architecture programme at the University of Pretoria was driven by the principles and beliefs of McHarg. The notion of 'stewardship of the land' which was instilled in their students, originates in this philosophy. In 1987, the programme became a full Department under Prof Michael Murphy, a visiting academic from Texas A&M University, followed by Prof Willem van Riet in 1988 who was responsible for establishing a Geographic Information Systems (GIS) laboratory in the newly formed Department. Not many African countries offered programmes in landscape architecture at the time and students from other African states, including Ghana, Malawi, Zambia, Botswana, and Namibia, enrolled at the Department, making it significant beyond the borders of South Africa.

The 1990s saw many changes and refinements. In 1991, Prof van Riet appointed Mr. Graham A. Young to guide the programme in a different direction (Figure 38.1). Design teaching had suffered over the years as the emphasis on environmental planning, with the aid of the GIS laboratory, took hold and became the focus of the Department. The mandate was to develop a balance between landscape design and environmental planning. To this end, the fourth-year curriculum was changed to shift the focus to urban design and construction detailing. The technical aspects were separated from the design studio and a series of construction courses were developed by Prof Piet Vosloo. This proved to be a huge success as students graduated with ability and confidence in landscape design and technical knowledge that had not been evident in the previous years.

The programme remained a four-year bachelor's degree during this time. However, in 1998, it was decided to change the degree to a five-year bachelor's programme to keep pace with international trends and standards. The circulation had just been completed, when the University mandated the Departments of Landscape Architecture and Architecture amalgamate into one department and that the programmes respond to government regulations by introducing a two-degree system. The landscape architecture programme responded to the new act[4] by

Figure 38.1 Senior lecturer Graham A. Young illustrating urban design principles to third-year architecture and landscape architecture students at the University of Pretoria.

Source: Photo by Marguerite Pienaar.

introducing a three-year BSc(LArch), followed by the second degree, a two-year professional master's by coursework ML(Prof). The latter allowed graduates to register as Professional Landscape Architects, after carrying out a two-year mentorship programme. This new system eventually included a third degree, which allowed a candidate to register as a Professional Senior Landscape Technologist after graduating with a BL(Hons).

The melding of the two departments, each with its own strong individual identity, was no easy task and in addition to architecture and landscape architecture, interior architecture was thrown into the mix, creating a new Department of Architecture, which contained three sister disciplines. Prof Schalk le Roux, the then Head of the Department of Architecture stated, "However, especially with the support and energy of Prof Mike Murphy and the maturity shown by members of staff, this was achieved and has proved to be an accomplishment of great value".[5] The strength in the arrangement was the mutual benefit that each programme brought to and received from the others, without sacrificing individual discipline identity.

Looking outward – international connections

Over the years, the Department forged ties with both professional and academic landscape architects, who brought an international perspective to the programme. Through the dedication and persistence of Dr Chris Mulder (landscape architect and developer), strong bonds were established between the University of Pretoria and Texas A&M University. Primarily through Prof Michael Murphy, who in addition to being appointed Head of Department for a short period, obtained his PhD, *Investigation of a process for developing a culturally and geographically relevant curriculum for landscape architecture education in South Africa*, from the University in 1999. Similarly,

Prof John Motloch, author of the book, *Introduction to Landscape Design*, formerly from Texas A&M and Head of the Department of Landscape Architecture at Ball State University, was awarded a PhD in 1991 for his thesis titled, *Delivery models for urbanization in the emerging South Africa*. In the mid-1990s, Prof Ed Fife, Head of the Department of Landscape Architecture, University of Toronto; Prof Glen Thomas from the Queensland University of Technology; and Prof Cristina Felsenhardt from Pontificia Universidad Catolica de Chile were guest lecturers. With the desire to raise the design profile of the landscape architecture programme (and as a counterpoint to the 'McHargian' approach to design and planning), the Department motivated to award Peter Walker an honorary doctorate from the University of Pretoria. The degree was conferred in 2003, in recognition of his profound influence on landscape architecture and the designed environment on a global scale.

Consistent with the staff's intent to look outward for international influence, the Department invited Mr Will Green, Head of the Department of Landscape Architecture, University of Rhode Island, and Ms Jocelyn Zanzot, from the University of Alabama, to take up a short-term residence at the Department, lecturing and working in the studio. In 2016 Mario Schjetnan, internationally renowned Mexican landscape architect, architect, urbanist, and academic inspired the students in seminars, lectures, and a workshop with the master's students. Adding to the significance of his visit was the recognition he received from IFLA in 2015 when he was presented with the Sir Geoffrey Jellicoe Award. This prestigious award is bestowed on a practitioner/academic 'whose work and achievements are respected internationally'. In 2018 Dr Martha Fajardo, a recognised Columbian landscape architect and former president of IFLA (2002–06) was invited as a guest of the landscape architecture programme. During her stay, she participated in student workshops, was an external examiner for our honours students, and delivered an open lecture.

Whilst the visits by international guests did not directly influence the education programme, they did bring a perspective that either reinforced what was being taught or challenged students to think beyond the curriculum. An example was Peter Walker's critique of McHarg's 'rational' approach, which was well entrenched in the Department. Walker challenged students to consider an approach that was not so much rational as perhaps phenomenological or abstract, and which fuelled the discussion on the relationship between landscape architecture, visual art, and spirituality (Walker 1997). This thinking began to permeate studio design projects and theory lectures, after his visit.

Research and publications

Over the years, research contracts have been completed at the university under the leadership of its landscape architectural staff. Historically these projects had little influence over the curriculum and were not the primary drivers of it. However, in recent years there has been a direct correlation between teaching, research, and publications (based directly on the research). It is now encouraged that in the post-graduate courses, publications are generated from studio projects and master's thesis work. The Department's research focus areas are continuously being refined to address relevant 'real world' problems of the day, that is, sustainability, green infrastructure, climate adaptation, resilient and smart cities, and spatial transformation and redress as contained in the South African National Development Goals. The research and its outcomes ultimately feed back into the undergraduate and post-graduate programmes through the formulation of studio briefs and thesis topics (Prof Chrisna du Plessis, Head of the Department of Architecture and Dr Ida Breed, Head of the Landscape Architecture programme, Department of Architecture, personal communication, 02 September 2021).

Dr Anton Rupert, founder of the South African Nature Foundation, who approached the university to complete a masterplan for the heritage town of Graaf-Reinet, commissioned the first research project in 1975. Prof Willem van Riet completed the project in collaboration with the Architecture and Town and Regional Planning Departments, using the GIS lab as a full environmental planning tool that reflected his training under McHarg. The Department of Environmental Affairs and Tourism (DEAT) who became aware of this work, later commissioned the Department in 1993 to begin one of the largest environmental research projects ever to be completed in South Africa: The Environmental Potential Atlas (ENPAT). The project provided DEAT and provincial environmental management agencies with a decision-making system with full geo-referenced data sets comprising biophysical, cultural, and social information. Dr Gwen Breedlove continued with the management of the project, from 1999 until she departed from the Department in 2007. During this period, she appointed research teams, comprised of students from the landscape architecture and architecture programmes, who conducted field trips to various parts of the country, including the natural areas of the Richtersveld, Mapungubwe, Pondoland, and carried out heritage research on South African cities (Dr G. Theron, personal communication, 02 September 2021).

In 2012 Dr Hennie Stoffberg, Mr Clinton Hindes, and Ms Liana Muller, former lecturers in the programme, edited two seminal books on landscape architecture in South Africa: *South African Landscape Architecture – A Reader* (2012a) and *South African Landscape Architecture – A Compendium* (2012b). The publications captured the history of the profession in South Africa and documented projects. Substantial contribution to the books came from personnel in the landscape architecture programme. The books were widely circulated in landscape architecture programmes at South Africa universities, where they were cited as recommended reading for theory courses.

Several PhDs have been awarded to staff for their research, which covered a variety of topics and reflected the nature of the research areas that the Department was grappling with at the time. These include:

- Prof Willem van Riet 1988 (DL) *An Ecological planning model for use in landscape architecture.*
- Dr Erika van den Berg 1990 (DL) *Indigenous tree species: identification and evaluation for use in ecological planning and landscape design.*
- Dr Gwen Breedlove 2002 (DL) *A systematics for the South African cultural landscape with a view to implementation.*
- Dr Finzi Saidi 2006 (DL) *Developing a curriculum model for Architectural Education in a Culturally Changing South Africa.*
- Prof Piet Vosloo 2008 (DL) *The determination of pertinent contract document requirements for landscape projects in South Africa.*
- Dr Ida Breed 2015 (DL) *Social production of ecosystem services through the articulation of values in landscape design practice in South Africa.*

Looking ahead[6]

The three-degree system continued through to 2018 when another change to the programme was mandated by the Dean of the Faculty of Engineering, Built Environment and IT (EBIT),[7] Prof Sunil Maharaj, when it was announced that the undergraduate programme in landscape architecture would close, and the last intake would be in 2020. The reasons given were that the continued low intake in numbers in the programme was unsustainable and not financially viable to the Faculty. This was indeed a dark day for undergraduate landscape architecture education in South Africa and despite appeals from the professional body (SACLAP) and the Institute of Landscape Architects (ILASA), along with emergency meetings pleading the case to keep the undergraduate degree, the decision was not reversed. The Dean, however, did concede that the programme could be reinstated when the numbers became viable. A bittersweet irony is that the undergraduate programme was abandoned in its 49th year.

Professional postgraduate programmes in landscape architecture (Bachelor and Master of Landscape Architecture) however, continue alongside Architecture and Interior Architecture, albeit under a new curriculum. To this end meetings were held in 2020 with the professional body, the advisory committee, and "Friends of the Department" that provided their input into the newly mandated post-graduate programme. Four broad areas of concern were to be considered:

- *Needs of the profession*: to create relevant, sustainable and transformed professional graduates.
- *Feeder programmes*: due to the absence of an undergraduate programme intake would focus on South African architecture programmes, as well as local and African landscape architecture undergraduate programmes.
- *The course aims to form professionals and advance research*: professional registration is indispensable for a course to have national standing and purpose. To deliver value to the programme and society, the focus of research on local and global challenges is essential.
- *Global challenges with local relevance:* the programme's curriculum and research would refer to the UN SDGs. It would also focus on a place-based design approach that is centred on the unique social-ecological qualities of a place.

The new programme aims to perpetuate a balance between ecological urban design and human considerations that recognise local social values and social justice issues.

The University of Cape Town[8]

The growing need for professional landscape architects in South Africa resulted in the development of a master's programme in landscape architecture (MLA) at the University of Cape Town. The MLA programme was initiated from within the programme of City and Regional Planning (CRP). It initially shared many courses with this programme and gradually separated to increasingly focus on the specific nature of landscape architectural design. The collaboration with CRP meant that the programme started with a strong focus on a combination of environmental planning and urban design, and with time-specific landscape architectural design content was added. Like the University of Pretoria, the programme evolved to accommodate the changing legislative requirements with input from academics and professionals. The programme was initiated in 2000 when a master's syllabus was created and convened by Bernard Oberholzer. Since then several staff changes have occurred:

- 2006–2008 – Finzi Saidi joins the programme and takes over programme convenorship
- 2008 – Acting convenorship by David Gibbs
- 2009 – Clinton Hindes takes over programme convenorship
- 2011–2014 – Liana Muller joins the programme
- 2014–2019 – Julian Raxworthy joins the programme
- 2018 – Redesign of curriculum and split of programme into a one-year Bachelor of Landscape Architecture BLA(Hons) and a one-year Master of Landscape Architecture (MLA)
- 2019 – Christine Price joins the programme

The programme continues to be guided by its vision, which references the

> dialogue between contemporary international landscape architectural history and theory, African cultures and innovations in landscape architecture and urbanism that can be reflected back to the world. . . . [The programme] encourage[s] students to design landscapes that provide rich aesthetic experiences, are socioeconomic and ecological catalysts, and function as fun, safe places that hope to bring communities together.[9]

Addressing local issues[10]

While Cape Town is physically located on the African continent, it simultaneously occupies both worlds of the Global North and Global South. The Global North and South are not geographical locations but ideological perspectives (Santos 2014, p. 10; Kerfoot and Hyltenstam 2017, p. 1). The Global South has, for centuries, been dominated by a single point of view that has resulted in, among many things, the justification of colonialism, patriarchy, exploitation of natural materials, and in South Africa, promulgated the system of apartheid. The effects of segregated spatial planning implemented during apartheid continues to persist, with increasing inequalities across neighbourhoods and communities in Cape Town. Situated in the realm of spatial design and urban fabric, the landscape architecture programme at UCT finds itself within this system of complexity "that questions the value systems underpinning the practice of landscape architecture" (Saidi 2012, p. 75). These questions land themselves in the choice and location of studio projects.

The landscape programme continues to build on the work of its academics who advocate for curricula to engage with local and contemporary social problems (Muller and Gibbs 2011; Saidi 2012). Situated in the School of Architecture Planning and Geomatics, the programme is also influenced by eminent academics actively involved in engaged scholarship (Winkler 2013), southern theory (Watson 2016), and revealing subaltern urban narratives (Toffa 2008). The landscape architecture programme maintains close connections with several disciplines within the school, regularly participating in collaborative and interdisciplinary studio and theory courses with the City and Regional Planning and Urban Design programmes. Although the specific studio sites or contexts may change from year to year, the programme deliberately engages in a range of Cape Town-based informal, urban, social, and ecological conditions to grapple with the rich, unexpected complexities of the built environment in the Global South. The programme endeavours to become "more sensitive to the everyday hardships faced by many residents in the cities we study" (Winkler 2013, p. 215).

In addition to local contemporary contexts, the landscape architecture programme engages in critical reflexivity. Through the programme, both lecturers and students are encouraged to reflect on their positionality and the impact of this on our work and research (Bourke 2014). Together, they explore the role of the landscape architect as a 'citizen professional' and the intersecting identities of the student, emerging professional, and citizen within local and contemporary spatial design contexts (McMillan 2017). The landscape architecture programme at UCT is also increasingly mindful that the dominant canon of landscape architecture is rooted in Eurocentric perspectives and that, not unlike most schools of architecture, it has inherited an educational system from the Global North (Saidi 2005, p. 3). Pedagogically the programme addresses this through the notion of 'recognition' (Archer and Newfield 2014; Bezemer and Kress 2016). Recognition[11] has played out in several ways in the landscape architecture programme: recognition not only changes 'orders of visibility' (Kerfoot and Hyltenstam 2017) to present alternative perspectives, but also improves access to those marginalised from the dominant perspectives (Archer and Newfield 2014).

The programmes at the University of Pretoria and the University of Cape Town, being in South Africa, one of the most culturally and environmentally diverse countries in Africa, are ideally placed to offer a wide range of relevant landscape architectural challenges to which students can apply creative expression, design exploration, strategic thinking, and technical resolution. Both programmes offer professional master's degrees required to meet the development needs in the country but in geographically diverse regions. For this reason, they complement each other as their combined teaching and research focus deals with the most pressing landscape planning and design issues encountered across Southern Africa.

However, with the undergraduate programme at the University of Pretoria recently being discontinued, there is now competition to attract students from related undergraduate programmes and the two technical universities that offer landscape architecture courses (i.e. Cape Peninsula University of Technology and Tshwane University of Technology).

Landscape architecture programmes in other African countries

This section sets a context for understanding landscape architecture education in other African countries. It briefly describes the nature of the programmes and the challenges they face in establishing a foothold for landscape architecture education and as a profession. In addition to the South African programmes, several institutions offer landscape architecture and related studies across the continent. The most established are in Kenya and Nigeria with other programmes located in Morocco, Tunisia, Tanzania,[12] and Ethiopia. Table 36.1 lists these programmes. The degree or diploma is listed as per the Institution's description. A diploma qualification usually emphasises a more practical approach to learning. Whilst a degree qualification emphasises an academic approach to the specific field of study. It should be noted that the programmes are accredited in their own countries, some by statutory legislation and others by professional bodies. Some, however, may not meet the minimum requirements described by the International Federation of Landscape Architects. Duration of the degree assumes that the prerequisite first degrees/diplomas have been completed as per the Institution's requirements.

Jomo Kenyata University of Agriculture and Technology, Nairobi, Kenya[13]

The first seeds of landscape architecture in Kenya were planted in the Architecture Programme at the University of Nairobi. A colleague of Prof P. G. Ngunjiri, Mr Frédèrique Grootenhuis, a qualified landscape architect,

> taught introductory courses in landscape architecture at the Nairobi University from 1973–1983. Even though at several IFLA seminars and conferences attention was paid to landscape architecture in Africa, the profession was far from being institutionalised. A Landscape Architecture Chapter was [however] established under the Kenya Association of Architects in 1981.
>
> (Duchart 2007, p. 12)

A syllabus for a landscape architecture undergraduate programme was finally mooted in 1994 and presented at an international conference, organised by IFLA in Nairobi, Kenya. The paper "Curriculum in Landscape Undergraduate Program for Africa" (Ngunjiri 1994) proposed a five-year undergraduate programme. The contents underwent several reviews by professional bodies, Kenyan practitioners, and the IFLA. In 2001, the Department of Landscape Architecture, under the chairmanship of Mr Samuel Kigondu, came in to being as one of the three departments that formed the newly established School of Architecture and Building Sciences (SABS). Prof Ngunjiri, an architect, landscape architect, and environmental planner was the founding Director of the School. He had qualified with an MLA from Berkeley, California, and became the first indigenous landscape architect in the Eastern African Region. The other sister departments are the Department of Architecture and Construction Management.

In 2012 after several reviews of the five-year bachelor's degree it was decided to change the programme to four years, in part, as a response to the paper "Towards Vision 2030: Developing a Responsive Landscape Architecture Curriculum for Kenya", presented by Toroitich and Saidi (2012) at the 49th International Federation of Landscape Architects Congress. It, however, remains under periodic review but so far runs as a four-year professional bachelor's degree. A positive consequence of the Department of Landscape Architecture being situated in the SABS is the cooperation between the two other Departments, particularly where specialist teaching and support staff are required. Students in the School have common foundation units including basic architectural communication techniques. In addition, the Department benefits from the interdisciplinary exchange within the School of Architecture and Building Sciences, and from other relevant fields such as Engineering, Environmental Sciences, and Horticulture among others, which provide for an enhanced landscape architectural learning experience.

The Department intake has grown steadily over the years and currently, student numbers stand at 200 with an average of 50 students per class across the four years. To date, the Department has graduated more than 300 landscape architects. In 2015, to meet the growing demand for post-graduate programmes, the Department introduced a Master of Landscape Architecture (MLA) and a Master of Environmental Planning and Management (MEPM). Initially teaching staff were drawn from architecture. However, staff members have increasingly gained higher qualifications (masters/PhD) in diverse fields including landscape architecture/planning, urban design/urbanism, environmental planning, and GIS. This has led to the curriculum emphasising a design with nature approach to accommodate human activities. There is therefore a strong focus on environmental planning, grading, and planting design.

The Department continues to grow and remain relevant. To this end, it has also sought collaboration with other programmes and has hosted two IFLA Africa Symposiums in the past ten years. Visiting staff/examiners have come from South Africa, Canada, and the United States of America. Graduates have revitalised the Landscape Architects Chapter of the Architectural Association of Kenya and have been active in IFLA Africa activities. In 2023, Kenya and Sweden will jointly host the IFLA World Congress. This has been planned jointly between the respective professional associations and drawing heavily from university staff teaching in landscape programmes in both host countries.

Landscape architecture education in Nigeria[14]

The quest for landscape architecture education in Nigeria is traceable to the 1969 Architects Registration Council of Nigeria Decree No. 10 Caps 20 Laws of the constitution of the Federal Republic of Nigeria. However, momentum for a landscape architecture programme was not realised until Zvi Miller's (IFLA President 1982–86) visit to Nigeria in 1985 (Obiefuna and Uduma-Olugu 2013). Miller was hosted by Mr Ige Fasusi (Individual IFLA Member at that time) and a handful of Nigerian landscape architects who later formed the core of the Society of Landscape Architects of Nigeria (SLAN). During his visit, a decision was taken to establish Master of Landscape Architecture programmes in the Department of Architecture at the University of Lagos, Lagos, and Ahmadu Bello University, Zaria. The immediate goal was to produce home-grown landscape architects who would be interested in teaching at Nigerian tertiary institutions. The many years of military dictatorship delayed the initiative until the advent of democratic governance in 2000, when the implementation of the plan could again be considered.

In 2007, a part-time MLA programme was finally approved in the Department of Architecture, University of Lagos, and in 2009 a similar programme was established at Ahmadu Bello University in 2009. The two programmes faced tremendous teething challenges. The first was attracting adequate students that met the architectural-based requirements, and the second was the lack of qualified landscape architects in academics who could teach in these programmes. This was compounded by the National University Commission's minimum requirement of a Doctor of Philosophy to lecture at Nigeria Universities. To meet this challenge SLAN members volunteered and still volunteer to teach on a part-time basis. Both programmes are conscious that landscape architecture should be grounded in theories and methodologies required to understand the complexity of the term 'landscape'.

Eight teaching modules were introduced.

- *Module 1: Landscape Architectural Studios*: Three design studios varying in magnitude from tot lots, through the hierarchy of parks to urban planning and landscape design and planning solutions to address environmental degradation.
- *Module 2: Communication Skills*: The aim is to develop imaginative and creative faculties and to gain confidence in working processes requiring communication skills.
- *Module 3: History and Theoretical Studies*: The emphasis in this module is the understanding of the history and theory of Landscape Architecture.
- *Module 4: Landscape Construction Technology*: Landscape construction prepares and implements technical planning documents that are needed to realise designed projects.
- *Module 5: Humanities*: These are courses aimed at exposing the student to the cultural, historical, psychological, anthropological, and sociological context within which manmade landscapes are created.
- *Module 6: Ecosystems and Ecological Processes*: Prerequisite to a thorough understanding of the environmental discipline in landscape architecture is knowledge of ecological principles.
- *Module 7: Management Studies*: This module aims to equip the student with management tools required for the coordination, control, administration, and management of project execution.

Addressing local issues

Adejumo's (2019) synopsis of critical Nigerian environmental issues is a window to evaluate the relevance of the existing curriculum in nation building. Included are massive crude oil pollution of fragile Niger Delta ecosystem and abandoned mined fields; gully erosion, desert encroachment, and climate change-driven incessant flooding; food insecurity and uncontrolled high population upsurge that triggers crisscrossing conurbations on agricultural precincts; and degradation of heritage landscapes, dearth of urban open space and collapse of urban developmental codes. The two programmes are not addressing these issues. The curriculum is too Western, meeting the needs of very few military and political elites. The fact is that the philosophy and principles used for urban and regional planning in post-independent Nigeria are still

Figure 38.2 Supervising lecturers, from Ahmadu Bellow University, Dr Maimuna Saleh-Bala and Mr Bartho Ekweruo collaborating on addressing UNDP-Sustainable Development Goal 3 on health and well-being in Zaria City, Nigeria.

Source: Photo by Ibrahim Kyari.

based on obsolete Global North approaches (Bolay 2015). Unfortunately, these ideologies and strategies do not accommodate indigenous landscapes value system, land tenure, rural livelihood concepts, and urban morphological worldviews. Typical examples are the total collapse of most urban parks and deeply held landscape values. Inherited Colonial spaces were used and functional until 1960, the year of Nigerian independence. Western value-driven design ingredients made them work for a colonial workforce.

The change in governance influenced cultural perception and reflected a change of values in green infrastructural development including open spaces. Whereas the relationship between people and the landscape in the Global North is based on material assets, this relationship is more spiritual in rural Nigeria. The belief system sees the landscapes as numinous environments needed to fulfil both spiritual and livelihood obligations without overexploitation.

To address these issues, the SLAN Education Committee resolved in 2016 that the National Universities Commission should take cognisance of these observations in a proposed two-tier landscape architecture education curriculum, which is to include a four-year bachelor's degree and a two-year Master of Landscape Architecture post-graduate degree. Derived from this process several suggestions for a people-centred landscape education curriculum (Papadopoulou and Lapithis 2011) were promoted. These include:

- Accommodation of Nigerian specific environmental ethics, geosophy, and land tenure systems in a theoretical framework.
- An inclusive regional planning and management studio that focuses on food security and rural development.

- Collaborative urban open spaces policies, planning, and design that focus on urban population needs.
- Climate-smart agriculture and flood risk management at urban and rural scales.
- Mainstreaming of heritage landscape preservation and sacred landscapes conservation with a capacity to accommodate pro-poor tourism.
- Urban interstitial spaces that accommodate worldview in the western public park system.

This proposal is already influencing the choice of MLA thesis in the two programmes and perhaps will move the landscape architecture curriculum in Nigeria away from a Global North pedagogy to one that emphasises the conscious relationship between human culture and natural processes in spatial configurations that transform space to 'place'.

North African countries

Morocco

In the francophone countries of North Africa, landscape architecture education follows more closely to the French tradition of *architecte paysagiste*. The landscape architecture programmes typically sit in agricultural institutions and are often removed from the art of landscape design. However, there have been changes to this approach and there are three institutions[15] that deliver what are either considered 'landscape architecture' degrees or at least produce graduates that are employed in landscape architecture offices (Carey Duncan, personal communication, 27 July 2021).

According to Prof Harrouni,[16] teaching of landscape architecture is quite recent in Morocco compared to other European and North American countries. In fact, it started thanks to the initiative of the founder and first director of the Hassan 2nd Institute of Agronomy and Veterinary Medicine (IAV Hassan II), who, in 1978, decided to send students abroad to specialise in landscape architecture. This visionary man established cooperation agreements with the two main French institutions that formed in this field at that time. These were the ENITHP (National School of Engineers of Horticulture and Landscape Techniques) in Angers and the ENSP (National School of Landscape Architecture) in Versailles. The two institutions were different but complementary. The first trained at the Application Engineer level (four years after the baccalaureate) with an emphasis on the technical aspects of landscape architecture. For Moroccan students, the training lasted one year after a general horticultural basis. The second trained at the Engineer level (six years after the baccalaureate) and mainly addressed the conceptual aspects of landscape architecture. The study programme consisted of four years after two years of general university studies. The objective was to allow the students to develop sufficient skills, both in terms of project design and project management for implementation'.

A recent development in landscape architecture education in Morocco is at the *Ecole d'Architecture de Casablanca*, a private school of architecture and landscape architecture, which offers 'diplomas in architecture and landscape architecture recognized by the Moroccan state'[17] and which according to its director is in line with the demand for the training of a new profile of landscape architect able to meet the challenges of new national dynamics in terms of development. It meets the needs for new profiles of professional, artistic, conceptual, scientific and technical, integrating naturally into multidisciplinary teams, intervening in space planning and complementing current profiles, horticultural or agronomic. The establishment of this sector alongside that of architecture within the establishment, will promote the creation of bridges between the two sectors, the crossing of knowledge and transdisciplinarity in use within the teams of specialists involved in these areas (Ms Carey Duncan, personal communication, 27 July 2021).

Tunisia

Landscape architecture education in Tunisia has come along a similar route to that of Morocco and only one institution, The Higher Agronomic Institute of Chott Mariem (ISA-CM), that offers landscape architecture education. According to Dr Ikram Saidane (Dr Ikram Saidane, personal communication, 11 June 2021),

> The landscape section was created in 1995 and the ISA-CM has issued landscape engineering diplomas since 1997. Before this date (since 1980) the ISA-CM delivered the horticultural engineering diploma with the mention 'specializing in landscape'. These engineers were authorized to practice the landscape profession.

Saidane suggests that 'things are much better structured today and the ISA-CM now delivers several landscape degrees corresponding to a different curriculum, either professional or research:

- Landscape Engineer (five-year curriculum)
- National License (three-year curriculum)
- A professional master's degree in landscape and planning systems
- A master's degree in research
- PhD in landscape'.

Higher education in landscape is recent, it has trained nearly 300 landscape engineers since the first promotion of graduates in 1997. However, it should be noted that the practice of landscape design is older and became visible during the French protectorate (1881–1956). During this period, numerous urban planning projects and the development of gardens and public spaces in large Tunisian cities such as Tunis, Sousse or Sfax, etc were conducted. These projects were carried out by French architects, town planners and gardeners such as the Belvédère park in Tunis by the Parisian landscape gardener Joseph Laforcade.[18] After this date, this profession was practiced by landscape (DPLG)[19] graduates of the National School of Landscape of Versailles.

It was in 2013 that the academic researchers who graduated from the ISA-CM joined the landscape academic team at the university. In 2021 the landscape department became an independent structure, until then it was affiliated with the horticulture department. The newly created landscape department, according to the directives of the Ministry of Higher Education, will have the task of initiating a reform of the education program by 2023. The new programme is envisaged to apply a skill-based approach that will consider the economic context, sectors of opportunity, future employers, and the professions of the future. This participatory approach will include the academic team, professionals and former graduates. It aims to offer a new education program adapted to new expectations, taking into account sustainable design, climate change, resilience, and renewable energies. It will also be a question of consolidating the relationship between the professional environment and the academic one.

Ethiopia

The Ethiopian Institute of Architecture, Building Construction and City Development (EiABC), located in Addis Ababa, is the only institution in Ethiopia that offers a landscape architecture programme. However, courses related to landscape architecture are available at other state universities (Dire-Dawa, Mekele, Hawassa, Bahir-Dar, Gonder, and Arbaminch) as part of the architecture curriculum. According to Aziza Busser, chair of the landscape architecture programme at the EiABC, a programme in landscape architecture for architects started in 2005. The programme morphed into an MSc in Environmental Planning and Landscape Design in 2007, guided and chaired by Professor Gerhard Albert and Dr Beate Quenteen, as part of a German Technical Cooperation (GTZ) funded project. A few years later, the two specialisations (landscape architecture and environmental planning) separated as demand grew. In October 2014, Aziza Busser, Kelly Leviker, with three other landscape architects from the faculty, Kalkidan Asnake, Wondifraw Fekadu, and Yusuf Zoheb, developed a curriculum for an MSc in Landscape Architecture, which launched in 2015. The new programme is driven by context, both academic and real-time experience in the country. One of the curriculum's main focuses is harmonising built environments with nature and keeping the impact on the landscape to the bare minimum. In this regard, it implies that "every public project will have to involve landscape architects to achieve the intended outcome" (Ms Aziza Busser, personal communication, 14 October 2021).

Seeking a post-colonial approach to education[20]

In his book on African Cities, Garth Myers (2011) speaks about the re-envisioning of teaching and learning that African academics in architecture and planning schools on the continent are engaged in. This process seeks to define a non-Eurocentric vision of African cities by 'moving' the practice and teaching of architecture (in its broadest sense which includes landscape architecture) and planning, beyond colonialism. Implied in this statement is that education and training provided in most African universities might be outdated and failing to respond to specific contextual and social-cultural issues of Africa.

Since 2015 with #DecolonizeEducation and #CurriculumMustFall campaigns by students in South Africa and other higher learning institutions across the Globe, there has been increased interest among academics in architecture and planning schools regarding the evolution and transformation of knowledge and teaching approaches in architecture and its related disciplines of planning landscape architecture and urban design. While some academics and their institutions have been quick to affirm that change must indeed happen in the curriculum design and planning disciplines, there are yet to be systematic approaches or empirical research that indicate that this change has begun across the African continent.

Most if not all higher education institutions in Africa carry a European heritage and it is not surprising that knowledge content and methods of delivery can be classified as Eurocentric in nature. Recent #Decolonize Education and #CurriculumMustFall debate have focused on decentring this Eurocentric approach in African institutions of higher learning. Apart from social sciences disciplines, developing a critical response to transformative educational agendas of architecture and planning schools is not an easy task as most if not all architects and planners have no formal training in how to initiate transformative pedagogical practices. Most academics in African schools of architecture, landscape architecture, and planning have attained their higher degrees, at master and PhD level, from universities outside their countries of origin.

However, despite the challenges faced by academics, they have developed innovative ways of making their curricula respond to the needs of the local environment. Many academics seek to make learning meaningful not only for themselves but also for their students in response to their societal challenges and opportunities. To further this process Dr Finzi Saidi initiated a dialogue with fellow academics from various institutions across Africa. The initiative was motivated first by the lack of critical cross-disciplinary dialogue on education in environmental programmes in Africa, and second by the need to seek solutions from academics who are actively engaged in the teaching process to better understand their attempts to create appropriate, responsive, and meaningful learning experiences for their students and in so doing 'unearth' innovative teaching strategies, which he calls 'bottom-up' approaches to curriculum transformation.

Dr Saidi invited five young and mid-career academics from five different universities on the African context and challenged them to share their grounded experience of teaching because he believed they understand, more than anyone else, the context of African cities, its context and the power they wield to influence the positive transformation of teaching and pedagogy in Africa. Further, they would be aware of contemporary students' demands in the university learning environments. The five academics were Dr Nelly Babere, a planner from ARDHI University in School of Social Sciences and Planning in Dar es Salaam; Caleb Toroitich, a landscape architect from the Jomo Kenyatta University of Agriculture and Technology (JUKAT) in Nairobi; Goabamang Lethugile, an environmental-researcher and planner at the Ba Isago University in Gaborone; Mathebe Aphane, an architect who previously taught at the Department of Architecture at the University of Pretoria, now at the GSA; and, Jabu Makhubu, urban designer at the University of Johannesburg in the Graduate School of Architecture in Johannesburg.[21]

Dr Babere's presentation, *Planning and Architecture Teaching Response to the Challenges of the City of Dar es Salaam,* focused on the complexities of teaching planning in one of the fastest-growing cities in Africa. She highlighted the issues that planners must navigate in the Global South that include poverty, informality, and climate change. She noted that the two critical challenges for planners that arise in the context of Dar es Salaam are the disparity between spatial planning and the actual situation on the ground, and the often-contested interests between government, international donors, and global influence that had a compounding effect on teaching and practice of planning. This suggests teaching strategies that are interdisciplinary for effective teamwork. Further, she argues that universities should have an open-minded approach to the teaching of planning that allows for different disciplines to collaborate and explore alternative ways to practice. She notes that it is important for planners to be taught to operate in a variety of institutional frameworks within society, not just the legislated practice.

Goabamang Lethugile's presentation, entitled *Examining the Implications Traditional Spatial Demarcation and Formation in Botswana on Teaching Architecture and Urban Design,* raised several questions in the way architecture, landscape architecture, and urban design is taught in the Global South against her indigenous spatial knowledge and experience in the city of Gaborone, Botswana. She observed that landscape architecture and planning programmes in universities lacked relatable local cultural and spatial references to enable students to develop innovative solutions to planning and landscape design in the contemporary urban context. She questioned whose landscape is celebrated when students are taught with landscape references from the Global North in their design exercises. She suggests initiating conversations that discuss what to include or exclude in design and planning-related programmes and emphasises the need to develop a pedagogy that meets the social-cultural and political needs of students.

Mathebe Aphane's presentation, entitled *Integrating the Spatial Concept of Township into the Architecture Curriculum at the University of Pretoria,* explored the spatial dynamic of townships in South Africa and how it could inform the curricula of architecture schools in the Global South. She explores how the education system can be transformed from high school level to empower and encourage young black students to take up environmental design programmes like architecture, landscape architecture, planning, and urban design as viable career paths. Aphane emphasises the need to expose young black students to black academics and professionals in the built industry and the need to teach them to embrace their past, context, and how to access local knowledge to address the challenges of contemporary African cities.

Caleb Toroitich's presentation, entitled *Developing a Responsive Landscape Architecture Curriculum for the City of Nairobi,* discussed developing responsive landscape architecture curricula in the context of the evolving city of Nairobi. Toroitich notes that government, political and professional institutions play a major role in determining what is taught in architecture and landscape programmes at universities. He further noted that international donor development programmes also influence what is taught as well as what happens in the industry. International donors, like the World Bank who are sponsors of large urban infrastructure programmes that is, implementation of Non-Motorized-Transport (NMT) in Nairobi, dictate what is built, how it is built, and who designs what is built. This has a knock-on impact on what schools of architecture and planning in universities teach their students. Toroitich notes that, during these developments, multiple societal problems are not addressed in the curricula relating to the spatial and environmental programmes in universities. He advocates for a shift in emphasis of schools, to address real-life urban issues in the Global South by creating curricula that are responsive to its context and ensures that the gap between what is taught and what is built in African cities is effectively bridged.

Jabu Makhubu's presentation, entitled, *Radical Teaching Practice at the University of Johannesburg – Africanizing the architecture curriculum,* discussed how politics play a role in the teaching of architecture and urban design in the Global South. He argues that African urban stories are often one-sided, and he advocates for imaginative teaching studios that allow for the untold African stories of the city to be expressed on platforms that allow students to narrate their version of the architecture, landscape, and urban design narratives. He argues that different explorations would allow for fluidity and the emergence of diverse topics of spatial interest (issues of race, gender, spirituality, etc.) in the studios that will ensure a radical transformation of the curriculum. What is needed he argues, is "a complete change from the teaching norm" – to be radical.

Many participants expressed the need to continue the conversation simply because there is a dearth of contemporary cross-disciplinary dialogue on education involving the fields of architecture, landscape architecture, planning, and urban design in the African context. Although each presenter spoke about their topic, context, and experience, most of the topics discussed had similarities and potential for synergies from which academics and practitioners can draw inspiration for the meaningful spatial transformation of African cities.

Looking to the future

Saidi[22] suggests that three key ideas frame the discussion around changing a Eurocentric approach to landscape architecture education in Africa. These are:

- Positionality – While teaching the landscape architecture canon we must reveal the geopolitical location of knowledge.
- Relationality – Transformation of the relationships established in the classroom. The diverse background of students should be recognised and valued to enrich the learning experience.
- Transitionality – To enable students to address the question of the meaning of the landscape architecture knowledge they are learning for their society.

And in answering the question, 'What issues should landscape architects address in Africa and South Africa?' He quotes Myers (2011) who suggest that five themes have emerged which should inform landscape architecture theory:

- Postcolonialism – overcoming colonial, in the inheritance of poverty, underdevelopment, and socio-spatial inequalities.

- Informality – dealing with informal sectors and settlements.
- Governance – governing justly.
- Violence – forging non-violent environments.
- Cosmopolitanism – Coping with globalisation.

Conclusion

The history of landscape architecture education in Africa dates to 1971 when the first programme was established at the University of Pretoria. Programmes dedicated to landscape architecture now exist in at least seven different countries. This chapter discusses how these have come into being and the influences that shaped their existence. It, however, makes no claims to being comprehensive in every area of landscape architectural education in Africa, a huge continent with many diverse cultures and approaches to development.

Like the profession, the pedagogy around the discipline is beginning to morph and respond to perceived environmental planning, social, and design needs across the continent. However, as we move towards the future, the challenge is to create new curricula that are driven by educators who understand the African context through a new 'indigenous' lens and who can begin to answer the challenge that Saidi posed to academics in the *South-North dialogues: A festival of the Minds*. What should we be teaching? Why do we teach what do; and how do we better teach the next generation of designers, architects, landscape architects, planners, and urban designers who will be able to respond effectively to our challenges and opportunities presented by the fast-changing complex, yet rich African urban context. Clearly, there is much new work needed in this direction.

Gareth Doherty, an Associate Professor of Landscape Architecture at the Harvard University Graduate School of Design (GSD) also asks a pertinent question within the context of the material discussed in this chapter. Writing in the inaugural issue of the *African Journal of Landscape Architecture*,[23] he states,

> What can African landscape teach landscape architecture in the rest of the world? The launch of this new journal offers an opportunity to consider this question and, to speculate on how landscape architects can work in an African context without being agents, however unwittingly, for the extension of colonialism which has had such a devastating impact on the continent. African landscape architects and their particular practices can contribute to the discipline of landscape architecture and enrich it, rather than by myopically applying western-centric landscape architecture without a critical translation.

These are questions which need urgent attention, but perhaps what is offered here is 'a useful box to put things in'.[24]

Notes

1 Contributors to this chapter include Dr Ida Breed, University of Pretoria, South Africa; Dr Christine Price, University of Cape Town, South Africa; Dr Finzi Saidi, University of Johannesburg, South Africa; Dr Tunji Adejumo, University of Lagos, Nigeria; Dr Dennis Karanja, Jomo Kenyata University of Agriculture and Technology, Kenya; Ms Carey Duncan, IFLA Africa Past President, Morocco; Dr Ikram Saidane, University of Sousse, Tunisia and Aziza Busser, Head of the Landscape Architecture programme, Ethiopian Institute of Architecture, Building Construction and City Development, Ethiopia.

2 Derived from a chapter by Graham A. Young, first published in Barker, A. ed. 2019. *Boukunde, 75th Anniversary of the Department of Architecture, University of Pretoria*. Pretoria: University of Pretoria, pp. 36–40. [Landscape Architecture]
3 An archive of Joane Pim's Papers 1939–1974 exists in the Historical Papers Research Archive, University of the Witwatersrand, Johannesburg, South Africa, 2017 (Inventory no. A882)
4 Act 45 of 2000 the Landscape Architecture Professions Act which established the South African Council for the Landscape Architecture Profession (SACLAP)
5 1971–2001, Landscape Architecture, University of Pretoria, Supplement to Urban Green File, Feb 2002.
6 Contribution by Dr Ida Breed, Head of the Landscape Architecture Programme, Department of Architecture, University of Pretoria
7 EBIT is the faculty within which the Department of Architecture and the landscape architecture programme reside.
8 Contribution by Christine Price PhD, Landscape Architecture Programme, UCT
9 UCT School of Architecture, Planning and Geomatics (2021) *Programmes of Study-Landscape Architecture*. [Online]. Available at: www.apg.uct.ac.za/apg/landscape-architecture (Accessed 26.06.2021)
10 References to the University of Cape Town (UCT) are from a contribution by Christine Price PhD, Landscape Architecture Programme, UCT.
11 See also: Price, C. 2020 *Redesigning landscape architecture in higher education: A multimodal social semiotic approach*. Doctoral thesis. University of Cape Town.
12 A section on Tanzania has not been included as, after many attempts to contact personnel at Ardhi University, landscape architecture programme, the author was not able to establish contact. However, through the University website (www.aru.ac.tz/pages/department-of-interior-design) it can be established that a four-year Bachelor of Science in Landscape Architecture is offered through the Department of Interior Design.
13 Contribution by Dr Dennis Karanja, Chairman of the Department of Landscape Architecture, JKUAT.
14 Contribution by Dr Adejumo, O.T., Department of Architecture, University of Lagos, Akoka. Lagos.
15 IAV Institut Agronomique et Veterinaire Hassan II; Ecole d'Architecture de Casablanca and Université Mohamed Premier, Oujda.
16 Prof Harrouni is the director of the Programme at IAV Institut Agronomique et Veterinaire Hassan II (The Hassan 2nd Institute for Agronomy and Veterinary Medicine).
17 Honoris United Universities 2021 *Ecole d'Architecture et de Paysage de Casablanca*. [Online]. Available at: https:// honoris.net/our-institutions/ecole-darchitecture-de-casablanca-eac-morocco/#1490888588232-ac863f6f-f240 (Accessed 29.07.2021).
18 Zhioua I. dans Les paysagistes dans le monde, Texte collectif concluant les Chroniques de Topia (2008-2014) de Pierre Donadieu, www.topia.fr.
19 DPLG is an abbreviation of Diplômés Par Le Gouvernement that means government graduates.
20 Extracted from "South-North Dialogues: A Festival of the Minds Webinar – Exploring Education in Design and Planning African Schools of Planning and Architecture: A Bottoms-Up Approach", Hosted by Dr Finzi Saidi, Jabu Makhubu and Dickson Adu-Agyei at the Graduate School of Architecture (GSA) University of Johannesburg, 17 September 2020.
21 Derived from "South-North Dialogues: A Festival of the Minds Webinar – Exploring Education in Design and Planning African Schools of Planning and Architecture: A Bottoms-Up Approach", Hosted by Dr Finzi Saidi, Jabu Makhubu and Dickson Adu-Agyei at the Graduate School of Architecture (GSA) University of Johannesburg, 17 September 2020.
22 'Re-Imagining Landscape Architecture in a Decolonized African Context', Presented to Master of Landscape Architecture Students, University of Cape Town, 19 August 2020.
23 Doherty, G. 2020 *Culture, Design and Two Yoruba Landscapes*. Issue 01. [Online]. Available at: www.ajlajournal.org/articles/culture-design-and-two-yoruba-landscapes (Accessed 20.07.2021).
24 Derived from Winnie-the-Pooh and misquoted by Thompson, C. W. (2013) 'Landscape perception and environmental psychology' in Howard, P. et al. (eds.) *The Routledge Companion to Landscape Studies*. London: Routledge, p. 49.

References

Adejumo, O. T. (2019) *Landscapes: Canvas of Civilization*. Nigeria: University of Lagos Press, University of Lagos, Lagos.

Archer, A. and Newfield, D. (2014) 'Challenges and Opportunities of Multimodal Approaches to Education in South Africa', in Archer, A. and Newfield, D. (eds.) *Multimodal Approaches to Research and Pedagogy: Recognition, Resources and Access*. London: Routledge, pp. 1–15.

Bezemer, J. and Kress, G. (2016) *Multimodality, Learning and Communication: A Social Semiotic Frame*. Abingdon: Routledge.

Bolay, J.-C. (2015) 'Urban Planning in Africa: Which Alternative for Poor Cities? The Case of Koudougou in Burkina Faso', *Current Urban Studies*, 3, pp. 413–431 [Online]. Available at: http://dx.doi.org/10.4236/cus.2015.34033 (Accessed 10 July 2021).

Bourke, B. (2014) 'Positionality: Reflecting on the Research Process', *The Qualitative Report*, 19(33), pp. 1–9.

de Santos, B. S. (2014) *Epistemologies of the South: Justice against Epistemicide*. London: Routledge.

Duchart, I. (2007) *Designing Sustainable Landscapes: From Experience to Theory, a Process of Reflection Learning from Case-Study Projects in Kenya*. Doctoral thesis, Wageningen University, The Netherlands.

Kerfoot, C. and Hyltenstam, K. (2017) 'Introduction: Entanglement and Orders of Visibility', in Kerfoot, C. and Hyltenstam, K. (eds.) *Entangled Discourses: South-North Orders of Visibility*. London: Routledge, pp. 1–15.

McMillan, J. (2017) 'I Understand That Infrastructure Affects People's Lives: Deliberative Pedagogy and Community-Engaged Learning in a South African Engineering Curriculum', in Shaffer, T. et al. (eds.) *Deliberative Pedagogy: Teaching and Learning for Democratic Engagement*. Ann Arbor: Michigan State University Press, pp. 159–168.

Muller, L. and Gibbs, G. (2011) *Reading and Representing the Cultural Landscape: A Toolkit*. Unpublished toolkit for the Association of African Planning Schools, Cape Town: University of Cape Town.

Myers, G. (2011) *African Cities: Alternative Visions of Urban Theory and Practice*. London and New York: Zed Books.

Ngunjiri, P. G. (1994) 'Curriculum in Landscape Undergraduate Program for Africa', International Federation of Landscape Architects – Education in Landscape Architecture in Africa Conference Proceedings, January, Nairobi, Kenya.

Obiefuna, J. and Uduma-Olugu, N. (2013) 'Evolution of Landscape Architectural Education: The Nigerian Experience', *International Journal of Sciences*, 2, pp. 28–32.

Papadopoulou, A. and Lapithis, P. (2012) 'Exploring Dimensions of Sustainable Design within the Architectural Curriculum', in *18th Annual International Sustainable Development Research Conference Proceedings*, Hull, 24–26 June.

Pim, J. (1971) *Beauty Is Necessary: Preservation or Creation of the Landscape*. Cape Town: Purnell.

Price, C. (2020) *Redesigning Landscape Architecture in Higher Education: A Multimodal Social Semiotic Approach*. Doctoral thesis, University of Cape Town, Cape Town.

Saidi, F. E. (2005) *Developing a Curriculum Model for Architectural Education in a Culturally Changing South Africa*. Doctoral thesis, University of Pretoria [Online]. Available at: https://repository.up.ac.za/bitstream/handle/2263/27969/Complete.pdf (Accessed 14 July 2021).

Saidi, F. E. (2012) 'Rethinking the Role of Landscape Architecture in Urban Cape Town', in Stoffberg, H., Hindes, C. and Muller, L. (eds.) *South African Landscape Architecture*. Pretoria: University of South Africa Press, pp. 75–88.

Stoffberg, H., Hindes, C. and Muller, L. (eds.) (2012a) *A Reader – South African Landscape Architecture*. Pretoria: University of South Africa Press.

Stoffberg, H., Hindes, C. and Muller, L. (eds.) (2012b) *A Compendium – South African Landscape Architecture*. Pretoria: University of South Africa Press.

Toffa, S. (2008) 'Cultural Landscapes in Terrains of Contestation: Space, Ideology and Practice in Bo-kaap, Cape Town', *Traditional Dwellings and Settlements Review*, 20(1), p. 91.

Toroitich, C. and Saidi, F. (2012) Towards Vision 2030: Developing a Responsive Landscape Architecture Curriculum for Kenya', *49th International Federation of Landscape Architects World Congress*, Cape Town, South Africa, 4–7 September.

Walker, P. (1997) *Peter Walker Minimalist Gardens*. Washington: Space Maker Press.

Watson, W. (2016) 'Shifting Approaches to Planning Theory: Global North and South', *Urban Planning*, 1(4), pp. 32–41.

Wiesmeyer, M. E. (2007) *Joane Pim South Africa's Landscape Pioneer*. Pinegowrie: The Horticultural Society.

Winkler, T. (2013) 'At the Coalface: Community-University Engagements and Planning Education', *Journal of Planning Education and Research*, 33(2), pp. 215–227.

39
EDUCATIONAL ECOSYSTEM ON LANDSCAPE IN LATIN AMERICA

Gloria Aponte-García, Cristina Felsenhardt, Lucas Períes, and Karla María Hinojosa De la Garza

Introduction

This chapter presents an overview of landscape architecture education in Latin America, a region characterized by heterogeneous development consequent with its cultural and physiographic diversity. From the broad spectrum of the diverse orientation, focus, level and reliability of educational programmes and courses, three examples in Chile, Argentina and Mexico that represent activities in the extremes of the region are discussed.

As published studies on landscape architectural education are quite scarce for Latin America, the information presented in this chapter is based mainly on the experience and previous work of the authors, including personal notes. By bringing four authors together, a consistent survey emerges. It is a region where many disciplines address landscape education in a dispersed way. Educational approaches, subjects and contents of landscape programmes and courses vary considerably. Many have different names, some even lacking the word 'landscape' in their title (Aponte-García 2015).

Context

Latin America is the most diverse of the five world regions that the International Federation of Landscape Architects (IFLA) defines. The region extends, in latitude terms, over more than the other four regions. It contains a wider variety of biomes, and consequently countless biocapacity. The greatest biodiversity is concentrated in South and Central America. This great natural diversity gave rise to multiple ancestral cultures. Being well aware of their dependence on nature, ancestral people considered themselves part of it and deeply rooted in the natural characteristics of their lands, confirming, in their own way the landscape theory about 'place forming people' (Swanwick 2002, 2009).

Complex and strong processes of conquest and colonization altered the extensive territory today known as Latin America. These changes simultaneously spoiled the physical richness, imposed new thoughts and led to a certain grade of homogeneity, principally through language and religion, the same causes that helped to deepen the difference between South, Central and North America. Nevertheless, México, located in the south of North America, remained an important part of Latin America. Several political and economic developments through history have contributed to a close relationship among Latin American countries, known as *Latin American Brotherhood*.

Undoubtedly colonization eroded the indigenous traditional *people-earth link* and has fostered the habit of looking for external models as a paradigm of what is the right way of living. As a consequence, the beginnings of intentional landscape interventions consisted in making gardens and parks, mainly in European styles that contributed to hide local identity. Nevertheless, the patrimonial traces of ancient cultures and their expressions, found in several nations, have recently become object of greater appreciation, research and inspiration as secure roots of the particular human-nature affinity. Numerous academic works that are being developed at present, in several university research groups, and other institutions and ONGs, from México (Maya and Azteca communities), Colombia (Arhuaco, Muisca, Uitoto and Tikuna communities) and Chile (Mapuche community) take into account this component as a fundamental part of the *New World* landscape identity.

The beginnings of design education and practice are marked by European and USA influences, through invited lecturers and professionals. Depending on where Latin American promoters of educational programmes had studied, the first opportunities were established through grants or universities exchange, for example with France, Italy, Spain and England. According to its natural and cultural diversity, education in landscape architecture in Latin America is also diverse. Interest in this branch of knowledge and professional activity started, with a few exceptions, during the second half of the twentieth century through courses offered within architecture faculties, schools or departments, actually as an extension of that profession. The dependence on architecture has continued in several programmes to the point that disagreements arise on the matter of subjects to be covered in official landscape architectural programmes.

Formal programmes in Latin America started to be structured and officially approved by national education authorities, almost simultaneously in Chile, Brazil and Argentina. According to the survey "International Opportunities in Landscape Architecture Education and Internship" collected by the IFLA Education and Academic Affairs Committee, around 40 official landscape architecture programmes are currently running in Latin America. Compared with a population of 629 million, there is one for 16 million inhabitants; while, according to the ASLA webpage, in the United States there is one programme for a population of 3.3 million. The Latin American subcontinent counts with 21 countries (including Puerto Rico), but many of them lack of a single landscape architecture programme, or some of those when they exist, are intermittent in their activity.

The region has experienced an outstanding dynamic around landscape architecture education in recent years. One example is the first Capacity Building course for educators carried out in Brazil in 2004, supported by local agencies, IFLA and UNESCO. Later on, in 2012 and 2015, two more but shorter Capacity Building courses were developed in Medellin, Colombia. Both were orientated mainly to countries that did not have landscape architecture programmes.

Representative examples of the development, present situation and coverage of education in landscape architecture in Latin America are presented through the wide and diverse panorama in three countries of the region: Chile, Argentina and México. Although there are also landscape architecture formal academic opportunities in Costa Rica, Puerto Rico, Colombia, Venezuela, Ecuador, Brazil and Paraguay, those are not enough to cover the extension, population and development need for the profession.

Landscape architecture education in Chile

Early teaching of landscape design in Chile is closely linked to French and German influence derived from the Paris World's Fair in 1889, which revealed prominent trends in architecture and garden architecture and which aroused the interest among traveling elites keen on realizing advances of modernity.

Another significant event occurred at the beginning of the twentieth century, as urban authorities from the Ministry of Housing and Urbanism invited foreign landscape designers, perhaps at the stage more public garden designers, like George Dubois, Oscar Prager, Guillaume Renner and Charles Thays to introduce the concept of public green space, a result of the democratization in the modernizing world. These guests also brought the first attempts to coach local landscape designers. A fundamental milestone is the influential work in the 1940s and 50s of the modernist and versatile Roberto Burle Marx in Brazil, who opens the gates (Felsenhardt 2021) of abstraction in landscape design, formalizing a more contemporary expression in South America and becoming an archetype for the region (Adams 1991). It is then, in the late 1950s, that the Universidad de Chile's School of Architecture establishes the first undergraduate degree programme in Landscape Design, beginning the country's initial formal steps in landscape instruction.

In the 1950s a different viewpoint with poetic and territorial roots emerges at the School of Architecture of the Universidad Católica in Valparaiso. Although not born from landscape principals as such, the school introduces a new relationship between its architecture education and the territory, the identity elements of geography as the opportunity to understand, plan and design the territory by its own natural law. The educational entity, formed by academics and students created a community that travelled across the country, and thaught the students 'on-site' the transcendence of macro space and landscape (Perez de Arce et al. 2003).

In 1954, and by people interested in design with natural elements, the Garden Club, was created, promoting and organizing workshops and flower exhibitions. From it a select group of self-taught individuals – mostly women – emerge, promoting Chile's landscape culture. They undertake design work together with architects, bring landscape strategies learned during their trips, and begin to teach brief landscape courses. In the late 1960s, private institutes begin offering short technical programmes thus educating some of the first landscape designers, once again mostly women, who contribute to designing public and private parks and gardens.

In the 70s, architecture undergraduate schools start to convey mixed visions that slowly generate new curriculum arrangements and integrate vegetation-related subjects as elements for urban space construction. A specific geographical cultural and local-climate approach also springs up at the Universidad del Norte, in Antofagasta. The approach is not generated from landscape methods, but from an undoubted sensitivity to the Chilean Northern xeriscape. Similar to this paradigm, the Universidad de Talca in 1999, started a School of Architecture based on a strong regional view, and on architecture focused on low impact construction within explicit landscape language considerations.

In the 1970s, the Architecture School at Pontificia Universidad Católica de Chile in Santiago generated a Department of Environment related to ecology, managing – unfortunately for a short time – to bring together scholars interested in interdisciplinary teaching approaches. Here, in the 80s, an important academic movement emerges to create a formal Specialization Programme in Landscape Architecture and Management. The materialization of holistic landscape concepts and its education starts based on a multidisciplinary approach. The understanding of landscape architecture commences to play an important role, making to converge technology, science and design with environmental systemic methodologies, reshaping the traditional spatial and symbolic perspectives.

Many years pass before the merging of a multidisciplinary students and academics programme is approved as completely mandatory for what finally is understood in Chile as landscape architecture. The characteristics of traditional self-contained programmes, especially of architecture, started to generate the need for an integral outlook, which is precisely the great value of the landscape architecture discipline.

In the 80s and due to the Education Reform introduced by the military regime, the unfortunate closing of the undergraduate degree programme in Landscape Design at Universidad de Chile takes place. Instead, the teaching of landscape architecture at university level is placed within the Schools of Architecture, including the education of architects dedicated to public space design and seminars that are part of a curriculum focused on an urban approach. The conceptual definition of landscape education as such begins, introducing the still sparse experience in Chile of ecologically systemic work.

In the 80s, within the same reform which allows the creation of private universities, and with the military dictatorship's expulsion of a group of academics from Universidad de Chile's school of Landscape Design, the private Universidad Central establishes the Ecology and Landscape undergraduate school in 1988, until today the only such programme in Chile. There, in 2004, with an important curricular modification, the agenda shifts to more present days landscape architectural methodology of education, thus completing the contemporary perspective.

The 80s are also a turning point in landscape architectural education, mainly because of the new Educational Reform that authorizes the opening of new private universities, and there were not enough academics. A large number of architects, along with some agronomists and geographers, travel to Europe and the USA – mainly UCL Berkeley, Harvard, Illinois, ETSAB in Barcelona and AA in England, to study landscape in master and doctoral programmes in design, sustainability and environment. Returning to Chile, they take positions at longstanding, as well as new universities and as professional landscape architects in public agencies, and private offices. It is when all the layers of efforts made until this time appear, and nourish the new programmes. At the same time, in 1989, the Chilean Institute of Landscape Architects (ICHAP) is created and becomes part of IFLA as an institution.

Since the end of the twentieth century, important programmes of landscape architecture arise boosting theory, research and publications, and the importance of multidisciplinary approaches. A more complex interpretation of landscape education begins, but maintaining the 'hard core' of each discipline. This leads to a reshaped more contemporary perspective in a quickly transforming world. Today, some research projects have won national public funds, though the overall amount of funded research is still modest. In addition to the scientific methodologies, new procedures from aesthetics and art are increasingly recognized. This is a breakthrough, as the previous perspectives were limited by the exclusive ones that rejected abstraction and subjectivity.

Today, reputable masters and diploma programmes and a number of seminars are offered at many universities, as well as short degree courses and seminars at the Professional Institutes. Landscape-specific courses are also immersed in the degrees of architecture, geography and agronomy; some of these are also offered in the south and north architecture undergraduate careers. Although the presence of landscape education topics is certainly increasing, the unfortunate part is that despite of the Chilean geography, very diverse climates and environments, formal landscape education is greatly concentrated in Santiago, the central region of the country (Figure 39.1).

Landscape-specific courses are also immersed in the degrees of Architecture, Geography and Agronomy, some of these are also offered in different places of the national territory. The impacts of climate change, with the unavoidable metamorphosis of places, are still a challenge in landscape education in Chile. Important matters like re-wildening of urban areas, the ecological finitude of the planet, an unreasonably excessive anthropisation of lands, seismic landscape,

Latin America

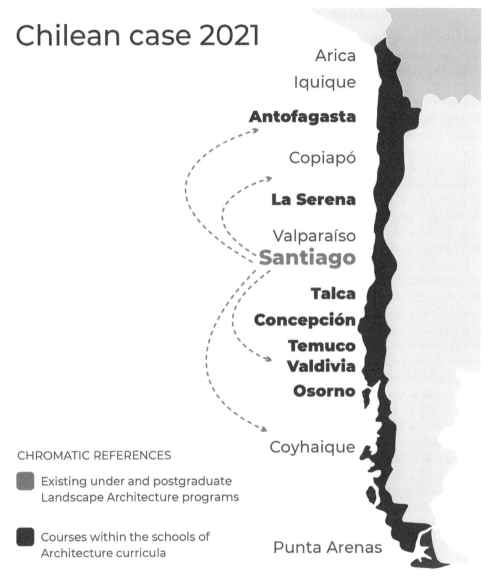

Figure 39.1 Landscape architecture education in Chile.
Source: Visualized by Felsenhardt (2021).

landscape as an "open window over geography" concepts, are topics that are still to be more powerfully integrated (Catalá Marticella 2017).

The influence of landscape architecture out view at territory and landscape has brought an important progress in the understanding of Chilean identity, undertaking territorial considerations, geographic processes as well as scenic approach. The Chilean field within Latin America has been growing into more complex outcome in landscape philosophy, becoming today an important landscape education locus within the continent.

Landscape architecture education in Argentina

Argentina is recognized as a pioneer nation in terms of disciplinary construction, regarding landscape education in Latin America. The influence of the French landscape designer Jules Carlos Thays (1849–1934) – naturalist and architect, as a disciple of Édouard François André (1840–1911) and Jean-Charles Adolphe Alphand (1817–1891), from whom he received professional training with the development of landscape projects for the cities of France – marks the beginning of the informal training of professionals, with the incursion of his professional practice throughout the country, being the designer of numerous public and residential parks, from the end of the nineteenth century.

The formal antecedents of education originate in the early twentieth century and in the context of agronomic sciences. The foundation of the chairs of *Parques y jardines* (Parks and gardens) of the Universidad de Buenos Aires in 1918, and of the Universidad Nacional de La Plata in 1931, establish the beginning of education history. The schools of architecture followed while the foray into engineering is more recent. This disciplinary development evolved from the traditional notion of landscape design to the current field of landscape architecture.

Regarding architectural programmes, in the mid-twentieth century landscape architecture was incorporated as a curricular activity in the study plan at the Universidad Nacional de Córdoba, 1956 (Budovski and Períes 2018). This innovation, for the time, which impacted the profile of professional training, stems from a search for renewal in urban and architectural planning approaches, with contextual, environmental and ecological principles. This was a paradigm shift developed in Córdoba. It is mainly based on the syncretism of international models: the harmonic composition of the *Beaux Arts*, the functionalist technicality of the *Bauhaus* and the architectural organicism of F. L. Wright (Naselli 1986). The precursor case was followed in 1964, by the subject *Paisajismo* (Landscaping) at the Universidad Católica de Córdoba and the optional subject *Diseño del Paisaje* (Landscape Design) in the Architecture programme of the Universidad de Buenos Aires, 1968. Then, most of the country's schools adhere to the proposal of inclusion of landscape content in architecture and urban planning programmes, which remained until present, even in the creation of new architecture programmes.

In specific training, the creation of the first postgraduate programme developed within the framework of the Landscape Institute of the Universidad Católica de Córdoba, in 1976, the specialization in landscape architecture, aimed at architecture professionals who are awarded the title Architect Specialist in Landscape Design. Subsequently, other postgraduate courses emerged in Buenos Aires and La Plata. The creation of a training programme enabling the diploma for professional practice was delayed until 1993 and arose from the integration of the schools of architecture and agronomy of the Universidad de Buenos Aires; the Bachelor's Degree in Landscape Planning and Design. The people who promoted the programmes also created the Centro Argentino de Arquitectos Paisajistas (CAAP) in 1971, the first Latin American association to join IFLA.

In the twenty-first century, landscape architectural training is diversified in terms of programmes at multiple academic levels. In the year 2000, the second degree in the country was created at the Universidad del Museo Social Argentino, under the name *Diseño del Paisaje* (Landscape Design). In 2004, the *Ingeniería en Paisaje* (Landscape Engineering) programme started at the Universidad Nacional de Catamarca. In 2002 the first master's degree was inaugurated at the Catholic University of Córdoba, with the name *Arquitectura Paisajista* (Landscape Architecture). In 2003, the specialization programme in *Diseño y planificación del paisaje* (Landscape Design and Planning) at the Universidad Nacional de Córdoba and the postgraduate seminar on *Arquitectura del Paisaje* (Landscape Architecture) at the Universidad Torcuato Di Tella were created.

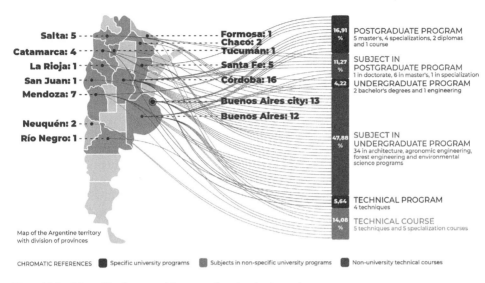

Figure 39.2 Map of landscape architecture education in Argentina.
Source: Visualized by Períes (2019).

Subsequently, three new specializations, four masters and two diplomas – the latter for graduate professionals – were created.

Other diploma, technical programmes and technical courses are available but not at university level. The current situation in landscape architecture education is presented in Figure 39.2.

Between 2000 and 2020 there were 19 landscape architecture university programmes (12 postgraduate, three undergraduate and four technical) and 42 subjects with specific content in university programmes of architecture, urban planning, environmental management and agronomic or forestry engineering (eight postgraduate subjects and 34 undergraduate subjects). This is significant because knowledge about landscape architecture is dispersed in multiple disciplines of professional training. In this regard, the singularity that all architecture programmes in the country have at least one landscape subject stands out; this establishes a distinctive feature for professional training in Argentina, which distinguishes it from other countries in the region. Finally, the educational platform is completed with 10 pre-university technical training courses. These activities propose a more practical and operational approach to landscaping, which is valuable for the training of human resources in terms of trade training.

Another relevant aspect is the geographical distribution. The territory of the Argentine Republic has an area of 2.78 million km² (Instituto Geográfico Nacional de la República Argentina 2020) and a longitudinal development of 4,361 km – between the northern and southern extreme boundaries of the continental surface. Given the territorial extent, the training offered is concentrated in the centre and east of the country, mainly the cities of Córdoba and Buenos Aires – the two metropolises with the largest population – in addition to La Plata. Not only do students from all over the country come to these cities, but they also come from neighbouring countries. This last factor is generally due to the condition of free public education and the reduced cost of postgraduate fees that favours foreigners due to the monetary exchange rate.

Currently, two degree programmes are in operation, one in engineering on landscape and one in bachelor's degrees in landscape design. Regarding the postgraduate programmes, there are nine: three master's degrees (cities of Córdoba, La Plata and Rosario), two specializations

(Buenos Aires City and Córdoba) and the Torcuato Di Tella postgraduate seminar; these data represent the discontinuity of six programmes. In technical programmes, the discontinuity of a technical title is recorded. The subjects included in the non-specific programmes remain constant. All of them are thematically specific to landscape architecture approaches, although the same name is not always used. This data is, in some way, reflected in the variation of content of the curricular plans. In some cases, there are approaches with a botanical, ecological and naturalistic orientation and, in others, urban-landscape planning and project planning are weighted higher. The variation in approaches is due to the institutional context of origin of the programmes, whether they are schools of architecture and urban planning or of applied and natural sciences. These particular profiles, in no case leave aside the holistic view around the multidisciplinary contents of landscape architecture.

Despite the considerable decrease in activities, the situation in Argentina is positive compared to the region. The offer is varied, and the ongoing programmes have extensive experience with specialist teachers and academic quality.

From the analysis of the collected information and the relationship and nature of the data, the condition of a 'dispersed' teaching model is concluded. This corresponds to the diversity of the educational offer in terms of: the variety of activities, the multiple academic levels, the thematic approaches and the territorial distribution. All this shows dispersion and distances itself from the models that prevail in other regions of the world. But, at the same time, a particular nature is built according to the local scope and possibilities. In this way, the approach to education, around planning and the landscape architecture project, develops with its own identity.

Landscape architecture education in México

Landscape architecture education in México started about 50 years ago with professionals of different disciplines studying, at some point in their training, landscape architecture abroad. Returning home, applied the new knowledge in the development of projects and their professional work. This process led to the development of education in landscape architecture in an informal way, however making contributions to the transformation in the professional landscape design culture. A slightly different case is architects from México City who went abroad to study for a postgraduate diploma or for taking design courses. These professionals are people close to the Mexican Schools of Architecture who, either by having been trained in or invited to collaborate with foreign institutions, have influenced the founding of the most important schools of landscape architecture in the country.

Three universities and one civil association have upgraded education in Mexico, the educators training professionals and promoting the profession. The first educational initiative in this discipline appeared in 1964 at the Universidad Nacional Autónoma de México (UNAM). For the first seminar, the Brazilian landscape designer Roberto Burle Marx and the American landscape architect Garret Eckbo were invited. Later, in 1967, the first course in landscape architecture starts within the Degree in Architecture at the same university (Larrucea 2010).

These developments had an impact on the training of the following generations of architects. Some of them continued their studies in landscape architecture at the University of California at Berkeley, and other universities abroad that offer different interpretations of landscape architectural concepts, including those offered by Robert Royston, Donald Appleyard and Ian McHarg. The influence of the University of Sheffield, UK, is significant due to the holistic vision of its Landscape Design Master's programme. The École Nationale D'Horticulture et Paysage, Versailles, France, offered knowledge and the first contact with

IFLA, when México was invited to become IFLA member with the assignment to create its own association to represent the country in that organization. Therefore the Sociedad de Arquitectos Paisajistas de México (SAPM) was founded on 18 August 1972. Later on, the landscape architecture firm SWA Group from Sausalito, California, gives impulse in 1985 to create the first landscape architecture undergraduate programme within the Architecture Faculty at UNAM.

Some architects graduated from UNAM also went to Aberdeen University, Scotland, to study and based on their experience these professionals with newly acquired knowledge returned home and got involved in the School of Architecture at the Universidad Autónoma de Baja California (UABC) that offered the country's first postgraduate programme in landscape architecture.

The objectives and human resources of that speciality were integrated to create a new Master in Architecture, where a research line in landscape architecture stands out. Consequently, UABC was the first Mexican university to offer a professional master's degree in landscape architecture. Resulting from academic restructuring, unfortunately this degree programme was eliminated from the institution in 2001.

At the same time, between 1983 and 1990, at the Universidad Autónoma Metropolitana, Azcapotzalco (UAM-A) in México City, a short course in landscape architecture at postgraduate level, is being taught continuously (Martínez 2017). A multidisciplinary group of academics developed these courses in the Environmental Department of UAM-A. These educators are graduated in different fields, including Landscape Architecture, Anthropology, History and Conservation. They had earned degrees from universities such as the Haute Ecole Charlemagne Gembloux, Belgium, the Università degli Studi di Genova, and at the UAM-A itself, with the participation of graduate masters from the UABC acting as lecturers.

A specialization in Environmental Design, containing a landscape architecture line is approved at UAM-A in 1990. Later on three post-graduated programmes in Planning and Conservation of Landscapes and Gardens are approved as specialization at both graduate and post-graduate levels. Since that time, this is the only opportunity for students to earn a doctoral degree in landscape architecture in Latin America.

Education in landscape architecture is offered in 13 of the 32 Mexican states. There are two undergraduate programmes in Chihuahua, one in Sinaloa and two diploma degree courses in Baja California, one undergraduate programme in San Luis Potosí and one in Jalisco, two master's programmes in Guanajuato and two master's programmes in Michoacán, one undergraduate programme, a specialization, two master's programmes and a doctoral programme in Mexico City, a specialization programme and a master's programme in Morelos, one undergraduate programme in Tabasco and one undergraduate programme in Guerrero, one master's programme in Veracruz and two master's programmes in Yucatán (Figure 39.3). The country offers 23 programmes: eight undergraduate and 15 postgraduate, two diploma courses, two specializations and ten graduate degree programmes, and the doctorate in Mexico City.

Several facts have facilitated the expansion of landscape architecture education in Mexico. First, the Society of Landscape Architects of Mexico formed and committed to supporting the discipline through workshops, conferences, seminars, congresses, student competitions and participation of its members in courses at undergraduate, graduate and postgraduate levels at national educational institutions.

Second, students do not have to pay tuition fees as the government supports them at universities such as UNAM and UAM. On the other hand, in the particular case of UAM-A, all students receive a scholarship because the institution belongs to the National Register of Quality

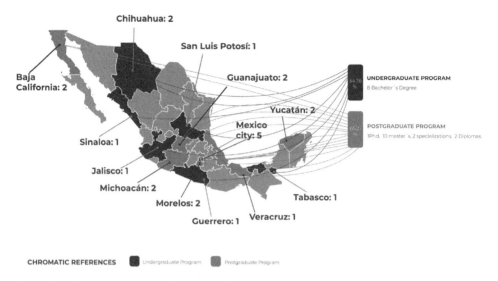

Figure 39.3 National coverage of landscape architecture education in México.
Source: Visualized by Hinojosa De la Garza (2021).

Postgraduate Courses (Padrón Nacional de Posgrados de Calidad) PNPC of the National Council of Science and Technology (Consejo Nacional de Ciencia y Tecnología, CONACYT).

Another determining factor for the success of landscape architecture studies in the capital city, is that students from all of the Mexican states attend and are interested in the matter. These students develop master plans and landscape proposals for various cities in the country, usually those where the students come from. In this way, once the students finalize their studies, besides obtaining the degree, they usually gain tangible proposals to spread the profession through presentation of those projects and also recommendation letters, or get work agreements. Sometimes they receive offers to teach courses or workshops that eventually become the basis for new undergraduate or master's programmes, thus contributing to this complex national educational ecosystem.

Conclusions

Important steps have been taken during last decades to develop landscape architecture education, in Latin America, as could be concluded from the three analyzed examples. Those are the outstanding countries with a number of programmes, teaching quality, academic exchange and economic support from the governments. It is actually a small coverage proportion for the region, composed of 35 countries of which only 17 landscape architecture associations belong to IFLA. This is one of the points that show the priority given to the profession in the region. From the IFLA member countries, three do not count with any programme at all, six count with one –although intermittent – and two countries count with two programmes each.

Relaying on the accelerated growing of virtual communication, new forms of interaction among Latin American landscape architects have led to active exchanges, including experience in education and standards. The work carried out by the multisectoral organization *Iniciativa Latinoamericana del Paisaje* (LALI – Latin American Landscape Initiative) has contributed to this line of articulation. The Landscape Institute of the Universidad Católica de Córdoba, Argentina, also promotes education in landscape architecture since its inception in 1973. Recently, the book *The Teaching and Research of Landscape in South America* (Períes et al. 2021) which forms a body representative of the current situation, in terms of the multiple ways of facing the landscape from different disciplinary approaches and their articulations. Frequent virtual dialogues have facilitated the recognition of interests, focus, emphasis and strengths of others and their own.

One topic that is being developed and strengthened is the recovery of ancestral principles that with shorter coverage, than modern Western thought, has been kept by the more than 800 indigenous groups in Latin America. They have maintained their traditions, particularly those that refer to a respectful relationship between people and *mother earth*.

Considering the growing importance of this discipline and the urgent need to prepare professionals to provide directions to the matter that it represents, the establishment of more programmes, on all academic levels, according to the development – or absence – of educational opportunities in each one of the region's countries, is called for.

Otherwise, although interdisciplinary method is part of landscape architectural thought, the principles of trans-disciplinary work could be applied more widely and their teaching at universities improved. Also, the promotion of and the renewed criteria of *working on challenges* bring to scene a wide scope to be faced from the landscape architecture education; it means priority on integral applied solutions, besides a concrete fundamental theory. It would be important for educators to take the lead in involving and giving landscape response to the current warnings about climate change and water crises, to the decline of biodiversity, for example by giving a more profound meaning to Nature Based Solutions (NBS) and to planning 'with' nature.

Finally, there is a need to build a formal network of educational institutions in landscape architecture. This network would include all Latin America and look for the future of the educational system, make suggestions for cooperative action, promote the establishment of formal educational opportunities prioritizing countries that lack landscape architecture educational opportunities in the region.

References

Adams, W. H. (1991) *Roberto Burle Marx: The Unnatural Art of the Garden*. New York: Museum of Modern Art New York.

Aponte-García, G. (2015) 'Educar hacia el paisaje en América Latina', *Bitacora arquitectura*, (31), pp. 56–6.

Budovski, V. and Períes, L. (2018) 'La trayectoria de la enseñanza del paisaje en Córdoba', in Períes, L. (Comp.) *La enseñanza del paisaje en Córdoba*. Córdoba: UCC.

Catalá Marticella, R. (2017) *La geografia com a narració descriptiva i com a construcció d'una pedagogia del mon. La Literatura Paisatgistica de Josep Pla*. Barcelona: Tesis Universidad de Barcelona, Facultad de Geografia e Historia, pp. 283–293. [Online]. Available at: https://77dialnet.unirioja.es/servlet/tesis?codigo=158513 (Accessed 18 July 2021).

Felsenhardt, C. et al. (2021) *Abriendo Territorios y Paisajes – Santiago en transición a la Cordillera de Los Andes' Book*, Finis Terrae (ed.). Santiago.

Larrucea Garritz, A. (2010) 'La arquitectura de paisaje en los 100 años de la UNAM. El reto de diseñar el paisaje mexicano', *Bitacora arquitectura*, (21), pp. 63–73. Available at: www.revistas.unam.mx/index.php/bitacora/article/view/25203/23691 (Accessed 7 October 2021).

Martínez Sánchez, F. A. (2017) 'Área de Investigación Arquitectura del Paisaje', *Diseñomas*, (0.1), pp. 17–34 [Online]. Available at: http://medioambientecyad.azc.uam.mx/index.php/dmasrevista/article/view/77 (Accessed 7 October 2021).

Naselli, C. A. (1986) 'El filo de la navaja: Carlos Alberto David, arquitecto paisajista argentino', *Summa*, 227, pp. 18–21.

Perez de Arce, R. and Perez, F. (2003) *Escuela de Valparaíso, Grupo Ciudad Abierta, Tanais Ediciones S.A.* Madrid: Tanais Ediciones.
Períes, L., Aponte-García, G. and Filla-Rosaneli, A. (eds.) (2021) *La enseñanza e investigación del paisaje en Sudamérica*. Córdoba: UCC.
Swanwick, C. (2002) 'Landscape Character Assesment, Guidance for England and Scotland', *The Countryside Agency and Scottish Natural Heritage*. Cheltenham: The Countyside Agency.
Swanwick, C. (2009) 'Society's Attitudes to and Preferences for Land and Landscape', *Land Use Policy*, (21), pp. S62–S75 [Online]. Available at: www.sciencedirect.com/science/article/abs/pii/S0264837709001112 (Accessed 17 September 2021).

40
CONCLUSIONS AND HOPEFUL PERSPECTIVES

Diedrich Bruns and Stefanie Hennecke

In this compendium, 60 authors report findings from six continents on stories of personal ambition, on political and professional ideology, social and political development, and mechanisms under which education emerges and thrives. These are foundations for the research community to grow on, and to develop the field of education studies further. Engaging in research is the more motivating and worthwhile as the volume of educational records and archives is expanding, and the numbers of study programmes, teachers, and students are increasing worldwide. The body of knowledge needed for model making and theory building is also growing, including educational concepts and structure, inclusive pedagogy, the appreciation of regional differences, and cultural diversity. By producing reports in greater numbers, landscape architecture researchers are supporting the field in gaining scholarly recognition and attention from outside.

There is much to learn. For, from reading the 40 chapters, it may appear that policymakers, managers, and teachers share a common understanding of the basic structures of programmes and curricula, of learning aims and outcomes, and of examination requirements in the field. Yet, after two centuries of landscape architectural education (including its 19th-century antecedents), many of the underlying ideas behind teaching and several core issues still have to be exposed. Debates persist about academization of and in an originally professional field allegedly requiring mainly practical training. About the balancing acts between arts and sciences, between generalist versus specialist, planning and design philosophies, worldly knowledge and local expertise. Furthermore, as landscape architecture is joining the ranks of disciplinary fields, we need clarity about its position in the greater discipline classification system. By way of contributing discipline specific approaches, methods and competences to interdisciplinary work, landscape architecture researchers are helping to define and orient the field in relation to others. In conducting educational studies, landscape architecture scholars will take an interdisciplinary and mixed-methods approach. They would ideally work in teams that include education, archival and library sciences, history and geography, sociology and psychology, and others. They would integrate the collection and analysis of various documents, biographies, discourses, and other sources for information on study programmes, curricula, course development, and learning practice.

This chapter discusses three crosscutting themes that we recognize as central to studying and furthering the field: (1) historical awareness and responsibility, (2) international networking, and (3) Education for Sustainable Development. The three themes link to perspectives of model and theory building. *Future research* will have opportunities to respond to a variety of questions regarding the genealogy of ideas and the transportation of ideologies via university education, regarding pedagogy, teaching and learning practices, and academic and personal development. It would also analyze and rank school policy and the work of study programme managers, for example by assessing educational performance, different qualities of study programmes and curricula, and so forth. More work needs to be done for education research to provide sound institutional and programmatic guidance to teachers, schools, accreditation agencies, and governments, for example on programme and course development, on pedagogic approaches and learning methods, and about teacher education.

Historical awareness and responsibility

To raise historical awareness, it is fundamental to collect chronological data, and evidence about the reasons and driving forces that motivate countries, universities, and organizations to engage in landscape architectural education, including the ideas and ideologies behind decisions on how we teach and learn (instructor and student values and thinking). One motivation for *countries* to provide education resources is the growing relevance of a field, for example, when it contributes to resolving societal problems, and when market demand finds expression in rising numbers of applicants to study programmes. For landscape architecture, both relevance and demand are growing as new tasks emerge during the last two centuries, for example in the course of industrialization and urbanization, and during periods of transition when politicians include the discipline into modernizing programmes. Research to date has focused on structural and resource responses of governments establishing study programmes, on teaching staff that is in touch with the profession. We are understanding how graduates qualifying to meet immediate practice requirements, their learning might be subject to meeting a variety of private and public interest. For example, how providing for recreation might link to maintaining industrial productivity. How design might contribute to raising real estate value. How regional and landscape planning paradigms interlace with nationalistic and protectionist politics, including classifications of design styles and even individual plant species as "invasive", even "alien". How governments see landscape design and planning supporting nation building and expansion, even colonization, or supporting post-war reconstruction, meeting environmental concerns about infrastructure projects, rehabilitating wasted and mined lands. From reading chapters in this volume, we are aware how students absorb a range of different ideas and ideology along with acquiring landscape and design knowledge. We learn, for example, how, following judgements made by a small number of scholars, students get to know a select number of emblematic designs valued as exemplary, ignoring most of the world's vernacular landscapes and landscape concepts.

In addition to politics, administration and the bourgeoisie, tourism and conservation, *universities* are important players. Their reasons to install landscape architecture education pertain to a variety of driving forces. For example, when competing for applicants at the end of the 19th century, it took European and North American institutes of higher education only a few decades to expand the range of their academic fields to include professional disciplines. During the 20th century, the numbers of universities to install landscape architecture grew rapidly in several parts of the world. Universities were not so quick in offering equal opportunity to staff and students, and in agreeing on titles for programmes and degrees. Tradition and institutional politics were in the way. For example, from analyzing staff and student rosters of European and North American universities, we learn how long it took for women to be admitted. How, even today, equal access and opportunity are still not the standard regarding gender, race, social, and other circumstances. Chronicling the naming of departments and study programmes reveals how replacing "garden" with "landscape" was a long process. Not the majority of schools but a small number of people, each with their own ideological baggage, drove it. Today, newly established programmes all over the world adopt landscape, including in countries that have no word for landscape in their vocabulary. In addition, replacing traditional degree titles with internationally accepted ones also took several decades to complete. One example is the engineering degree. European universities used to award degrees with the term "engineer" in the title, for example in Germany *Diplom-Ingenieur*, in Poland *inżynier ogrodnik* (gardening engineer), in the Netherlands Agricultural Engineer. It took the political pressure of the Bologna Process to create unified higher education standards, including cycles and degrees, a system now applied around the globe.

Further driving forces pertain to the formation of knowledge societies. Knowledge growth is associated with social status. The professional field enjoys a high status and priority at universities in many parts in the world. In Asia, for example in China, South Korea, and Thailand, universities only admit candidates with the best grades to study landscape architecture. Some of the universities themselves are striving to acquire a new subject such as landscape architecture in the course of educational policy renewal and competition with each other. The field is considered as one with its finger on the pulse of time. Supporting a university that offers education in fields that are considered modern is also relevant for regional policy and competition. In addition, there is a strong ranking of universities, in several countries, which is important for the career prospects of graduates.

The role of *professional organizations* as driving force is threefold. Firstly, associations act as institution that exert pressure on governments and universities to establish study programmes; secondly, they act as political supporters of higher education initiatives; and thirdly, they serve as reservoirs from which universities recruit teachers. In some regions, graduation from a professional programme is a prerequisite for becoming a member of professional associations, which in turn is required for practicing one's profession. Historically, associations striving for the identity of their discipline and an education tailored to it, aimed to keep landscape architecture apart from landscape gardening, an activity that is, particularly in the Anglo-Saxon world, perceived as associated with certain (frowned upon) activities and style. In order to make explicit their claim to a wide professional range with a design competence equal to architecture, North American universities were the first to rename their courses of study from Landscape Gardening to Landscape Architecture, such as the Massachusetts Agricultural College in Amherst in 1903 and the University of Illinois, Champaign, in 1907. Universities in Europe follow suit, again in slow progress, for example, in Reading in 1926 and Manchester in 1934 (UK), in Warsaw (Poland) in 1930, and in Wageningen (Netherlands) in 1948. Reasons for professional organizations around the world to introduce the term "architect" relate to the forming, at the end of the 19th century, of international professional organizations.

International networks, culturally contextualized education

International networks initially develop through personal contacts and travel. They develop further through formal correspondence and publications, for example through the dissemination of books, and through contributions to specialist journals. Graduates of one school take on teaching at another and carry pedagogical ideas and canonized knowledge from one place to the next across national borders, such as Louis in Belgium, Moen in Norway, Cabral in Portugal, Rebhuhn in Romania, and so forth. Comparing stories collected in this volume regarding national and international exchange, and the debates carried on through such exchange, we observe how networking relationships existed between design teachers since the early 19th century, marking the beginning of a continuous formation of networks between educators and schools. These are paths of inspiration and mutual support in building knowledge, but also mechanisms of colonization, and of marginalization of local traditions.

From the turn of the century onwards, regular exchange takes place, beyond the networking of individuals, through systematically preparing and partaking in professional meetings, conferences, and congresses. Comparing study programmes presented in this volume suggests that educators will have learned about content and qualifications standards via exchange. While jealously guarding realms of academic freedom it seems that, over the years, people began to abide by agreed upon educational agendas, including curricular structure and minimum learning requirements. Academic and professional organizations in the UK and the USA were among the first to agree on educational syllabus and minimum standards. International organizations followed suit, such as the European Council of Landscape Architecture Schools (ECLAS), which issued their Guidance on Education in 2010. With its Standing Committee on Education and Academic Affairs (EAA), the International Federation of Landscape Architects (IFLA) aims to advance education globally and inclusively, thereby broadening the Common Ground, and recognizing Africa, Americas, Middle East, Asia-Pacific, and Europe as geographically and culturally distinct areas.

International networking and agreements on standards and qualification has led countries and institutions to promote exchange. The idea of international learning itself, however, is comparatively old. Modelled after the Académie de France à Rome (founded in 1666 to offer a three-, four- or five-year scholarship to select French artists), the American Academy in Rome was established at the turn of the 20th century. When it added landscape architecture to the programme in 1914–15, this might be the first institution to offer to (a select group of) students opportunities to study abroad. International student and teacher exchange has become common practice ever since. A range of organizations offers resources for the running of international courses, workshops and projects, and others learning formats such as summer schools. Students cherish the opportunity to compare design cultures in different parts of the world. Long before and at the latest since the pandemic of 2020–22, the use of internet-based learning platforms has been a fixture in everyday teaching and learning practice.

Education for Sustainable Development

Bounding beyond the confined space of a single discipline (A), and taking advantage of international networking (B), we appreciate the greatness of our common world and take on the sustainability challenge (C). UNESCO and ministries of education worldwide are working, within the framework of the World Action Programme 'Education for Sustainable Development' (ESD), to promote education to achieve Sustainable Development Goals (SDGs). Individual states implement the Action Programme. They involve educational and municipal institutions, private educators, and education organized through professional associations and commercial enterprises. The core question for landscape students is how we might design landscapes to meet, as the Brundland Report states in 1987, "the needs of the present without compromising the ability of future generations to meet their own needs"?

Understanding landscape design as referring not simply to scenery and the physical environment but also to community of place, and process of shaping ideas about people's material and social surroundings, students will ask many more questions that education research will need to address. How can teachers facilitate that we learn what access do different sections of the population have to landscape information and design? Where and how does social injustice manifest itself in the everyday landscape, and might it be overcome? What can we do to strengthen general landscape awareness? Studies into education that revolves around these and similar questions will help teachers who are preparing students to meet the challenges of inclusive ways of opinion-forming and democratic forms of decision-making about landscape. These are learning objectives, for example, of the ERASMUS+ project 'Landscape Education for Democracy'.[1] The project serves to illustrate how landscape architecture education might use international networks to develop the sustainability story further. As part of a strategic partnership, educators from several universities collaborate on establishing and offering courses on participatory and inclusive landscape design. As ESD includes appreciation of cultural and social diversity, students learn what different people associate with things in their surroundings, determine what plants and buildings, forest and commercial areas and other things mean to them. Learning consecutively and building up knowledge, students gradually become sensitive to people's experience and to perception of landscape, learn to communicate how people perceive landscape with all of our senses. Based on findings from education studies discloses how adding methods of inclusive and community-based learning to standard didactic repertoires helps generate forms of landscape knowledge that complement specialist knowledge. Ministries and educational institutions can work harder than before to include democratic principles and the rule of law into their sets of learning objectives and into curricula and study programmes. Didactic principles follow suit that combine theory and practice, learning by doing, including inter- and transcultural learning.

Perspectives

Advancing the field of Education Research, we are looking at three hopeful perspectives that connect to historical awareness, international networking, and Education for Sustainable Development. These three perspectives together connect to theory building. The perspectives are hopeful also in the way we, taking guidance from the landscape theorist François Jullien, deliberately relocate positions from where we look at things. Thus including views representing different cultures, countries, schools, and people who are involved in education. Including educational views from different parts of the global public and sciences, each with their own ideological roots and collective memories. Paying respect to the various landscape and design traditions of world. Freeing our imagination, from the shackles of established narratives, and the confines of history defined by self-appointed academic gatekeepers.

Critical historical analysis aims to compare and contrast differing sets of ideas, ideals, and ideologies that students absorb along with acquiring landscape and design knowledge. Studies would, for example, explore how governments foster education that includes reflection on practice to recognize power structures, to uphold specific ethical standards, and that prepares professionals abide by rules of conduct enshrined in legal statutes, and follow democratic principles and rules of law. They would explore how universities, in hiring staff and accepting students, abide by statues of human rights. How they pay respect to regional design cultures, where the term "landscape" does not exist. Where using authentic formats of learning and graduating might better apply rather than adopting globalized ones. Studies would aim to shed light on questions of national and regional policy, professional and cultural identity, progress in educational practice. They would place ideas in historical context (institutional transition, social change) and so forth. For example, a question might be what roles professional organizations play in facilitating exchanges between education and practice. How and how quickly are professional practice challenges and changes in policy and legal regulations reflected in educational practice, and, vice versa, how do findings from research and teaching become professional knowledge. Similar questions might be raised as to the role of national politics and government. In addition, researchers might engage in studies about the roles and effects of different schools of thought, for example about the relationship between art and science, design and planning, large and small scale, nature and culture. They might start developing genealogical models, for example by tracing histories of ideas that several generations of students incorporate into their thinking repertoire. In studio and history classes, the consumption of landscape and art-historical works might lead students to form typological genealogies that, in turn, influence student's landscape perception and values regarding landscape, the role of ecology, the classification of design styles, the connotations associated with different plants, and so forth. Genealogical studies are still in their infancy in educational research. Reviewing studies in this volume, it appears students build up a stock of landscape images (and preferences), for example by means of field visits, perusing richly illustrated books. This repertoire they constantly expand with new examples, thus contributing to the growth of a professionally conditioned reference list the coverage of which circles much of the same canon, often including projects declared iconic.

We bear an international responsibility as we address the further development of educational structures and educational content. The purpose of future educational research on the history of ideas will be to overcome the boundaries of the litany that formed over decades. To open up the view, future studies would include all regions of the world cherishing their rich and often vernacular landscape and design legacy. In the same vein, studies would look at how people around the world use a multiplicity of words to give expression to their perception and understanding of landscape, art, and design. Scholars would ask how education in a landscape design discipline can support our ability to communicate about the terms at the core of our identity. For example, to Mediterranean, Arabian, and Asian regions having some of the oldest design and design education traditions in history, the idea of "landscape" is alien, as a word and as an idea. The term origins in Europe and immigrants introduced landscape to the Americas and, as a technical (geographical) term via Russia to Asia before the advent of landscape architecture. For people to give expression in their own words to individual facets of the concept of landscape, the languages of the world have a variety of specific terms available. In education and professional practice, we continue using "landscape" knowing how it spread in the course of colonial processes, and how it refers to the environment changed by humans, as designed space, as place and polity, and as an influential factor in the shaping of ideas in a just social and physical environment. Are we not striving, at the same time, towards promoting the plurality of possibilities to speak about the human environment with alternatives to Western bias? How rigid or flexible are the concepts of landscape and the concepts of dealing with people and nature conveyed with it? Can education research help shed light on developing adaptations or the use of alternative concepts?

International networking helps broaden the view and the geographic scope of studies, of collecting and reporting research, and of exposing and discussing findings. What began in the early 19th century with cross-border exchange of teachers continued in the 20th century with international meetings and has developed into regular conferences of large organizations, publication series, and a system of mutual recognition of study programmes and degrees. There is an international dimension to the debates about the discipline's position between arts and sciences, about academic education versus practical training, generalist versus specialist. International initiatives are underway to build up the common body of knowledge, for example through cooperation in the documentation and archiving of the knowledge pool, and by strengthening specialist journals and the publication of books through cooperation in the reporting of newly acquired knowledge. International programmes and initiatives also foster responsible forms of knowledge storage, retrieval, and transfer. Aiming at collecting and archiving knowledge resources, education institutes cooperate in building up systems for landscape design documentation. Storing and sharing of disciplinary knowledge includes ways to record and make available implicit, tacit, and explicit knowledge.

Using common web-based platforms, conferences and publication, academic institutions in the Global North and South continue to exchange concepts and ideas on landscape architecture education and will discuss the impact of colonizing strategies in the past until present days. Research is needed in how universities in Africa, Latin America, parts of Asia and the Arabian World will meet the many challenges of decolonization mentioned in contributions to this volume. In addition, international communication and frameworks offer benchmarking opportunities to teachers and institutions, such as the ranking of schools against education and knowledge achieved worldwide, while at the same time preserving their own identity and values. The role of ranking is based, at least partly, on students who hope for good opportunities on the market, but partly also on politics and administration who place value on numerical measure, particularly where numbers are recorded and translate into educational success. With regards to institutions and people's interests in status, it would be important to raise the question how rankings and efforts of optimization come to the fore, and which effects a focus on economic criteria and ranking has on learning and educational outcome, on the value of local anchoring of a degree programme.

Regarding social justice and inclusiveness, educational studies are needed about the ways teachers and students are engaging with diverse communities as an opportunity to develop culturally contextualized learning, thereby addressing challenges of cross-cultural understanding. As institutions of higher education have set up landscape architecture programmes in all parts of the world, studies might be launched with the purpose to learn how connected programmes and people are in education practice, whether teachers work independently or with the help of external support, and how either way effects learning practice and outcomes. Studies might also look into cooperation in educational practice as the forms of collaborative teaching spread across institutional and regional borders, for example in the form of teacher and student exchanges, international courses, workshops and projects, through the use of globally operated learning platforms, and more. In addition, as graduates of one school take on teaching at another, it would be important to know, with respect to agenda building, how people carry pedagogical ideas and content-related ideals and ideologies from one place to the next, including transfers across national borders and institutional barriers. From chapters in this book we have several examples that suggest how international exchange facilitates the spread of ideas. However, a systematic genealogical analysis would reveal which roles institutes of higher education have played and are still playing, and what policy they are using regarding the distribution of ideas and ideologies through teaching.

Sharing knowledge across borders is an important service in many ways, such as fostering mutual understanding and peacekeeping; it might also, and again, carry colonial elements. In the past, several countries have, through their foreign services, established schools and educational institutions abroad. For example, the American University of Beirut (AUB) was founded as a college of higher learning in 1920 to educate academic and political leaders in the region. Landscape architecture was added in 2000. German and Norwegian institutions support Palestinian universities. However, in several regions of the world, a contextualized school of landscape architecture continues to be a challenge. When installing study programmes, universities that initially deferred to North American or European influences have since adopted a policy to deal with local challenges and values. As an example, China represents the case of countries where landscape architecture is successfully regaining regionally specific cultures. Wondering what would have happened if continuous development had occurred along the traditional route of design education is pure speculation: we do not know if the country might have developed its own version of landscape architecture by now. What did happen is that, initially, China, like other countries that experienced abrupt breaks, including those of their education and design traditions, started to look for orientation in the wider world. During the cultural break, traditions such as Fengshui were questioned, even banned. More recently, cultural qualities and technological achievements that are specific traditional values are being increasingly revisited. Future studies might provide insight into ways of revisiting history from different points of view, and looking at the "mending" of what might be interpreted as "broken" regional and other traditions. An appreciation might be important of differences between varieties of traditions, including those of the elite and of local everyday practice, the latter rarely considered in a disciplinary canon.

Addressing sustainability challenges, education research might contribute to crossing disciplinary boundaries, because landscape architecture stands at the crossroads of sustainability, weaving threats of ecology, socio-economy, and governance into the sustainability knowledge-fabric. Studies would aim to connect global, regional, and local dimensions and cultures in an internationally defined framework. Education research might investigate where to locate landscape architecture in the canon of disciplines when we include the boundary-spanning role that the discipline plays in pursuing Sustainable Development Goals.

Overlooking two centuries of education history, the volume of evidence is growing to levels needed to *develop theory*, for instance by systematically comparing conceptual, structural, methodological, and pedagogic models from different regions and times. By extracting similarities and differences, and paradigms and paradigm shifts, for example on the role of art and science subjects and their respective weights in study programmes and place in curricula. By analyzing concepts of design and landscape education that evolve differently in the Global North, South, East, and West. By categorizing curricula and clusters of subjects, thus contributing to the formation of structural models which the global community will discuss and assess as to their regional applicability. Studies might also compare which approaches and methods appear to work best in different situations of course development and learning.

For building theory by and through landscape architecture education research, the community would set an agenda of combining inductive and deductive research, with deductive approaches beginning with a theory that is then tested, and with inductive research used to collect information to develop theory. For example, starting with cognitive models of teaching and learning, and of problem-solving models in particular, pertinent theories might be tested by running a number of experimental and observational studies. From case studies collected over several decades, researchers are in a good position to know how students construct their learning from two sources, what they are being taught and their own experiences, and how they are likely to retain information when receiving positive reinforcement such as offered during studio sessions. In the future, researchers might look for commonalities in the ways teachers and students think and interact (pertaining to cognitive process of learning), and how they learn and grow while forming connections between different kinds of experience and information. New developments require revision of models and theory. To monitor and evaluate changes that occur over periods of months and years, conducting longitudinal studies would help looking at different variables over an extended period of time, such as for example at the ones that characterize in distance and hybrid learning that are adding new dimensions to learning practice.

Note

1 Ruggeri, D. and Fetzer, E. (2019) 'Landscape Education for Democracy. Methods and Methodology', In_bo, Università di Bologna 10 (4), 18–33. https://in_bo.unibo.it/issue/view/816. (16.03.2022).

INDEX

Aachen 148–150
Abel, L. 243, **245**, 246–247, 251
Abercrombie, P. 166
academic community 46–51, 63, 72, 347
academic degree 104, 145, 156, 243, 245–246, 250–251
academic freedom 17, 40–41
academization 15, 421
Academy of Fine Arts: Amsterdam 189; Copenhagen 140; Vienna **244**, 246–247, 249
accreditation 172, 225, 237–240, 311–312
accreditation body 63, 93, 308, 422
accreditation standards 16
accreditation system 333
accredited 7, 93, 104, 166, 169, 236, 239, 308, 365, 370, 375, 397
acculturation 124
Adams, T. 166
Adejumo, T. 399, 406
adjunct teachers 373
admission 104, 130, 147, 227, 256, 258, 260, 266, 357
aesthetics 17, 20, 122, 188–190, 194, 204–206, 224, 235, 276, 283, 315, 318, 320, 412
affiliation 64, 102, 105, 113, 127, 223, 228, 248
African American 21, 93, 105, 122–126, 130–131
Agar, M. 110, 112, 115, 165
Agricultural College (including Agricultural University) 55, 101, 102, 105, 111, 122–124, 126, 139, 143, 145–146, 152, 155, 189–192, 204, 207, 280, 423
Agricultural College Act *see* Morrill Act
agriculture 20, 147, 234–227, 239–242; engineering 106, 113, 177, 192, 306–309, 311; experiment stations 55, 125–126, 397, 404, 406
agro-colonial enterprises 22
Alphand, A. 293, 414

American: American Baptiste Home Mission Society 123; American Civil War 122, 124; American Missionary Association 123; American Society of Landscape Architects (ASLA) 20, 60, 115, 121, 126, 128, 165, 319; American University of Beirut 304, 308–309, 312, 427; Anglo-American 390
Ammann, G. 254
ancestral 7, 26, 337–338, 409, 419
ancien regime 22
André, R.-É. 293
Angers 275, 292, 299, 300
Aničić, B. 211, 217, 220
anti-racist 18, 26
apartheid 366, 396
Aphane, M. 404
apprenticeship 32–38, 104, 156, 165, 215, 234, 247, 256, 258, 260, 276, 316, 332–333
architecte(s) paysagiste(s) 178, 189, 293–295, 401
architecture: architectural education 4, 127, 147–148; programmes 20; school(s) 121, 130, 147–148, 172, 277, 373, 401, 404, 407, 411
architecture and urbanism 217, 220, 230, 239, 308, 396
archival 6, 54–60, 117, 164, 201, 245
Arnold, Z. 211–213, 220
art: academy 117, 146–148; college 143, 145–148, **146**; historian 54, 148; history 54–55, 135, 189, 192–195
artistic 105, 136, 143, 148–149, 167, 173, 178, 193, 216–217, 224, 227, 246, 250, 261, 295, 401; artistic admission examination 147; artistically 146–147, 203, 298, 320
Arts and Crafts 140, 224, 254
arts and sciences 116, 368, 421, 426
Ås 1, 54, 101, 107, 110, 135, 141
Asnake 403

Index

Aspinall, P. 21
assessment methods 83, 84, 88
assimilation 124, 130–131
Assunto, R. 283
atelier 107, 147, 276, 296–297, 300, 326–327, 332
Audias, A. 295
AudioVisual Lab 269
Australasian Educators in Landscape Architecture (AELA) 374

Babere, N. 404
bachelor degree 279, 288, 309
Bagatti Valsecchi, P- F. 282
Bailey, L. H. 122
Barillet-Deschamps, J.-P. 293
Barth, E. 104, 110–113, 143, 145–146, **146**, 148, 150, 152n3, 155
Bartman, E. 353
Barton, C. 21
beaux-arts 127, 295
Behrens, P. 147, 153n16, 254
Beijing Forestry College 315, 318–319
benchmark 2, 66, 427
Berlin 43, 45, 101, **103**, 104, 106–107, 110–114, 137–139, 143–148, 150, 152, 155–161, 177, 193, 196n9, 201, 212–213, 215, 231, 248, 250, 254, 256, 264, 289n13
bi-cultural governance 366
Bijhouwer, J. T. P. 103, 106, 108, 110–111, 113, 187, 189–190, 192–193, 195
biodiversity 299, 310–311, 383, 409, 419
biographical studies 5, 55
biography, biographies 109, 150, 198, 223, 421
BIPOC 130
Bir Zeit University (BZU) 301
Bloch, M. 17, 27
Blois 275, 292, 299
blood and soil 157
Boffa, F. 368–369
Bogdanowski, J. 351, 353
Bois de Boulogne 293
Bologna Process 8, 47, 63, 229, 258–259, 284, 423
Boone, K. 21
Bordeaux 275, 292, 299
Boston 129
botanist 147, 149
botany 105, 106, 112–113, 116, 129, 137, 147, 166–167, 176, 178, 183, 189, 194, 199, 218–219, 248, 253, 255, 257, 259, 268, 283; as "feminine science" 129
Botha, R. 390–391
Bratislava 233, 235–237, 239, 241
Breed, I. 393–394, 406–407
Breedlove, G. 394
British ruling oligarchy 22

Brno 234, 236, 240
Brussels 43, 47, 110, 114–115, 196n7, 201
Bucharest 199–201, 203–207
Budapest 55, 59, 223–226, 228, 230–232
building code 357, 362
Bund Schweizerischer Garten- und Landschaftsarchitekten (Swiss Association of Garden- and Landscape Architects) 262
Burle-Marx, R. 391, 411, 416
Bush School 379
Busser, A. 403, 406
Bussey Institute 127

Cabral, F. C. 103, 107, 108, 110–111, 113–114, 176–179, 184, 423
Cambridge 103, 113, 114, 128, 129
campus 104, 107, 115, 122–123, 126, 137, 140, 261, 300, 310, 311, 343, 374–375, 380
campus design 122–123, 126
canon 21; historical 18; imperialist 22; Western 21
canonization 13, 106
carbon landscape 376
carceral landscapes 19
career changers 375
Carolina rose *see* Cherokee rose
cartographic 55, 380–384
Carver, G. W. 125
Casa Pia 110, 176, 181–183
case studies 3, 16, 21, 27, 35, 49, 57, 60, 83, 93, 428
Castner, E. 112
cemetery, cemeteries 108, 110, 115, 140, 177, 201, 225, 282, 381
Centre national d'Étude et de Recherche du Paysage (CNERP) 296
Centre of Excellence 374
Centro Argentino de Arquitectos Paisajistas (CAAP) 414
Challenger, C. 366, 368–369
Chamber of: Agronomists and Foresters 279; Architects 223, 225, **244**, 278–9; engineers 357; town planners 357
Charlottenburg 112, 143, 148
Châtelaine Horticultural School 253, 254
chemistry 107, 116, 167, 178, 189, 199, 220, 255, 317
Cherokee rose 125
Chilean Institute of Landscape Architects (ICHAP) 412
Chinese garden design 315, 320
Chinese painting 320
Chinese Society of Landscape Architects 319
Chiusoli, A. 283–284
Choisy, A. 293
Ciriani, E. 296
city garden director 148, 150
city planning 20, 126, 278

civic art 127
civic association 130
civil engineering 87, 113, 207, 239, 309, 355, 356
Clément, G. 263, 300
Cleveland, H. W. S. 122
climate adaptation 26, 267, 270, 393
climate change 91, 267, 301, 338–339, 354, 357, 363, 365, 375–376, 378–380, 388, 399, 402–404, 412, 419
climate emergency 378–379, 387–388
climatology 268
code of practice 369
cognitive skills 88, 89
College of Applied Arts 147
college of art 148
colonialism 396, 403
colonization 178, 337, 409–410, 422, 424, 427
Colvin, B. 109, 114–115, 165–167, 172–173
Comité de l'art des jardins 292
Committee on Education 114–115, 424
community-based 15, 66, 68, 93, 96, 230, 425
comparative analysis 7, 68, 102
computer-aided design 321
conceptual models 6, 83, 85, 88
conceptual thinking 81–82, 84, 86–88
Conder, J. 328
Confederate monument 18
Confucius 322
conservation: heritage conservation 55, 298–299, 301; nature conservation 178, 256, 310–311
conservatism 18
constructivism 33
containment 376
continuing education 1
conversations (as study method) 6, 17, 27, 75, 110, 404
Copenhagen 106, 138
Corajoud, M. 296
Correspondence (as study method) 6, 14, 117, 198, 304, 390, 423
Council of Educators in Landscape Architecture (CELA) 14, 341, 367, 374
Council of Europe 40, 62–63, 74, 353
countryside 106, 109, 167, 192, 359
course development 77, 421–422, 428
course structure 96, 105, 164, 170–171
Covid-19 366, 374–376
Craft 123, 126
Cramer, E. 254, *260*
Critical Path' 370
critical race theory (CRT) 18
critical reflexivity 396
crosscutting themes 7, 338, 421
Crowe, S. 165–166, 172
Crystal Palace School of Gardening 110, 165
Cultural Revolution 319

culture, cultural: ancestral 7, 337, 409; diversity 309, 410, 421; erasure 18; heritage 35, 209; history 257, 263; landscapes 21; Native American 124; wars 17–18
cultures and precedents 21
culture wars 17–18
curriculum, curricula 19, 27, 368, 372, 403; design 20; integrated 27; professional 19
Cursol livre de arquitectura paisagista 111–113
cycles 1, 63, 423

Dahlem *144*, 144–146, 148, 150, 152n2, 152n4, 250, 254
Damec, J. 237
Darcel, J. 107, 293
Dauvergne, P. 296
De Chirico, G. 282
decolonization 7, 26, 104, 337–338, 427
defence landscape 157
deforested 22
democracy 50, 230, 338, 340–341, 345–346, 425; *Democracy Index* 341, 345–348
dendrology 129, 235, 239
densification 6, 31, 33, 35, 37
depopulat(ion) 22
design competition 107, 137, 212, 241
design culture 7, 416, 424–425
design examples 13, 126
DesignLab 374
Design studios 20
design training 106, 116
Deutsche Gesellschaft für Gartenkunst (DGfG, German Society for Garden Art) 145, 147–148, 152n3, 152n6
didactic, didactically 1, 6, 55, 107, 138, 261, 265, 268, 284, 343, 358, 361, 363, 425
diploma 1, 103, 106, 110, 115, 144, 158, 167, 173, 178, 205, 213, 262, 295, 299, 309, 310, 338, 361, 365, 368, 370, 397, 402, 412, 415, 417
disaster 19, **67**, 251, 311, 331, 337
disciplinary field 5, 13, 96, 103, 115, 117, 176, 321, 421
discrimination 125, 127, 130–131
dissertation 2, 21, 58, 224, 331, 368, 386
doctoral: degree 71–72, 146, 227, 248, 270, 330–331, 417; programme 21, 299, 412, 417; schools 230, 298
Doherty, G. 406–407
domestic sphere 123, 129–130
Downing, A. J. 111, 113, 122
Downing, M. F. 43, 45, 110, 122
draughtsmanship 106, 112, 116
drawing: freehand 59, 107, 117, 129, 137; hand 59, 194–195; nature 189, 195; skills 137, 139, 193, 195
Dresden 148, 150, 152

dual system 102, 104, 109, 117, 299
Du Bois, W. E. B. 123
Duchêne, A. 295
Duncan, C. 401, 406
du Plessis, C. 393
Duprat, F. 293–295
Düsseldorf 147–148

earthquake 33–31, 327, 337–338, 366, 375
Echtermeyer, T. *144*, 153n4
ECLAS Conference 14, 44–45, 239
ECLAS statutes 46
École Nationale Supérieure de Paysage (ENSP) 55, 60, 111, 263, 292, 296
ecological: landscape design 312; ecological networks 299; ecological planning 332, 394
ecology (including landscape ecology) 14, 64, 68, 116, 173, 179, 219, 229, 235, 239, 244, 250–251, 255, 257, 259, 263, 268, 282, 284, 296, 298, 299, 301, 304, 307, 312, 321, 330, 341, 346, 368, 370, 411–412, 426, 428
Ecosystem Management 276, 304, 309, 311
Ecosystem Services 82, 85, 230, 394
Education Committee, Committee on Education 44, 63, 101, 104, 110, 113, 116, 167, 400
Education for Sustainability 91
Education for Sustainable Development (ESD) 6, 14, 62, 65, 91, 333, 421, 424–425
education history 2, 5, 7, 102, 105, 233, 326, 341, 414, 428
education policy 63, 104, 337
educational agendas 13, 101, 109, 229, 251, 275, 403, 424
educational syllabus *see* syllabus
Eidgenössische Technische Hochschule Lausanne (EPFL) 263, 266
Eidgenössische Technische Hochschule Zürich (ETH Zurich) 260–270
Eisgrub 105, 111, 234, 247
elective subjects 370
Eliot, C. 127
Ellis, C. 21
emergency remote teaching (ERT) 375
empirical 3, 6, 33, 57, 403
Encke, F. 147, 150, 153n16
Enclosures Acts 22
English landscape garden(s) 22
enslaved Africans 22
Environmental Potential Atlas (ENPAT) 394
environmental: conservation 333, 392; crises 250–251; humanism 17; Integrated Management (IEM) 391; sciences 179, 267, 330, 398
equal opportunity 101, 102, 105, 122–123, 131, 277, 422
Erasmus Programme 43, 44, 47, 63
Esch, A. **245**, 247–248, 251

Estrela garden 110, 179, 180–181
ethnicities 338, 366
Eurocentric 13, 18, 396, 403, 405
Euroleague for Life Sciences (ELLS) 374
European: Common Market 366; Higher Education Area 47, 63, 258; Landscape Convention 40, 63, 74, 305, 342, 347, 358; Landscape Education Exchanges (ELEE) 43; Qualification Framework 73, 105; Single Act 43, 62; Single Market 40, 43, 51; Union 43
European Council of Landscape Architecture Schools (ECLAS) 14, 45, 73, 239–240, 305, 341, 345, 374, 424
European Federation for Landscape Architecture (EFLA) 14, 43–44, 62–63, 354
European Landscape Convention (ELC) 63, **79**, 82, 358, 362
evicted cottagers 22
examination 17–13, 59, 93, 109, 112–113, 147, 164, 169, 235, 256, 421
excursion 116–117, 137, 177, 188, 195, 213, 269; *see also* field trip; field visit

Fajardo, M. 393
Fariello, F. 280–281
Farrand, B. J. 127–128
Fasusi, I. 398
Federation of Landscape Architecture Associations 361
Fekadu, F. 403
feminism 127
Feriancová, L. 237
field trip 55, 113, 177, 269, 287, 358, 394
field visit 49, 138, 426
Fife, E. 393
Filipsky, K. P. 250
Fine Arts 20, 64, 69, 105, 138, 166–167, 189, **218**, 220, **244**, 246–247, 249, 307, 308
fire 337, 366, 375, 378, 383
flood 337, 338, 366, 375, 380–381, 399, 401
floriculture 235–236
Forestier, J.-C.N. 295
former communist countries 337, 340–341, 344–347
Freedom house 341, 345, 347–348
Frissell, H. B. 131
Fröbel School, Fröbel Kindergarten 179, 180–181
Froebel, O. 253–254
Froebel, T. 104, 253
Frost, H. A. 128
Fuchs, L. 107, 110, 193
Fukuba, H. 329–330

Gál, P. 237, 239
Galí-Izard, T. 266

Index

Garden Architecture 233–237, 239; *Architetti giardinieri* 280–281; Landscape and Garden Architecture 237, 239

garden: antebellum estate 124; art 55, 59, 104–106, 110–116, 135–139, 143–145, 155–156, 188–190, 192–193, 200–203, 215–217, 233–235, 246–248, 253, 262, 276, 279, 280–281, 292–296, 317–320; Chinese 315–320; English Landscape 22, 315; flower 126, *128*; historic 110, 225; history 144, 147–148, 148–149, 150; school 123, 130, 205; vegetable 123, 125–126

garden design 234–236; exhibition 149; garden and landscape design 102, 114, 155, 199, 237, 251, 276, 326, 332; garden designer 104, 107, 109, 111, 113, 143, 145–150, 155, 157–159, 166, 201–203, 207, 253–254, 326, 411; *Gartenkunst* 112, 137, 143, 146–147, 152n8

Gärtnerlehranstalt 4, 110, *144*, *145*, 146–148, 150, 152n4

Geisenheim 111, 150

gender 105, 128, 130, 168, 248, 405, 422

gender bias 128–129

genealogy, genealogical 2, 3, 422, 426, 427

Generalplan Ost 158–163

geography 40, 116, 189, 218, 266, 283, 301, 357, 411–412, 421

geology 107, 116, 173, 199

Giacomini, V. 283

Gibbs, D. 395–396

Ginsburg R. 21

Giovannoni, G. 278

Girot, C. 20–21, 263–265, 269

GIS, GIS laboratory 391, 394

Global Challenge University Alliance 376

Global North 396, 400–401, 404

global 'oil shock' 366

Global South 396, 404–405

global sustainability agenda 1, 6, 66

governance 14, 67, 338, 366, 399, 406

grading 106, 109, 116, 123, 127, 398

green: area planning 83–84, 86, 87; belts 333; infrastructure 82; space 235; space design **244**, 249–250

Green, W. 393

Gregotti, V. 296

Grootenhuis, F. 397

Groth, P. 19

Groton 103, 112, 128, 129

Gurlitt, C. 148

Hackett, B. 173, 321

Haines, J. 128

Hammerbacher, H. 148

Hannah-Jones, N. 18

Hanover 148

Hartogh Heys van Zouteveen, Hendrik François 110, 188–190, 192–193, 195

Harvard: approach 20; School of Landscape Architecture 20; university 20, 105, 111, 121, 126, 321, 406

Henrici, K. 148–149

heritage: conservation 55; garden 54; protection 60; sites 55; value 60

Hertzka, Y. 111, 247–248

Higgins, C. 73

Hindes, C. 394–395

historical 6, 54, 58, 421–422, 425; actors (practitioners, researchers, and critics) 25; constructs 17; empathy 22; evidence 25; narrative(s) 16, 17, 18; questions 22; scholarship 16, 19, 21, 26

history: books 20; course 16–20, 26, 57–60; critical 18; curricular 20; of landscape architecture education in Poland 337, 350; restorative 16; speculative (abductive) history 25; survey 16, 21; -theory seminars 21

Hochschulkonferenz Landschaft (HKL) 43

Hoemann, R. 146–148, 150

Hollein, H. 278

Honda, S. 330–331

Hongo, T. 330–331

Hood, W. 21

horticultural engineering 227, 275, 402

Horticultural Society of Zagreb 211–216

horticultural training 189, 254, 300

Hrubík, P. 235, 237

Hruška, E. 237, 239

Hubbard, H. V. 20

Hubbard, T. K. 20

human rights 62, 157, 311, 381, 425

human-nature: affinity 410; relationships 82

hydrology 107, 116, 268

'Hyperdensity' design studio 381

ideologies 7, 13, 17, 31, 251, 277, 400, 422, 425, 427

Ihm, E. M. 248

inauguration 1, 43, 207, 237

inclusive 1, 3, 16, 44, 50, 63, 66, 94, 131, 279, 337, 338, 380, 421, 424, 427

independence: of the landscape architecture profession 127; of women 127–128

indigenous knowledge 276, 398, 400, 406

Indigenous peoples *see* Native Americans

inequality 122, 131

institutes, academic: Hampton Institute 105, 123–125, 130–131; Hassan 2nd Institute of Agronomy and Veterinary Medicine (IAV Hassan II) 401, 407; Higher Agronomic Institute of Chott Mariem (ISA-CM) 402; Higher Horticultural Institute for Women 105; horticultural institute 144–145, 146, 147, 149,

152; Institute for Garden Design 156–158, 166; Institute for Landscape and Garden Design 158, 160; institute of technology 145–146, 148; *Institut für Landschaft und Urbane Studien* (Institute of Landscape and Urban Studies) 266; *Institut für Orts-, Regional- und Landesplanung* (Institute for Local, Regional and National Planning - ORL) 262; Massachusetts Institute of Technology 128; *Silva Tarouca Research Institute of Landscape and Ornamental Gardening* 234; Swiss Federal Institute of Technology Zurich 102; Tuskegee Normal and Industrial Institute (Tuskegee Institute Tuskegee University) 122–123, 130–131; Uehara Landscape Research Institute 332; *see also* research institute

institutes, professional: Australian Institute of Landscape Architects (AILA) 369; Bucharest Institute for Planning 204; Chilean Institute of Landscape Architects (ICHAP); Ethiopian Institute of Architecture, Building Construction and City Development (EiABC) 402–403; Institute of Landscape Architects of South Africa (ILASA) 390–391, 394–395; Landscape Institute (LI) 43, 109, 164, 166; New Zealand Institute of Landscape Architects (NZILA) 369; Town Planning Institute 165–166, 172

institutional: legend 20; pedagogy 19; perspectives 19

Instituto Superior de Agronomia 110–111, 113, 177–178

integrative thinking 6, 81–82, 84, 86–88, 89

Intensive Programmes 43

interdisciplinary 19; research 4, 322

international: collaboration 376; conferences 374; developments 365; networks 374; partners 366; partnerships 374; relationships 366; students 373–374; teaching 374

International Federation of Landscape Architects (IFLA) 14, 42, 114, 227, 237, 240, **244**, 281, 287, 305, 319, 369, 391, 397–398, 424

intersections (types of: e.g. race, diversity, cultural landscape, and design justice) 21

interviews (as study method) 5, 33, 35, 83, 109, 223, 253, 261, 279, 292, 315

Iran Society for Landscape Professionals (ISLAP) 304

Iron Curtain 7, 42, 337, 340–341, 343, 345, 347

Ivy League 128, 131

iwi 372

Jackman, A. 369
Jackson, J.B. 19, 111, 122
Japan Garden Association 331
Japanese Garden 318, 326–329, 331
Jeffrey, J. T. 167
Jeglič, C. 211–215, 220

Jellicoe, G. 20, 109, 111, 114, 116, 167, 172, 315, 393
Jellicoe, S. 20
Jenkins, G. 172
Jie, H. 321
Joffet, R. 295
Jones Allen, D. 21
Journal of Landscape Architecture (JoLA) 47, 237
justice 7, 16, 18, 21, 27, 37, 101, 104, 131, 305, 337, 338, 395, 424, 427
Juyuan, W. 317

Karlsruhe 149–150, 152n13
Kavka, B. 235–236
Kemmer, E. 146
Kemp, E. 123
Kendi, I. X. 18
Kienast, D. *153*, 259, 262–263, *271*
Kigondu, S. 398
Klaić, S. 212
Kodaira, Y. 329
Kodoň, M. 237, *238*, 239
Koenig, H. 145–146, 156
Kongjian, Y. 315, 321–322
Königliche Gärtner-Lehranstalt 143
Köstritz 111, 152n2, 254
Krakow School of Landscape Architecture 353
Kreis, W. 147–151, 153n16
Krzywda-Polkowski, F. **351**, 353
Kunstakademie 147
Kunstgewerbeschule 147, 149

Laeuger, M. 149–150, *151*, 152n13, 152n14, 153n16
Laforcade, J. 402
Landbouwhogeschool 110
Landbrukshøgskole 71, 101, 103, 111, 135
land-grant colleges 105, 121–123; institutions 20
landscape: aesthetics 235; 'Apostles' 368; assessment 373, 375; concepts 277, 422; construction 116, 169, 219, 246, 248, 250, 259, 342, 399; designer 19, 85, 102, 114, 158–60, 329–32, 411, 414, 416 (*see also* garden and landscape design); landscape drawing *237–238*, *240*; education 237, 239; Education for Democracy (LED) 50, 425; engineer 107, 137, 141, 229, 237, 239, 275, 292, 299, 307–308, 312, 402, 414; engineering 107, 137, 141, 237, 239, 299, 307–308, 402, 414; history agenda 21; history (equitable) 26; Landscape@Lincoln 365, 376; literacy 26; management **244**, 250; painting 189; scholarship 26, 46, 312, 333; style 139, 144, 246, 329; urban 370
Landscape Architecture Accreditation Board (LAAB) 93, 311
Landscape Architecture Forum 350, **352**, **359**, 361, 363

Landscape Journal 21
Landwirtschaftliche Hochschule 143, 146, 152n4
Lange, W. 139
L'architecture d'aujourd'hui 282
Lassus, B. 263, 296–300
Latinne, René 110, 113–115
Laugier, M.-A. 279
Lawin, L. 353
Le Roux, Schalk 392
LE:NOTRE 374
LE:NOTRE Institute 50
LE:NOTRE Project 47, 50–51, 110
LE:NOTRE.org 48–49
Leadership in Energy and Environmental Design (LEED) 2, 9
Lear, J. 388
learning: method 3, 66, 422; outcomes 83, 84–85, 87–88; practice 3, 5, 240, 421, 424, 427, 428
Lebanese Landscape Association (LELA) 308
Leder, W. *118, 254*
Lednice 105, 111, 233–237, 240–242, 247
Lenné, P. J. 107, 110, 138, 143, 157
Lethugile, G. 404
Leutscher, P. 390
Leveau, T. 295
Leviker, K. 403
Levins Morales, A. 17, 25
Lewis, P. 16
Liangyong, W. 317, 322
Lincoln College 365, 369
Lisbon city council 176–177, 179, 183–184
longitudinal studies 3, 96, 428
Low, J. M. 128
Lowthorpe School of Landscape Architecture 103, 112, 128–130, 129
Lullier 253, 262
Lynch, K. 296

Machovec, J. 236
Majdecki, L. 353
Makhubu, J. 404–405, 407
mana whenua 372
Māori 365–366, 372, 376
Mareček, J. 236
Margiochi, F. S. 107, 179–183
Martinet, H. 293
Massachusetts Agricultural College 122, 126
Mattern, Hermann 148
Maurer, Erich 146
Mawson, E. P. **111**, 165
Mawson, T. H. 106, 139, 165–166, 224
McHarg, I. 321, 379, 391–394, 416
Mertens, E. 104, 253
Mertens, O. 254
Mertens, W. 254
meteorology 107, 116
Meyer, G. 15, 155, 247

Meyer, K. 157–159
Michigan Agricultural College (Michigan State University) 122
Migge, L. 139, 254
military 20, 124, 157, 159, 399, 412
Miller, Z. 398
Milner, E. 111, 166
minimum requirements 64, 103, 351, 397, 399, 423
minimum standards 116, 424
Ministry of Agriculture 147
Ministry of Commerce 147
Ministry of Education 279, 284, 286
Misvær, H. M. 107, 111, 137
Mitchell, W.J.T. 22
Mlyňany Arboretum 235
Mőcsényi, M. 223–228
Moen, O. L. 103, 107, 110–112, 135, 137–141, 423
Montégut, J. 295
Morrill Act 22, 122–123
Motloch, J. 393
Mulder, C. 392
Muller, L. 394–396
municipal garden departments 143
municipal garden intendants 144
Murphy, M. 391–392
Muthesius, H. 147, 152n6
Myers, G. 403

National Socialism 7, 155–160; National Socialist 147, 248
Native Americans 105, 121–122, 124, 125, 130–131; culture of 124; and deprivation of political power 131; education of 124; gardening traditions of 125; land of 122; suffering of 125
natural disasters 267
natural resources 20, 64, 321, 353, 368
natural-science 40, 72, 107, 116, 173, 203, 216, 246, 248, 251, 255, 263, 267–268, 281, 347, 353–354, 415
Nature Based Solutions 82
Neal, D. 319
neo-liberal public policies 366
Network of European Landscape Architecture Archives (NELA) 6, 14, 54, 61
Newton, N. 20
New York City 121
New Zealand Experiment 366, 370–373
Ngunjiri, P. G. 397–398
Niemirski, W. 353
Nitra 233, 235, 237–242
Nivet, H. 293
non-Western 21
Norges Landbrukshøgskole 71, 101, 103, 111, 135
Novák, Z. **351**, 353

nurseries 113, 146–147, 177, 179, 199, 201, 212, 215–216

Oberholzer, B. 395
Oeschberg Horticultural School 254, **255**, *260*
Ogawa, J. 328
Ogrin, D. 2, 21, 110, 211, 217, 220
Oldham, J. 391
Olmsted, F. L. 103, 106–107, 111, 121–122, 125–127, 131n4, 136, 315
Olmsted, F. L. Jr. 126
open access digital resource 374
open entry 373
open space: conservation of 126; design of 211–212, **219**, 224, 229–230, 282, 287
oppression 16, 18
Ormos, I. 223–229
ornamental horticulture 235
Otruba, I. 236
Otto, H. 190
outdoor art 122, 126
Owens, H. 391

painting 106, 116–117, 189, 224, 234, 294, 316–318, 320, 328
pandemic 51, 117, 354, 357, 366, 374, 376, 388, 424
paradigm shift 36, 350, 381, 414, 428
Paris 110–111, 114, 127, 167, 179, 201, 212–213, 221n6, 263–264, 266, 289n13, 293–294, 296, 301n2, 301n5, 410
park: city park 110, 212, 331; national park 19, 203, 319, 374; parks 123, 233–235, 237, 239, 241–242
partnership: with indigenous people 339, 365, 372, 376; intercultural 312; international 373–374, 376; strategic 435
paysagiste: *architecte 293–296, 401*; *concepteur* 292, 299; *Diplômé DE 299*
pecan trees 124–125
pedagogical tradition 20
pedagogic approach 14, 15, 43, 53, 81, 88, 96, 106, 118, 422
pedagogy 1, 3, 4, 6, 57, 92, 107, 261, 374, 380, 400, 404, 406, 422; conceptual 81–88; inclusive 421; institutional 19; landscape 372, 387; online 375; for sustainability 91–97
Peets, E. 127
Pejchal, M. 237
People's Council 204
Performance Based Research Fund (PBRF) 373
periodical 5, 136, 141, 176, 183
Piacentini, M. 280
Piccinato, L. 280
Pim, J. 390, 407
planning: countryside 106–107, 109; Earthscape 319, 321; for Environmental Conservation 391; green area 83–89; integrated 89; *integrative urban* 82; landscape 2, 35, 81–82, 89, 106, 165, 173, 207, 220, 227, 237, 239, 244, 246, 248, 250–251, 257–258, 279, 284, 300, 321, 327, 333, 342, 350, 361, 375, 414, 416, 422; and the naming of the profession 126; open space 106–107; and Property Valuation 375; reclamation 262; regional planning 107, 109, 116, 166, 173, 195, 227, 250, 262, 391, 394–395, 399–400; urban planning 81, 189, 192–193, 195, 237; *see also* spatial planning
plant: communities 191, 193; knowledge 190, 259; material 49, 106, 116, 127; sociology 190, 192, 248, 268
planting design 78, 112, 114, 116, 127, 129, 187, 205, 237, 269, 328, 398
Pocock, C. 376
Polak, E. 211
Politecnico di Milano 283, 287
Politecnico di Torino **286**, 287
pomiculture 146
Porcinai, P. 118, 280–283, 288n7, 289n14, 341
post-colonial education 390
post-graduate 1, 58, 63, 123, 173, 393, 398, 417
Potager du Roi 111, 275, 292, 295–300
Potsdam 143, 145, 147, 152n2
practical exercise 214, *295*
Prague 212, 233–236, 240–241
Price, C. 395, 406–407
professional: association 5, 63, 101, 150, 209, 233, 247, 263, 295, 331, 361, 398, 423–424; chamber 104, 361; history 17; identity 20; practice 60, 130; recognition 104, 169, 293; requirement 63, 108, 300; training 41, 44, 110, 127–128, 164, 206, 224, 234, 414–415; years 373
professoriate 17
programme manager 6, 107, 265, 277, 422
Proskau 152, 152n2
Průhonice 234–236
public monuments 19
publication (types) 21

qualification requirement 101, 103
qualitative: analysis 176, 187, 279; interview 35; research 3, 384
quantitative: analysis 150, 279; approach 93; studies 6
Quenteen, B. 403
questionnaire 6, 304

racism 122, 130–131; systemic 130–131
racist construct 18
racist worldview 18
ranking 1, 96, 147, 423, 427
Rapperswil 60
Raxworthy, J. 395

reconciliation 7, 104, 157, 337–338, 388
recreation 20, 32–33, 35, 109, 117, 227, 318–319, 359, 370, 422
Regia Scuola Superiore di Architettura (Royal School of Architecture) 280
relational models 81–85, 87–88
remote learning and teaching 376
reporting (of research findings) 3–4, 229, 426
representation 58–59, 60
Rerrich, B. 107, 223–225
research; design 57, 235, 280, 380, 384; by drawing 58; and education 57, 58; laboratory 230, 331, 391; methods 5, 6, 57, 176, 278; questions and investigative methods 22; skills 374
research institute 108, 234, 235, 322, 333
resilience 81–82, 84–85; community 376
revisionist 18
Riley, R. 19, 21–22, 37
Riousse, A. 295
Rogers, E. B. 21
Rotkvić, A. 212
Royal Horticultural Department 201, 207
Rui, Y. 322
Rumelhart, M. 298
Rupert, A. 394
Ruys, M. 190

Sadovnictví 233, 235–236, 241
Saidane, I. 402, 406
Saidi, F. 395–396, 398, 403–407
Saint-Maurice, J.-C. 296
Schjetnan, M. 393
Scholz, J. 235–236
Scholtz-Richert, A. Z. 353
School of: Cambridge of Architecture and Landscape Architecture 128–129; *Crystal Palace of Gardening*; Engineers of Horticulture and Landscape Techniques (ENITHP) 401; Harvard Graduate of Design 406; *Höhere Gartenbauschule für Frauen* Vienna 247; *Höhere Obst- und Gartenbauschule* Eisgrub/Lednice 247; Horticultural and Gardening in Köstritz 111, *152*, 254; of Landscape Architecture (SOLA) 374; Lowthorpe 103, 112, 128, 129; National of Landscape of Versailles 402; Pennsylvania of Horticulture for Women *128*; *Regia Scuola Superiore di Architettura* (Royal of Architecture) 280; Rhode Island of Design 113, 393; *Scuola Superiore di Architettura* (of Architecture) 280; *see also* university
Schultze Naumburg, P. 150, 153n16
Schweizer, J. 254
Scuola Superiore di Architettura (School of Architecture) 280
Seeck, F. 148, 152n8
Seissel, S. 212

self-studies (self-training) 121, 127, 130
Semper, G. 262
Sereni, E. 283
Sgard, J. 296, 300
shared power 22
Sharp, T. 166
Shurcliff, A. 128
Silesia 152
Šimek, P. 237
Simon, J. 296
Sinatra, J. 379
SITES (The Sustainable SITES Initiative) 92
Sitte, C. 148
sketching 33, 78, 79, 116, 169, 193, 195
social change 20; construction 17; justice 18, 37, **79**, 131, 395, 427
social sciences 107, 219, 297, 312, 403–404
Sociedad de Arquitectos Paisajistas de México (Society of Landscape Architects of Mexico) 417
societal challenge 64, 270, 338, 363, 403, 405, 422
Société nationale d'horticulture de France 292
Society of Landscape Architects of Nigeria (SLAN) 398–400
sociology 6, 69, 116, 190, 192, 218, 248, 421
socio-political 57, 165, 211, 250, 337, 350, 362
soft skills 84
soil bioengineering 262
soil science 107, 116, 137, 192, 255, 268, 368
Sørensen, C. T. 138
Sörrensen, W. 148
South-North dialogues 406–407
spatial awareness 59
spatial planning 126, 189, 216–217, 239, 262–264, 266, 309, 353, 357–358, 361–362, 396, 404
specialist education 106, 165
Spooner, P. 379
Springer, L. A. 187–190, 193, 197
state interference 18
Stern, C. 256, 259, 262
Stilgoe, J. 19
Stoffberg, H. 391, 394
Storm (as natural hazard) 75, 337
strategic adaptions 365
structural models 102, 104, 106, 428
structure(s) of knowledge 20
student design competition 241
student enrolment **367**, 375
student project 6, 31, 35, 59, 376
studio: course 83, 87–88; teaching 3, 310, 343, 368, 374
Studley Horticultural and Agricultural College for Women 128
subaltern histories 21
subdivisions: design 127
Sudell, R. 166

437

suffrage movement 127
summer school 138, 172, 424
Supuka, J. 235, 237
surveying 106, 116, 121, 129, 165, 167, 169, 173, 189, 293, 317, 320, 379
surveys 3, 5, 21, 83, 121, 327
sustainability 9, 81–82, 84–85; agenda 1, 6, 66; development 365; education 3, 14; principles 14, 69; Sustainable Development Goals (SDGs) 14–15, 64–65, 400, 424, 428; grey and green infrastructure 376; teaching 15, 92, 94, 96; urban planning 81
Sutton, A. 390
Swanley Horticultural College 105, 112, 115, 123, 128, 165, 172
syllabus 43, 106, 113, 115–116, 146, 167, 172, 281, 351, 395, 397, 424
system change 337–338, 350, 363

Tamura, T. 330–331
teacher: history 16; teacher-student interaction 2, 118; teacher training 107, 123; teaching 235–237, *238–239*, 241; teaching communities 71–77; teaching competence 107, 150; teaching department of the Berlin Museum of Applied Art 148; teaching methods 88
Technische Hochschule 145–149, **148**, 177, 261
Tentori, F. 282
tenure(d) 17
Te Wai Pounamu - New Zealand South Island 365
territorial planning 237, 239
Terror 337, 339, 366
Thacker, C. 20–21
Thematic Network 47, 49, 63
therapeutic design 376
Thomas, G. 393
topography 107, 116, 178, 201, 383
Toroitich, C. 398, 404–405
tourism 276, 296, 322, 370, 374, 394, 401, 422
Trip, J. 148
Tuinbouwschool 110–111
Tuning Project 47, 63
Turri, E. 283

Uehara, K. 330–331
UK model 368, 370
Ungar, P. 212
Union of Universities for the Development of Studies in Landscape Architecture 354
university: Ahmadu Bello 399; Ancona 283; Ba Isago 404; Balamand 308; Bari 284; Beijing Agriculture 317; Beijing Forestry 315, 318, 321–323; Beijing University 321; Berkeley 128; Berlin University 110, 112–113; Bir Zeit University (BZU) 304, 309; Black 122–126; Bodenkultur Wien (BOKU) 59, 60, **244–245**, 245–246, 248–251; Bologna 283; Budapest 55; Cape Peninsula 397; Chiba 330; Columbia 121; Cornell 103, 111, 122, 126, 128; Cracow University of Technology **351–352**, **355**, **359–360**; Dar es Salaam 404, 407; Delft 113; Dresden 330; Florence 284, 287; Genoa 284, 287; Hannover 148, 155, 254; Harvard 20, 105, 111, 121, 126–127, 129, 130–131, 321, 406; Illinois 122, 128; Iowa State 122; Johannesburg 404–407; *Jomo Kenyata* 397, 404, 406; Kassel 254, 263, 321; King Abdulaziz 304; Lagos 399, 406–407; Lincoln 338, 365, 370, 373–374, 376; Lisbon 101, 176; Ljubljana 44; Manchester 44, 105, 111, 114, 167, 169, 173; Milan 284, 287; Munich 254, 330; Naples 284; Palermo 278, 287; Pennsylvania 165, 379, 391; Pretoria 404, 406–407; Rapperswil 60; Reading 101, **103**, **111**, 113, 167; Rome 280, 283, 287–288; Thames Polytechnic, Greenwich 43; Tokyo 316, 328, 330–332; Tshwane 397; Tsinghua 317, 321–322; Turin 284, 287; Tuscia, Viterbo 288; Wageningen University 44, 101, 103, 106, 110, 111–112
Unterrichtsanstalt des Kunstgewerbemuseums 148
urban: bias 121, 130; expansion 108, 110, 193, 321, 327; green infrastructure 82–83; metabolism 85
urbanism 237, 239, 241
urbanization 121, 126, 130, 339, 393, 422
urban-nature concepts 81–84, 86–88

Vacherot, J. 295
van den Berg, E. 394
van der Swaelmen, Louis 193
van Riet, W. 391, 394
van Zouteveen, Hartogh Heys 187–190, *191*, 192–193, 195, 197
vegetation, vegetation studies 107, 113, 116, 150, 183, 192, 200, 205, 236, 381, 411
Vereinigung Österreichischer Gartenarchitekten, VÖGA **245**, 247
vernacular landscape 19, 21, 26–27, 116, 131, 277, 306, 422, 426
vernacular practices 21, 124, 131; and West African planting 125
Versailles 55, 60, 263, 292–301, 401–402, 416
Vienna 43, 45, 58–59, 112, 212–213, 234, 243, **244**, 246–251
Vilvoorde 110, 114–115, 193
vocational schools 6, 102–103, 107, 215, 333
vocational training 103, 130, 144, 253, 259, 260, 333, 361
Vogt, G. 264
volunteer work 130
von Brandis, A. 149
von Engelhardt, W. 143, 147–148, 150, 152n5, 152n7, 153n16
Vosloo, P. 391, 394

Vouk, V. 211–212, 214, 220
Vreštiak, P. 235, 237

Wageningen 44, 187–190, *190*, 192–193, 195
Wagner, B. 236
Walker, P. 393
Ward Thompson, C. 21
Ware, I. 279
Warsaw School of Landscape Architecture 353
Warsaw University of Life Sciences – SGGW **351–352**, **355**, 357, **359–360**
Washington, B. T. 123, 125
water management 75, 116, 220
Waugh, F. A. 126
Way, T. 17, 26
web-databases 49
Weihenstephan 111, 150, 254
Weimar Republic 152, 155–157
Werkbund 149–150, 153n16
Werkmeister, H. 391
Western: civilization 155–156; -ness 19, 27; perspective 21; thought 419
white supremacy 123–124, 131
White, E. 167
White, M. 166
Whiteness 19, 27
Wieler, A. L. 149

Wiepking, H. F. 152n3, 156–160, 248
Wikis 27, 48–49
Williston, D. A. 122, 126
Woess, F. 243, **244**, 248–251
Women 21, **245**, 246–248, 250, 411; admission 20, 104–105, 112, 422; and early landscape architectural education 127–130, 167, 179, 181, 301; and extension service 126; and Indigenous horticultural practices 125; and landscape gardening 123
Woolley, G. 172
working group 47, 49
World Wide Web 45, 48

Xiangrong, W. 321
Xiaoxiang, S. 276, 315, 317, 322

Yamagata, A. 329, 332
Ye, Y. 319
Young, G. 392, 407
Youngmann, P. 173

Zanzot, J. 393
Zhi, C. 316, 322
Zielonko, A. **351**, 353
Zoheb, Y. 403
Zrinjevac park 212

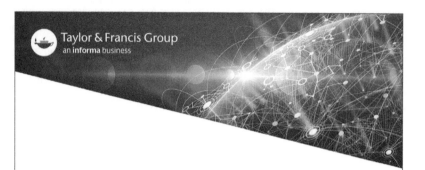